The chemistry of the
sulphonium group
Part 1

Edited by

C. J. M. STIRLING

School of Physical and Molecular Sciences
University College of North Wales
Bangor

1981

JOHN WILEY & SONS

CHICHESTER – NEW YORK – BRISBANE – TORONTO

An Interscience® Publication

British Library Cataloguing in Publication Data:

The chemistry of the sulphonium group.—(The
    chemistry of functional groups).
    Part 1
    1. Sulphonium compounds
    I. Stirling, Charles James Matthew
    II. Series
    547′.065   QD341.S8   80-40122
ISBN 0 471 27769 X
ISBN 0 471 27655 3 Set of 2 vols

Typeset by Preface Ltd, Salisbury Wilts.
Printed in the United States of America.

FOR E. AND L.—

ANOTHER LINK BETWEEN WALES AND ISRAEL
הקדשה עבור א., ל., והקשר בין ויילס וישראל
CYSWLLT ARALL RHWNG CYMRU AC ISRAEL

The chemistry of the
sulphonium group
Part 1

THE CHEMISTRY OF FUNCTIONAL GROUPS

A series of advanced treatises under the general editorship of
Professor Saul Patai

The chemistry of alkenes (2 volumes)
The chemistry of the carbonyl group (2 volumes)
The chemistry of the ether linkage
The chemistry of the amino group
The chemistry of the nitro and nitroso groups (2 parts)
The chemistry of carboxylic acids and esters
The chemistry of the carbon–nitrogen double bond
The chemistry of amides
The chemistry of the cyano group
The chemistry of the hydroxyl group (2 parts)
The chemistry of the azido group
The chemistry of the acyl halides
The chemistry of the carbon–halogen bond (2 parts)
The chemistry of the quinonoid compounds (2 parts)
The chemistry of the thiol group (2 parts)
The chemistry of the hydrazo, azo and azoxy groups (2 parts)
The chemistry of the amidines and imidates
The chemistry of cyanates and their thio derivatives (2 parts)
The chemistry of diazonium and diazo groups (2 parts)
The chemistry of the carbon–carbon triple bond (2 parts)
Supplement A: The chemistry of double-bonded functional groups (2 parts)
Supplement B: The chemistry of acid derivatives (2 parts)
The chemistry of ketenes, allenes and related compounds (2 parts)
The chemistry of the sulphonium group (2 parts)

Contributing authors

K. K. Andersen   Department of Chemistry, University of New Hampshire, Durham, New Hampshire 03824, USA

M. R. F. Ashworth   Lehrstuhl für Organische Analytik, Universität des Saarlandes, Fachbereich 15.3, 66 Saarbrucken, W. Germany

G. C. Barrett   Oxford Polytechnic, Headington, Oxford OX3 0BP, England

L. F. Blackwell   Department of Chemistry, Biochemistry and Biophysics, Massey University, Palmerston North, New Zealand

E. Block   Department of Chemistry, University of Missouri–St. Louis, St. Louis, Missouri 63121, USA

J. D. Coyle   Chemistry Department, The Open University, Milton Keynes MK7 6AA, England

D. C. Dittmer   Department of Chemistry, Syracuse University, Syracuse, New York 13210, USA

A. Gavezzotti   Istituto di Chimica Fisica e Centro CNR, Università di Milano, Milan, Italy

J. Grimshaw   Department of Chemistry, David Keir Building, Queen's University of Belfast, Belfast BT9 5AG, N. Ireland

A. C. Knipe   School of Physical Sciences, The New University of Ulster, Coleraine, Co. Londonderry, Northern Ireland BT42 1SA

P. A. Lowe   The Ramage Laboratories, Department of Chemistry and Applied Chemistry, University of Salford, Salford M5 4WT, England

G. A. Maw   Department of Biochemistry, Glasshouse Crops Research Institute, Littlehampton, Sussex, England

T. Numata   Department of Chemistry, University of Tsukuba, Sakura-mura, Ibaraki, 305 Japan

S. Oae   Department of Chemistry, University of Tsukuba, Sakura-mura, Ibaraki, 305 Japan

B. H. Patwardhan   Department of Chemistry, Syracuse University, Syracuse, New York 13210, USA

I. C. Paul   Department of Chemistry, School of Chemical Sciences, University of Illinois, Urbana, Illinois 61801, USA

E. F. Perozzi   Ethyl Corporation, 1600 West Eight Mile Road, Ferndale, Michigan 48220, U.S.A.

R. Shaw   Chemistry Department, Lockheed Palo Alto Research Laboratory, Lockheed Missiles & Space Company, Inc., 3251 Hanover Street, Palo Alto, California 94304, USA

H. J. Shine   Department of Chemistry, Texas Tech University, Lubbock, Texas 79409, USA

J. Shorter   Department of Chemistry, The University, Hull HU6 7RX, England

M. Simonetta   Istituto di Chimica Fisica e Centro CNR, Università di Milano, Milan, Italy

T. Yoshimura   Department of Chemistry, University of Tsukuba, Sakura-mura, Ibaraki, 305 Japan

Foreword

Our collaboration on production of this volume on the chemistry of the sulphonium group arose from a sabbatical visit from Wales to Israel.

There is remarkably little review literature on the chemistry of this group, one of the most versatile in organic chemistry, and we have been most fortunate in the group of contributors who have brought this volume to fruition. Chapters on nuclear magnetic resonance spectroscopy and mass spectroscopy did not materialise in time for publication.

We hope that assembly of so much up to date information on this functional group will stimulate further exploration of its wide-ranging and fascinating chemistry.

The preface to the series contains the promise of further volumes on the chemistry of sulphur-containing compounds. We are in the early stages of planning further volumes in the series on the chemistry of other sulphur-containing functional groups, notably those concerned with sulphonyl and sulphinyl groups.

Bangor   May 1980   CHARLES STIRLING
Jerusalem   SAUL PATAI

The Chemistry of Functional Groups
Preface to the series

The series 'The Chemistry of Functional Groups' is planned to cover in each volume all aspects of the chemistry of one of the important functional groups in organic chemistry. The emphasis is laid on the functional group treated and on the effects which it exerts on the chemical and physical properties, primarily in the immediate vicinity of the group in question, and secondarily on the behaviour of the whole molecule. For instance, the volume *The Chemistry of the Ether Linkage* deals with reactions in which the C—O—C group is involved, as well as with the effects of the C—O—C group on the reactions of alkyl or aryl groups connected to the ether oxygen. It is the purpose of the volume to give a complete coverage of all properties and reactions of ethers in as far as these depend on the presence of the ether group but the primary subject matter is not the whole molecule, but the C—O—C functional group.

A further restriction in the treatment of the various functional groups in these volumes is that material included in easily and generally available secondary or tertiary sources, such as Chemical Reviews, Quarterly Reviews, Organic Reactions, various 'Advances' and 'Progress' series as well as textbooks (i.e. in books which are usually found in the chemical libraries of universities and research institutes) should not, as a rule, be repeated in detail, unless it is necessary for the balanced treatment of the subject. Therefore each of the authors is asked *not* to give an encyclopaedic coverage of his subject, but to concentrate on the most important recent developments and mainly on material that has not been adequately covered by reviews or other secondary sources by the time of writing of the chapter, and to address himself to a reader who is assumed to be at a fairly advanced post-graduate level.

With these restrictions, it is realized that no plan can be devised for a volume that would give a *complete* coverage of the subject with *no* overlap between chapters, while at the same time preserving the readability of the text. The Editor set himself the goal of attaining *reasonable* coverage with *moderate* overlap, with a minimum of cross-references between the chapters of each volume. In this manner, sufficient freedom is given to each author to produce readable quasi-monographic chapters.

The general plan of each volume includes the following main sections:

(a) An introductory chapter dealing with the general and theoretical aspects of the group.

(b) One or more chapters dealing with the formation of the functional group in question, either from groups present in the molecule, or by introducing the new group directly or indirectly.

(c) Chapters describing the characterization and characteristics of the functional groups, i.e. a chapter dealing with qualitative and quantitative method of determination including chemical and physical methods, ultraviolet, infrared, nuclear magnetic resonance and mass spectra: a chapter dealing with activating and directive effects exerted by the group and/or a chapter on the basicity, acidity or complex-forming ability of the group (if applicable).

(d) Chapters on the reactions, transformations and rearrangements which the functional group can undergo, either alone or in conjunction with other reagents.

(e) Special topics which do not fit any of the above sections, such as photochemistry, radiation chemistry, biochemical formations and reactions. Depending on the nature of each functional group treated, these special topics may include short monographs on related functional groups on which no separate volume is planned (e.g a chapter on 'Thioketones' is included in the volume *The Chemistry of the Carbonyl Group*, and a chapter on 'Ketenes' is included in the volume *The Chemistry of Alkenes*). In other cases certain compounds, though containing only the functional group of the title, may have special features so as to be best treated in a separate chapter, as e.g. 'Polyethers' in *The Chemistry of the Ether Linkage*, or 'Tetraaminoethylenes' in *The Chemistry of the Amino Group*.

This plan entails that the breadth, depth and thought-provoking nature of each chapter will differ with the views and inclinations of the author and the presentation will necessarily be somewhat uneven. Moreover, a serious problem is caused by authors who deliver their manuscript late or not at all. In order to overcome this problem at least to some extent, it was decided to publish certain volumes in several parts, without giving consideration to the originally planned logical order of the chapters. If after the appearance of the originally planned parts of a volume it is found that either owing to non-delivery of chapters, or to new developments in the subject, sufficient material has accumulated for publication of a supplementary volume, containing material on related functional groups, this will be done as soon as possible.

The overall plan of the volumes in the series 'The Chemistry of Functional Groups' includes the titles listed below:

The Chemistry of Alkenes (two volumes)
The Chemistry of the Carbonyl Group (two volumes)
The Chemistry of the Ether Linkage
The Chemistry of the Amino Group
The Chemistry of the Nitro and Nitroso Groups (two parts)
The Chemistry of Carboxylic Acids and Esters
The Chemistry of the Carbon–Nitrogen Double Bond
The Chemistry of the Cyano Group
The Chemistry of Amides
The Chemistry of the Hydroxyl Group (two parts)
The Chemistry of the Azido Group

The Chemistry of Acyl Halides
The Chemistry of the Carbon–Halogen Bond (two parts)
The Chemistry of Quinonoid Compounds (two parts)
The Chemistry of the Thiol Group (two parts)
The Chemistry of Amidines and Imidates
The Chemistry of the Hydrazo, Azo and Azoxy Groups
The Chemistry of Cyanates and their Thio Derivatives (two parts)
The Chemistry of Diazonium and Diazo Groups (two parts)
The Chemistry of the Carbon–Carbon Triple Bond (two parts)
Supplement A: The Chemistry of Double-bonded Functional Groups (two parts)
Supplement B: The Chemistry of Acid Derivatives (two parts)
The Chemistry of Ketenes, Allenes and Related Compounds (two parts)
The Chemistry of the Sulphonium Group (two parts)

Titles in press:

Supplement E: The Chemistry of Ethers, Crown Ethers, Hydroxyl Groups and their Sulphur Analogs

Future volumes planned include:

The Chemistry of Organometallic Compounds
The Chemistry of Sulphur-containing Compounds
Supplement C: The Chemistry of Triple-bonded Functional Groups
Supplement D: The Chemistry of Halides and Pseudo-halides
Supplement F: The Chemistry of Amines, Nitroso and Nitro Groups and their Derivatives

Advice or criticism regarding the plan and execution of this series will be welcomed by the Editor.

The publication of this series would never have started, let alone continued without the support of many persons. First and foremost among these is Dr Arnold Weissberger, whose reassurance and trust encouraged me to tackle this task, and who continues to help and advise me. The efficient and patient cooperation of several staff-members of the Publisher also rendered me invaluable aid (but unfortunately their code of ethics does not allow me to thank them by name). Many of my friends and colleagues in Israel and overseas helped me in the solution of various major and minor matters, and my thanks are due to all of them, especially to Professor Z. Rappoport, Carrying out such a long-range project would be quite impossible without the non-professional but none the less essential participation and partnership of my wife

The Hebrew University
Jerusalem, ISRAEL

SAUL PATAI

Contents

1. General and theoretical aspects   1
   M. Simonetta and A. Gavezzotti

2. The structural chemistry of sulphonium salts   15
   E. F. Perozzi and I. C. Paul

3. Analysis and determination   79
   M. R. F. Ashworth

4. Thermochemistry of the sulphonium group   101
   R. Shaw

5. Photochemistry of sulphonium compounds   107
   J. D. Coyle

6. Electronic spectra   123
   G. C. Barrett

7. Electrochemistry of the sulphonium group   141
   J. Grimshaw

8. Isotopically labelled sulphonium salts   157
   L. F. Blackwell

9. Electronic effects of the sulphonium group   187
   J. Shorter

10. Stereochemistry and chiroptical properties of the sulphonium group   229
    K. K. Andersen

11. Synthesis of sulphonium salts   267
    P. A. Lowe

12. Reactivity of sulphonium salts   313
    A. C. Knipe

13. Cyclic sulphonium salts   387
    D. C. Dittmer and B. H. Patwardhan

14. Organosulphur cation radicals   523
    H. J. Shine

15. Heterosulphonium salts   571
      S. Oae, T. Numata and T. Yoshimura

16. Synthetic applications of sulphonium salts and sulphonium ylides   673
      E. Block

17. The biochemistry of sulphonium salts   703
      G. A. Maw

    Author Index   771

    Subject Index   833

The Chemistry of the Sulphonium Group
Edited by C. J. M. Stirling and S. Patai
© 1981 John Wiley & Sons Ltd.

CHAPTER 1

General and theoretical aspects

M. SIMONETTA and A. GAVEZZOTTI

Istituto di Chimica Fisica e Centro CNR, Università di Milano, Milan, Italy

I.	INTRODUCTION	1
II.	QUANTUM CHEMICAL DESCRIPTION OF THE SULPHONIUM GROUP.	2
	A. Simple picture of sulphur valence in the sulphonium group . . .	2
	B. Survey of *ab initio* calculations	3
	1. Acyclic compounds	3
	2. Cyclic compounds	5
	C. Extended Hückel calculations on sulphonium compounds . . .	6
	1. Geometries and parameterization	6
	2. Sulphonium ($\overset{+}{S}H_3$)	8
	3. Trimethylsulphonium [$\overset{+}{S}(CH_3)_3$]	12
	4. Triphenylsulphonium [$\overset{+}{S}(C_6H_5)_3$]	13
	5. Conclusions	13
III.	REFERENCES	14

I. INTRODUCTION

As a member of the second row of the Periodic Table, sulphur was not considered in the early applications of quantum chemical methods, the first heteroatoms to be included in the calculations obviously being nitrogen, oxygen, and fluorine. This is in general true both for semiempirical methods (the parameterization of which was extended to second-row elements only in subsequent stages of application) and for *ab initio* methods (for which the larger number of electrons posed initially strict limitations on the feasibility of the calculations). Today, however, the situation is radically changed, and there is no longer such a severe separation between the first row and the remainder of the Periodic Table. Moreover, the interesting bonding capabilities of sulphur, an element that forms a variety of compounds escaping simple dot-and-dash description, have made it an extremely interesting target for theoretical chemists. Some simple facts about the valence properties of sulphonium sulphur are presented in Section IIA.

A number of reviews dealing with calculations on sulphur-containing compounds have appeared. A very careful and critical review was contributed by Fabian[1];

1

Fabian has also given an updated account of the subject[2], which had been reviewed previously[3] and is dealt with continuously under specific headings in a periodic publication[4]. Therefore, no attempt will be made here to review once more the general aspects of the quantum chemistry of sulphur compounds; only some very recent publications, dealing specifically with matters of importance in the theoretical description of the sulphonium group, have been considered, and these are reviewed in Section IIB. Since these are few, we present in Section IIC a series of Extended Hückel calculations on SR_3 compounds, where R is H, CH_3, or C_6H_5. The out-of-plane bending and torsional barriers, and the shapes of some relevant molecular orbitals, have been obtained. These results may be helpful in casting some light on the main electronic features of sulphonium compounds.

II. QUANTUM CHEMICAL DESCRIPTION OF THE SULPHONIUM GROUP

A. Simple Picture of Sulphur Valence in the Sulphonium Group

The parent compound for sulphonium compounds can be considered the simple ion **1**, analogous to hydronium, OH_3; by substitution of H atoms with organic groups, one obtains sulphonium derivatives (**2**); strictly analogous are the sulphonium ylides (**3**) and the alkoxysulphonium group (**4**).

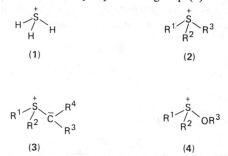

(1)   (2)

(3)   (4)

In all of these species, sulphur is formally trivalent and bears a positive charge. The simplest way of describing this electronic arrangement is to consider the five electrons of the valence shell of S^+ as being distributed in three bonds and one lone pair of electrons:

There can be little doubt that the configuration of sulphonium groups should be pyramidal, on the basis, for instance, of the need of mutual avoidance of valence shell electron pairs (VSEPR theory[5]; note, however, that the VSEPR description of first-row hydrides and their hybridization properties have been discussed by frozen-orbital calculations[6], and the results seem to infer that VSEPR should be applied with care to small molecules).

One possible and very simple way of describing the stabilization of the pyramidal structure with respect to the planar structure for **1** is in terms of perturbation interaction between HOMO and LUMO[7]. For AH_3 molecules, the relevant orbitals are shown in Figure 1; it can be seen that HOMO and LUMO are orthogonal in

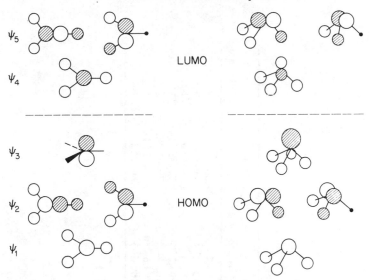

FIGURE 1. Relevant valence MOs for planar (left) and pyramidal (right) AH_3 molecules.

the planar structure, but can mix as the molecule assumes a pyramidal conformation leading to a stabilization energy

$$SE = 2\langle\psi_{HOMO}|\mathscr{H}|\psi_{LUMO}\rangle/(E_{HOMO} - E_{LUMO})$$

This point will be discussed further in Section C2.

B. Survey of Ab Initio Calculations

1. Acyclic compounds

The only sulphonium compound that has received considerable attention in the field of quantum chemical calculations is the parent ion SH_3. Large gaussian basis set LCAO MO SCF calculations gave, for instance[8], an equilibrium distance of 2.57 bohr for S—H and an equilibrium HSH angle of 98.7°. The barrier opposing the symmetrical inversion was computed to be 30.4 kcal/mol. For the substituted compounds (2), a wide variety of X-ray crystallographic evidence confirms the pyramidal structure (see Table 1); from these data, it is also clear that the RSR angles are always significantly smaller than the tetrahedral angle, even in the presence of bulky substituents. This suggests a hybridization with a p contribution larger than in sp^3, and the average angle of about 100° is in good agreement with the calculated value for SH_3. It is therefore surprising that, in some extended Hückel (EH)[15] and ASMO–SCF[16] calculations, a planar C_{3v} structure has been considered the most likely for trimethylsulphonium chloride.

The question may arise of how important the contribution of d orbitals to the description of bonding in sulphur compounds can be. This is a much debated question, and will not be discussed here in detail since the sulphonium group has never been considered specifically in this respect; the reader is referred to the

TABLE 1. Calculated and experimental properties of R_3S^+ groups[a]

R_1	R_2	R_3	RSR' angles	R—S bond lengths	Reference
H	H	H	98.7	1.36	8
CH_3	CH_3	CH_3	101–104	1.76–1.87	9
CH_3	neo-C_5H_{11}	neo-C_5H_{11}	96–105	1.76–1.90	10
CH_3	CH_2COOH	$C_2H_4C(COO^-)NH_3^+$	99.2–103.8	1.797–1.814	11
CH_3	$CH_2C(CH_3)=C(CH_3)_2$	$C(CH_3)_2CH(CH_3)_2$	101.1–107.5	1.799–1.882	12
CH_3	CH_3	$C_2H_4OOC(C_6H_5)$ (cyclohex)	99.6–101.6	1.784–1.811	13
Compound (7)			100–106	1.80–1.82	14

[a]Distances in Å and bond angles in degrees.

reviews mentioned earlier[1-3] for further discussion of this point. It may be mentioned that minimal basis set calculations on H_2S showed[17] that the contribution of 3d sulphur orbitals to the bonding molecular orbitals is extremely small, and that many molecular properties could be calculated with good accuracy without including these orbitals in the basis set; in what concerns positively charged sulphur, overlap between C 2p and S 3d orbitals has been explicitly invoked[18] in the explanation of n.m.r. data for the sulphonium ion:

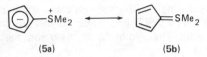

In another example, in the calculation (by the Molecules in Molecules method) of intramolecular electron transfer for the transition from the ground state **5a** to the excited state **5b** in the system[19]:

(5a)   (5b)

one of the π electrons of the rings was considered to transfer to an empty 3d orbital on the sulphur atom.

The electronic structures of ylides are written as resonance hybrids of ylides and ylenes. For the parent compound[20]

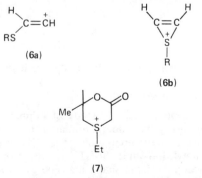

ylide   ylene

a gauche conformation of the ylide is calculated to be more stable[20].

2. Cyclic compounds

Considerable interest has been aroused by the electronic structure and properties of the system **6a** ↔ **6b**, and by the consequent problem of the most stable conformation of **6b**. Although the strain due to the smallness of the ring in this last compound makes the problem peculiar, the pyramidal structure at S survives in cyclic compounds, as exemplified by the crystal structure of **7** (see Table 1).

(6a)

(6b)

(7)

In a first study of the relative thermodynamic stabilities of **6a** and **6b** (R = H)[21], non-empirical SCF MO wave functions were obtained with full geometry optimization. Three different basis sets (31 sp, 53 sp, 77 spd) of contracted gaussian-type functions were used. Optimization was achieved by varying one geometrical parameter at a time for both the β-thiovinyl cation and the thiirenium ion; from the energy difference between the two absolute minima, the cyclic structure was found to be more stable (by 1–14 kcal/mol, depending on the basis set). The two minima were then connected through the multidimensional energy surface (with the assumption that all atoms move in a synchronized fashion), and the energy was calculated along this trajectory using the smallest basis set. A conversion barrier of 12.84 kcal/mol for **6b** → **6a** and of 8.96 kcal/mol for the reverse process were obtained.

The most stable conformation at sulphur in **6b** was found to be pyramidal; a very high barrier to inversion (72.9 kcal/mol) was calculated. The destabilization of the planar structure was considered to be related to violation of the Hückel rule (four π electrons in the ring), but also to the strained geometry of the ring, since test calculations on $\overset{+}{S}H_3$ showed that only when one HSH angle was kept at 38.78° (the optimized CSC bond angle for the thiirenium cation) did such a large barrier appear, while the symmetrical inversion mode was hindered by a barrier of only 9.8 kcal/mol.

The results summarized so far were refined in subsequent calculations with split-valence (4–31G) basis sets and more extensive geometry optimization[22]. The planar–pyramidal inversion for **6b** and the s-*cis* to s-*trans* inversion for **6a** were considered, and the four tautomers were connected through the energy surface by a procedure similar to that described above. Relative stabilities and barriers for the various transformations were thus obtained. The same calculations were carried out for the related systems $C_2H_2\overset{+}{O}H$, $\overset{+}{S}H_3$ and $\overset{+}{O}H_3$. The essential conclusions of the preceding paper[21] for $C_2H_2\overset{+}{S}H$ were unchanged; the analysis of the full energy hypersurface revealed, however, a number of easy paths for the pyramidal inversion in thiirenium via β-thiovinyl open structures.

An experimental confirmation of some of the results obtained by quantum-mechanical calculations came from an X-ray crystal structure analysis of **8**[23], which revealed a pyramidal structure at sulphur, with C—S—CH$_3$ angles of 106.0° and 106.3°.

(8)

C. Extended Hückel Calculations on Sulphonium Compounds

1. Geometries and parameterization

Since no systematic quantum-chemical study of sulphonium compounds could be found in the literature, we have undertaken an extended Hückel calculation on three sample derivatives, the sulphonium ion, $\overset{+}{S}H_3$, the trimethyl derivative, $\overset{+}{S}(CH_3)_3$, and the triphenyl derivative, $\overset{+}{S}(C_6H_5)_3$. The purpose of these calculations was to give some indications of the shape and energy of some relevant molecular orbitals, and on the energetics of the deformation from a planar to a pyramidal conformation.

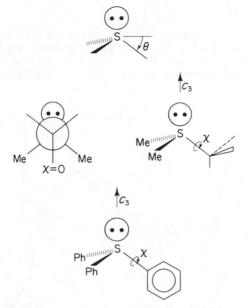

FIGURE 2. From top to bottom: definition of the bending coordinate θ; $\dot{S}(CH_3)_3$, definition of the methyl torsion angle χ; $\dot{S}(C_6H_5)_3$, definition of the torsion angle χ (at $\chi = 0$ the phenyl ring eclipses the direction of the lone pair). C_3 is a three-fold symmetry axis.

Figure 2 shows some geometrical details. The bond lengths were assumed to be as follows: S—H 1.36 Å (as resulting from *ab initio* calculations[8]); S—C 1.8 Å (average of the X-ray results, see Table 1); C—H 1.1 Å; and C—C 1.40 Å. The methyl groups were considered to be tetrahedral, and the phenyl rings were assumed to be rigid, planar hexagons. Besides the planar–pyramidal inversion coordinate, θ, only one torsional degree of freedom was allowed for methyl or phenyl derivatives (see Figure 2). The EH parameters were standard ones (see Table 2).

TABLE 2. Valence orbital ionization potentials (VOIPs) and Slater orbital exponents used in the Extended Hückel calculations

	Atom	VOIP, eV	Slater exponent
S[a]	3s	−20.0	2.122
	3p	−11.0	1.827
	3d	− 8.0	1.50
C	2s	−21.4	1.625
	2p	−11.4	1.625
H	1s	−13.6	1.3

[a]Reference 24.

2. Sulphonium ($\overset{+}{S}H_3$)

The energy profile for the symmetric deformation of this ion is shown in Figure 3. With or without d orbitals in the basis set, a pyramidal structure ($\angle HSH \approx 95°$) is predicted to be the most stable; the calculated barrier to inversion is 25 kcal/mol with d orbitals and 18 kcal/mol without d orbitals. These results can be compared

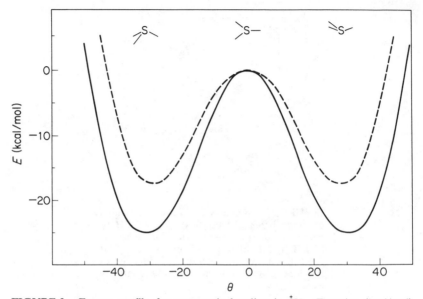

FIGURE 3.   Energy profile for symmetric bending in $\overset{+}{S}H_3$. Energies (kcal/mol) relative to $\theta = 0$. Full line, S 3s, 3p, 3d; broken line, S 3s, 3p basis set.

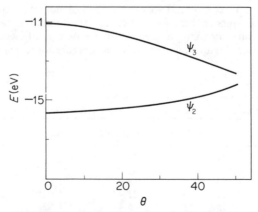

FIGURE 4.   Energy variations for levels ψ_2 and ψ_3 of $\overset{+}{S}H_3$ (see Figure 1) as a function of the bending coordinate θ. S 3s, 3p basis set.

with the *ab initio* values ($\angle$HSH = 98.7° and barrier = 30.4 kcal/mol from the large basis set calculations in reference 8). Figure 4 shows the variation in energy of some relevant molecular orbitals along the bending coordinate; it can be seen that the minimum at $\theta \approx 30°$ is obtained as a balance between the energy increase of the four electrons in level ψ_2 and the energy decrease of the two electrons in ψ_3. The behaviour of the HOMO illustrates well the HOMO–LUMO interaction argument proposed to explain the stabilization of the pyramidal structure[7]. The inclusion of d orbitals affects the magnitude of the calculated barrier, but does not change the geometric prediction of $\angle$HSH $\approx 95°$, and affects only slightly the overall shape and the energy trends of the MOs shown in Figure 4.

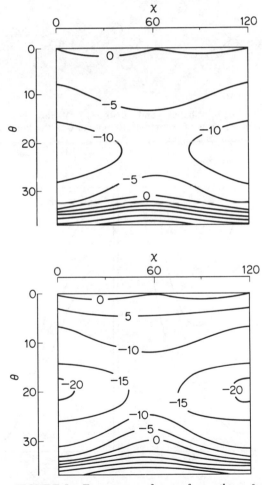

FIGURE 5. Energy surface for the θ bending/methyl torsion in trimethylsulphonium. Energies in kcal/mol relative to $\theta = \chi = 0$. Top, without; bottom, with S 3d orbitals.

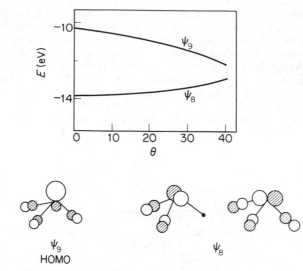

FIGURE 6. Shapes and energies of the highest occupied MOs of trimethylsulphonium. θ is the bending coordinate (see Figure 2).

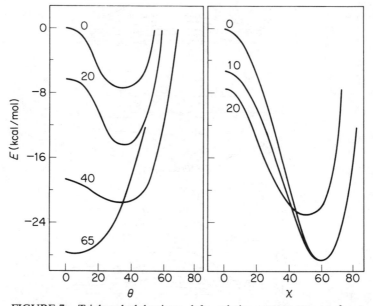

FIGURE 7. Triphenylsulphonium: left, relative energy curves for θ bending at various values of χ (the numbers above each curve); right, the same for χ torsion at various values of θ (the numbers above each curve). S 3s, 3p basis set. Energies in kcal/mol relative to $\theta = \chi = 0$.

TABLE 3. Some calculated properties of sulphonium derivatives (EH results)

	θ_{min} (degrees)	χ_{min} (degrees)	∠HSH or ∠CSC	Barrier (kcal/mol)	E(HOMO) (eV)	E(LUMO) (eV)	Charge on S (electrons)	Charge on H or C (electrons)	Bond overlap population (H—S or C—S)
$\overset{+}{S}H_3$									
with d orbitals	33	—	93	25	-12.27	-9.30	+1.20	-0.07	0.7877
without d orbitals	30	—	97	18	-12.26	-0.76	+1.19	-0.06	0.7106
$\overset{+}{S}(CH_3)_3$									
with d orbitals	20	0	109	24	-11.16	-8.12	+0.94	-0.13	0.8311
without d orbitals	20	0	109	13	-10.94	-0.35	+1.21	-0.15	0.6724
$\overset{+}{S}(C_6H_5)_3$									
without d orbitals	~0	60	~120	—	-10.62	-8.37	+1.47	0.0	0.7561

3. Trimethylsulphonium [$\overset{+}{S}(CH_3)_3$]

As expected, trimethylsulphonium also has a pyramidal conformation (see Figure 5); the barrier to inversion is calculated to be 24 kcal/mol with d orbitals and 13.5 kcal/mol without d orbitals. Since a three-fold symmetry axis was imposed on this molecule, the torsional degree of freedom χ (Figure 2) was the only other degree of freedom that could be varied. Figure 5 shows also the influence of χ; as could have been predicted by simple steric arguments, $\chi = 0°$ is always preferred.

The strict similarity between $\overset{+}{S}H_3$ and its trimethyl derivative is confirmed by the results shown in Figure 6. There the shapes of the highest occupied orbitals of trimethylsulphonium are shown, together with their variation in energy along the bending deformation coordinate. The analogy with the situation for sulphonium (see Figure 4) is evident.

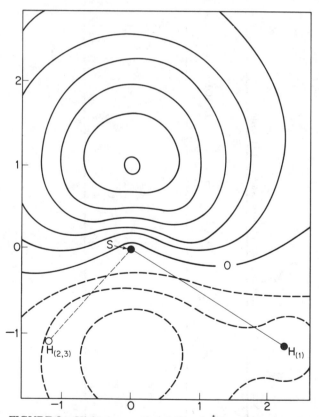

FIGURE 8. Highest occupied MO of $\overset{+}{S}H_3$. Section in the plane containing the S atom, $H_{(1)}$ (closed circle) and bisecting the $H_{(2)}SH_{(3)}$ angle. The wave function is plotted at 0.05 intervals. 0 denotes a nodal plane. Coordinates in a.u.

4. Triphenylsulphonium $[\overset{+}{S}(C_6H_5)_3]$

This ion adopts a slightly bent conformation ($\theta \approx 0°$), as can be seen from the results shown in Figure 7. The need for mutual avoidance of the bulky groups, which would favour small values of χ, is largely balanced by conjugation effects that favour maximum coplanarity of the phenyl rings, resulting in a large χ value for the minimum energy conformation ($\chi \approx 60°$). From Figure 7 it can be seen that at small values of χ a large θ bend is possible, with a sizeable barrier to inversion; as χ increases, θ bending becomes less and less favoured, and correspondingly the barrier becomes smaller and smaller. At large χ values a shallow minimum near $\theta = 0°$ allows a large amplitude vibration of the molecule along the θ coordinate.

5. Conclusions

Table 3 gives some Extended Hückel results for sulphonium compounds, including charges and bond overlap populations. These results are generally reasonable, and the comparison with *ab initio* results for $\overset{+}{S}H_3$ is very favourable. The success of EH in reproducing the shape of this compound suggests that the results for the substituted compounds are also reliable. The good performance of EH in what concerns the energies of bending and torsional deformations is, anyway, well known.

The shape of the HOMO of sulphonium and trimethylsulphonium is shown in Figures 8 and 9. The 'lone pair' character of this orbital is clear, although some

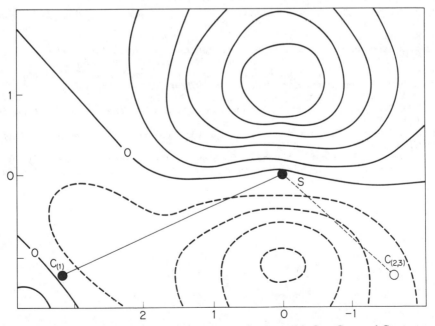

FIGURE 9.   Same as Figure 8 for trimethylsulphonium with $C_{(1)}$, $C_{(2)}$, and $C_{(3)}$ atoms instead of $H_{(1)}$, $H_{(2)}$, and $H_{(3)}$.

bonding tails extend towards the substituents. The energy of this orbital, and its shape, are only very slightly modified by the inclusion in the basis set of the sulphur d orbitals; only a small contribution from a d_{z^2} orbital appears. The atomic charges show the expected trends, corresponding to a molecular dipole with the positive end placed very near to the sulphur atom, which bears a positive charge of about unity.

III. REFERENCES

1. J. Fabian, in *Sulfur in Organic and Inorganic Chemistry*, Vol. 3 p. 39 (Ed. A. Senning), Marcel Dekker, New York, 1972.
2. J. Fabian, in *Organic Compounds of Sulphur, Selenium and Tellurium* (Ed. D. H. Reid), Specialist Periodical Report, Vol. 3, p. 728, The Chemical Society, London, 1975.
3. D. T. Clark, in *Organic Compounds of Sulphur, Selenium and Tellurium* (Ed. D. H. Reid), Specialist Periodical Report, Vol. 1, p. 1, The Chemical Society, London, 1970.
4. *Organic Compounds of Sulphur, Selenium and Tellurium* (Ed. D. H. Reid), Specialist Periodical Report, Vols. 1–4, The Chemical Society, London, 1970–77.
5. R. Gillespie and R. Nyholm, *Quart. Rev. Chem. Soc.*, **11**, 339 (1957).
6. J. Jarvie, W. Willson, J. Doolittle and C. Edmiston, *J. Chem. Phys.*, **59**, 3020 (1973).
7. W. Cherry, N. Epiotis and W. T. Borden, *Accounts Chem. Res.*, **10**, 167 (1977).
8. R. E. Kari and I. G. Csizmadia, *Can. J. Chem.*, **53**, 3747 (1975).
9. P. Biscarini, L. Fusina, G. Nivellini and G. Pelizzi, *J. Chem. Soc. Dalton Trans.*, 664 (1977).
10. B. T. Kilbourn and D. Felix, *J. Chem. Soc. A*, 163 (1969).
11. J. W. Cornforth, S. A. Reichard, P. Talalay, A. L. Carrell and J. P. Glusker, *J. Amer. Chem. Soc.*, **99**, 7292 (1977).
12. W. Barnes and M. Sundaralingam, *Acta Crystallogr.*, **B29**, 1868 (1973).
13. J. J. Guy and T. A. Hamor, *J. Chem. Soc. Perkin Trans. II*, 467 (1975).
14. E. Kelstrup, A. Kjaer, S. Abrahamsson and B. Dahlen, *J. Chem. Soc. Chem. Commun.*, 629 (1975).
15. K. Ohkubo and H. Kanaeda, *Bull. Chem. Soc. Jap.*, **45**, 11 (1972).
16. K. Ohkubo and H. Kanaeda, *J. Chem. Soc. Faraday Trans. II*, **68**, 1164 (1972).
17. S. Polezzo, M. P. Stabilini and M. Simonetta, *Mol. Phys.*, **17**, 609 (1969).
18. M. C. Caserio, R. E. Pratt and R. J. Holland, *J. Amer. Chem. Soc.*, **88**, 5747 (1966).
19. K. Iwata, S. Yoneda and Z. Yoshida, *J. Amer. Chem. Soc.*, **93**, 6745 (1971).
20. M. Whangbo and S. Wolfe, *Can. J. Chem.*, **54**, 949 (1976).
21. I. G. Csizmadia, A. J. Duke, V. Lucchini and G. Modena, *J. Chem. Soc. Perkin Trans. II*, 1808 (1974).
22. I. G. Csizmadia, F. Bernardi, V. Lucchini and G. Modena, *J. Chem. Soc. Perkin Trans. II*, 542 (1977).
23. R. Destro, T. Pilati and M. Simonetta, *J. Chem. Soc. Chem. Commun.*, 576 (1977).
24. R. Hoffmann, J. Fujimoto, J. R. Swenson and C. Wan, *J. Amer. Chem. Soc.*, **95**, 7644 (1973).

The Chemistry of the Sulphonium Group
Edited by C. J. M. Stirling and S. Patai
© 1981 John Wiley & Sons Ltd.

CHAPTER **2**

The structural chemistry of sulphonium salts

EDMUND F. PEROZZI

Ethyl Corporation, 1600 West Eight Mile Road, Ferndale, Michigan 48220, USA

and

IAIN C. PAUL

Department of Chemistry, School of Chemical Sciences, University of Illinois, Urbana, Illinois 61801, USA

I.	INTRODUCTION	16
II.	STRUCTURES THAT HAVE BEEN STUDIED BY X-RAY METHODS	21
	A. Sulphonium Salts	21
	B. Sulphonium Ylides	27
	C. Aminosulphonium Salts	30
	D. Sulphilimines	31
	E. Selenium Compounds	34
III.	MOLECULAR DIMENSIONS OF SULPHONIUM SALTS AND RELATED COMPOUNDS	35
	A. Molecular Dimensions of Sulphonium Salts	38
	B. Molecular Dimensions of Sulphonium Ylides	47
	C. Molecular Dimensions of Aminosulphonium Salts	50
	D. Molecular Dimensions of Sulphilimines	55
	E. Molecular Dimensions of Selenium Analogues of Sulphonium Salts	64
	F. Other Features of Intramolecular Geometry of Sulphonium Salts and Related Compounds	66
IV.	INTERMOLECULAR INTERACTIONS IN SULPHONIUM SALTS AND RELATED COMPOUNDS	68
V.	ACKNOWLEDGEMENTS	74
VI.	REFERENCES	76

I. INTRODUCTION

This chapter surveys the geometrical information available on the sulphonium group. Among the types of information considered will be the bond lengths and angles involving the sulphur atom, torsion angles around bonds involving the sulphur atom, questions of planarity and non-planarity of the sulphonium group, and geometric aspects of the way molecules containing sulphonium groups interact with other molecules or ions. Virtually all of the information presented was obtained from the results of X-ray crystallographic investigations on the solid state of molecules containing the sulphonium group. There do not appear to be any electron diffraction studies or microwave studies on sulphonium salts in the vapour phase. In addition to discussions of sulphonium salts, there is some description of molecules containing the selenonium group.

In molecules that have been studied by X-ray crystallography, four categories of sulphonium groups or of groups closely related to sulphonium groups can be recognized. These are: the sulphonium group in normal sulphonium salts where sulphur is bonded to three carbon atoms **1**, the case where the sulphonium group is part of an ylide with the negative charge residing on a carbon atom bonded to the sulphur **2**, the case where one of the atoms bonded to sulphur is a nitrogen atom rather than a carbon atom, i.e. aminosulphonium, azasulphonium, or sulphiminium salts **3**, and the case where that nitrogen atom bears a negative charge and is involved in ylide formation, i.e. sulphilimines **4**. The fifth category

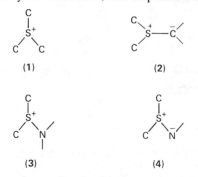

described is that of selenonium salts. In this last case, the number of compounds studied is so small that it is of little value for them to be categorized as for the sulphur compounds. Three other classes of molecule can be thought of as related to sulphonium salts. These are the sulphuranes **5**, thiathiophthenes **6**, and dithifurophthenes **7**. However, as the resemblance is not very great, and many aspects of their geometries have been discussed recently[1-3], they are not considered further in this chapter.

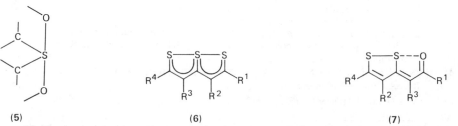

As the number of compounds studied in the first five categories described above is not very large, an attempt is made to include all of the available data in the survey. There is a brief description, by category, of the various structures that have been determined in each group; in this description, any unusual features of the crystal or of the analysis are discussed. Then there is a comparative survey of various aspects of the molecular geometry. Finally, any information on the capability of the molecules for intermolecular interactions that can be obtained from an examination of their crystal packing is described.

Many of the X-ray analyses reported in this chapter have been carried out in the past 10–15 years and have therefore mostly been based on intensity data measured on a diffractometer equipped with a counter. Much of the early work was based on visual estimates of intensities recorded on photographic film. At a later stage, photometers were used to measure the intensities that were integrated on to film. In general, better accuracy is obtained from counter measurements, although the parallel method of data acquisition associated with film data is better suited to a rapidly decaying crystal than is the serial method associated with counters. However, for a variety of reasons, including the quality of the crystals, the means of data collection, possible decay of the crystal, the possible existence of molecular disorder in the crystal, and the extent of refinement, the precision of the analyses covered in this chapter does vary widely.

The crystallographic R-factor, which is defined as $\Sigma \| F_{obs} | - | F_{calc} \| / \Sigma | F_{obs} |$, has been traditionally used as indicator of the agreement between the observed and calculated structure factors and, indirectly, as a measure of the quality of the analysis. There are other measures of the quality of the analysis or refinement, e.g. the size of the estimated standard deviations for the structural parameters, the value for the standard deviation of an observation of unit weight, or the 'goodness of fit'[4], and the internal consistency of bond lengths that should be chemically equivalent. Some of these are definitely superior to the R-factor, but the R-factor remains widely quoted and is still a fairly useful measure of the level of reliability of the analysis. With photographic data, R-factors less than 0.13–0.14 indicate reasonably well refined structures, whereas R-factors of about 0.20 imply either that the measured intensities are inaccurate or that there is something significantly wrong with the structural model, although probably not to the extent that the overall molecular structure is incorrect. When the reflection data are measured by counters, usually on an automatic diffractometer, R-factors below 0.05 can be obtained in careful work.

Another useful guide to the quality of an analysis can often be obtained by the treatment and behaviour of the hydrogen atoms. If these atoms were clearly located from a difference map, i.e. a map calculated with the difference between the measured and calculated structure amplitudes as coefficients, then it is unlikely that there are serious errors in the data. Another indicator of a good model is if the hydrogen atoms were included in the final refinements and their positions and thermal parameters were varied and the results gave chemically reasonable values. If the hydrogen atoms are not well defined, their inclusion in the refinement often leads to a chemically unreasonable molecular geometry. A common technique with reasonably well defined hydrogen atoms is to include them in the refinement but to hold the thermal parameters constant. If the hydrogen atoms cannot be located from a difference map, they can often be positioned using standard criteria. For example, a hydrogen atom attached to a carbon atom in a benzene ring is placed at a distance of 1 Å from the carbon and such that it lies in the plane of the benzene ring and bisects the external C—C—C angle. The inclusion of hydrogen atoms is

important both for their contribution to the quality of the analysis and as a guide to
its precision. For example, if hydrogen atoms of a methyl group are not included in
the analysis, there will be a tendency for the C—C(methyl) bond to be lengthened
artificially, as the refinement process will tend to place the methyl carbon further
out in order to compensate partly for the electron density of the hydrogen
(Figure 1).

The estimated standard deviations quoted in an X-ray diffraction paper are
usually obtained from the elements of the inverse matrix of the normal equations
relating the atomic parameters in the least-squares procedures. Other things, such
as the precision of the measured data, being equal, the larger the number of
independent measurements, the smaller will be the estimated standard deviations.
For a more detailed discussion of these procedures, see reference 5. However, such
estimated standard deviations are based on an assumption that the errors in the
data follow a normal distribution, i.e. that they are subject only to random errors.

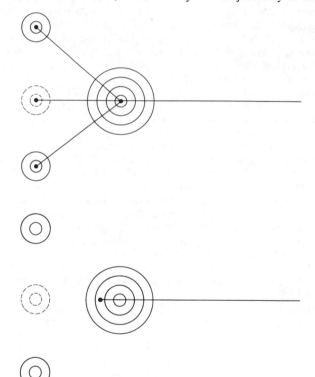

FIGURE 1.   Effect of omitting hydrogen atoms from
the model on the length of a C—C bond. If hydrogen
atoms are included in model, the correct length is
obtained (upper drawing). If hydrogen atoms are *not*
included in the model, there will be a tendency for an
artificial lengthening of the C—C bond as the carbon will
be placed in a position where it can partially take
account of the electron density due to hydrogen atoms
(lower drawing).

Unfortunately, there are several sources of systematic error in X-ray intensity measurements, some of which can be difficult to correct exactly. In the case of data collected by film, errors often arise due to variations in the response of the X-ray film, the difficulty of measuring integrated intensities, particularly when the intensity of the reflection was distributed in different shapes at different parts of the film, and because of the problems of scaling the films of different nets together. While the advent of automatic diffractometers employing counters has removed many of these problems, errors can also arise due to the difficulty of treating such corrections that arise from absorption or extinction of the scattered X-ray beams. Many of the effects of these types of errors would not be reflected in the value of the R-factor or in the estimated standard deviations that would be obtained from the least squares refinement.

In all crystals above 0 K, there are vibrations of the atoms. The simplest way to treat this vibration is to consider it as isotropic and apply an exponential term to the atomic scattering factor. The scattering factor at temperature T, f_T, is given by

$$f_T = f_0 \exp(-B\sin^2\theta/\lambda^2)$$

where f_0 is the scattering factor for the atom at rest, B is the isotropic temperature factor in Å^2, 2θ is the scattering angle, and λ is the wavelength. In general, a more appropriate treatment is to describe the thermal ellipsoid by a tensor, which is defined by six parameters; three of these are related to the vibration along the three crystallographic axes (b_{11}, b_{22}, b_{33}) and the other three are cross terms (b_{12}, b_{13}, b_{23}):

$$f_T = f_0 \exp - (b_{11}h^2 + b_{22}k^2 + b_{33}l^2 + 2b_{12}hk + 2b_{13}hl + 2b_{23}kl)$$

In this expression, h, k, and l are the three crystallographic indices and refer to the three axes, a, b, and c. An illustration of a molecule where the vibrations of the non-hydrogen atoms were described in anisotropic terms is shown in Figure 2.

An inappropriate treatment of the thermal motion of the atoms can lead to errors in the bond lengths. For example, if a methyl carbon atom vibrates significantly in a direction normal to an $S-CH_3$ bond, then there will be an artificial shortening of the measured $S-C$ distance as the maximum of the peak due to the carbon atom will correspond to the mid-point of the two extreme positions (Figure 3). These problems and others associated with thermal motion can be lessened by collecting the intensity data at a lower temperature.

While the molecular dimensions obtained in an X-ray diffraction experiment conducted under optimum conditions (good crystals, good data collection facilities, and adequate computing facilities) should be fairly accurate, e.g. about 0.005 Å for a $C-C$ (or $C-N$) bond, it should be kept in mind that the molecules in the crystal are in a high state of aggregation, that they are almost always required by the nature of crystals and their symmetry to adopt a uniform conformation, and that intermolecular interactions can play a more important role in determining the molecular conformation than they would in the gas phase or in solution.

The space group to which a crystal belongs describes the symmetry operations that relate the molecules in three-dimensional space. Some of these symmetry operations involve rotation operations which do not transform an object into its mirror image; others involve centres of inversion, rotation–inversion axes, or mirror reflections which do result in a transformation of an object into its mirror image. Certain space groups, which will be designated here as chiral space groups, contain a molecule in just one enantiomorphic form, while others can contain the molecule as a racemate. As individual molecules are discussed, the type of space group in

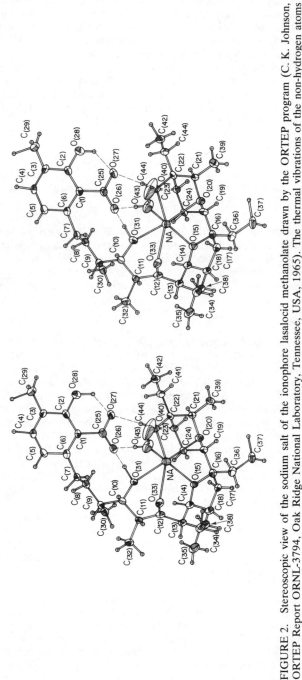
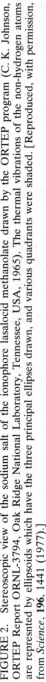

FIGURE 2. Stereoscopic view of the sodium salt of the ionophore lasalocid methanolate drawn by the ORTEP program (C. K. Johnson, ORTEP Report ORNL-3794, Oak Ridge National Laboratory, Tennessee, USA, 1965). The thermal vibrations of the non-hydrogen atoms are represnted by ellipsoids which have the three principal ellipses drawn, and various quadrants were shaded. [Reproduced, with permission, from *Science*, **196**, 1441 (1977).]

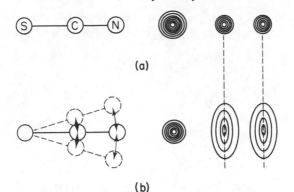

(a)

(b)

FIGURE 3.   Effects of thermal vibrations normal to a chemical bond upon the bond length determined by X-ray diffraction. The figure shows the effect on a linear group such as the thiocyanate group resulting in a shortened bond length. (Reproduced, with permission, from *The Chemistry of the Thiocyanate Group and Derivatives, Part I,* (Ed. S. Patai), Wiley–Interscience, 1978, pp. 69–129.)

which they crystallize will be stated. A fuller description of space groups, their symmetry, and the conditions that they impose on molecules can be found in reference 6.

II. STRUCTURES THAT HAVE BEEN STUDIED BY X-RAY METHODS

A. Sulphonium Salts 1

Of the five categories, simple sulphonium salts have been the subjects of more X-ray analyses than have members of the other categories. The earliest analysis was carried out in 1959 by Zuccaro and McCullough[7] on trimethylsulphonium iodide **8**. Since then there have been studies on dimethylphenylsulphonium perchlorate **9**[8], 2,3-dimethyl-2-butenyl-1,1,2-trimethyl propylmethylsulphonium 2,4,6-trinitrobenzenesulphonate **10**[9], 2-[cyclohexyl(phenyl)acetoxy]ethyldimethylsulphonium (hexasonium) iodide **11**[10], *R*-{[(3*S*)-3-amino-3-carboxypropyl](carboxymethyl)methylsulphonium} 2,4,6-trinitrobenzesulphonate **12**[11], *S*-methyl-L-methionine chloride hydrochloride **13**[12], 1-methyl-1-thioniacyclohexane iodide **14**[13], *cis*- and *trans*-4-*t*-butyl-*S*-methylthianium perchlorate **15** and **16**[14], *S*-ethyl-β-thia-δ-caprolactone tetrafluoroborate **17**[15], 1-acetonyl-1-thionia-5-thiacyclooctane perchlorate **18**[16], phenylsulphoxonium (phenylphenoxtinium) iodide **19**[17], 1,4-dithioniabicyclo[2.2.2]octane tetrachlorozincate **20**[18], and *trans*-1-thioniabicyclo[4.4.0]decane bromide **21**[19]. In addition to the above structures, there have been some analyses where the geometry of the sulphonium group was incidental to the main purpose of the study. Included in this group are analyses of methyldineopentylsulphonium triiododineopentylsulphonium methyl zincate **22**[20] and of trimethylsulphonium mercuritriiodide **23**[21] and bistrimethylsulphonium mercuritetraiodide **24**[22], where the purpose of the investigations was the structures of the anions, and the analyses of two modifications of water clathrates **25**[23,24] which enclosed tri-*n*-butylsulphonium fluoride but in a highly disordered fashion.

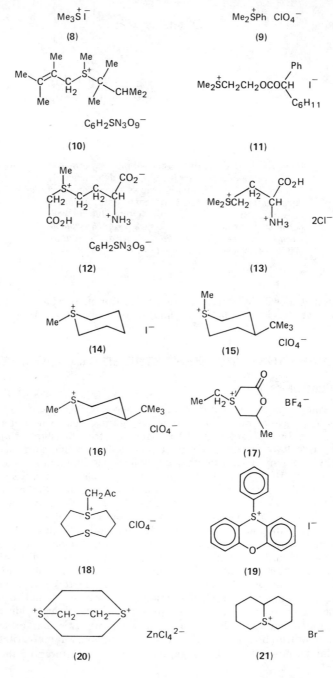

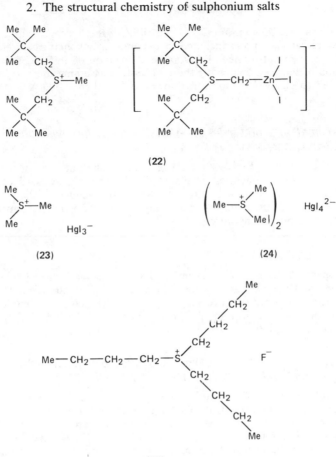

(22)

(23)

(24)

(25)

1. Trimethylsulphonium iodide 8[7]

There is some ambiguity about the correct space group for this molecule. The model was refined in $P2_1$ (chiral), but the structure closely approaches C_s (mirror) symmetry through the sulphur and one of the carbon atoms, and the authors do not make a definite choice of $P2_1$ over $P2_1/m$ (racemic), which would require exact C_s symmetry. The model was refined to an R-factor of 0.069 on about 350 non-zero reflections that were collected photographically. The sulphur and iodine atoms were refined with anisotropic thermal parameters, the carbon atoms were included with final B values of 5.0 Å, and no hydrogen atoms were included. The standard deviation of the S—C bond lengths ranged from 0.06 to 0.12 Å.

2. Dimethylphenylsulphonium perchlorate 9[8]

The space group is $P2_1/c$ (racemic). The model was refined to an R-factor of 0.14 on 823 non-zero reflections collected photographically. As a super-lattice appeared when the crystal was cooled, a greater number of reflections could not be obtained at lower temperatures. The positions of the hydrogen atoms of the phenyl group were calculated and included in the model, but their positional and thermal parameters were not refined. The standard deviation of an S—C bond is 0.02 Å. A correction of the coordinates for thermal motion assuming the two ions act as rigid bodies led to changes in position that were less than 0.006 Å in the sulphonium ion.

3. 2,3-Dimethyl-2-butenyl-1,1,2-trimethylpropylmethylsulphonium 2,4,6-trinitrobenzenesulphonate 10[9]

The space group is $P\bar{1}$ (racemic). The model was refined to an R-factor of 0.057 on 3504 non-zero reflections measured on a diffractometer. All of the hydrogen atoms were located from a difference map, were included in the model, and their positional, but not their thermal, parameters were refined. The standard deviations of the S—C bond lengths were 0.003–0.004 Å.

4. Hexasonium iodide 11[10]

The space group is $P2_1/c$ (racemic). The model was refined to an R-factor of 0.063 on 2443 non-zero reflections collected on a diffractometer. Hydrogen atoms were located from a difference map, but were included in the model at calculated positions and were not refined. The estimated standard deviations of the S—C bond lengths were 0.009–0.010 Å.

5. R-{[(3S)-3-Amino-3-carboxypropyl](carboxymethyl)methyl-sulphonium} 2,4,6-trinitrobenzenesulphonate 12[11]

The space group is $P1$ (chiral). There were two independent molecules in the asymmetric unit. The crystals displayed a tendency for a marked anisotropic decay after 40–45 h of exposure to X-rays; three crystals were used to assemble the data. The model was refined to an R-factor of 0.070 on 3719 reflections collected on a diffractometer. The hydrogen atoms except those of the amino group were located from a difference map, and were included in the refinement with fixed idealized positional parameters and isotropic thermal parameters equal to those of the atoms to which they were attached. The estimated standard deviations of the S—C bond lengths were 0.006–0.007 Å.

6. S-Methyl-L-methionine chloride hydrochloride 13[12]

The space group was $P2_1$ (chiral). The model was refined to an R-factor of 0.022 on 1684 non-zero reflections collected on a diffractometer. The hydrogen atoms were located from a difference map and included in the calculations with fixed parameters ($B = 4.0$ Å^2). The estimated standard deviations of the S—C bond lengths were 0.002–0.003 Å.

7. 1-Methyl-1-thioniacyclohexane iodide 14[13]

The space group is $Pna2_1$ (racemic). The crystals decomposed readily and four crystals were used to assemble the data. The final model was refined to an R-factor of 0.057 on 884 non-zero reflections measured on a semi-automatic diffractometer. The hydrogen atoms bonded to the ring carbon atoms were included in the model at their calculated positions but apparently were not refined. The standard deviations of the individual atoms or bond lengths were not included in the paper, but the average standard deviation was about 1% for bond lengths, suggesting a value of about 0.02 Å for a C—S bond.

8. cis-4-t-Butyl-S-methylthianium perchlorate 15[14]

A full description of this analysis and of that of the corresponding *trans*-isomer has not been published, so only a minimal number of experimental details are available. The space group is $P2_12_12_1$ (chiral). The model was refined to an *R*-factor of 0.063 on 875 non-zero reflections collected on a diffractometer. The hydrogen atoms were included in the refinement with fixed parameters. The standard deviation of the S—C bond length was 0.010 Å.

9. trans-4-t-Butyl-S-methylthianium perchlorate 16[14]

The space group is *Pbca* (racemic). The model was refined to an *R*-factor of 0.104 on 890 non-zero reflections collected on a diffractometer. The hydrogen atoms were included in the refinement with fixed parameters. The standard deviations of the S—C bond lengths were 0.01–0.02 Å.

10. S-Ethyl-β-thia-δ-caprolactone tetrafluoroborate 17[15]

Only a preliminary account of this study is available. The space group is $P2_1/c$ (racemic). The model was refined to an *R*-factor of 0.13 on 795 non-zero reflections collected on a diffractometer. Apparently the hydrogen atoms were included in the final calculations, but no details were given. The standard deviations of the S—C bond lengths were 0.03–0.04 Å.

11. 1-Acetonyl-1-thionia-5-thiacyclooctane perchlorate 18[16]

The space group is *Pbca* (racemic). The model was refined to an *R*-factor of 0.107 on 1117 independent reflections collected photographically. The hydrogen atoms were located from a difference map and included in the early cycles of refinement, but they were held constant in the later cycles. The standard deviations of the S—C bond lengths were 0.01–0.02 Å.

12. Phenylsulphoxonium (phenylphenoxtinium) iodide 19[17]

The space group is $P2_1/n$ (racemic). The model was refined to an *R*-factor of 0.106 on 800 reflections recorded on film and estimated visually. Hydrogen atoms were not located or included in the model, and only isotropic thermal parameters were employed for the other atoms. The standard deviation of an S—C bond length was about 0.04 Å.

13. 1,4-Dithioniabicyclo[2.2.2]octane tetrachlorozincate 20[18]

The space group of this compound is in some doubt. The author[18] presents three possible models: an ordered one in the space group *Pnma*, a disordered one in *Pnma*, and an ordered one in the space group $Pn2_1a$; both space groups would correspond to a racemate. The author concluded that a disordered structure in

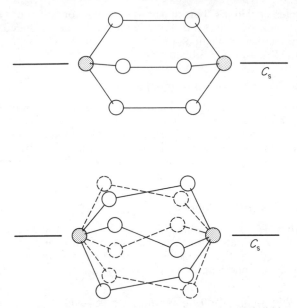

FIGURE 4. Two possible structures for 1,4-dithio-
niabicyclo[2.2.2]octane tetrachlorozincate **20**. (a)
Ordered structure with C_s crystallographic symmetry. (b)
Two-fold disordered arrangement of two individual
structures that each lack C_s crystallographic symmetry.

Pnma was probably correct. The cation has crystallographic C_s symmetry with both
sulphur atoms lying on the mirror plane, but all carbon atoms occupying different
sites in the two disordered molecular orientations (Figure 4). As pointed out by the
author[18], a standard X-ray analysis does not enable one to make a distinction
between this model with static disorder or one where the disorder is due to
a dynamic process. The disordered model was refined (with isotropic thermal
parameters for the carbon atoms) to an R-factor of 0.110 on 908 non-zero
reflections collected on a semi-automatic diffractometer. Hydrogen atoms were
included in all models and both positional and thermal parameters for the hydrogen
atoms were refined. The standard deviations of the S—C bond lengths ranged from
0.009 to 0.017 Å.

14. trans-1-Thioniabicyclo[4.4.0]decane bromide **21**[19]

The space group is $Cmc2_1$ (racemic). The molecule has C_s symmetry in the
crystal, with the mirror plane passing through the sulphur atom and relating one
six-membered ring to the other. The final model was refined to an R-factor of
0.044 on 603 non-zero reflections (four strong reflections were removed from the
last calculations) collected on a diffractometer. Although the hydrogen atoms were
included in the calculations with fixed parameters ($B = 5.0$ Å^2), the hydrogen
coordinates were not given in the paper. The standard deviation of the S—C bond
length was 0.008 Å.

15. Methyldineopentylsulphonium triiododineopentylsulphonium methyl zincate **22**[20]

This structure was obtained from an attempt to prepare a structure with the group $R_2\overset{+}{S}CH_2ZnCH_2\overset{+}{S}R_2$. The actual structure determined is unusual in that there is a sulphonium group in both the cation and the anion. The main aim of the analysis was to determine the gross structure of the species obtained. The space group is $P2_1/c$. The final R-factor was 0.095 for 1263 reflections collected on a semi-automatic diffractometer. Hydrogen atoms were not included in any calculations. The standard deviation of an S—C bond length was 0.06 Å.

16. Trimethylsulphonium mercuritriiodide **23**[21]

The purpose of this analysis and that of bismethylsulphonium mercuritetraiodide was to determine the structure of the anions which are used as heavy atom derivatives in protein structure analysis. The space group is $P2_1/c$ (racemic). The model was refined to an R-factor of 0.12; the number of reflections, measured on a semi-automatic diffractometer, was not given. Hydrogen atoms were not included. The standard deviations of the S—C bond lengths were not given, but the range of the lengths was 1.5–2.0 Å.

17. Bistrimethylsulphonium mercuritetraiodide **24**[22]

The space group was $Pna2_1$ (racemic). The R-factor was 0.09 on data collected photographically, but the methyl groups could not be located owing to the dominating influence of the mercury and iodine atoms. There was even uncertainty as to which of two possible peaks was occupied by the sulphur atoms; accordingly, they may not be ordered. Hence no useful information on the geometry of the sulphonium group was obtained.

18. Tri-n-butylsulphonium fluoride hydrate **25**[23,24]

Two forms of this clathrate have been studied by X-ray methods. One is cubic and the space group is either $P\bar{4}3n$ or $Pm3n$[23]. In the cubic clathrate structure, a disordered arrangement for the tri-n-butylsulphonium atoms was postulated to be present in the voids formed by the tetrakaidecahedra. No useful information on the geometry of the sulphonium group could be obtained from the analysis.

The other form is monoclinic (space group $P2_1/m$)[24]. In this form the structure of the sulphonium group was rather better defined. The model was refined to an R-factor of 0.13 on 3708 non-zero reflections measured photographically. The t-butyl groups exhibit disorder so no detailed information on the geometry of the cation was obtained. Two views of the crystal structure are given in Figures 5 and 6. In Figure 5 the dodecahedra of water molecules are shown by lines with oxygen atoms at the vertices. In between layers of dodecahedra are layers of anions and other water molecules; these layers also contain large and irregular cavities in which are disordered arrangements for the cations. A closer view of the arrangements of the cations is shown in Figure 6.

B. Sulphonium Ylides 2

Four sulphonium ylides have been examined by X-ray structure analysis and all four analyses were carried out in the last decade. The structures determined were

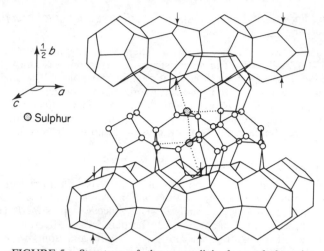

FIGURE 5. Structure of the monoclinic form of the tri-*n*-butylsulphonium fluoride hydrate clathrate **25**. The *n*-butyl chains are omitted for clarity. The vertices of the dodecahedra represent oxygen atoms. Those oxygen atoms not belonging to a dodecahedral sheet are indicated as circles. The solid lines represent hydrogen bonds; the dotted lines indicate the nearest S---S and S---O distances. [Reproduced, with permission, from *J. Chem. Phys.*, **40**, 2800 (1964).]

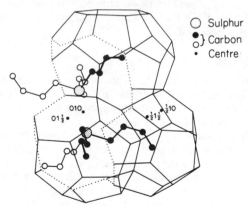

FIGURE 6. Irregular cavity with cation pair in the monoclinic form of tri-*n*-butylsulphonium fluoride hydrate clathrate **25**. Of the disordered carbon atoms only the ones with the highest weight are shown. Only one half of the cavity is shown; this is filled with the shaded carbon chains. The second part is related by the centre (0,1,0) and is filled with the unshaded carbon chains. The two parts share the dotted edges. [Reproduced, with permission, from *J. Chem. Phys.*, **40**, 2800 (1964).]

2-dimethylsulphuranylidenemalononitrile **26**[25], 2-dimethylsulphuranylidene-1,3-indanedione **27**[26], dimethylsulphonium cyclopentadienylide **28**[27], and 7-chloro-1,3,5-trimethyl-5*H*-pyrimido[5,4-*b*][1,4]benzothiazine-2,4(1*H*,3*H*)-dione **29**[28].

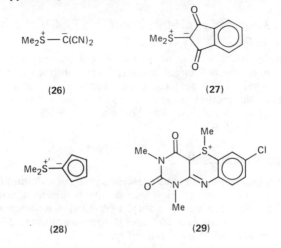

(26)   (27)

(28)   (29)

1. 2-Dimethylsulphuranylidenemalononitrile **26**[25]

The space group is *Pna*2$_1$ (racemic). The crystals darkened slowly during exposure to the X-rays, but no decrease in the intensities was noted. The model was refined to an *R*-factor of 0.048 on 660 non-zero reflections measured on a diffractometer. Hydrogen atoms were located from a difference map and included in the refinement, although their isotropic thermal parameters were held constant. A correction was made to the bond lengths to allow for thermal motion; the corrections to the S—C bond lengths were 0.008–0.010 Å.

2. 2-Dimethylsulphuranylidene-1,3-indanedione **27**[26]

The space group is *P*2$_1$2$_1$2$_1$ (chiral). An absorption correction was applied to the intensity data. The model was refined to an *R*-factor of 0.028 on 1170 non-zero reflections collected on a diffractometer. Hydrogen atoms were located from a difference map and included in the refinement with positional and isotropic thermal parameters being varied. The thermal motion of the molecule was analysed but no corrections were applied to the bond lengths. The standard deviation of the C—S bond length was 0.003 Å.

A second set of data on **27** was collected at −150°C[29]. The model was refined to an *R*-factor of 0.039 on 1264 non-zero reflections collected on a diffractometer. The model included hydrogen atoms, and their positional and thermal parameters were varied. The standard deviations of the C—S bond lengths were 0.002–0.004 Å.

3. Dimethylsulphonium cyclopentadienylide **28**[27]

The space group is *P*$\bar{4}$2$_1$c (racemic). The model was refined to an *R*-factor of 0.072 on 561 non-zero reflections collected on a diffractometer. The hydrogen atoms of the methyl groups were located from a difference map, those on the cyclopentadiene ring were positioned. All hydrogen atoms were included in the refinement, but the isotropic thermal parameters were held constant. The standard deviations of the S—C bond lengths were 0.007–0.009 Å.

4. 7-Chloro-1,3,5-trimethyl-5H-pyrimido[5,4-b][1,4-b]benzothiazine-2,4(1H,3H)-dione 29[28]

The space group is $P2_1/n$ (racemic). The model was refined to an R-factor of 0.058 on 1699 non-zero reflections collected on a diffractometer. The hydrogen atoms on the phenyl ring were located and included in the refinement, but the hydrogen atoms of the methyl groups could not be located because of either large thermal motion or positional disorder. The standard deviations of the S—C bond lengths were 0.004–0.005 Å.

C. Aminosulphonium Salts 3

Two molecules (aminosulphonium, azasulphonium, or sulphiminium salts) having an uncharged nitrogen atom forming one of the bonds to a sulphonium sulphur have been studied; they are *trans*-4-*t*-butyl-1-[*N*-ethyl-*N*-*p*-toluenesulphonylamino]-1-thioniacyclohexane fluoroborate 30[30] and dehydromethionine 31[31]. In both cases, the analyses are recent and of good reliability.

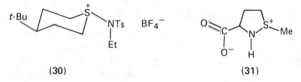

(30)   (31)

1. trans-4-t-Butyl-1-(N-ethyl-N-ptoluenesulphonylamino)-1-thioniacyclo-hexane fluoroborate 30[30]

It proved difficult to obtain a crystal of 30 suitable for X-ray analysis. The space group was $P\bar{1}$ (racemate). The data were collected with a diffractometer and the model of the structure was refined to an R-factor of 0.068 on 2212 non-zero reflections. The hydrogen atoms were located from a difference map and their coordinates and isotropic thermal parameters were varied in the refinement. Difficulty was encountered in the treatment of the BF_4^- anion as it exhibited considerable disorder which could be either static or dynamic in origin. The unlikely possibility that the BF_4^- groups were ordered and that there were two crystallographically independent cations in the space group $P1$ was considered but was rejected by the authors. However, the problems with the BF_4^- group should not affect greatly the precision of the dimensions involving the sulphur atom. The standard deviations of the S—C bond lengths were 0.005 Å.

2. Dehydromethionine 31[31]

In the case of 31, the data were also collected on a diffractometer. The space group was $P2_1/n$ (racemate). The model was refined to an R-factor of 0.044 on 972 non-zero reflections. Hydrogen atoms were located from difference maps and their coordinates and isotropic thermal parameters were included in the final refinement. As the position of H(6), the hydrogen bonded to nitrogen, is critical to an

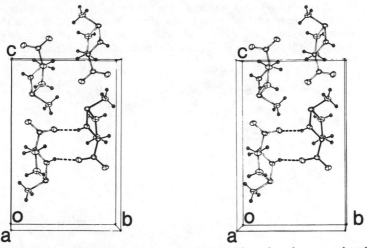

FIGURE 7. Stereoscopic view of the packing in the crystal of dehydro-methionine **31**. The bonds of the reference molecule (i.e. that whose coordinates are given in the paper and from which the entire structure can be generated by symmetry transformations) are shaded in black. The N—H---O hydrogen bonds are shown by discontinuous lines. The two S⋯O contacts involving the sulphur atom of the reference molecule are with the oxygen atom of the carboxyl group above it and with an oxygen atom in the carboxyl group that is involved in the hydrogen bonding to the reference molecule.

assessment of the geometry of the nitrogen atom, the behaviour of this atom upon the refinement is of importance. The isotropic thermal parameter was low [2.2(7) Å^2], the N—H bond was short [0.72(3) Å], but not unusually so, and the C—N—H and S—N—H angles were about equal [105(2)° and 107(2)°].

The N—H group forms a centrosymmetrically related pair of hydrogen bonds with an oxygen [O(2)] of the carboxylate group. The N—H---O(2) angle is 147(3)°, deviating considerably from linearity. However, it can be seen from the packing diagram (Figure 7) that if the hydrogen were in a more planar arrangement around nitrogen, the N—H---O(2) angle would be even less than 147°. The standard deviations of the S—C bond lengths were 0.003–0.004 Å.

D. Sulphilimines 4

There has been considerable interest in the structures of sulphilimines **4**, i.e. sulphur ylides with a nitrogen atom bearing the negative charge, particularly from the groups of Cameron (Glasgow) and Kálmán (Budapest). The structures that have been determined are as follows:

SS-dimethyl-*N*-methylsulphonylsulphilimine **32**[32,33];
SS-dimethyl-*N*-*p*-toluenesulphonylsulphilimine **33**[34];
S-*n*-propyl-*S*-phenyl-*N*-*p*-toluenesulphonylsulphilimine **34**[35];
SS-diphenyl-*N*-*p*-toluenesulphonylsulphilimine **35**[36];
SS-dimethyl-*N*-benzoylsulphilimine **36**[37];
SS-dimethyl-*N*-trichloroacetylsulphilimine **37**[38];
SS-diethyl-*N*-dichloroacetylsulphilimine **38**[39];

a penicillin derivative **39** that contains a *p*-toluenesulphonylsulphilimine group[40]; two S—N ylides where part of an aromatic ring is interposed between the sulphur and nitrogen atoms **40**[41] and **41**[42]; and a compound **42** where the nitrogen of the sulphilimine group is bonded to a second positively charged sulphur atom[43].

$$Me_2\overset{+}{S}-\overset{-}{N}SO_2Me$$

(32)

$$Me_2\overset{+}{S}-\overset{-}{N}Ts$$

(33)

(34)

$$Ph_2\overset{+}{S}-\overset{-}{N}Ts$$

(35)

$$Me_2\overset{+}{S}-\overset{-}{N}COPh$$

(36)

$$Me_2\overset{+}{S}-\overset{-}{N}COCCl_3$$

(37)

$$Et_2\overset{+}{S}-\overset{-}{N}COCHCl_2$$

(38)

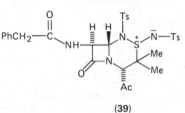

(39)

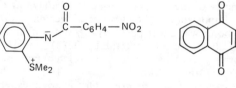

(40)

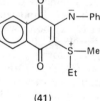

(41)

$$Me_2\overset{+}{S}-\overset{-}{N}-\overset{+}{S}Me_2 \quad Br^-$$

(42)

1. SS-*Dimethyl*-N-*methylsulphonylsulphilimine* **32**[32,33]

The original paper on this compound described an analysis based on photographic data[32]. The space group is $P2_1/c$ (racemate). The model was refined to an *R*-factor of 0.106 on 1027 non-zero reflections measured by using a photometer. The crystals decomposed slowly upon irradiation. The two methyl carbon atoms attached to sulphur had a great deal of thermal motion and the thermal parameters showed considerable anisotropy. Consequently, it proved impossible to locate the hydrogen atoms. The standard deviations of the S—C bond lengths were 0.03 and 0.07 Å, that of the S—N bond was 0.009 Å; the high values for the S—C bonds are due to the large thermal motion.

More recently, Kálmán and his group have re-collected diffractometer data on this compound[33]. Although a description of this analysis has not been published, some results are available. The R-factor on this new analysis is 0.04. The new standard deviation of the S—N bond length is 0.002 Å.

2 SS-*Dimethyl*-N-p-*toluenesulphonylsulphilimine* **33**[34]

The analysis of this compound was also based on photographic data. The space group is $P2_1/c$ (racemate). The model was refined to an R-factor of 0.107 on 1085 non-zero reflections. Hydrogen atoms were not included in the model. The standard deviations of the S—C bond lengths were 0.012–0.013 Å and that of the S—N bond length was 0.008 Å.

3. S-n-*Propyl*-S-*phenyl*-N-p-*toluenesulphonylsulphilimine* **34**[35]

The space group is $P2_1/a$ (racemate). The model was refined to an R-factor of 0.087 on 1621 non-zero reflections measured on a diffractometer. Significant crystal decomposition was noted. The positions of the hydrogen atoms were located from a difference map, but were not published in the paper[35]. The standard deviation of the S—C bond length was 0.009 Å and that of the S—N bond length was 0.007 Å.

4. SS-*Diphenyl*-N-p-*toluenesulphonylsulphilimine* **35**[36]

For this compound also, the crystals tended to decompose in the X-ray beam. The space group is $P2_1/c$ (racemate). Intensity data were collected photographically and measured on a photometer. Data were obtained from several crystals in order to minimize crystal decomposition. The model was refined to an R-factor of 0.095 on 1758 non-zero reflections. Hydrogen atoms were included in the refinement. The standard deviation of the S—C bond length was 0.008 Å and that of the S—N bond length was 0.007 Å.

5. SS-*Dimethyl*-N-*benzoylsulphilimine* **36**[37]

The space group is $P2_1/c$ (racemate). The model was refined to an R-factor of 0.034 on 1481 non-zero reflections collected on a diffractometer. Hydrogen atoms were included in the model. The standard deviations of the S—C bond lengths were 0.003–0.004 Å and that of the S—N bond length was 0.002 Å.

6. SS-*Dimethyl*-N-*trichloroacetylsulphilimine* **37**[38]

The space group is $P2_12_12_1$ (chiral). The crystals were stable to exposure to X-rays. The structural model was refined to an R-factor of 0.060 on 826 non-zero reflections collected on a diffractometer. All of the hydrogen atoms were included in the model and, in a rather unusual procedure, were given the anisotropic thermal parameters of the atoms to which they were bonded. The standard deviations of the S—C bond lengths were 0.010–0.012 Å and that of the S—N bond length was 0.007 Å.

7. SS-*Diethyl*-N-*dichloroacetylsulphilimine* **38**[39]

There is only a preliminary account of this structure analysis. The space group is $P2_1/n$ (racemate). There was some crystal decomposition. The structure model was refined to an R-factor of 0.103 on 963 non-zero reflections collected photographically. Hydrogen atoms were included in the model, although the exact treatment was not described. The standard deviation of the S—C bond length was 0.013 Å and that of the S—N bond length was 0.010 Å.

8. A β-*lactam-fused ylide derived from methyl* 6β-*phenyl-acetamidopenicillinate* **39**[40]

The space group is $P2_1$ (chiral). The structural model was refined to an R-factor of 0.048 on 3105 non-zero reflections collected on a diffractometer. The positions of the hydrogen atoms were calculated, included in the model, but were not refined. The standard deviation of the one S—C bond length was 0.005 Å and those of the two S—N bond lengths were 0.004–0.005 Å.

9. N-(p-*Nitrobenzoyl*)-2-*iminophenyldimethylsulphur(IV)* **40**[41]

The space group is *Pbca* (racemic). The structural model was refined to an R-factor of 0.040 on 2209 non-zero reflections collected on a diffractometer. The positions of the hydrogen atoms were located from difference maps and were included in the refinement. The standard deviations of the S—C bond lengths were 0.002–0.003 Å.

10. *The* 2-*anilino*-3-*ethylmethylsulphonium ylide of* 1,4-*naphthoquinone* **41**[42]

The space group is *Cc* (racemic). The structural model was refined to an R-factor of 0.073 on 1342 non-zero reflections collected on a diffractometer. All of the hydrogen atoms, except those for the methyl groups, were located from a difference map and were included in the model, but were not refined. The standard deviations of the S—C bond lengths were 0.006–0.008 Å.

11. [(CH₃)₂S]₂N⁺Br⁻·H₂O **42**[43]

The space group is $P2_1/n$ (racemic). As the solvated crystals were believed to be somewhat unstable, the intensity data were collected at −70°C on a diffractometer. Difficulties were encountered owing to the poor quality of the crystals and to the problems in maintaining a constant temperature. The model was refined to an R-factor of 0.085 on 349 non-zero reflections. Hydrogen atoms were not located or included in the model. The standard deviations of the S—C bond lengths were 0.05–0.06 Å and that of the S—N bond length was 0.04 Å.

E. Selenium Compounds

Among selenium compounds, there are two studies on simple selenonium salts[44,45] and two on selenium ylides[46,47]. There is a very early partial analysis of triphenylselenonium chloride **43**[44] and a more recent one of trimethylselenonium iodide **44**[45].

The two selenium ylides whose structures have been determined are

$$Ph_3\overset{+}{Se}\ Cl^-\qquad\qquad\qquad Me_3\overset{+}{Se}\ I^-$$

$$(43)\qquad\qquad\qquad\qquad (44)$$

tetrahydroselenophenium 4,4-dimethylcyclohexylide-2,6-dione **45**[46] and diacetylmethylenediphenylselenurane **46**[47]. There is also another paper in the Russian literature[48] that reports cell data for a number of selenium ylides related to **45**.

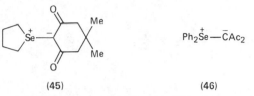

$$Ph_2\overset{+}{Se}-\overset{-}{C}Ac_2$$

$$(45)\qquad\qquad\qquad\qquad (46)$$

The paper on triphenylselenonium chloride established the space group ($Pna2_1$, racemate) and the positions of the selenium and chlorine atoms, but the carbon atoms were not located. Hence no information on the molecular geometry was obtained.

1. Trimethylselenonium iodide **44**[45]

Intensity data were collected photographically on crystals kept in capillaries at temperatures of -15 to $-20°C$. The space group is $Pnma$ (racemate) and the selenonium group occupies a site of C_s symmetry in the crystal. The structure was refined to an R-factor of 0.056 on 580 non-zero reflections. Anisotropic thermal parameters were introduced for the selenium and iodine atoms, and isotropic ones from the two crystallographically independent carbon atoms; no attempt was made to locate or include hydrogen atoms in the structural model. The standard deviations of the Se—C bond lengths were 0.016–0.023 Å.

2. Tetrahydroselenophenium 4,4-dimethylcyclohexylide-2,6-dione **45**[46]

Intensity data were collected photographically. The space group is $P2_1/a$ (racemate) and the structure was refined to an R-factor of 0.115 on 403 non-zero reflections. Only isotropic temperature factors were included and no hydrogen atoms were located or included in the model. The standard deviation of the Se—C bond length was 0.02 Å.

3. Diacetylmethylenediphenylselenurane **46**[47]

The space group is $P2_1/c$ (racemate). The model was refined to an R-factor of 0.088 on 2370 non-zero reflections measured on a diffractometer. All hydrogen atoms were included in the model, but the positions of several of them were held at fixed positions during the least-squares refinement. The standard deviation of the Se—C bond length was 0.008–0.009 Å.

III. MOLECULAR DIMENSIONS OF SULPHONIUM SALTS AND RELATED COMPOUNDS

In this section, we again adhere to our division of the sulphur compounds we are discussing into four categories **1–4** with the selenium compounds making up a fifth category.

TABLE 1. Molecular dimensions involving the sulphur atom in sulphonium ions in which the sulphur atom is bonded to three carbon atoms

Compound	Atom Nos. for C—S bonds[a]	C—S bond distances with standard deviations (Å)	Atom Nos. for C—S—C angles[a,b]	C—S—C angle (degrees)	Distance (Å) of S from plane of three C atoms	Reference
9	$C_{(1)}$—S	1.82(2)[c,d]	$C_{(7)}$—S—$C_{(8)}$	102(1)	0.759	8
	$C_{(7)}$—S	1.81(2)	$C_{(1)}$—S—$C_{(8)}$	105(1)		
	$C_{(8)}$—S	1.83(2)	$C_{(1)}$—S—$C_{(7)}$	103(1)		
10	$C_{(7)}$—S	1.799(4)	$C_{(8)}$—S—$C_{(14)}$	107.5(3)	0.747	9
	$C_{(8)}$—S	1.830(4)	$C_{(7)}$—S—$C_{(14)}$	105.1(3)		
	$C_{(14)}$—S	1.882(3)	$C_{(7)}$—S—$C_{(8)}$	101.1(3)		
11	$C_{(16)}$—$S_{(1)}$	1.811(9)	$C_{(17)}$—$S_{(1)}$—$C_{(18)}$	101.6(5)	0.821	10
	$C_{(17)}$—$S_{(1)}$	1.789(10)	$C_{(16)}$—$S_{(1)}$—$C_{(18)}$	99.6(5)		
	$C_{(18)}$—$S_{(1)}$	1.784(9)	$C_{(16)}$—$S_{(1)}$—$C_{(17)}$	101.0(5)		
12	$C_{(3)}$—$S_{(6)}$	1.814(6)	$C_{(\varepsilon)}$—$S_{(\delta)}$—$C_{(\gamma)}$	103.8(3)	0.801	11
		1.797(6)[e]		102.5(3)	0.795	
	$C_{(\gamma)}$—$S_{(\delta)}$	1.801(6)	$C_{(\varepsilon)}$—$S_{(\delta)}$—$C_{(3)}$	101.9(3)		
		1.801(7)		102.3(3)		
	$C_{(\varepsilon)}$—$S_{(\delta)}$	1.786(6)	$C_{(\gamma)}$—$S_{(\delta)}$—$C_{(3)}$	99.2(3)		
		1.782(6)		100.8(3)		
13	$C_{(4)}$—$S_{(1)}$	1.800(2)	$C_{(5)}$—$S_{(1)}$—$C_{(6)}$	101.6(1)	0.786	12
	$C_{(5)}$—$S_{(1)}$	1.778(3)	$C_{(4)}$—$S_{(1)}$—$C_{(6)}$	102.6(1)		
	$C_{(6)}$—$S_{(1)}$	1.787(2)	$C_{(4)}$—$S_{(1)}$—$C_{(5)}$	102.2(1)		
14	$C_{(1)}$—S	1.804[f]	$C_{(5)}$—S—$C_{(6)}$	106.4	0.758	13
	$C_{(5)}$—S	1.807	$C_{(1)}$—S—$C_{(6)}$	104.8		
	$C_{(6)}$—S	1.799	$C_{(1)}$—S—$C_{(5)}$	100.1		
15	$C_{(2)}$—S	1.795(10)	$C_{(6)}$—S—$C_{(11)}$	104.5(5)	0.771	14
	$C_{(6)}$—S	1.816(10)	$C_{(2)}$—S—$C_{(11)}$	105.3(5)		
	$C_{(11)}$—S	1.777(10)	$C_{(2)}$—S—$C_{(6)}$	98.9(4)		
16	$C_{(2)}$—S	1.80(2)	$C_{(6)}$—S—$C_{(11)}$	102.2(7)	0.831	14
	$C_{(6)}$—S	1.81(1)	$C_{(2)}$—S—$C_{(11)}$	102.2(8)		
	$C_{(11)}$—S	1.83(2)	$C_{(2)}$—S—$C_{(6)}$	97.5(7)		

No.[a]	C—S bond	Length (Å)[c]	C—S—C angle (°)[b]		ratio	No.[a]
17	C(1)—S(1)	1.80(4)	C(4)—S(1)—C(6)	104(2)	0.761	15
	C(4)—S(1)	1.81(3)	C(1)—S(1)—C(6)	100(1)		
	C(6)—S(1)	1.82(3)	C(4)—S(1)—C(1)	106(1)		16
18	C(2)—S(1)	1.84(2)	C(8)—S(1)—C(9)	100.7(6)	0.806	
	C(8)—S(1)	1.80(1)	C(2)—S(1)—C(9)	99.9(7)		
	C(9)—S(1)	1.82(1)	C(2)—S(1)—C(8)	104.9(7)		17
19	C(1)—S(1)	1.834[f,g]	C(7)—S(1)—C(18)	102.4	0.738	
	C(7)—S(1)	1.779	C(1)—S(1)—C(18)	108.6		
	C(18)—S(1)	1.778	C(1)—S(1)—C(7)	101.8		18
20	C(1D)—S(1)	1.808(10)[h]			—	
	C(3DA)—S(1)	1.807(16)				
	C(3DB)—S(1)	1.753(16)				
	C(2D)—S(2)	1.794(9)				
	C(4DA)—S(2)	1.794(17)				
	C(4DB)—S(2)	1.834(15)				
21	C(5)—S	1.806(8)	C(1)—S—C(5')	98.1(4)	0.825	19
	C(1)—S	1.872(8)	C(5')—S—C(5)	106.9(4)		
	C(5')—S	1.806(8)[i]	C(1)—S—C(5)	98.1(4)		

[a] These are the atom numbers used in the original paper.

[b] The C—S—C angles are listed such that an angle is on the same horizontal line as the bond opposite it.

[c] The figures in parentheses refer to the estimated standard deviations in the length or angle. For more information, see the first two sections.

[d] In the molecule of 9, the C(1)—S bond is that to the phenyl ring.

[e] In the crystal of 12, there were two crystallographically independent molecules. The values for both molecules are given.

[f] The standard deviation in atom position and that for bond length and angle were not included in the original paper; hence, it was not possible to include the value.

[g] In the crystal of 19, the C(1)—S(1) bond is to the single non-fused phenyl ring.

[h] There are two sulphonium groups involving S(1) and S(2) in the molecule of 20. The results presented here correspond to the ordered model discussed in reference 18.

[i] There is a crystallographic mirror plane passing through the cation in 21. As a result of this symmetry element the C(5)—S and C(5')—S bonds and the C(1)—S—C(5) and C(1)—S—C(5') angles are identical.

A. Molecular Dimensions of Sulphonium Salts 1

The bond lengths and angles involving the sulphonium sulphur atoms obtained in the various X-ray diffraction studies are listed in Table 1. In general, the sulphonium salts exhibit pyramidal geometry with the sulphur atom lying approximately 0.8 Å above the base of the pyramid defined by the three bonded carbon atoms; the C—S̈—C angles are all slightly less than tetrahedral. A stereoscopic view of a typical sulphonium cation, that of **9**, is shown in Figure 8. Considering all of the structures included in Table 1, the mean value for a C—S$^+$ bond length is 1.806 Å with a range from 1.753(16) Å to 1.882(3) Å; the mean value for a C—S̈—C bond angle is 102.5° with a range from 97.5(7)° to 108.6°. There are insufficient data on $C_{(aromatic)}$—S$^+$ dimensions to justify separating them from the $C_{(sp^3)}$—S$^+$ dimensions. If only the most accurately determined structures are considered, i.e. those of **10**[9], **12**[11], and **13**[12], then the average C—S$^+$ bond length is 1.805 Å and the range is from 1.778(3) to 1.882(3) Å. For bond angles from these three structures, the mean value is 102.6° and the range is from 99.2(3)° to 103.8(3)°. These average values for the C—S$^+$ lengths agree well with the value of 1.81 Å for the sum of the covalent radii of $C_{(sp^3)}$ and sulphur[49] and with the lengths in the range 1.77–1.82 Å found for the C—S bond in thiols[50]. It is clear from the data in Table 1, particularly when one considers the well determined structures, that there is considerable variation in both the C—S$^+$ bond length and

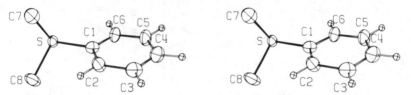

FIGURE 8.   Stereoscopic view of a single cation of dimethylphenylsulphonium perchlorate **9**, drawn from the coordinates presented in reference 8.

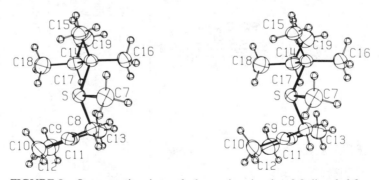

FIGURE 9.   Stereoscopic view of the cation in the 2,3-dimethyl-2-butenyl-1,1,2-trimethylpropylsulphonium 2,4,6-trinitrobenzenesulphonate structure **10**, drawn from the coordinates given in reference 9. It corresponds to the opposite configuration from that shown in the figures in that paper. The crystal contains equal numbers of molecules with both configurations. For the H4 $C_{(17)}$ atom, we used coordinates −0.361, −0.340, and 0.835.

the C—$\overset{+}{S}$—C angles and that the actual values obtained vary considerably with molecular environment. In their highly accurate study on the 2,3-dimethyl-2-butenyl-1,1,2-trimethylpropylsulphonium salt **10**, Barnes and Sundaralingam[9] noted that the C—S[+] bond length increases with increasing bulk of the substituent attached to sulphur. A view of the sulphonium salt **10** is shown in Figure 9. The S—$C_{(7)(methyl)}$ bond length is 1.799(4) Å, the S—$C_{(8)(butenyl)}$ bond length is 1.830(4) Å, while the $\overset{+}{S}$—$C_{(14)(trimethylpropyl)}$ bond length is 1.882(3) Å, the longest $\overset{+}{S}_{(sulphonium)}$—C bond reported in Table 1. It is of interest to compare the variations in the lengths of the C—S[+] bonds found for **10** with those found for **12** and **13**[11,12], the analyses of which are of comparable accuracy. No great variation in C—S[+] bond length is found in the structures of **12** and **13**. If one examines the nature of the groups attached to sulphur in the molecules of **12** and **13** (Figures 10 and 11), one would not expect any significant intramolecular strain along the C—S[+] bond as a consequence of steric bulk in these molecules. Concomitant with the C—S[+] bond lengthening that results from the presence of a bulky substituent, there is a contraction of the C—$\overset{+}{S}$—C angle not involving that bond. The effect is once again clearly seen in the molecule **10**[9].

In their paper on the structure of dimethylphenylsulphonium perchlorate (**9**), Lopez-Castro and Truter[8] stated that the values of the C—$\overset{+}{S}$—C angles of less than tetrahedral were due to the repulsion between the lone pair of electrons on sulphur and the atoms bonded to sulphur.

Another factor which appears to influence the molecular geometry, particularly the size of the C—$\overset{+}{S}$—C angles, is incorporation of the sulphur atom into a ring. In the molecules of **14**[13], **15**[14], and **16**[14], where the sulphur atom is part of a non-fused six-membered ring, the intra-ring C—$\overset{+}{S}$—C angle is notably smaller than the *exo*-ring angles. Inclusion of the sulphur in the lactone ring in **17**[15] or in the fused ring system in **19**[17] does not lead to such predictable results. However, the main

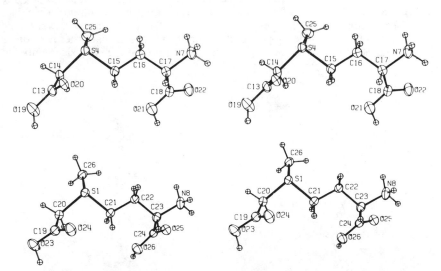

FIGURE 10. Stereoscopic views (a and b) of the two crystallographically independent cations in the structure of R-{[(3S)-3-amino-3-carboxylpropyl](carboxymethyl)methyl-sulphonium} 2,4,6-trinitrobenzenesulphonate **12**, drawn from the coordinates in reference 11.

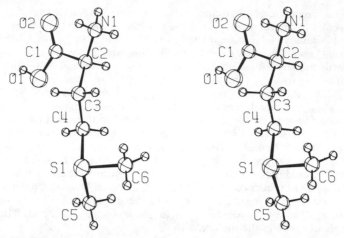

FIGURE 11. Stereoscopic view of the cation of S-methyl-
L-methionine chloride hydrochloride 13, drawn from the
coordinates in reference 12.

aim of the analysis of 17[15] was not to provide accurate geometry, and there could
be other intra-molecular steric influences in the case of 19 to cause the unusual
pattern of C—$\overset{+}{S}$—C bond angles. Inclusion of the sulphur atom in an
eight-membered ring in the molecule of 18[16] results in the intra-ring C—$\overset{+}{S}$—C angle
[104.9(7)°] being larger than the two exo C—S—C angles [99.9(7)° and
100.7(6)°]. The sulphonium cation trans-1-thioniabicyclo[4.4.0]decane 21[19]
(Figure 12) is interesting in that steric effects due to eclipsed groups along the
central C—S⁺ bond and the effects due to incorporation of the sulphur atom into
six-membered rings are both operative. The central $C_{(1)}$—S⁺ bond is very long
[1.872(8) Å], but the angle opposite it is, in fact, larger than the two other
C—$\overset{+}{S}$—C angles, presumably because of their incorporation in the six-membered
rings.

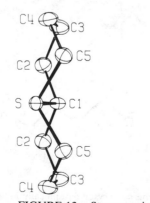
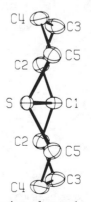

FIGURE 12. Stereoscopic view of the cation of trans-1-
thioniabicyclo[4.4.0]decane bromide 21, drawn from the
coordinates in reference 19.

Also given in Table 1 is the perpendicular distance of the sulphur atom from the plane of the three carbon atoms to which it is bonded. When all structures in Table 1 are included, the average distance is 0.784 Å with a range from 0.738 to 0.831 Å. If only the molecules **10**, **12**, and **13** are considered, the mean value is 0.776 Å and the range is from 0.747 to 0.801 Å. The distance above the plane is a function of the bond lengths and angles, so that variations in them lead to variation in the distance above the plane.

Newman projections down a number of the $\overset{+}{S}$—C bonds, where there is some conformational flexibility in the alkyl group attached to carbon, are shown in Figure 13. For convenience in these and subsequent discussions, we refer to the 'unoccupied lobe' of the sulphonium salt as the position that would be occupied by a group to make an ideal trigonal arrangement of groups as one looks along one of the $\overset{+}{S}$—C bonds; however, no such group is shown in the Newman projections. We shall also refer to the carbon atoms as α, β, etc., in order of their position from the sulphonium sulphur atom (e.g. in **47**).

$$\overset{\diagdown}{\underset{\diagup}{\overset{+}{S}}} - C_{(\alpha)} - C_{(\beta)} - C_{(\gamma)}$$

(47)

In the molecule **10**, where there is a primary (methyl), a secondary (dimethylbutenyl), and a tertiary (trimethylpropyl) group attached to sulphur, we find a completely staggered arrangement for the somewhat symmetrical trimethylpropyl and methyl groups (Figure 13a, b) and an arrangement where the secondary C—C bond approaches the *anti* position to $C_{(7)}$ for the dimethylbutenyl group (Figure 13c). This arrangement where the secondary C—C bond is highly *anti* to one of the S—C bonds is also found for the secondary groups attached to S$^+$ in **11**, **12** (two molecules), **13**, **17**, and **18** (Figure 13d–k).

The two sulphonium salts which have phenyl rings that are free to rotate and are attached to the S$^+$ atom are **9** and **19**. The torsion angles about the $\overset{+}{S}$—C (phenyl) rings are different in the two cases (Figure 14a, b). In the case of **9**, where the other two substituents are methyl groups, the phenyl ring roughly bisects the $C_{(methyl)}$—S—$C_{(methyl)}$ angle (Figure 14a), whereas in **19**, where the other two substituents are part of a fused ring system, the phenyl ring makes roughly equal angles with the two $\overset{+}{S}$—$C_{(methyl)}$ bonds but does not intersect the $C_{(methyl)}$—$\overset{+}{S}$—$C_{(methyl)}$ angle.

There are four molecules, **14**, **15**, **16**, and **17**, where the sulphonium sulphur atom is part of an isolated six-membered ring. In the molecules of **14** and the *trans* form of S-methylthianium perchlorate **16** (Figure 15), the S-methyl groups are in equatorial positions and the six-membered rings have fairly regular chair conformations. In these instances the torsion angles around the S—$C_{(ring)}$ bonds are such that the exocyclic $\overset{+}{S}$—$CH_{3(methyl)}$ group is arranged almost exactly *anti* to the $C_{(\alpha)}$—$C_{(\beta)}$ bonds (Figure 16a–d). In the S-ethyl-β-thia-S-caprolactone structure **17**, the S-ethyl substituent is axial but the conformation of the ring is irregular (Figure 17). Hence the torsion angles around the $\overset{+}{S}$—$C_{(ring)}$ bonds have the other S—$C_{(ring)}$ bond in an orientation that is rather close (~24°) to eclipsing the $C_{(\alpha)}$—$C_{(\beta)}$ bond (Figure 16e, f). In the *cis* isomer of S-methylthianium perchlorate **15** (Figure 18), the ring is in the chair conformation, but the S-methyl substituent is axial. This results in an orientation about the $\overset{+}{S}$—$C_{(ring)}$ bonds with $C_{(ring)}$—$\overset{+}{S}$—$C_{(ring)}$—$C_{(ring)}$ torsion angles of ~45° and with the $C_{(\alpha)}$—$C_{(\beta)}$ bond lying between the other $C_{(ring)}$—$\overset{+}{S}$—$C_{(methyl)}$ angle (Figure 16g, h).

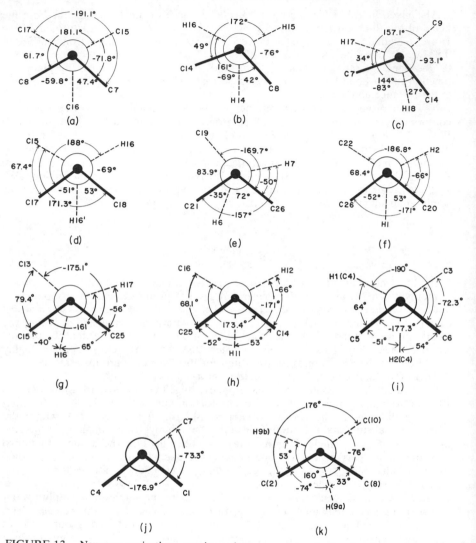

FIGURE 13. Newman projections, or views along several bonds showing the torsion angles, in various sulphonium salts. The torsion angle A—B—C—D is considered positive if, when looking along the B—C bond, atom A has to be rotated clockwise to eclipse atom B. For example, in Figure 13a, which is a Newman projection along the S—C$_{(14)}$ bond in **10**, the atom C$_{(8)}$ has to be rotated 61.7° to eclipse atom C$_{(17)}$. (a) View along S—C$_{(14)}$ bond in **10**. (b) View along the S—C$_{(7)}$ bond in **10**. (c) View along the S—C$_{(8)}$ bond in **10**. (d) View along the S—C$_{(16)}$ bond in **11**. (e) View along the S$_{(1)}$—C$_{(20)}$ bond in one of the independent molecules of **12**. (f) View along the S$_{(1)}$—C$_{(21)}$ bond in the same molecule of **12**. (g) View along the S$_{(4)}$—C$_{(14)}$ bond in the second molecule of **12**. (h) View along the S$_{(4)}$—C$_{(15)}$ bond in the second molecule of **12**. (i) View along the S—C$_{(4)}$ bond in **13**. (j) View along the S—C$_{(6)}$ bond in **17**. (k) View along the S$_{(1)}$—C$_{(9)}$ bond in **18**. The signs of the torsion angles that we calculate [and the configuration of the molecule that we draw (Figure 9) from the coordinates given in reference 9] are opposite those given in that paper.

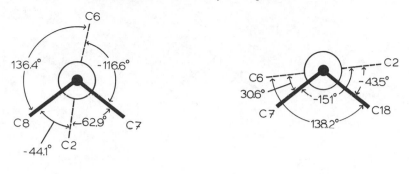

(a)   (b)

FIGURE 14.   Torsion angles about the S—$C_{(1)}$ bond in **9**. (b) Torsion angles about the S—$C_{(1)}$ bond in **19**.

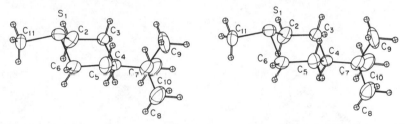

FIGURE 15.   Stereoscopic view of the cation in *trans*-4-*t*-butyl-*S*-methylthianium perchlorate (**16**), drawn from the coordinates that were supplementary material to reference 14. We changed the *z*-coordinates for $H_{(8\gamma)}$, $H_{(9\beta)}$, and $H_{(9\gamma)}$ to 0.024, -0.036, and -0.074, respectively.

For the remaining $\overset{+}{S}$—C bonds (the bonds in **20** will not be considered on account of uncertainties in the actual ring conformation), a number of different environmental factors are operative. In **18**, the sulphonium sulphur is part of an eight-membered ring which has an approximate boat–chair (BC) conformation (Figure 19). The arrangements of $\overset{+}{S}$—C bonds about the two $\overset{+}{S}$—$C_{(ring)}$ bonds are different (Figure 20a, b); in both instances the $C_{(\alpha)}$—$C_{(\beta)}$ bond is almost *anti* to the exocyclic $\overset{+}{S}$—$C_{(acetonyl)}$ bond but in one case (Figure 20a) the $C_{(\alpha)}$—$C_{(\beta)}$ bond is close to the 'unoccupied lobe' of the sulphonium group, while in the other it is close (43°) to the other $\overset{+}{S}$—C ring bond (Figure 20b). In the molecule of **21** (Figure 12), despite the incorporation of the sulphur atoms into two fused six-membered rings, we find an almost ideally staggered arrangement for the $C_{(\alpha)}$—$C_{(\beta)}$ bonds with respect to the C—$\overset{+}{S}$—C groups; the $C_{(\alpha)}$—$C_{(\beta)}$ bonds are almost exactly *anti* to $\overset{+}{S}$—$C_{(ring)}$ bonds (Figure 21a, b). In the structure of **19** (Figure 22), where the sulphur atom is involved in a tricyclic ring system, a reasonably staggered arrangement for the C—$\overset{+}{S}$—C—C groups is still possible (see Figure 23a, b).

With the exception of one of the structures, in all cases where there is an $\overset{+}{S}$—$C_{(methyl)}$ group and where the hydrogen atoms of the methyl group have been included in the model, we find a very close approach to the fully staggered arrangement illustrated by the structure of **10** shown in Figure 13b. The exception occurs for the $\overset{+}{S}$—CH_3 groups in the two molecules of **12**. In one of the molecules

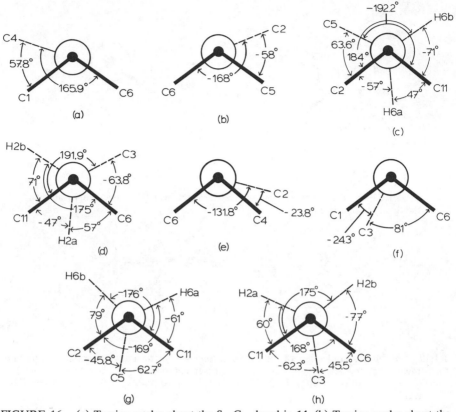

FIGURE 16.   (a) Torsion angles about the S—C$_{(5)}$ bond in **14**. (b) Torsion angles about the S—C$_{(1)}$ bond in **14**. (c) Torsion angles about the S—C$_{(6)}$ bond in **16**. (d) Torsion angles about the S—C$_{(2)}$ bond in **16**. (e) Torsion angles about the S$_{(1)}$—C$_{(2)}$ bond in **17**. (f) Torsion angles about the S$_{(1)}$—C$_{(4)}$ bond in **17**. (g) Torsion angles about the S—C$_{(6)}$ bond in **15**. (h) Torsion angles about the S—C$_{(2)}$ bond in **15**.

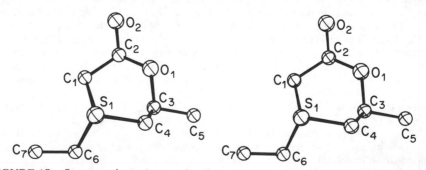

FIGURE 17.   Stereoscopic view of the cation of *S*-ethyl-β-thia-δ-caprolactone tetrafluoroborate **17**, drawn from coordinates sent to us by Dr B. Dahlén.

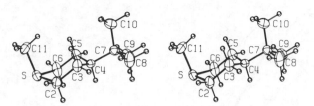

FIGURE 18.   Stereoscopic view of the cation in *cis*-4-*t*-butyl-*S*-methylthianium perchlorate **15**, drawn from the coordinates that were supplementary material to reference 14. We changed the *x*-coordinate of $H_{(3\alpha)}$ to 0.031.

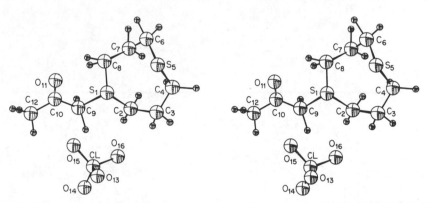

FIGURE 19.   Stereoscopic view of the structure of 1-acetonyl-1-thionia-5-thiacyclooctane perchlorate **18**. [Reproduced, with permission, from *J. Chem. Soc. B.* 1603 (1970).]

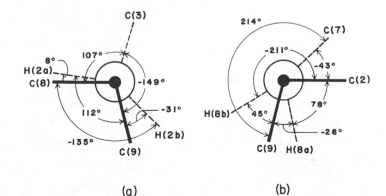

(a)   (b)

FIGURE 20.   (a) Torsion angles about the $S_{(1)}$—$C_{(2)}$ bond in **18**. (b) Torsion angles about the $S_{(1)}$—$C_{(8)}$ bond in **18**.

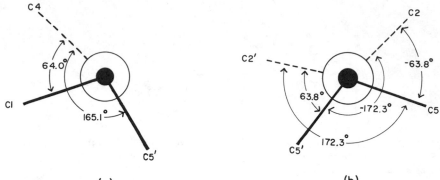

(a)   (b)

FIGURE 21.   Torsion angles about the S—C$_{(5)}$ bond in **21**. (b) Torsion angles about the S—C$_{(1)}$ bond in **21**.

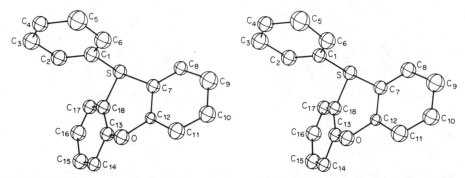

FIGURE 22.   Stereoscopic view of the structure of the cation of phenylsulphoxonium iodide **19**, drawn from the coordinates in reference 17.

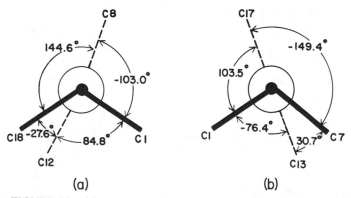

(a)   (b)

FIGURE 23.   (a) Torsion angles about the S$_{(1)}$—C$_{(7)}$ bond in **19**. (b) Torsion angles about the S$_{(1)}$—C$_{(18)}$ bond in **19**.

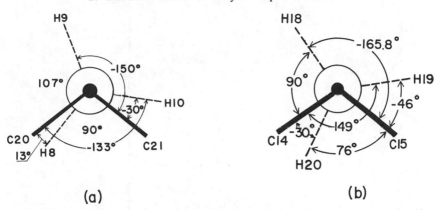

(a)   (b)

FIGURE 24.   (a) Torsion angles about the $S_{(1)}$—$C_{(26)}$ bond in one of the molecules of **12**. (b) Torsion angles about the $S_{(4)}$—$C_{(25)}$ bond in the other molecule of **12**.

there is a close approach to a staggered arrangement with one of the C—S—C—H torsion angles being 13° (Figure 24a), while in the other molecule of **12** the torsion is approximately half way between fully staggered and fully eclipsed (Figure 24b). There is no apparent environmental cause for this unusual effect.

A comparison of the conformations of the two crystallographically independent molecules of **12** at the sulphonium group is of interest (Figure 10a, b, Figure 13e–h, and Figure 24a, b). Around two of the C—S⁺ bonds there is a systematic difference in twist (~17° about the S̟—CH$_{3(methyl)}$, ~4–5° about the S̟—CH$_2$CO$_2$H), while the angles around the other S—CH$_2$ bond are almost identical in the two molecules (Figure 13e–h).

B. Molecular Dimensions of Sulphonium Ylides 2

The bond lengths and angles involving the sulphur atoms in sulphonium ylides in the structures studied by X-ray diffraction methods are given in Table 2. In this category of compound, it is fortunate that all four studies are recent and accurate. The geometry of the sulphonium sulphur atom in the ylides resembles closely that in the normal sulphonium salts. The arrangement is pyramidal with the pyramid being slightly lower (~0.73 Å) than in the normal sulphonium salts. This lowering probably arises because the S̟—C̄ ylide bond is significantly shorter than the other two S̟—C bonds in the ylides. The arrangement of the three atoms around the ylide carbon atom is very close to planar. The mean value of the S̟—C̄ ylide bond length is 1.715 Å with a range from 1.707(7) to 1.721(4) Å. These values do suggest considerable double bond character in the S̟—C̄ ylide bond. The ylide can be represented as a hydrid of structures **48** and **49**. The pyramidal nature of the arrangement of atoms around sulphur suggests contributions from **48**, while the length of the S—C bond (and the smaller pyramid compared with normal sulphonium ions) indicates contributions from **49**. Depending on environment, the

(48)   (49)

TABLE 2. Molecular dimensions involving the sulphur atom in sulphonium ions that are part of a sulphur—carbon ylide

Compound	Atom Nos. C—S bonds[a]	C—S bond distances with standard deviations (Å)	Atom Nos. for C—S—C angles[a,b]	C—S—C angle (degrees)	Distance (Å) of S from plane of the C atoms	Reference
26	$C_{(2)}$—$S_{(1)}$	1.719(8)[c,d] / 1.730[e]	$C_{(3)}$—$S_{(1)}$—$C_{(4)}$	101.7(7)	0.737	25
	$C_{(3)}$—$S_{(1)}$	1.800(10) / 1.810	$C_{(2)}$—$S_{(1)}$—$C_{(4)}$	103.0(5)		
	$C_{(4)}$—$S_{(1)}$	1.831(10) / 1.841	$C_{(2)}$—$S_{(1)}$—$C_{(3)}$	107.3(7)		
27	$C_{(2)}$—$S_{(12)}$	1.707(3)[d] / 1.710(3)[f]	$C_{(13)}$—$S_{(12)}$—$C_{(14)}$	99.9(1) / 100.1(1)	0.727	26 29
	$C_{(13)}$—$S_{(12)}$	1.784(3) / 1.785(4)	$C_{(2)}$—$S_{(12)}$—$C_{(14)}$	105.4(1) / 105.4(1)		
	$C_{(14)}$—$S_{(12)}$	1.787(3) / 1.795(2)	$C_{(2)}$—$S_{(12)}$—$C_{(13)}$	106.8(1) / 106.9(1)		
28	$C_{(1)}$—S	1.712(8)[d] / 1.790(7)	$C_{(6)}$—S—$C_{(7)}$	99.8(4)	0.717	27
	$C_{(6)}$—S	1.769(9)	$C_{(1)}$—S—$C_{(7)}$	107.0(4)		
	$C_{(7)}$—S		$C_{(1)}$—S—$C_{(6)}$	106.4(4)		
29	$C_{(10)}$—S	1.721(4)[d] / 1.765(4)	$C_{(5)}$—S—$C_{(13)}$	102.2(2)	0.772	28
	$C_{(5)}$—S	1.803(5)	$C_{(10)}$—S—$C_{(13)}$	104.1(2)		
	$C_{(13)}$—S		$C_{(5)}$—S—$C_{(10)}$	100.4(2)		

[a]These are the atom numbers used in the original paper.
[b]The C—S—C angles are listed in the Table such that an angle is on the same horizontal line as the bond opposite it.
[c]The figures in parentheses refer to the estimated standard deviation in the length or angle.
[a]This is the ylide (i.e. S—C) bond.
[e]In the refinement of **26**, a correction for the effect of thermal vibration was applied to the bond lengths. The second numbers for each length correspond to the 'corrected' values.
[f]A set of data was collected for **27** at −150°C[29]. The results of a refinement on this data are presented below those from the room temperature data.

S=C bond length can range from 1.60 to 1.72 Å[51]. In their paper on **28**, Adrianov and Struchkov[27] argued that the Ṡ—C̄ bond length was indicative of an 80% contribution from the ylide structure (such as in **48**) and a 20% contribution from the double-bonded structure (as in **49**).

The mean value for the other two C—S[+] bonds is 1.791 Å with a range from 1.765(4) to 1.831(10) Å. These values are slightly smaller than those reported for the well determined normal sulphonium ions (1.805 Å with a range of 1.778 to 1.882 Å), but are clearly compatible with a formulation of these bonds as largely $C_{(sp^3)}$—S[+] single bonds. A stereoscopic view of a typical sulphonium ylide **27** is shown in Figure 25. In these averages we have used the lengths for **26** that were uncorrected for thermal vibrations and we have used the results from the room temperature data for **27**; in both instances the values are almost the same for the different sets of results. In the molecules of **26**, **27**, and **28**, which have considerable structural similarity, the bond angle opposite the ylide bond is smaller than the other two. The mean value of the angle opposite the ylide bond in these three molecules is 100.5° with a range from 99.8(4)° to 101.7(7)°, while the average of the two angles is 106.0° with a range from 103.0(5)° to 107.3(7)°. The structure of **29** is considerably different from the others in having the Ṡ—C̄ ylide bond incorporated into a fused ring system (Figure 26). The S—C̄ ylide bond is still the shortest of the three S—C bonds, but it is slightly longer than the ylide bonds found in the other structures. However, the trend found in **26**, **27**, and **28** for the bond angles is not observed in **29**. The bond angle opposite the ylide bond is not the smallest of the three angles around the sulphur atom, being 102.2(2)° compared with the value of 100.4(2)° for the angle which is within the six-membered ring.

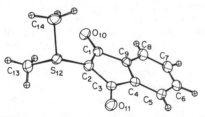 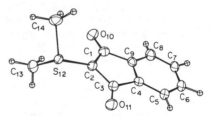

FIGURE 25. Stereoscopic view of the ylide 2-dimethylsulphuranylidene-1,3-indanedione **27**, drawn from the coordinates in reference 26.

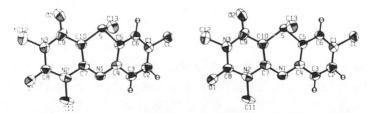

FIGURE 26. Stereoscopic view of the ylide 7-chloro-1,3,5-trimethyl-5*H*-pyrimido-[5,4-*b*]-[1,4]benzothiazine-2,4(1*H*,3*H*)-dione **29**. As only three hydrogen atoms were located in the analysis, they were excluded from the picture. [Reproduced, with permission, from *J. Amer. Chem. Soc.*, **94**, 908 (1972).]

For the molecules **26**, **27**, and **28**, the mean distance of the sulphur atom from the plane defined by its three bonded atoms is 0.727 Å with a range from 0.717 to 0.737 Å. These values are very significantly smaller than the corresponding ones for the normal sulphonium ions and do suggest a significant flattening of the pyramid in the ylides, probably due to contributions from structures such as **49**. In the case of **29**, the distance of the sulphur atom from the plane is considerably greater (0.772 Å). In all four of the molecules **26**, **27**, **28**, and **29**, there appears to be a slight, but significant, displacement of the ylide carbon atom from the plane defined by its three bonded neighbours. In **26** the distance is 0.020 Å, in **27** it is 0.041 Å, in **28** it is 0.015 Å, and in **29** the corresponding distance is 0.057 Å. While all of these distances are small, they do suggest a slight non-planarity of the ylide carbon atom.

In their study on dimethylsulphonium cyclopentadienylide **28**, Adrianov and Struchkov[27] point out that a structure with localized charges (e.g. as in **48**) is supported by the approximately equal C—C bond lengths in the cyclopentadienyl ring.

Christensen and Witmore[25] proposed that the geometry of the sulphur ylide **26** suggests that there was some delocalization from unshared electrons on the ylide carbanion into the 3d orbitals on sulphur.

Views of the Newman projections along the $\overset{+}{S}$—$\overset{-}{C}$ ylide bond for **26**, **27**, **28**, and **29** are shown in Figure 27. There is a close resemblance in the distribution of the torsion angles in **26**, **27**, and **28**. The dimethyl groups are arranged such that they are roughly equally disposed on either side of the plane of the four S—C$\diagup^{\textstyle C}_{\diagdown C}$ atoms of the ylide group. Alternatively, one could say that the 'unoccupied lobe' of the approximate tetrahedron around sulphur approximately eclipses one of the C—C bonds attached to the ylide carbon atom. From these pictures, one can also see the out-of-plane bending that occurs at the ylide carbon, particularly in the case of **27**. The situation in **29** is different. Here, the two groups attached to the sulphur atom have much less flexibility than in the dimethylsulphonium compounds and this results in a very approximately eclipsed arrangement (torsion angle 23.2°) for the $\overset{+}{S}$—C$_{(ring)}$ bond with one of the C$_{(ylide)}$—C bonds. A stereoscopic view of the molecule of **29** is shown in Figure 26, from which it can be seen that the $\overset{+}{S}$—C$_{(methyl)}$ bond is almost at right-angles to the remainder of the molecule. Another consequence of the presence of the ring system in **29** is the near eclipsing (torsion angle 19.9°) of the C$_{(5)}$—C$_{(4)}$ bond with the $\overset{+}{S}$—$\overset{-}{C}$$_{(10)}$ ylide bond (Figure 28a). In the molecules of **26**, **27**, and **28**, where the methyl groups attached to the sulphonium sulphur atom are not constrained to any particular torsional conformation, the ideal, completely staggered, arrangement is found in every case. An example is shown in Figure 28b. The hydrogen atoms of the methyl group could not be located in the analysis of **29**.

C. Molecular Dimensions of Aminosulphonium Salts 3

The bond lengths and angles in the two examples of aminosulphonium salts (**30** and **31**) are given in Table 3. Both analyses are comparatively recent and are accurate. Stereoscopic views of the two molecules are shown in Figures 29 and 30. In this category, it would appear to be relatively meaningless to give average values as the two structures have significant differences from each other. In both **30**[30] and **31**[31] there is the usual pyramidal structure around the sulphur atom, although the

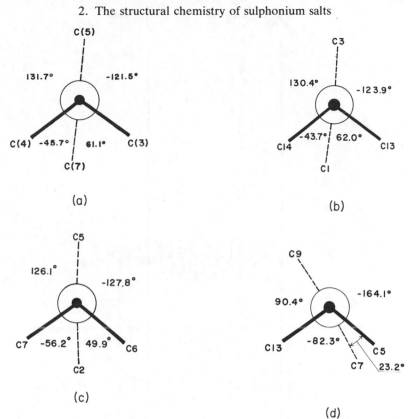

FIGURE 27. Torsion angles about the S—C ylide bonds in **26**, **27**, **28**, and **29**. (a) Torsion angles about the $S_{(1)}$—$C_{(2)}$ bond in **26**. (b) Torsion angles about the $S_{(12)}$—$C_{(2)}$ bond in **27**. (c) Torsion angles about the S—$C_{(1)}$ bond in **28**. (d) Torsion angles about the S—$C_{(10)}$ bond in **29**.

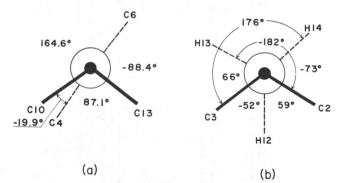

FIGURE 28. (a) Torsion angles about the S—$C_{(5)}$ bond in **29**. (b) Torsion angles about the $S_{(1)}$—$C_{(4)}$ bond in **26**.

Edmund F. Perozzi and Iain C. Paul

TABLE 3. Molecular dimensions involving the sulphur atom in aminosulphonium salts

Compound	Atom Nos. for X—S bonds[a]	X—S bond distances with standard deviations (Å)	Atom Nos. for X—S—C angles[a,b]	X—S—C angle (degrees)	Distance (Å) of S atom from plane of three atoms	Reference
30	N—S$_{(2)}$ C$_{(8)}$—S$_{(2)}$ C$_{(12)}$—S$_{(2)}$	1.644(5)[c] 1.789(5) 1.789(5)	C$_{(8)}$—S$_{(2)}$—C$_{(12)}$ N—S$_{(2)}$—C$_{(12)}$ N—S$_{(2)}$—C$_{(8)}$	100.6(3) 107.5(3) 105.9(3)	0.704	30
31	N—S C$_{(1)}$—S C$_{(5)}$—S	1.679(3) 1.834(4) 1.802(3)	C$_{(1)}$—S—C$_{(5)}$ N—S—C$_{(5)}$ N—S—C$_{(1)}$	100.6(2) 101.4(1) 96.6(1)	0.832	31

[a] These are the atom numbers used in the original paper. In this instance X = C or N.
[b] The X—S—C angles are listed such that an angle is on the same horizontal lines as the bond opposite it.
[c] The figures in parentheses refer to the estimated standard deviations in length or angle.

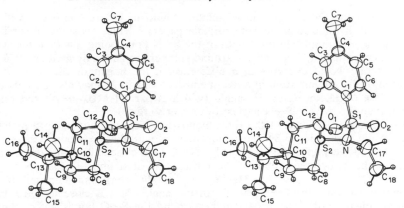

FIGURE 29. Stereoscopic view of the cation of the *trans*-4-*t*-butyl-1-[*N*-ethyl-*N*-*p*-toluenesulphonylamino]-1-thioniacyclohexane fluoroborate **30** structure, drawn from the coordinates in reference 30.

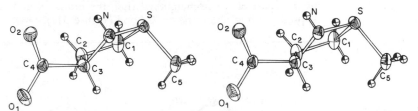

FIGURE 30. Stereoscopic view of the dehydromethionine **31** molecule, drawn from the coordinates in reference 31.

height of the pyramid varies from 0.704 Å for **30**, which is among the flatter pyramids considered in this chapter, to 0.832 Å for **31**, which is among the higher ones.

The $\overset{.}{S}$—N bond lengths are 1.644(5) Å in **30** and 1.679(3) Å in **31**. Both values are clearly shorter than the sum of the covalent radii for single bonds (1.74 Å) given by Pauling[49]. There seems to be a difference in the hybridization of the nitrogen atoms in the molecules of **30** and **31**. In **30**, the hybridization appears to be intermediate between sp^2 and sp^3; the nitrogen atom lies 0.19 Å out of the plane defined by its three bonded atoms, one of which is a sulphur of a sulphonamide. The $S_{(sulphonium)}$—N—$S_{(sulphonamide)}$, $S_{(sulphonium)}$—N—C, and $S_{(sulphonamide)}$—N—C bond angles are 114.5(3)°, 121.9(3)°, 119.6(3)°, respectively, giving a sum of 356°. We have already discussed the location of the $H_{(N)}$ hydrogen atom in **31** (see Section IIC). The exact position of this atom is essential to an assessment of the hybridization of nitrogen in the molecule of **31**. If one uses the position of the hydrogen given by Glass and Duchek[31] (and our analysis of the situation would support this), the nitrogen is definitely sp^3 hybridized. The nitrogen atom lics 0.40 Å out of the plane of its three neighbours and the sum of the three bond angles involving nitrogen is 321.5°. The crucial difference between the two structures may be the presence of the sulphonamido group in **30**, which imparts more sp^2 character to the nitrogen atom in that molecule. The shorter $\overset{.}{S}$—N distance and the flatter pyramid at sulphur found in **30** are consistent with this difference in hybridization. In both original papers[30,31] there is some discussion of bonding theories to explain the length of the S—N bond being less than the sum of the covalent radii. The bond angles around sulphur are generally greater in **30** than in **31**, as one would expect with a difference in the height of the pyramid; the inclusion of the sulphur atom in a five-membered ring in **31** may also be important. In the case of **30**, the C—S—C angle (which is part of a six-membered ring) is smaller than the two C—S—N angles; this is not so for **31**.

The torsion angles around the three bonds involving sulphur in both molecules are shown in Figure 31. In **30**, the arrangement around the S—N bond (Figure 31a) is such that the very approximate plane S—N—C—S is close to bisecting the C—S—C angle, or the 'unoccupied lobe' of the tetrahedron at sulphur is close to eclipsing the N—S$_{(sulphonamide)}$ bond. In **31**, because of the constraint imposed by the five-membered ring, the C—S—N—C torsion angle (21°) is close to the eclipsed position (Figure 31d). The arrangements around the S—C bonds where there is some conformational flexibility (i.e. where the carbon atom is part of a methyl group or participates in a six-membered ring) result in completely staggered arrangements (Figure 31b–d); the anomalous appearance of Figure 31c is presumably due to an incorrect location for H$_{(12)}$. In the case of **31**, where the carbon attached to sulphur is part of a five-membered ring, the atoms N and C$_{(2)}$ are held in an almost eclipsed position about the S—C bond (Figure 31f).

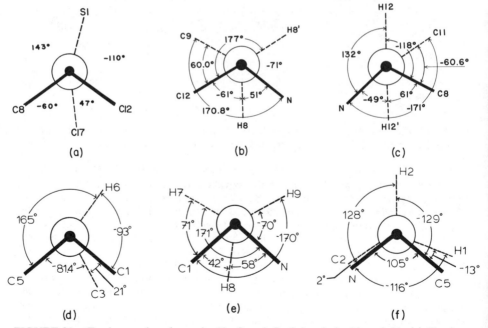

FIGURE 31. Torsion angles about the N—S and C—S bonds in **30** and **31**. (a) Torsion angles about the S$_{(2)}$—N bond in **30**. (b) Torsion angles about the S$_{(2)}$—C$_{(8)}$ bond in **30**. (c) Torsion angles about the S$_{(2)}$—C$_{(12)}$ bond in **30** [the position of H$_{(12)}$ is fairly clearly incorrect but the coordinates in reference 30 were used]. (d) Torsion angles about the S—N bond in **31**. (e) Torsion angles about the S—C$_{(5)}$ bond in **31**. (f) Torsion angles about the S—C$_{(1)}$ bond in **31**.

Cook *et al.*[30] pointed out that the lone pair of electrons on the nitrogen atom is at approximately right-angles to the sp³ (or unoccupied) lobe of the sulphonium sulphur atom. They suggested that this geometrical arrangement is due to lone pair–lone pair repulsion. Glass and Duchek[31] also recognized that the angle between the unoccupied lobe on sulphur and the lone pair of electrons on the sp³-hybridized nitrogen atom in **31** is virtually 90°.

D. Molecular Dimensions of Sulphilimines 4

The bond lengths and angles in sulphilimines **4** are given in Table 4. Again, the sulphur atom has a pyramidal arrangement of the three atoms bonded to it, as shown in a typical structure, that of **37** (Figure 32). The $\dot{S}$—N bond lengths range from 1.592(5) to 1.673(10) Å with a mean value of 1.637 Å. However, there is a clear distinction between the lengths in molecules where the N⁻ atom is attached to another sulphur atom, as in sulphonyl derivatives (i.e. **32, 33, 34, 35, 39**, and **42**), and those where the N⁻ is attached to the carbon atom of a carbonyl group, i.e. **36, 37**, and **38**. In the first category, the range of lengths is from 1.592(5) to 1.64(4) Å or, if one wishes to exclude **42** because of low accuracy, to 1.636(8) Å with a mean of 1.625 Å (including **42**) or 1.617 Å (excluding **42**). In the second category, the range of $\dot{S}$—N bond lengths is from 1.659(2) to 1.673(10) Å with a mean value of 1.666 Å. The $\dot{S}$—C$_{(aliphatic)}$ bond lengths for the molecules **32–39** range from 1.776(3) to 1.858(5) Å with a mean value of 1.803 Å, while the $\dot{S}$—C$_{(aromatic)}$ bond lengths range from 1.769(8) to 1.799(9) Å with a mean of 1.781 Å. These values do not differ greatly from those found in the other categories of sulphonium salts and are definitely compatible with the $\dot{S}$—C bonds being considered as single bonds. In the molecules of **32–35** and **39**, where the N⁻ is attached to another sulphur atom, the $\bar{N}$—S bond lengths range from 1.598(8) to 1.620(7) Å with a non-sulphonium mean value of 1.610 Å. Thus the bond orders of the two S—N bonds in these molecules are roughly equivalent. In these five molecules, the $\dot{S}$—$\bar{N}$—S bond angle ranges from 113.4(5)° to 116.3(2)° with a mean of 115.0°, while in **36–38** the range is from 110.0(6)° to 112.6(8)° with a mean of 111.0°.

The two groups of molecules, **32–35** and **39** on the one hand and **36–38** on the other, can also be grouped somewhat according to the height of the pyramid. For the first group the height ranges from 0.741 to 0.774 Å, while for the second it is from 0.775 to 0.805 Å. Another distinguishing feature between the groups is in the distribution of bond angles around the S⁺ atom. In the first group of molecules, the bond angle opposite the $\dot{S}$—N bond is always the smallest of the three angles, whereas in the second group this distinction is not observed.

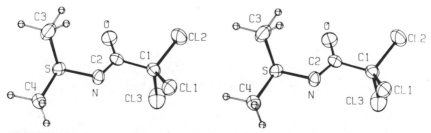

FIGURE 32. Stereoscopic view of the molecule *SS*-dimethyl-*N*-trichloroacetyl-sulphilimine **37**, drawn from the coordinates in reference 38.

TABLE 4. Molecular dimensions involving the sulphur and nitrogen atoms in sulphilimines

Compound	Atom Nos. for X—S or N—Y bonds[a]	X—S or N—Y bond distances with standard deviations (Å)	Atom Nos. for X—S—C and S—N—Y angles[a,b]	X—S—C or S—N—Y angle (degrees)	Distance (Å) of S atom from plane of three atoms	Reference
32	N—S(IV)	1.633(9)[c] 1.626(2)[d]	C$_{(2)}$—S(IV)—C$_{(3)}$	98.0(1.9) 98.2(2)	0.741	32, 33
	C$_{(2)}$—S(IV)	1.74(7)	C$_{(3)}$—S(IV)—N	102.6(7) 101.2(1)		
	C$_{(3)}$—S(IV)	— 1.74(3)	C$_{(2)}$—S(IV)—N	106.4(1.7) 105.1(2)		
	N—S(VI)	1.581(10) 1.603(2)	S(IV)—N—S(VI)	116.2(6)		
33	N$_{(1)}$—S$_{(2)}$ C$_{(8)}$—S$_{(2)}$ C$_{(9)}$—S$_{(2)}$ N—S$_{(1)}$	1.636(8) 1.801(13) 1.794(12) 1.591(8)	C$_{(8)}$—S$_{(2)}$—C$_{(9)}$ N$_{(1)}$—S$_{(2)}$—C$_{(9)}$ N$_{(1)}$—S$_{(2)}$—C$_{(8)}$ S$_{(2)}$—N$_{(1)}$—S$_{(1)}$	98.4(5) 102.2(5) 104.3(5) 113.4(5)	0.774	34
34	N—S$_{(1)}$ C$_{(21)}$—S$_{(1)}$ C$_{(31)}$—S$_{(1)}$ N—S$_{(2)}$	1.620(7) 1.780(9)[e] 1.820(9) 1.618(7)	C$_{(21)}$—S$_{(1)}$—C$_{(31)}$ N—S$_{(1)}$—C$_{(31)}$ N—S$_{(1)}$—C$_{(21)}$ S$_{(1)}$—N—S$_{(2)}$	97.4(4) 103.6(4) 103.6(4) 115.7(4)	0.774	35
35	C$_{(21)}$—S$_{(1)}$ C$_{(31)}$—S$_{(1)}$ N—S$_{(2)}$ N—S$_{(1)}$	1.628(7) 1.769(8)[e] 1.799(8)[e] 1.598(8)	C$_{(21)}$—S$_{(1)}$—C$_{(31)}$ N—S$_1$—C$_{(31)}$ N—S$_{(1)}$—C$_{(21)}$ S$_{(1)}$—N—S$_{(2)}$	101.0(4) 103.8(4) 103.7(4) 113.4(5)	0.774	36
36	N—S$_{(2)}$ C$_{(8)}$—S C$_{(9)}$—S N—C$_{(7)}$	1.659(2) 1.779(4) 1.776(3) 1.344(3)	C$_{(8)}$—S—C$_{(9)}$ C$_{(9)}$—S—N C$_{(8)}$—S—N S—N—C$_{(7)}$	101.0(2) 99.5(1) 104.2(1) 110.4(1)	0.775	37
37	N—S S—C$_{(3)}$ S—C$_{(4)}$ N—C$_{(2)}$	1.667(7) 1.791(12) 1.782(10) 1.320(10)	C$_{(3)}$—S—C$_{(4)}$ N—S—C$_{(4)}$ N—S—C$_{(3)}$ S—N—C$_{(2)}$	99.9(5) 98.5(4) 104.7(5) 110.0(6)	0.791	38
38	N—S C$_{(3)}$—S C$_{(5)}$—S	1.673(10) 1.837(13) 1.788(13)	C$_{(3)}$—S—C$_{(5)}$ N—S—C$_{(5)}$ N—S—C$_{(3)}$	99.3[f] 99.2(5) 103.8(5)	0.805	39

	Bond length (Å)	Angle (°)		Ref
39	N—C(2) 1.344(16)	S—N—C(2) 112.6(8)	0.753	40
	N(1)—S(1) 1.592(5)g	N(2)—S(1)—C(1) 97.0(2)		
	N(2)—S(1) 1.702(4)	N(1)—S(1)—C(1) 100.9(2)		
	C(1)—S(1) 1.858(5)	N(1)—S(1)—N(2) 107.6(2)		
	N(1)—S(2) 1.613(4)	S(1)—N(1)—S(2) 116.3(2)		
40	C(1)—S 1.772(2)	C(14)—S—C(15) 100.4(1)	0.772	41
	C(14)—S 1.793(3)	C(1)—S—C(15) 105.0(1)		
	C(15)—S 1.789(3)	C(1)—S—C(14) 102.5(1)		
	C(1)—C(2) 1.405(3)			
41	C(2)—N(1) 1.385(3)	C(2)—N(1)—C(13) 120.4(1)	—	42
	C(1)—S 1.724(6)	C(Me)—S—C(E1) 99f		
	C(E1)—S 1.842(8)	C(1)—S—C(Me) 109		
	C(Me)—S 1.812(8)	C(1)—S—C(E1) 108		
	C(1)—C(2) 1.423(8)			
	C(2)—N 1.298(7)			
42	N—S(1)i 1.63(4)	C(2)—N—C(B1) 122	0.807	43
	C(1)—S(1) 1.82(5)	C(1)—S(1)—C(2) 96(2)		
	C(2)—S(1) 1.78(5)	C(2)—S(1)—N 101(2)	0.739	
	N—S(2)i 1.64(4)	C(1)—S(1)—N 103(2)		
	C(3)—S(2) 1.84(5)	C(3)—S(2)—C(4) 102(2)		
	C(4)—S(2) 1.76(6)	N—S(2)—C(4) 105(2)		
		N—S(2)—C(3) 103(2)		
		S(1)—N—S(2) 111(2)		

[a] These are the atom numbers used in the original paper. In this instance X = C or N, while Y is the other atom bonded to the sulphilimine nitrogen.

[b] The X—S—C angles are listed such that an angle is on the same horizontal line as the bond opposite it. The S—N—Y angle is on the line opposite the N—Y bond.

[c] The figures in parentheses refer to the estimated standard deviations in length or angle.

[d] The lower figures are the values reported for some of the dimensions in **32** from a new analysis (see ref. 33).

[e] This corresponds to a C(aromatic)—S bond.

[f] The standard deviation was not given in the original paper.

[g] In the structure of **39**, there are two S—N bonds. The S(1)—N(1) bond corresponds to the ylide bond.

[h] In the manuscript sent by the authors[42], we were unable to reproduce exactly from the coordinates the quoted lengths and angles. Hence, we have used the values in the manuscript for the dimensions but do not include any additional calculations.

[i] In the molecule of **42**, there are two independent $\overset{\displaystyle C}{\underset{\displaystyle C}{>}}\overset{+}{S}-\bar{N}$ groups.

Kálmán[32] was the first to point out that the distribution of S—N bond lengths in **32** suggested considerable delocalization in the Ṡ—N̄—S region. Cameron *et al.*[34] suggested that this delocalization takes place in part by an interaction between an empty d-orbital of the S⁺ atom and a lone pair of electrons on the nitrogen atom. These authors noted that in the Ṡ—N̄—S molecules, the 'S^{IV}—N bond is significantly longer than the S^{VI}—N bond'. Subsequent work has shown this to be true, with the exception of the penicillin derivative **39** where additional influences may be important. They also noted[34] that all of these S—N bonds are much shorter than the S—N single bond length of 1.76(2) Å found in sulphamic acid[52] or than the sum of the single-bond covalent radii for sulphur and nitrogen (1.74 Å)[49]. The S—N distances found in the Ṡ—N̄—S molecules are also significantly shorter than those found in (C)₂Ṡ—N(R)₂ compounds [1.644(5) and 1.679(3) Å][30,31], while those found in the Ṡ—N̄—C compounds are comparable [1.659(2)–1.673(10) Å]. Cameron *et al.*[37] attempted to explain the differences in dimensions found in ylides stabilized by sulphonyl groups and those stabilized by carbonyl groups by indicating that delocalization of the negative charge between nitrogen and the adjacent carbonyl group could take place only via a p_π–p_π overlap, whereas d-orbital participation would be possible in the case of the sulphonyl compounds. A further factor which will affect the amount of negative charge on nitrogen is the extent to which delocalization takes place within the sulphonyl and carbonyl groups. These authors[37] concluded that, whichever effect predominates, delocalization is more efficient with carbonyl groups, as in these compounds (**36–38**) the Ṡ—N̄ bonds are significantly longer. As a result of the analyses on **36**, **37**, and **38**[37–39], it appears that the nature of the other group attached to the carbonyl carbon atom, whether electron withdrawing as —CCl₃ or —CHCl₂, or aromatic as —C₆H₅, has little influence on the bond lengths. In a related examination on the effect of phenyl *vs.* methyl substituents on either sulphur in molecules of type **50**, no significant

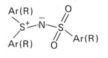

(50)

differences or trends were noted[34,37]. Kálmán and Párkányi[33] explained the S—N—S angle in **32** being less than 120° as due to repulsion of the lone pair of electrons on nitrogen with the two N—S bonds. Kálmán *et al.*[36] have also noted that there is sometimes a significant difference in length between the two Ṡ—C bonds when there were two identical substituents. This difference was noted for **32**[32], **35**[36], and **38**[39]. It was also pointed out that the effect had been observed in other identically substituted sulphur compounds, e.g. in the sulphur ylide **26**[25] and in dimethyl sulphoxide[53]. Kálmán *et al.*[36] presented evidence that the difference in Ṡ—C$_{(phenyl)}$ bond lengths in **35** may be due to the differences in the angles (56.8° and 39.8°) that the two phenyl rings make with the plane defined by the C—Ṡ—C atoms. A stereoscopic view of a single molecule of **35** is shown in Figure 33. The structure of **42**[43] is somewhat anomalous in that it can be thought of as a 'double ylide'. The molecular dimensions of each half of the molecule are in general agreement with those found in the sulphonyl-stabilized structures. Unfortunately, the accuracy of the analysis is poor and small differences would not be noted. A stereoscopic view of the structure of the cation in **42** is shown in Figure 34.

In many of these compounds there is a close intramolecular approach between the S⁺ atom and an oxygen atom of the group that stabilizes the negative charge on

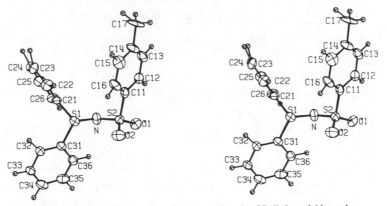

FIGURE 33. Stereoscopic view of the molecule *SS*-diphenyl-*N-p*-toluene-sulphonylsulphilimine **35**, drawn from the coordinates in reference 36. The coordinates of H$_{(36)}$ had to be changed to correspond to a reasonable position for this atom.

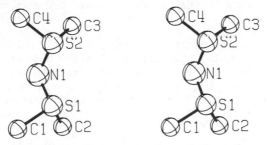

FIGURE 34. Stereoscopic view of the cation [(CH$_3$)$_2$S]N$^+$ **42**, drawn from the coordinates in reference 43.

nitrogen **51**. It is apparent that in all of the carbonyl-stabilized compounds the carbonyl group is arranged *trans* to the nitrogen lone pair (assuming trigonal

(51)

geometry at nitrogen); this arrangement then leads to a *cis* S---O contact of approximately 2.7–2.8 Å (see Table 5). There is undoubtedly a repulsive component because, as was pointed out by Kálmán *et al.*[39], the N—C—O angles in **36**, **37**, and **38** are increased to 125.9(1)°, 130.7(8)°, and 130.1(1.1)°, respectively. In the sulphonyl compounds, there is more flexibility, yet the S---O distances range from 2.86 to 3.01 Å (Table 5). All of these distances are much less than the sum of the van der Walls radii of sulphur and oxygen (3.25 Å)[54,55]. The approach of the sulphur and oxygen atoms can be seen in the views of **37** (Figure 32) and **35** (Figure 33).

TABLE 5. The non-bonded $\overset{+}{S}$---O distances in various sulphilimines[a]

Compound	Atoms in contact	Distance	Reference
32	$S_{(1)}$---$O_{(2)}$	3.01(1)	32
33	$S_{(2)}$---$O_{(1)}$	2.940(7)	34
34	$S_{(1)}$---$O_{(1)}$	2.988(7)	35
35	$S_{(1)}$---$O_{(2)}$	2.908(8)	36
36	S----O	2.708(2)	37
37	S----O	2.769(8)	38
38	S----O	2.836[b]	39
39	S----O	2.861(5)	40
40	S----N[c]	2.734(2)	41

[a]These distances were calculated by us from the coordinates in the original paper.
[b]No standard deviations are available from the analysis of **38**.
[c]In the molecule of **40**, it is a nitrogen atom rather than a oxygen that is involved in the intramolecular contact.

The structures of **40** and **41** are rather different from the others described in this section. In both instances, the authors[41,42] stated that they could be thought of as $\overset{+}{S}$—$\overset{-}{N}$ ylides with a C=C bond imposed between the S^+ and N^- atoms. However, inspection of the lengths of the C—C bond indicates that it can be considered to have only partial double bond character, and we also find that the C—N bond, especially in **41**, has substantial double bond character. It is also of interest to note that in **41** the negative charge on nitrogen is not stabilized by a sulphonyl or carbonyl group, as in all other structures discussed in this section except **43**. There is also a difference in the $\overset{+}{S}$—C(=C—N) bond length in **40** and **41**; the value for **41** is 1.724(6) Å, which suggests considerable double bond character and is virtually the same as the values reported for the $\overset{+}{S}$—C$_{(ylide)}$ bond lengths. Indeed, a structure of the type shown in **52** may be appropriate for **41**. The

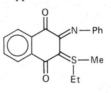

(52)

bond lengths in the 'quininoid' ring in **41** are strongly suggestive of such a structure (Figure 35), with most of the ring C—C bonds approaching $C_{(sp2)}$—$C_{(sp2)}$ single-bond values. The molecule of **40** is rather different. The $\overset{+}{S}$—C(=C—N) bond length is considerably longer (1.772 Å) and the C—C lengths in the phenyl ring suggest aromatic character.

Newman projections around the $\overset{+}{S}$—$\overset{-}{N}$ bond in the sulphonyl-stabilized compounds are shown in Figure 36. In all cases the sulphonyl sulphur atom is in a position that is close to eclipsing the 'unoccupied lobe' of the sulphonium group. However, the orientation is never exactly symmetrical with the torsion angles being as different as 146.5° and 111.1°. The projections along the $\overset{+}{S}$—$\overset{-}{N}$ bond in the carbonyl-stabilized compounds **36**, **37**, and **38** are shown in Figure 37. In **36** and **37**, the N—C bond is almost *exactly* *anti* to one of the $\overset{+}{S}$—C bonds, thus indicating a

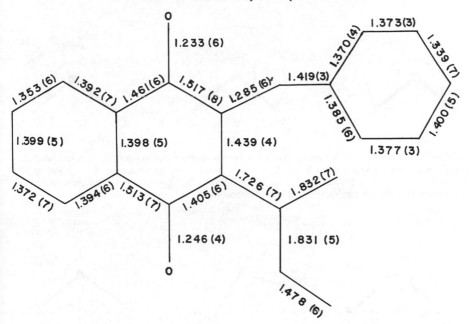

FIGURE 35.   Bond lengths in the molecule of the 2-anilino-3-ethylmethylsulphonium ylide of 1,4-naphthoquinone **41**. (Reproduced with the permission of Dr. F. M. Lovell of Philips Laboratories.)

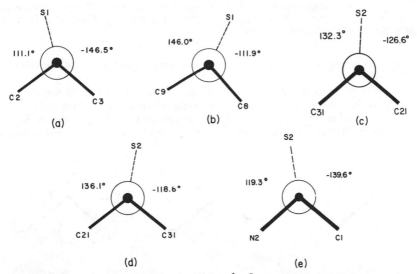

FIGURE 36.   Torsion angles around the $\overset{+}{S}$—$\overset{-}{N}$ bond in sulphonyl-stabilized sulphilimines. (a) Torsion angles around the $S_{(2)}$—N bond in **32**. (b) Torsion angles around the $S_{(2)}$—$N_{(1)}$ bond in **33**. (c) Torsion angles around the $S_{(1)}$—N bond in **34**. (d) Torsion angles around the $S_{(1)}$—N bond in **35**. (e) Torsion angles around the S—$N_{(1)}$ bond in **39**.

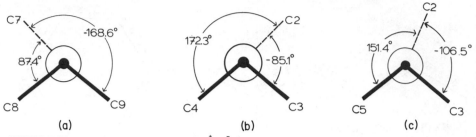

FIGURE 37. Torsion angles along the Ŝ—N̄ bonds in the carbonyl-stabilized sulphilimines. (a) Torsion angles about the S—N bond in **36**. (b) Torsion angles about the S—N bond in **37** (the values we calculate for the torsion angles in **37** have the opposite sign from those given in the paper). (c) Torsion angles about the S—N bond in **38**.

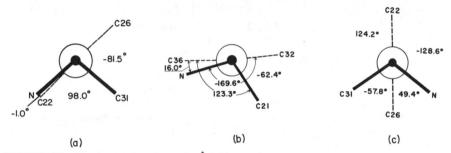

FIGURE 38. Torsion angles about the Ŝ—C(phenyl) bonds in **34** and **35**. (a) Torsion angles about the S$_{(1)}$—C$_{(21)}$ bond in **34**. (b) Torsion angles about the S$_{(1)}$—C$_{(31)}$ bond in **35**. (c) Torsion angles about the S$_{(1)}$—C$_{(21)}$ bond in **35**.

fully staggered arrangement if one considers the 'unoccupied lobe' as equivalent to an S—C bond.

The Newman projections around the Ŝ—C$_{(phenyl)}$ bonds in **34** and **35** are shown in Figure 38. In **34**, one of the C—C bonds of the phenyl ring almost eclipses the S—N bond. A similar arrangement is found for one of the phenyl rings in **35**, but the other phenyl ring is arranged such that one of the C—C bonds virtually bisects

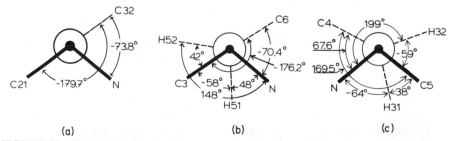

FIGURE 39. Torsion angles about the S—C$_{(ethyl or propyl)}$ bonds in **34** and **38**. (a) Torsion angles about the S$_{(1)}$—C$_{(31)}$ bond in **34**. (b) Torsion angles about the S—C$_{(5)}$ bond in **38**. (c) Torsion angles about the S—C$_{(3)}$ bond in **38**. The positions of the hydrogen atoms in **34** were not given in the paper.

the N—S—C bond angle (Figure 38c). In the molecules **34** and **38**, where there are ethyl or *n*-propyl groups attached to sulphur, the torsion angles are shown in Figure 39. In all three cases, the $C_{(\alpha)}$—$C_{(\beta)}$ bond is close to *anti* to the S—C bond. In all cases where there is an S—CH_3 group and where the hydrogen atoms have been located in the analysis, the three hydrogen atoms of the methyl group adopt an arrangement that is very close to being completely staggered with respect to the C—S—N bond angles (see Figure 40 for an example).

In Figure 41 are shown the torsion angles about the S—$C(CH_3)_2$ bond in **39**, the S—N bond in **39**, the S—C(=C) bond in **40**, and the two S—N bonds in **42**. In the case of the S—$C(CH_3)_2$ bond in **39** (Figure 41a), there is an almost completely

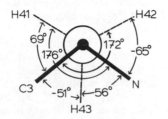

FIGURE 40. Torsion angles about the S—$C_{(4)(methyl)}$ bond in **37** showing the staggered arrangement of atoms.

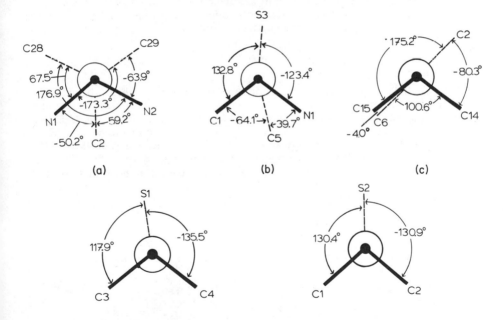

(a)   (b)   (c)

(d)   (e)

FIGURE 41. Torsion angles about a number of bonds in **39**, **40**, and **42**. (a) Torsion angles about the $S_{(1)}$—$C_{(1)}$ bond in **39**. (b) Torsion angles about the $S_{(1)}$—$N_{(2)}$ bond in **39**. (c) Torsion angles about the $S_{(1)}$—$C_{(1)}$ bond in **40**. (d) Torsion angles about the $S_{(2)}$—N bond in **42**. (e) Torsion angles about the $S_{(1)}$—N bond in **42**.

staggered arrangement, while the arrangement for the Ṡ—N bond in **39** (Figure 41b) is very similar to that found for the related bond in **30** (see Figure 31a). One of the C—C bonds in the quinoid ring in **40** virtually eclipses one of the S—CH$_3$ bonds (Figure 41c). In the molecule of **42**, the arrangement around the Ṡ—N bonds (Figure 41d, e) is similar to that found in the sulphonyl-stabilized compounds (**32–35** and **39**), although in the case of the torsions about one of the bonds, the N—Ṡ bond is virtually symmetrical with respect to the two Ṡ—C bonds.

E. Molecular Dimensions of Selenium Analogues of Sulphonium Salts

As there are only three compounds to be considered under this heading, we include all the structural data in Table 6, although two of the structures are ylides and one is a selenonium salt.

The Sė—C bond lengths in the trimethylselenonium salt **44**[15] agree with the value of 1.94 Å[49] for the sum of the single-bond covalent radii for carbon and selenium. The values reported for the Sė—C bonds in the ylide **45**[46] are longer than the sum of the covalent radii, while the length of the Sė—C ylide bond is significantly shorter. Surprisingly, in **46**[47] all three Sė—C bond lengths are approximately equal with the Sė—C ylide length being intermediate between the other two. It should be recognized that the other two Sė—C bonds in **46** involve phenyl rings. The sum of the double-bond covalent radii for carbon and selenium is 1.73 Å[49], hence the Sė—C ylide lengths are clearly intermediate between single and double bonds. Wei *et al.*[47] interpreted the bonding in **46** as involving the structures **53** and **54**, with a very significant component from **53**. In all three structures, there

$$Ph_2Sel = C(Ac)_2 \quad\longleftrightarrow\quad Ph_2\overset{+}{Se} - \overset{-}{C}(Ac)_2$$

$$(53) \hspace{5em} (54)$$

is the pyramidal configuration familiar in sulphonium salts, with the height of the pyramid varying from 0.951 Å in **44** to 0.780 Å in **46**. In both **45** and **46**, the angle opposite the Se—C ylide bond is significantly smaller than the other two C—Sė—C angles. A stereoscopic view of the structure of **46** is shown in Figure 42.

The arrangement of the groups around the Sė—C̄ ylide bonds in **45** and **46** is shown in Figure 43. The C—C bonds are arranged such that they bisect the C—Sė—C angle. A slightly unsymmetrical arrangement can be noted for both

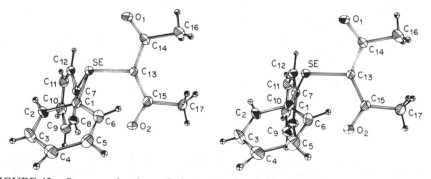

FIGURE 42. Stereoscopic view of the molecule of diacetylmethylenediphenylselenurane **46**. [Reproduced, with permission, from *J. Amer. Chem. Soc.*, **96**, 4099 (1974).]

TABLE 6. Molecular dimensions involving the selenium atom in selenonium salts and ylides

Compound	Atom Nos. for C—Se bonds[a]	C—Se bond distances with standard deviations (Å)	Atom Nos. for C—Se—C angles[a,b]	C—Se—C angles (degrees)	Distance (Å) of Se atom from plane of C atoms	Reference
44	C(1)—Se	1.962(23)[d]	C(2)—Se—C(2')[c]	99.1(7)	0.951	45
	C(2)—Se[c]	1.946(16)	C(1)—Se—C(2')	97.9(1)		
45	C(5)—Se[e]	1.88[f]	C(1)—Se—C(4)	90	0.91	46
	C(4)—Se	1.99	C(1)—Se—C(5)	104		
	C(1)—Se	2.01	C(4)—Se—C(5)	105		
46	C(13)—Se[e]	1.906(8)	C(1)—Se—C(7)	100.8(3)	0.780	47
	C(1)—Se	1.898(9)	C(7)—Se—C(13)	105.0(3)		
	C(7)—Se	1.926(9)	C(1)—Se—C(13)	107.5(3)		

[a] These are the atom numbers used in the original paper.
[b] The C—Se—C angles are listed in the Table such that an angle is on the same horizontal line as the bond opposite it.
[c] In the molecule **44**, there is C_s crystallographic symmetry that relates C(2) to C(2').
[d] The figures in parentheses refer to the estimated standard deviations in the length or angle.
[e] This is the Se—C ylide bond.
[f] There are no standard deviations quoted for the atomic positions in the original paper. The range of Se—C bond lengths was 0.02 Å and for C—Se—C angles 2°.

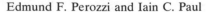

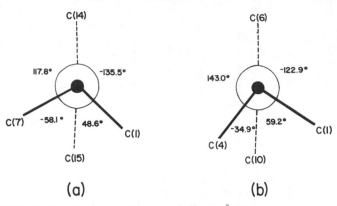

(a)   (b)

FIGURE 43. Torsion angles around the Sȅ—C̄ ylide bonds in **46** and **45**. (a) Torsion angles around the Se—C$_{(13)}$ bond in **46**. (b) Torsion angles around the Se—C$_{(5)}$ bond in **45**.

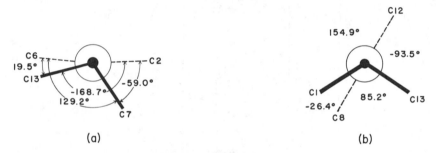

(a)   (b)

FIGURE 44. Torsion angles about the Se—C$_{(phenyl)}$ bonds in **46**. (a) Torsion angles about the Se—C$_{(1)}$ bond in **46**. (b) Torsion angles about the Se—C$_{(7)}$ bond in **46**.

molecules. The arrangement of the two phenyl rings about the Sȅ—C bonds in **46** is shown in Figure 44. The C$^-$ ylide atom lies 0.017 and 0.036 Å from the plane defined by its three bonded neighbours in **46** and **47**, respectively. Indeed, while both **45** and **46** could be thought of as having formal C_s symmetry, the configurations found in the crystal depart very significantly from that ideal symmetry. In the case of **46**, this deviation involves both the disposition of the acetylacetonate group with respect to the Sȅ—C bond and also the orientation of the two phenyl rings with respect to the very approximate plane that passes through the Sȅ—C ylide bond and the acetylacetonate group (Figure 42).

F. Other Features of Intramolecular Geometry of Sulphonium Salts and Related Compounds

There were many purposes for the X-ray investigations of sulphonium salts described in the previous sections. They ranged from the early work where the object of study was the gross geometry of the sulphonium group, to later investigations where the finest details of conformation and of molecular geometry were sought. There have been few systematic investigations of a range of compounds, the most notable being those of Cameron and Kálmán and co-workers on compounds of type **4**. In the course of the various studies reported in this chapter, a number of interesting points have emerged that do not properly belong in the previous sections, and these are described here.

While there are no examples of sulphonium salts crystallizing with precise $C_{2v}(3m)$ symmetry in the crystal, in general they do approach this geometry around sulphur with the typical pyramidal arrangement described earlier. The analyses of the salts that had three identical substituents, those of **8**[7], **20**[18], **23**[21], **24**[22], and **25**[23,24] were all plagued with difficulties regarding the establishment of the symmetry of the sulphonium group. In **8** and **20**, there may be crystallographic $C_s(m)$ symmetry in the cations, although this cannot definitely be concluded from the X-ray work, while the other analyses were of insufficient accuracy to draw conclusions regarding the symmetry of the cations.

The groups around the sulphur atoms in the molecules **10**, **12**, **17**, **29**, **31**, **34**, **39**, and **40** have the potential for creating a chiral sulphonium ion. However, only in **12** and **30** is the group in a chiral environment and in each case there are other asymmetric centres in the molecule. Thus, there are no examples among those discussed above of asymmetrically substituted sulphonium salts that are resolved spontaneously upon crystallization.

Several investigations were directed at the determination of the conformational preference for substituents on sulphonium sulphur when it is part of a six- or five-membered ring. The studies of Gerdil[13] on **14**, of Eliel et al.[14] on **15** and **16**, of Kelstrup et al.[15] on **17**, of Gusev and Struchkov[17] on **19**, and of Cook et al. on **30**[30] revealed an equatorial preference for substituents at sulphonium sulphur when it is part of a six-membered ring, except in the case of cis-4-t-butyl-S-methylthianium perchlorate (**15**)[14], where the methyl group is axial. Presumably the presence of the bulky t-butyl group in **15** and the preference of the six-membered ring for the chair conformation (Figure 18) forces the $\overset{+}{S}$—CH_3 group into the axial orientation. Studies of $\overset{+}{S}$—CH_3 groups in six-membered rings in solution[56] had led to a belief that, in general, they adopted equatorial orientations and the accumulated X-ray evidence tends to support this belief. The conformation of the dehydromethionine molecule (**31**) is of interest[31]. Here, the sulphur atom is part of a five-membered ring (Figure 30) which adopts an envelope conformation with $C_{(3)}$ lying 0.53 Å out of the approximate plane through the other four atoms. The $\overset{+}{S}$—CH_3 group adopts a definite axial orientation. Unfortunately, there are no other X-ray data available on the conformational preference of $\overset{+}{S}$—CH_3 groups when the sulphur is part of a five-membered ring.

The results of Del Re et al.[12] on the conformation of **13** in the crystal (Figure 11) were at considerable variance with those reported by the same authors[12] for **13** from the results of calculations and of solution n.m.r. studies. On the basis of both of these studies, a more folded structure with a closer approach of the sulphonium group to the carboxylate group had been predicted.

The study by Johnson et al.[16] on the dithiacyclooctane derivative **18** revealed a transannular $\overset{+}{S}$---S contact of 3.121(5) Å, less than twice the van der Waals radius of sulphur (3.50 Å)[54], and also an $\overset{+}{S}$---O intramolecular distance of 2.81(1) Å (see Figure 19). This $\overset{+}{S}$---O distance is comparable to those found in sulphilimines (Table 5). The arrangement of the sulphur and oxygen atoms around the central sulphonium sulphur atom in **18** is shown in the stereoscopic drawing in Figure 45 and can be thought of as a distorted trigonal bipyramid. The study by Deutsch[18] on **20**, although hampered by disorder, gave a reliable value of 3.130(4) Å for the

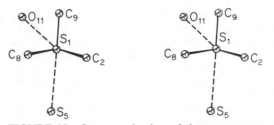

FIGURE 45. Stereoscopic view of the arrangement of atoms around $S_{(1)}$ in the dithiacyclooctane derivative **18**. The discontinuous lines represent interactions between non-covalently bonded atoms. [Reproduced, with permission, from *J. Chem. Soc. B*, 1603 (1970).]

intra-ring $\overset{+}{S}$---S distance. As the vacant p-orbitals of the sulphur atoms would be pointing outwards from the ring system, any interaction between the two sulphur atoms would have to be through bonds rather than through space.

There has been interest[57] in the effect of exocyclic substituents on the size of the C—C—C bond angles in phenyl rings. The conclusion has been drawn[57] that when there are electron-withdrawing groups (such as nitro) the bond angles are decreased below 120°, and when there are electron-releasing groups the C—C—C bond angles are increased above 120°. For sulphonium derivatives, the evidence is not completely conclusive. In the early analysis of dimethylphenylsulphonium perchlorate **9**[8], the C—C—C phenyl angle at the sulphonium group was 123(2)° and in the 'free' phenyl ring in phenylsulphoxonium iodide **19**[17] the C—C—C(phenyl) angle at the sulphonium group was 129(3)°. In **19** and in the ylide **29**, the C—C—C angles in the fused phenyl rings ranged from 123.1(4)° to 127(3)°. In the sulphilimine **4** class of molecules, there are four phenyl rings attached to S⁺ groups [in the molecules **34**, **35** (two phenyl rings), and **40**]; the C—C$_{(S)}$—C phenyl angles were 121.6(9)°, 120.8(8)°, 121.3(7)°, and 122.9(2)°. It does therefore appear that the presence of a bonded exocyclic sulphonium group increases the C—C$_{(S)}$—C angle; it is also probable that the increase is less in the case of sulphilimines than in the normal sulphonium salts.

IV. INTERMOLECULAR INTERACTIONS IN SULPHONIUM SALTS AND RELATED COMPOUNDS

It is interesting to ask whether the sulphonium group participates in any form of intermolecular electrostatic interaction and, if so, whether there is some directional preference for this interaction. There is considerable evidence for a relatively weak electrostatic interaction in a number of the structures examined, but it is clearly not a necessary condition for crystallization of a sulphonium salt as there are a number of structures where no such interaction takes place. Further, while many of the interactions do occur in the general direction of the 'unoccupied lobe', it is clear that there is no very specific directional requirement. In a number of the sulphonium salts that have been examined by X-ray methods, the groups attached to sulphur as sufficiently bulky that a close approach by an anion or polarizable group would be prohibited.

If we take values of 1.75 and 1.40 Å for the van der Waals radii of sulphur and oxygen, respectively[54,55], distances of less than 3.15 Å would appear to indicate strong interactions.

The shortest intermolecular S---O contacts are found in the structure of **31**. A view of the packing of **31** is shown in Figure 7. There is an Ṡ---O$_{(1')}$ contact of 3.006(3) Å. The authors[31] felt that this might indicate a significant interaction. They pointed out that the two C—O bond lengths in the ionized carboxyl group were unequal [1.283(4) and 1.259(4) Å] and that the longer of the two bonds was involved in the S---O contact. However, unequal C—O lengths are normal in zwitterionic amino acids[58], and the authors[31] failed to comment on another Ṡ---O interaction in **31** of 3.127 Å that involves O$_{(2)}$. This latter contact is between the two molecules that are held together by hydrogen bonding and involves the O$_{(2)}$ that participates in the hydrogen bond. While it is unlikely that these Ṡ---O interactions are sufficiently strong to alter C—O bond lengths by several hundredths of an Ångström, it is interesting that the arrangement of the five atoms (the three bonded atoms and the two oxygen atoms in other molecules) around sulphur is approximately trigonal bipyramidal with the C$_{(methyl)}$ group and O$_{(2)}$ in the molecule at $2 - x$, $1 - y$, $1 - z$ in the apical positions (Figure 7). Other short Ṡ---O contacts are found in the crystal of **40**[41] [Ṡ---O of 2.98 Å (Figure 46)], in the structure of **12**[11] [Ṡ---O of 3.115 Å (Figure 47)], and in the structure of **41**[42] [Ṡ---O of 3.19 Å]. In the last instance the oxygen atom involved in the contact lies at the corner of a very distorted tetrahedron around sulphur (Figure 48).

Short Ṡ---O contacts are not, however, always present in crystals of sulphonium ions. In the relatively simple structure of dimethylphenylsulphonium perchlorate (**9**)[8], where there would appear to be many opportunities for Ṡ---O contacts, the shortest Ṡ---O contact found in the crystal (Figure 49) is 3.23 Å.

A number of relatively short contacts have been noted that involve methyl groups attached directly to the sulphonium sulphur and other polarizible groups. For this type of contact, the sum of the van der Waals radii would be ~3.40 Å. A contact of 3.309 Å was found[9] in the structure of **10**, where packing in the crystal

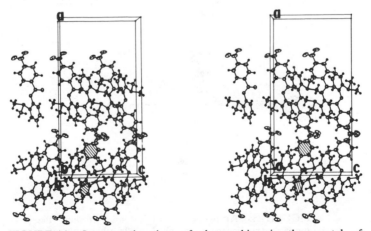

FIGURE 46. Stereoscopic view of the packing in the crystal of *N*-(*p*-nitro-benzoyl)-2-iminophenyldimethylsulphur(IV) **40**. The phenyl rings in the reference molecule are shaded. The Ṡ---O ($-x$, $\frac{1}{2} + y$, $\frac{1}{2} - z$) and S---S ($-x$, $1 - y$, $1 - z$) contacts are shown by dashed lines. This figure was drawn from the coordinates in reference 41.

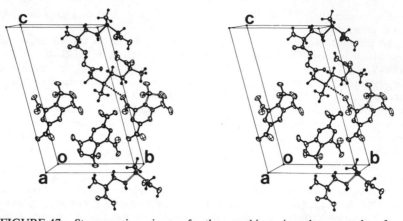

FIGURE 47. Stereoscopic view of the packing in the crystal of *R*-{[(3*S*)-3-amino-3-carboxypropyl](carboxymethyl)methylsulphonium} 2,4,6-trinitrobenzenesulphonate **12**. The two crystallographically independent reference molecules are shaded in black. There was some uncertainty as to which carboxylic acid groups were ionized; all possible carboxylic acid hydrogen atoms are included in the drawing. This figure was drawn from the coordinates in reference 11.

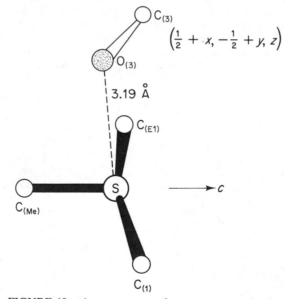

FIGURE 48. Arrangement of atoms around the sulphur atom in the 2-anilino-3-ethylmethylsulphonium ylide of 1,4-naphthoquinone **41**. The S---$O_{(1)}$ contact of 3.19 Å is shown as an unshaded bond. (Reproduced with the permission of Dr. F. M. Lovell of Philips Laboratories.)

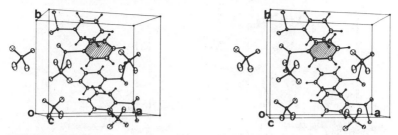

FIGURE 49. Stereoscopic view of the packing in the crystal of dimethylphenyl-sulphonium perchlorate **9**. The phenyl ring of one of the cations in the unit cell is shaded. This cation was obtained by adding 1 to the y-coordinates and 1 to the z-coordinates given in reference 8. The reference perchlorate anion is at the lower left hand corner. There are S---O contacts of 3.23 Å and 3.32 Å between the reference (shaded) sulphonium sulphur and $O_{(3)}$ at $x, \frac{1}{2} - y, \frac{1}{2} + z$ (behind) and with $O_{(2)}$ at $-x, \frac{1}{2} + y, \frac{1}{2} - z$ (to the left). The picture was drawn from the coordinates in reference 8.

was dominated by stacking of the trinitrobenzenesulphonate groups. In the sulphonium ylide **28**[28], there is a $C_{(13)}$---$O_{(1)}$ contact of 3.15 Å (see Figure 50) involving the methyl group attached to the sulphonium sulphur.

There are few reports of short Ṡ---Ṡ intermolecular contacts. Considered purely as an electrostatic interaction, one would, of course, not expect such an interaction. An exception is the molecule **40** (Figure 46)[41], where, in addition to the short Ṡ---O contact noted above, there is an Ṡ---Ṡ contact of 3.48 Å; this value is approximately equal to twice the van der Waals radius for sulphur[54,55]. In a structure (**11**) where the sulphur atoms would appear to be relatively free to approach each other (they have two methyl substituents and are properly oriented (Figure 51)[10], the Ṡ---Ṡ contact is 4.073(3) Å.

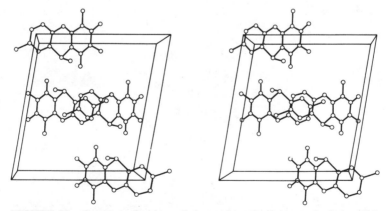

FIGURE 50. Stereoscopic view of the packing in the crystal of the ylide 7-chloro-1,3,5-trimethyl-5H-pyrimido[5,4-b][1,4]benzothiazine-2,4 (1H, 3H)-dione **29**. The view down b has a horizontal and c vertical with the origin at the back lower right corner. There is a short contact of 3.15 Å between $C_{(13)}$ in the reference molecule (the molecule at the centre right of the cell) and $O_{(2)}$ in the molecule at $\frac{1}{2} - x, \frac{1}{2} + y, \frac{1}{2} - z$. [Reproduced, with permission, from *J. Amer. Chem. Soc.*, **94**, 908 (1972).]

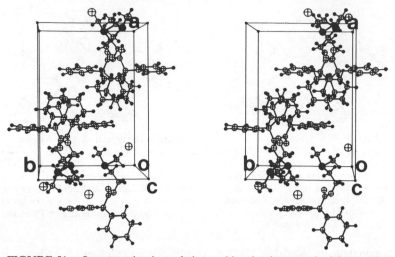

FIGURE 51. Stereoscopic view of the packing in the crystal of hexasonium iodide **11**. Sulphur atoms are shaded in black. The cation of the reference molecule is near the origin and extends in the *a* direction. The S---S contact of 4.073 Å is between two sulphur atoms separated along the *b*-axis. This picture was drawn from coordinates in reference 10.

More definite indications of interactions involving the sulphonium sulphur atom are found where there are halide anions in the crystal. The work of Battelle *et al.*[59] has shown that when divalent selenium is incorporated in a six-membered ring, it can form interactions to halogen in an axial direction. In the structure of

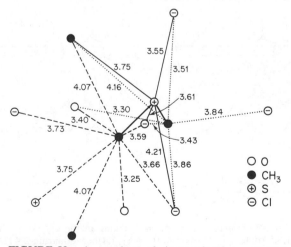

FIGURE 52. A number of intermolecular contacts involving the $-\overset{\cdot}{S}-(CH_3)_2$ group in the *S*-methyl-L-methionine chloride hydrochloride **13** structure. [Reproduced, with permission, from *Acta Crystallogr.*, **B33**, 3289 (1977).]

S-methyl-L-methionine chloride hydrochloride $\mathbf{13}^{12}$, there were a number of contacts of the sulphur and its immediate surrounding atoms with halogens (see Figure 52). The authors[12] felt that these contacts, the shortest involving $\overset{+}{S}$---Cl is 3.55 Å, were indicative of a Coulombic interaction and they postulated that there would be a polarization effect on the cation which would weaken the S—C bonds. However, inspection of Table 1 reveals that even among the accurately determined structures, the lengths of the $\overset{+}{S}$—C bonds in **13** are certainly not longer than those in other comparable structures.

In the structures of 1-methyl-1-thioniacyclohexane iodide $\mathbf{14}^{13}$ and phenylsulphoxonium iodide $\mathbf{19}^{17}$, there are $\overset{+}{S}$---$\bar{I}$ contacts of 4.287 and 3.85 Å, respectively. The packing of these structures is shown in Figures 53 and 54.

Perhaps the clearest example of an intermolecular interaction occurs in the trimethylselenonium iodide **44** structure[45], where there is an $\overset{+}{Se}$---I contact of 3.77 Å; this value is well below the sum of the van der Waals radii[54] of 4.15 Å. A stereoscopic view of the packing is shown in Figure 55, from which it can be seen that there are two $\overset{+}{Se}$---$\bar{I}$ contacts that lie in the crystallographic mirror plane that contains one of the $\overset{+}{Se}$—CH$_3$ bonds and the bisector of the other two. The two $\overset{+}{Se}$---$\bar{I}$ distances are 3.77 and 4.07 Å. The three methyl carbon atoms and the two iodine ions describe an approximate trigonal bipyramid around the selenium atom

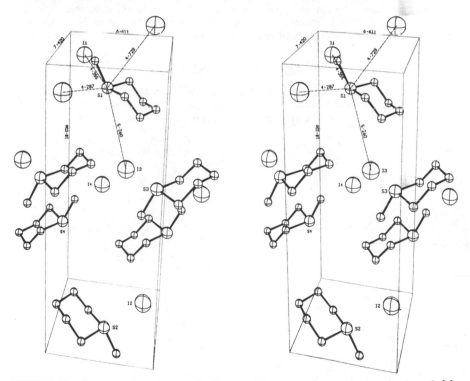

FIGURE 53. Stereoscopic view of the packing in the crystal of 1-methyl-1-thioniacyclohexane iodide **14**. The origin is at the upper left rear corner and the a-axis is vertical, the b-axis is horizontal, and the c-axis is toward the viewer. Various $\overset{+}{S}$---$\bar{I}$ contacts are shown. [Reproduced, with permission, from *Helv. Chim. Acta*, **57**, 489 (1974).]

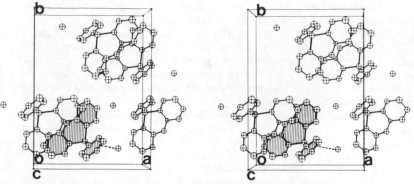

FIGURE 54. Stereoscopic view of the packing of phenylsulphoxonium iodide **19**. The rings in the reference molecule are shaded. The S---I contact of 3.85 Å is shown by a discontinuous line. This picture was drawn from the coordinates in reference 17.

That such interactions are not common to all selenonium salts can be seen from the packing of the selenium ylide **46**[47] (Figure 56), where there are no short intermolecular contacts involving selenium and the arrangement resembles that of a molecular crystal. It is probable that the strong Se---I interaction in **44** is enhanced by the symmetrical substitution of selenium by the relatively small methyl groups.

V. ACKNOWLEDGEMENTS

The authors acknowledge their great indebtedness to Dr Eileen Duesler and Ms Hi-Shi Chiang for their assistance in the preparation of this manuscript. Miss Becky McSwine and Miss Tina Tuminella drew several of the figures. The stereoscopic

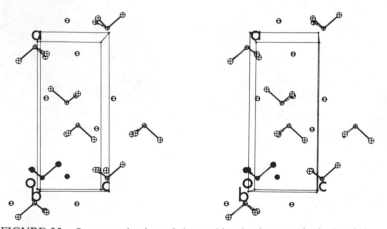

FIGURE 55. Stereoscopic view of the packing in the crystal of trimethyl-selenonium iodide **44**. The reference molecule is shaded black and the two iodide ions that make Se---I contacts of 3.77 and 4.07 Å lie in the plane at $y = \frac{1}{4}$ and are below to the left and right, respectively. There are crystallographic mirror planes at $y = \frac{1}{4}$ and $y = \frac{3}{4}$. This picture was drawn from the coordinates in reference 45.

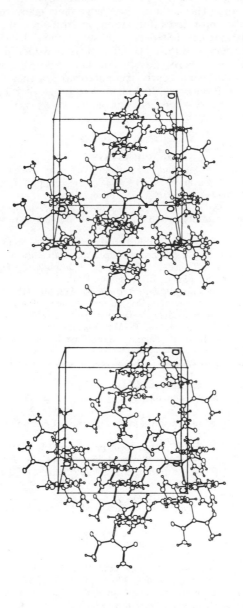

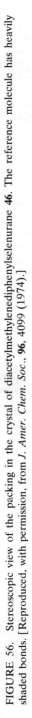

FIGURE 56. Stereoscopic view of the packing in the crystal of diacetylmethylenediphenylselenurane **46**. The reference molecule has heavily shaded bonds. [Reproduced, with permission, from *J. Amer. Chem. Soc.*, **96**, 4099 (1974).]

pictures were drawn using the ORTEP program, written by Dr C. K. Johnson, Oak Ridge National Laboratory. The authors also thank their many colleagues around the world who generously provided information on structures of sulphonium salts and related compounds that had not been published. Those to whom gratitude is due include Arild Christensen, Birgitta Dahlén, Raymond Gerdil, Jenny Glusker, Tom Hamor, Alajos Kálmán, F. M. Lovell, and Fernando Mazza. Joel Bernstein gave assistance to providing some early computerized searches of the literature. Finally, thanks are due to Professor J. C. Martin who provided E.F.P. with facilities and hospitality in his laboratory where most of the work was carried out.

VI. REFERENCES

1. J. C. Martin and E. F. Perozzi, *Science*, **191**, 154 (1976).
2. R. J. S. Beer, in *Organic Compounds of Sulphur, Selenium and Tellurium*, Specialist Periodical Reports, Vol. 2, Chemical Society, London, 1973, pp. 497–510.
3. I. C. Paul, *Kém. Közl*, **46**, 245 (1976).
4. The goodness of fit is defined as $[\Sigma w |F_{obs} - F_{calc}|^2/(m - n)]^{1/2}$, where w are the weights, m is the number of independent observations, and n is the number of variable parameters in the model.
5. A good general reference to crystallographic procedures is G. H. Stout and L. H. Jensen, *X-Ray Structure Determination: A Practical Guide*, Macmillan, New York, 1968.
6. N. F. M. Henry and K. Lonsdale (Editors), *International Tables for X-Ray Crystallography*, Vol. I, Kynoch Press, Birmingham, 1965.
7. D. E. Zuccaro and J. D. McCullough, *Zeit. Kristallogr.*, **112**, 401 (1959).
8. A. Lopez-Castro and M. R. Truter, *Acta Crystallogr.*, **17**, 465 (1964).
9. W. Barnes and M. Sundaralingam, *Acta Crystallogr.*, **B29**, 1868 (1973).
10. J. J. Guy and T. A. Hamor, *J. Chem. Soc. Perkin Trans. II*, 467 (1975).
11. J. W. Cornforth, S. A. Reichard, P. Talalay, H. L. Carrell and J. P. Glusker, *J. Amer. Chem. Soc.*, **99**, 7292 (1977).
12. G. Del Re, E. Gavuzzo, E. Giglio, F. Lelj, F. Mazza and V. Zappia, *Acta Crystallogr.*, **B33**, 3289 (1977).
13. R. Gerdil, *Helv. Chim. Acta*, **57**, 489 (1974).
14. E. L. Eliel, R. L. Willer, A. T. McPhail and K. D. Onan, *J. Amer. Chem. Soc.*, **96**, 3021 (1974).
15. E. Kelstrup, A. Kjaer, S. Abrahamsson and B. Dahlén, *J. Chem. Soc. Chem. Commun.*, 629 (1975).
16. S. M. Johnson, C. A. Maier and I. C. Paul, *J. Chem. Soc. B*, 1603 (1970).
17. A. I. Gusev and Yu. T. Struchkov, *Zh. Strukt. Khim.*, **12**, 1120 (1971); *J. Struct. Chem. (USSR)*, **12**, 1042 (1971); *Kristallografiya*, **18** 525 (1973); *Soviet Physics (Crystallography)*, **18**, 330 (1973).
18. E. Deutsch, *J. Org. Chem.*, **37**, 3481 (1972).
19. M. Matsui, T. Watanabé, F. Miyoshi, K. Tokuno and T. Ohashi, *Acta Crystallogr.*, **B32**, 3157 (1976).
20. B. T. Kilbourn and D. Felix, *J. Chem. Soc. A*, 163 (1969).
21. R. H. Fenn, *Acta Crystallogr.*, **20**, 20 (1966).
22. R. H. Fenn, *Acta Crystallogr.*, **20**, 24 (1966).
23. G. A. Jeffrey and R. K. McMullan, *J. Chem. Phys.*, **37**, 2231 (1962).
24. P. T. Beurskens and G. A. Jeffrey, *J. Chem. Phys.*, **40**, 2800 (1964).
25. A. T. Christensen and W. G. Witmore, *Acta Crystallogr.*, **B25**, 73 (1969).
26. A. T. Christensen and E. Thom, *Acta Crystallogr.*, **B27**, 581 (1971).
27. V. G. Adrianov and Yu. T. Struchkov, *Izv. Akad. Nauk SSSR*, **26**, 687 (1977); *Bull. Acad. Sci. USSR*, **26**, 624 (1977).
28. J. P. Schaefer and L. L. Reed, *J. Amer. Chem. Soc.*, **94**, 908 (1972).
29. A. T. Christensen, personal communication (1977).
30. R. E. Cook, M. D. Glick, J. J. Rigau and C. R. Johnson, *J. Amer. Chem. Soc.*, **93**, 924 (1971).

31. R. S. Glass and J. R. Duchek, *J. Amer. Chem. Soc.*, **98**, 965 (1976).
32. A. Kálman, *Acta Crystallogr.*, **22**, 501 (1967).
33. The results of a new refinement of this structure by Kálmán and Párkányi are reported in *J. Chem. Soc. Perkin Trans. II*, 1322 (1977).
34. A. F. Cameron, N. J. Hair and D. G. Morris, *J. Chem. Soc. Perkin Trans. II*, 1951 (1973).
35. A. Kálmán and K. Sasvári, *Cryst. Struct. Commun.*, **1**, 243 (1972).
36. A. Kálmán, B. Duffin and Á. Kucsman, *Acta Crystallogr.*, **B27**, 586 (1971).
37. A. F. Cameron, F. D. Duncanson and D. G. Morris, *Acta Crystallogr.*, **B32**, 1998 (1976).
38. A. Kálmán, K. Sasvári and Á. Kucsman, *Acta Crystallogr.*, **B29**, 1241 (1973).
39. A. Kálmán, K. Sasvári and Á. Kucsman, *J. Chem. Soc. Chem. Commun.*, 1447 (1971); A. Kálmán, personal communication.
40. A. F. Cameron, I. R. Cameron, M. M. Campbell and G. Johnson, *Acta Crystallogr.*, **B32**, 1377 (1976).
41. A. F. Cameron, F. D. Duncanson and D. G. Morris, *Acta Crystallogr.*, **B32**, 2002 (1976).
42. F. M. Lovell and D. B. Cosulich, *Abstracts of American Crystallographic Association, Summer Meeting*, p. 104 (1971); F. M. Lovell, personal communication (1977).
43. A. M. Griffin and G. M. Sheldrick, *Acta Crystallogr.*, **B31**, 893 (1975).
44. J. D. McCullough and R. E. Marsh, *J. Amer. Chem. Soc.*, **72**, 4556 (1950).
45. H. Hope, *Acta Crystallogr.*, **20**, 610 (1966).
46. V. V. Saatsazov, R. A. Kyandzhetsian, S. I. Kuznetsov, N. N. Madgesieva and T. C. Khotsyanova, *Dokl. Akad. Nauk SSSR*, **206**, 1130 (1972).
47. K.-T. H. Wei, I. C. Paul, M.-M. Y. Chang, and J. I. Musher, *J. Amer. Chem. Soc.*, **96**, 4099 (1974).
48. V. V. Saatsazov, R. A. Kyandzhetsian, S. I. Kuznetsov, N. N. Magdesieva and T. C. Khotsyanova, *Izv. Akad. Nauk SSSR, Ser. Khim.*, 671 (1973).
49. L. Pauling, *The Nature of the Chemical Bond*, 3rd Edition, Cornell University Press, Ithaca, N.Y., 1960, p. 224.
50. I. C. Paul, in *The Chemistry of the Thiol Group, Part I* (Ed. S. Patai), Wiley, London, 1974, pp. 111–151.
51. P. L. Johnson and I. C. Paul, *J. Chem. Soc. B*, 1296 (1970).
52. R. L. Sass, *Acta Crystallogr.*, **13**, 320 (1960).
53. R. Thomas, C. B. Schoemaker and K. Ericks, *Acta Crystallogr.*, **21**, 12 (1966).
54. L. Pauling, *The Nature of the Chemical Bond*, 3rd Edition, Cornell University Press, Ithaca, N.Y., 1960, pp. 257–264.
55. There is general agreement that Pauling's value (1.85 Å) for the van der Waals radius of sulphur is too high. A more appropriate value is probably 1.75 Å; see, for example, P. L. Johnson, K. I. G. Reid and I. C. Paul, *J. Chem. Soc. B*, 946 (1971).
56. J. B. Lambert, C. E. Mixan and D. H. Johnson, *Tetrahedron Lett.*, 4335 (1972).
57. A. Domenicano, A. Vaciago and C. A. Coulson, *Acta Crystallogr.*, **B31**, 221 (1975).
58. J. F. Griffin and P. Coppens, *J. Amer. Chem. Soc.*, **97**, 3496 (1975).
59. L. Batelle, C. Knobler and J. D. McCullough, *Inorg. Chem.*, **6**, 958 (1967).

The Chemistry of the Sulphonium Group
Edited by C. J. M. Stirling and S. Patai
© 1981 John Wiley & Sons Ltd.

CHAPTER **3**

Analysis and determination

M. R. F. ASHWORTH

Lehrstuhl für Organische Analytik, Universität des Saarlandes,
Saarbrücken, W. Germany

I.	INTRODUCTION	80
II.	SUMMARY OF ANALYTICAL METHODS	81
III.	CHEMICAL METHODS	82
	A. Ion Combination	83
	1. Anionic surface-active agents	83
	2. Dichromate	83
	3. Dyes	84
	4. Heteropoly acids	84
	5. Hexachloroplatinate	85
	6. Perchlorate	85
	7. Picrate	85
	8. Reineckate	86
	9. Tetraiodobismuthate	86
	10. Tetraphenylborate	86
	11. Triiodide	87
	B. Reduction	87
	1. Polarography	87
	2. Chemical reducing agents	88
	a. Ascorbic acid	88
	b. Cadmium	88
	c. Chloromolybdate(III)	88
	d. Chromium(II)	88
	e. Dithionite	88
	f. Iron(II)	89
	g. Tin(II)	89
	h. Titanium(III)	89
	i. Vanadium(II)	89
	C. Degradation with Acid or Alkali	89
	D. Other Reactions of the Sulphonium Ion	90
	1. Oxidative methods	90
	2. Substitution	90
	E. Reactions of Other Groups in the Molecule	91

IV.   PHYSICAL METHODS   91
      A. Spectroscopic Methods   91
      B. Separation Methods   92
            1. Paper chromatography   92
            2. Thin-layer chromatography   93
            3. Ion exchange   94
            4. Column and liquid chromatography   95
            5. Electrophoresis   95
            6. Gas–liquid chromatography   95
V.   REFERENCES   95

I. INTRODUCTION

There are few analytical methods which apply to sulphonium compounds in general. This chapter is devoted to those members of the compound class which have received particular analytical attention. Three broad groups may be distinguished:

(i) Sulphonium compounds of biological interest, i.e. those derived from methionine and typified by *S*-methylmethionine and *S*-adenosylmethionine sulphonium compounds:

$$Me_2\overset{+}{S}-CH_2-CH_2-CH(NH_2)-COOH$$

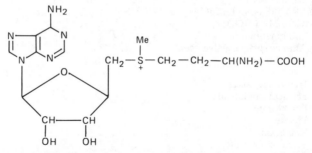

The isolation and purification of these compounds from natural sources involve steps which can be regarded as analytical, meriting at least brief reference here.

(ii) Sulphonium compounds containing a long-chain group and therefore surface-active. Some of these have been determined quantitatively by methods analogous to those used for certain quaternary ammonium compounds. Procedures for the latter compounds are also included in this chapter where it is reasonably sure that they would apply equally to similar sulphonium compounds.

(iii) Thiazine dyes (phenothiazonium compounds), such as methylene blue. These have sulphonium character seen in the canonical forms:

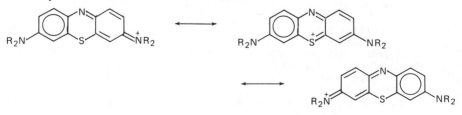

TABLE 1. Thiazine dyes

| (Trivial) Name | Substituents on nitrogens | | Other substituents |
	Position 3	Position 7	
Lauth's violet (thionine)	H, H	H, H	
Azure C	H, H	H, CH$_3$	
Azure A	H, H	CH$_3$, CH$_3$	
Azure B	H, CH$_3$	CH$_3$, CH$_3$	
Methylene blue	CH$_3$, CH$_3$	CH$_3$, CH$_3$	
Toluidine Blue C	CH$_3$, CH$_3$	CH$_3$, CH$_3$	CH$_3$ at 2–
Methylene Green	CH$_3$, CH$_3$	CH$_3$, CH$_3$	NO$_2$ at 4–
New Methylene Blue N	H, C$_2$H$_5$	H, C$_2$H$_5$	CH$_3$ at 2– and 8–

Much analytical work has been carried out on these dyes, justifying inclusion here. Table 1 gives some of the best known.

II. SUMMARY OF ANALYTICAL METHODS

Detection, isolation, identification and quantitative determination of sulphonium compounds depend on the following principles:

III. Chemical Methods
   A. Ion Combination

The sulphonium ion can combine with suitable, mostly large, anions to yield products which are coloured and/or poorly soluble in the reaction medium, which is usually aqueous.

$$R_3S^+ + An^- \longrightarrow R_3\overset{+}{S}An^-$$

This may be exploited quantitatively in several ways:

(a) direct titration with the reagent containing the large anion;
(b) indirect titration, generally back-titration, of unused anion;
(c) gravimetric determination of the precipitated product;
(d) colorimetric determination of the product.

Many reagent anions have been used to determine phenothiazonium compounds and surface-active sulphonium salts, and in the isolation and identification of methionine sulphonium compounds.

   B. Reduction
   Sulphonium ions are reduced to sulphides:

$$R_2\overset{+}{S}R' + 2e + H^+ \longrightarrow R_2S + R'H$$

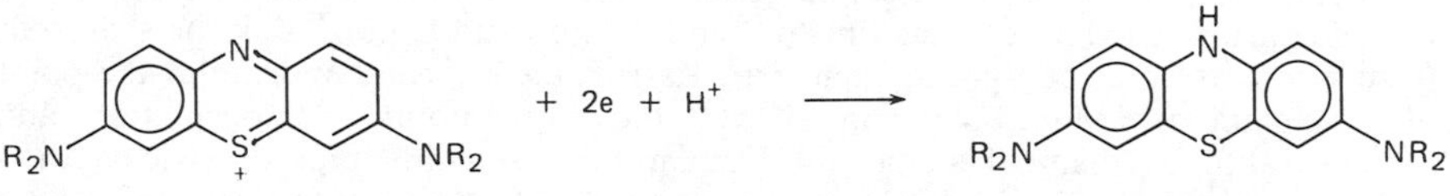

leuco—compound

Reduction can be with standard reagents or polarographically.

C. Degradation with Acid or Alkali

The initial reaction appears to yield sulphide:

$$R_2\overset{+}{S}R' + OH^- \longrightarrow R_2S + R'OH$$

Further reactions can occur subsequently, such as elimination of water from the alcohol or, with adenosylmethionine, fission of other bonds.

D. Other Reactions

Few examples are to be found in this 'miscellaneous' group. They include oxidation and substitution and apply only to the phenothiazonium cation.

E. Reactions of Other Groups in the Molecule

Strictly these fall outside the scope of the chapter but titrations with acid and reactions with ion exchangers in the OH^- form are mentioned:

$$An^- + H^+ \longrightarrow HAn$$

$$An^- + Exch.-OH \longrightarrow Exch.--X + OH^-$$

IV. Physical Procedures

A. Spectroscopic Methods

U.v., i.r., m.s., and n.m.r. studies of sulphonium compounds have been extensive and are dealt with elsewhere in this volume. There are few analytical applications.

B. Separation Methods

Chromatographic procedures have been widely applied to the phenothiazonium and methionine sulphonium compounds. The easy visual detection of the former simplifies and encourages use of the method.

III. CHEMICAL METHODS

A. Ion Combination

The analytical information can be conveniently classified according to reagents in alphabetical sequence.

1. Anionic surface-active agents

Considerable quantitative use has been made of ion combination between cationic and anionic surface-active agents. It enables the one type of agent to be determined with the help of the other as reagent. Most cationic agents in published work have been quaternary ammonium compounds containing one larger (C_{12}–C_{18}) group. However, it is reasonable to expect that similar sulphonium compounds will react analogously and thus be determinable. A short account of the procedure is given here. As indicated in the summary of methods, most ion combination methods are titrimetric or colorimetric. The anion-active reagents have generally been sulphonates or sulphates.

a. *Titration procedures in a single phase*. Physical end-point determination has been frequently used here. Thus Preston[1] and Preston and Epton[2] took the sharp fall in surface tension as end-point; Lambert[3], Karrman *et al.*[4], and Wickbold[5] the point of maximum turbidity; Ullmann *et al.*[6] took the point of flocculation and disappearance of foam; Tanaka and Tamamushi[7] used an amperometric method and Yoda[8] noted the position of minimum conductivity, titrating with silver dodecylsulphate.

Titration to colour or fluorescence changes after adding an indicator has also been used. These indicators have large cations or anions which combine with the anion or cation, respectively, taking part in the titration reaction. The indicators are therefore either liberated or taken up just before or after the equivalence point, depending on the direction of titration. This causes change of colour or fluorescence (Table 2).

Thus, Hartley and Runnicles[9] used bromophenol blue and took a colour between blue and purple as end-point; Saltor and Alexander[10] used pinacyanol bromide, obtaining a colour change from pink–purple to blue–purple; Doležil and Bulandr[11] used eosin, which yielded yellow–green fluorescence at the end-point, so titration was to the point of maximum fluorescence.

b. *Titration procedures in two phases.* More scope in end-point indication is obtained by adding a second solvent immiscible with the usually aqueous reaction mixture. Chloroform or dichloroethane is popular because their high density accelerates the separation of the layers. An indicator is then chosen which is soluble in only one of the phases and this can often be regulated by pH. The ion-combination products are generally soluble in the other phase. This provides end-points based on colour intensification or diminution in one or other layer. Epton[12] and Harada and Kimura[13], for example, used methylene blue indicator and took as end-point equal blue intensities in the aqueous and chloroform layers. In the Antara method[14], movement of blue into the water layer was taken. Disappearance of colour from chloroform was the end-point, using dichlorofluorescein[15], eosin[16], and bromophenol blue[17]. Carkhuff and Boyd[18] used dimethylaminoazobenzene as indicator and took the first colour in the chloroform layer as the end-point.

2. Dichromate

Precipitation with excess of standard dichromate solution and determination of unused reagent in the filtrate has been a standard quantitative procedure for determining methylene blue, for example by Dister[19], Davidson[20], and Moraw[21] and in the Pharmacopoeias of Switzerland[22], the GDR[23] and GFR[24], all of which have the same procedure. The final stage involves acidification of the filtrate with hydrochloric or sulphuric acid, addition of potassium iodide and, after a few

TABLE 2. Titration of surface-active compounds using indicators

Cationic agent, e.g. $=\overset{.}{N}=$ or $=\overset{.}{S}-$	Titrated with	Anionic agent, e.g. RSO_3^- or $ROSO_3^-$
combines with titrand ion	Indicator with large anion (e.g. bromophenol blue)	Indicator displaced and liberated by last trace of titrant
	Indicator with large cation (e.g. methylene blue, pinacyanol bromide)	Indicator combines with first trace of titrant after the equivalence point

minutes, titration of the iodine liberated from the unused reagent, using thiosulphate. Ferrey[25] proposed a gravimetric method.

3. Dyes

Basic dyes, such as the thiazines, which contain large cations, are precipitated by the large anion of acid dyes. The considerations in Section IIIA1 apply here, with the titrand or titrant serving as indicator. Numerous examples are in the literature. For instance, Pelet and Garuti[26] titrated methylene blue with some acid dyes, such as carmine, Pyramine Orange, Cotton Brown, and, best of all, Crystal Ponceau, which formed a product in the stoichiometric ratio of 1:2 (Crystal Ponceau:methylene blue). They located the end-point by spotting on to paper but also indicated the possibility of doing this by measurement of electrical resistance of the solution. This general principle has been extended to determine cationic surface-active materials, e.g. by Klevens[27], who titrated with dyes such as indophenol, eosin, Sky Blue FF and benzopurpurine to changes of colour or fluorescence. These changes are not specified by Klevens, nor do his examples include sulphonium salts although they must be titratable in the same way. An interesting, more recent variant is the titration by Schiffner and Borrmeister[28] of cationic dyes in acetate buffer pH 4.7 with a solution of the commercial optical brightener Blankophor G Extra High Conc. This yielded a non-fluorescent, poorly soluble product with the dyes. Their end-point was the appearance of permanent fluorescence from the first trace of titrant in excess.

Colorimetric application of the principle appears more usual. Examples are again of quaternary ammonium compounds but with the justified expectation of application to analogous sulphonium compounds. For instance, Auerbach[29] determined quaternary ammonium germicides by shaking the aqueous sample with sodium carbonate and bromophenol blue, extracting the coloured product with dichloroethane (in a later publication[30], he recommended benzene instead) and measuring the extract colorimetrically. Colichman[31] modified Auerbach's method by using no extraction solvent and a definite excess of bromophenol blue. Mukherjee[32] extracted into chloroform, and measured at 606 nm in 0.01 M sodium hydroxide or at 416 nm in 0.01 M hydrochloric acid. Levine[33] also used bromophenol blue to determine various 'onium' compounds, including bis (dimethylsulphonium) dibromide and hexamethylenebis(diethylsulphonium) diiodide. He reacted these compounds, containing one chain of at least four carbon atoms, in a solution containing sodium carbonate and dipotassium hydrogen phosphate and measured light absorbance at 600 nm.

Baldinus[34] suggested that dimethyldodecylsulphonium iodide or others with long chains could be determined by reaction with methyl orange in a buffer of pH 4 or 10, extraction of the product into chloroform, addition of methanolic hydrochloric acid and determination of absorbance at 525 nm.

4. Heteropoly acids

Heteropoly acids yield poorly soluble products with many bases. Ogawa[35] titrated methylene blue and other basic dyes in >1 M hydrochloric acid with tungstosilicic acid, using an amperometric end-point. A similar method was used by Mareš and Stejskal[36] who, however, took larger samples and precipitated in weaker hydrochloric acid (ca. 0.2 M). They reported a stoichiometric ratio of tungstosilicic acid to methylene blue of 1:3. Burgess et al.[37] evaluated methods for purifying and quantitatively determining certain basic dyes (including methylene blue) and preferred precipitation with tungstosilicic acid. Colour reaction with molybdophosphoric acid is among detection methods quoted by Ivanova et al.[38] for quaternary ammonium and sulphonium compounds, e.g. dimethyl β-bromoethylsulphonium bromide, after paper chromatography.

Precipitation with tungstophosphoric acid has been a stage in the isolation, concentration, and eventual quantitative determination of S-methylmethionine- and S-adenosylmethioninesulphonium compounds. McRorie et al.[39] used this method to isolate the former from cabbage. Ogawa et al.[40] obtained the same salt from drugs by precipitation with tungstosilicic acid before using a colorimetric method for determination (Section IIIE). The principle has been used for the S-adenosyl compound by several investigators, notably Schlenk and co-workers[41–46]. In most procedures, precipitation with tungstophosphoric acid is carried out in the cold filtering after 1–2 h. The precipitate is dissolved in 1:1 acetone–water and free tungstophosphoric acid is extracted into isoamyl alcohol–diethyl ether, leaving the S-adenosyl compound in the aqueous layer. Parks[47] and also Schlenk et al.[43] used a similar procedure to isolate S-adenosylethioninesulphonium compounds.

5. Hexachloroplatinate

Hexachloroplatinates are standard derivatives for characterizing bases by melting point. Examples of the procedure applied to S-methylmethioninesulphonium and 2-carboxyethyldimethylsulphonium compounds from asparagus and algae, respectively, come from Challenger's work.[48,49]

6. Perchlorate

Perchlorate was at one time a standard precipitant for methylene blue in the US Pharmacopeia. It is recommended for this in the Japanese Pharmacopeia[50], using precipitation with saturated potassium perchlorate and gravimetric estimation. Moraw[51] and Maurina and Neulon Deahl[52] studied this method of assay and the latter workers proposed a procedure for precipitation of methylene blue with sodium perchlorate and then gravimetry. Bock and Hummel[53] checked the purity of triphenylsulphonium chloride gravimetrically as the perchlorate.

7. Picrate

François and Séguin[54] determined methylene blue by mixing an aliquot of an approximately 1% solution of sample with twice its volume of 0.5% aqueous picric acid and weighing the precipitated picrate. Direct titration was preferred by Bolliger[55] for solutions of methylene blue containing neither strong acid nor base; he added chloroform and titrated with 0.01 N picric acid to disappearance of blue from the aqueous layer. Maurina and Neulon Deahl[52] studied various methods for the assay of medicinal methylene blue; picric acid precipitation and gravimetry was preferred to back-titration of unused picric acid with sodium hydroxide. Kračmár[56] titrated a thiophenium salt (Arfonad) in aqueous buffer solution of pH 5–7.5, using amperometric end-point indication. Ashworth and Walisch[57] used a polarovoltric end-point in the titration of methylene blue with picric acid in dilute acetic acid. Weiner[58] employed a photometric end-point and while his examples were quaternary ammonium compounds, application to sulphonium compounds appears probable.

Jureček et al.[59] converted dialkyl sulphides into sulphonium salts by reaction with p-bromophenacyl bromide, dissolved the bromides in water and precipitated with saturated aqueous picric acid. The picrates, after crystallization from water, served for characterization of the original sulphides. Veibel and Nielsen[60] prepared the picrates in the same way and then titrated them with perchloric acid reagent (Section IIIE) to determine equivalents and hence identify the original sulphides.

Challenger and co-workers[48,49] also prepared picrates to isolate sulphonium salts from naturally occurring material (Sections IIIA5 and 8).

8. Reineckate

Reineckate ion, $[Cr(NH_3)_2(SCN)_4]^-$, usually as the ammonium salt, is another standard precipitant for organic bases, including quaternary ammonium salts and hence probably sulphonium compounds. This precipitation is another of the stages in isolating S-adenosylmethionine and -ethionine from yeasts, e.g. by Schlenk and co-workers[43,61] or Janicki et al.[62] The usual yeast extract was treated with excess of ammonium reineckate, left for several hours and then filtered. The product in acetone–sulphuric acid was submitted to ion exchange on a cation-active resin.

Challenger and co-workers[48,63] adopted reineckate precipitation for isolating 2-carboxyethyldimethylsulphonium compounds from algae. Seno and Nagai[64] applied the procedure to determine S-methylmethioninesulphonium chloride (vitamin U). They used a large excess of ammonium reineckate in 0.1 M hydrochloric acid. After 1 h at 0°C, the precipitate was filtered, dissolved in acetone, and the absorbance measured at 520–535 nm. The absorbance of the filtrate at the same wavelength can also be checked to give the amount of unused reagent anion.

9. Tetraiodobismuthate

Dragendorff reagent, containing tetraiodobismuthate ion, BiI_4^-, has found general use for detecting bases and quaternary ammonium compounds in chromatography and other applications. Application to sulphonium compounds is highly probable and Ivanova et al.[38] refer to its use in paper chromatography. A quantitative method for quaternary ammonium and sulphonium compounds and tertiary amines is due to Buděšinský and Körbl[65]. They prepared a Bi–EDTA reagent containing potassium iodide from the disodium salt of EDTA, potassium iodide and bismuth trichloride or trinitrate. An aliquot of the reagent was mixed with the sample and dilute hydrochloric acid. The ammonium or sulphonium cation evidently reacts with the tetraiodobismuthate anion, which then liberates an equivalent amount of EDTA. They titrated this with a Th^{4+} reagent at pH 2–3 to methylthymol blue indicator.

10. Tetraphenylborate

This reagent yields poorly soluble sulphonium salts and was used in the direct titration of sulphonium salts by Uno et al.[66] and Nezu[67]. The former workers titrated in hydrochloric acid with methyl orange as indicator. It functions as described under Anionic surface-active agents (Section IIIA1). Removal of the last trace of sulphonium compound by added reagent evidently liberates methyl orange from its ion combination product with the sulphonium salt and this gives a red end-point. Nezu titrated solutions of sulphonium (also selenonium and phosphonium) salts with 0.01 M sodium tetraphenylborate to an amperometric end-point using a dropping mercury cathode and an applied voltage of 1.8 V. Kirsten et al.[68] carried out direct and indirect potentiometric titrations of quaternary ammonium salts in acetate buffer, usually of pH 4.6, with 0.1 M aqueous sodium tetraphenylborate but no sulphonium example was reported.

11. Triiodide

Most triiodides of organic bases are poorly soluble in aqueous solvents. Direct and indirect titrations of quaternary ammonium compounds have been based on this observation. The rare but successful application to other cations, such as sulphonium, indicate that greater use would be possible. Methylene blue and other basic dyes were titrated by Pelet and Garuti[69] at about 0.1% concentration in water with 0.5% iodine in aqueous potassium iodide using external starch–iodide indicator. Most dyes reacted with two atoms of iodine. Sabalitschka and Erdmann[70] used excess of triiodide reagent and back-titrated the supernatant liquid with thiosulphate to a blue colour. This was due to decomposition of the first trace of methylene blue triiodide after all the excess of triiodide reagent had been removed. Indirect titration procedures were also described by Gurmendi and Guevara[71] and by Sabalitschka and Erdmann[72]. The stoichiometry of the reaction was variously reported by these authors, and Holmes[73] studied the factors influencing it. DuBois[74] titrated high molecular weight quaternary ammonium, phosphonium and sulphonium salts to starch indicator.

B. Reduction

Polarographic and chemical methods have been applied to coloured thiazine dyes (phenothiazonium compounds).

1. Polarography

Much polarographic work on sulphonium compounds has been devoted to measurement of physical data and is therefore only potentially analytical.

Colichman and Love[75] studied polarography of trimethyl- and p-cresyldimethylsulphonium iodides and methosulphates, finding well defined reduction waves in neutral ethanol or aqueous phosphate buffer of pH 8 with quaternary ammonium salts as supporting electrolyte. They reported a linear relation between diffusion current and concentration in the region 5×10^{-4} to 5×10^{-3} M. Shinagawa et al.[76] found that triphenylsulphonium chloride gave a two-step reduction wave at -1.5 V (vs. S.C.E.) in 0.1 M potassium chloride. Wave heights were proportional to concentration. Fujita et al.[77] later identified the products of electrolytic reduction as benzene, biphenyl and diphenyl sulphide using gas chromatography. Half-wave potentials were measured by Lüttringhaus and Machatzka[78] on many salts in phosphate–borate buffer at pH 7.4, and by Savéant[79] on sulphonium salts with α-carbonyl groups, e.g. phenacylsulphonium compounds, $RCOCH_2—\overset{+}{S}R'_2$. The latter work also included the mechanism of electrochemical reduction and rupture of the C—S link to yield the corresponding ketone (e.g. $C_6H_5COCH_3$) was found. Shimanskaya et al.[80] studied the polarography of methionine and S-methylmethioninesulphonium iodide in ethanol containing tetraethylammonium iodide, observing reduction waves at -1.97 and -2.05 V (vs. S.C.E.). Pozdeeva et al.[81] described the polarographic determination of

10^{-2}–10^{-4} M solutions of S-methylmethioninesulphonium chloride (vitamin U) and other salts such as the iodide, gallate, orotate, salicylate and glycyrrhizate. They used acid media, 0.05–0.1 M in tetraethylammonium or tetrabutylammonium iodide, and suggested that the patented method[82] could be used on powders, pills, and mother liquors.

Bozsai-Bolda and Sata-Masonyi[83] investigated the polarography of stain solutions. For methylene blue they found a half-wave potential at ca. −0.1 V in Michaelis buffer of pH 9.8, wave height being proportional to concentration in the range from 10 to 40 mg per 100 ml. The method appeared suitable for determining methylene blue in microbial stain solutions.

2. Chemical reducing agents

Most of the chemical reductions have been applied to the quantitative determination of methylene blue and some date back to the beginning of this century. Titration procedures, mostly direct, predominate and, as expected, the usual end-point is decoloration, since the leuco-compound produced is colourless. The occasional indirect procedure involves either back-titration of unused reagent or titration of the leuco-compound.

The examples quoted represent only a selection since methylene blue has been much studied. Information is again classified according to reagents in alphabetical order.

a. *Ascorbic acid*. Růžička and Kotouček[84] determined several thiazine dyes (thionine, methylene blue, methylene green) by potentiometric titration with 10^{-3} M ascorbic acid. The sample was in aqueous or ethanolic, 2 M hydrochloric or sulphuric acid at 60–70°C under carbon dioxide.

b. *Cadmium*. Treadwell[85] used an indirect titrimetric method for determining methylene blue. He reduced it with cadmium, prepared by electrolytic precipitation, in dilute sulphuric acid under carbon dioxide and then titrated the leuco-compound potentiometrically with iron(III).

c. *Chloromolybdate(III)*. Sigi and Babu[86] reduced methylene blue quantitatively with chloromolybdate(III) in 6–8 M hydrochloric acid; this yields the leuco-dye and molybdenum(V), although in acid of lower concentration molybdenum(VI) is also formed. They titrated the chloromolybdate with methylene blue to a potentiometric end-point and also visually to the first blue given with excess dye.

d. *Chromium(II)*. Tandon and Mehotra[87] and Podolenko[88] titrated methylene blue and other dyes directly with chromium(II) reagents as well as with titanium(III) in acid solution. Davidson[89] used excess of chromium(II) sulphate in sulphuric acid at 50°C, back-titrating potentiometrically with dichromate.

e. *Dithionite*. Šrámek[90] carried out the ascending paper chromatography of many water-soluble dyestuffs as their leuco-compounds, using as developing solvent a mixture of water, 1% Turkey Red Oil, sodium hydroxide, pyridine and sodium dithionite. He included thiazine dyes among the examples. The colours appeared on drying and exposure to the air, facilitating detection.

Munemori[91] has given a coulometric procedure for titrating indigo carmine and methylene blue with dithionite, generated at a mercury pool cathode in a cell containing hydrogen sulphite:

$$2H_2SO_3 + H^+ + 2e \longrightarrow HS_2O_4^- + 2H_2O$$

Reduction was best at pH 3–5 and in 0.01 M hydrogen sulphite. He used a photometric end-point, recording the absorbance of the reaction mixture at 610 nm.

f. *Iron(II)*. Methylene blue was determined by Hein *et al.*[92] by adding Rochelle salt and concentrated ammonia solution, warming to 35–40°C in atmosphere of nitrogen and titrating with iron(II) ammonium sulphate to decoloration.

g. *Tin(II)*. Shcherbachev[93] heated methylene blue with a measured excess of tin(II) chloride reagent under carbon dioxide and back-titrated with standard dichromate to reappearance of a blue colour.

h. *Titanium(III)*. This is the most reducing agent for dyes, including methylene blue. The earliest work on the direct titration of methylene blue by titanium(III) is that of Knecht[94], followed by Knecht and Hibbert[95,96], Atack[97], Kikuchi[98], Koo Wha Lee[99], Tandon and Mehotra[87], and Podolenko[88]. Knecht and Hibbert[100] used it to determine residual methylene blue in a quantitative determination of glucose by reduction using the dye in measured excess. Gál[101] similarly determined ascorbic acid by reducing with excess of methylene blue and back-titrating unused dye with titanium(III). Pummerer *et al.*[102] titrated thiazines, and Burgess *et al.*[103] included titanium(III) reduction among methods studied for the purification and quantitative determination of basic dyes, including methylene blue. The method is that recommended by the British Pharmacopoeia[104] for determination of methylene blue. A sample in dilute hydrochloric acid is titrated under carbon dioxide with 0.1 M titanium(III) chloride to the colour change from blue to reddish grey. Reduction normally proceeds smoothly but some workers use 60–70°C and others add tartrate or fluoride as catalysts.

i. *Vanadium(II)*. Both Banerjee[105] and Matrka and Ságner[106] determined methylene blue and other dyes by direct titration with vanadium(II) sulphate to decoloration or potentiometrically. They also used an indirect procedure with excess of reagent and back-titration with ferric alum.

C. Degradations with Acid on Alkali

Identification of the products of breakdown of sulphonium salts by acids and/or alkalis helps to establish their structure and shows their presence. Degradation to dimethyl sulphide, identified as the mercurichloride, for example, has served to show the presence of the dimethylsulphonium group in *S*-methylmethionine sulphonium or 2-carboxyethyldimethylsulphonium compounds from natural products such as algae (Challenger and co-workers[48,63]), green tea (Kiribuchi and Yamanishi[107]), or fresh tomatoes (Wong and Carson[108]).

Fission of *S*-adenosylmethionine and related compounds has also been used for identification and analysis. In solution at pH 3.5–4, fission occurs between the sulphur atom and the xylose moiety. More vigorous conditions, such as alkali at pH 11–13 or *ca.* 0.1 M acid, at temperatures up to 100°C, cause more extensive degradation. *S*-Adenosylmethionine can give, for example, adenine, homoserine, *S*-ribosylmethionine,5'-methylthioadenosine, thiomethylribose, and methionine. Parks[47] identified analogous products by paper chromatography in his investigations of the isolation of *S*-adenosylethionine. Zappia *et al.*[44] prepared and studied analogous compounds, e.g. *S*-inosylmethionine, and used paper and thin-layer chromatography to identify the degradation products.

In quantitative applications of degradation, Ackman and Dale[109] determined 2-carboxyethyldimethylsulphonium chloride (dimethyl-β-propiothetin) using cold 7.5 M sodium hydroxide:

$$Me_2\overset{+}{S}-CH_2CH_2COOH + OH^- \longrightarrow Me_2S + HOCH_2CH_2COOH$$

The volatile dimethyl sulphide produced was determined by gas chromatography.

This method could be used successfully in the presence of many other amino acids, such as cysteine, cystine, homocystine, arginine, and methionine.

Bezzubov and Gessler[110] adopted a similar procedure for S-methylmethionine-sulphonium compounds (vitamin U) in plants. They used aqueous solutions of pH 9.7–10 at 97°C. Dimethyl sulphide was removed in a stream of nitrogen, trapped in toluene at −78°C and determined by gas chromatography.

Wronski and Mazurkiewicz[111] treated sulphonium salts (5–50 μM solutions in dimethylformamide) with thiourea and potassium iodide to yield S-alkylthiouronium salts:

$$R_3\overset{+}{S} + S{=}C(NH_2)_2 \longrightarrow R_2S + \left[R{-}S{-}\underset{NH_2}{\overset{\displaystyle C{=}NH_2}{|}} \right]^+$$

These were hydrolysed with sodium hydroxide to give urea and thiol, which was titrated with 2-hydroxymercuribenzoate.

D. Other Reactions of the Sulphonium Ion

1. Oxidative methods

Methylene blue has been determined by various oxidative methods. Chernavina[112] added dilute sulphuric acid (1:2) to solutions of dyes, including methylene blue, heated to 100°C and titrated with 0.1 M chloramine T to decoloration. This procedure was claimed to be cheaper and simpler than titration with titanium(III). Wojtas[113] likewise acidified solutions of methylene blue with sulphuric acid and added excess of potassium permanganate at the boiling point. Unused reagent was then back-titrated with oxalic acid. Muralikrishna and Ramanadham[114] carried out potentiometric titrations of methylene blue with ammonium hexanitratocerate (<0.1 M) in 0.5–1.5 M nitric acid. An average of 28 equivalents of cerium(IV) were reduced by each mole of the dye and the equation:

$$C_{16}H_{18}N_3SCl + 14\,O \longrightarrow C_{12}H_6N_3O_4SCl + 4\,HCO_2H + 2\,H_2O$$

was formulated. The authors evidently did not attempt to isolate any reaction products or suggest what oxidation has occurred, but the product corresponds to a dinitrophenothiazine from oxidation of the four dimethylamino groups, each methyl

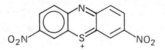

group giving formic acid. Muralikrishna et al.[115] based a spectrophotometric method on this reaction. They also oxidized with cerium(IV) and extracted the product with nitrobenzene. The extract was analysed spectrophotometrically at 545 nm.

Atherden[116] tested for amines (including methylene blue) with ammonium hexanitratocerate in nitric acid, which gives a colour change after 1 min at 50°C.

2. Substitution

Růžička[117] titrated 10^{-3} M solutions of thiazine dyes such as thionine, methylene blue, methylene green, and Azure A in 50% ethanol with freshly prepared saturated bromine water and observed the changes in potential. Thionine and methylene blue gave tetrabromo substitution products, the others dibromo compounds. Růžička quoted this as a possible method of quantitative determination.

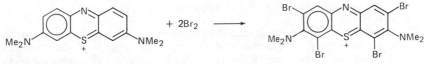

$+ 2Br_2 \longrightarrow$

Kalinowski and Czlonkowski[118] titrated methylene blue coulometrically in 2% potassium bromide solution, using a dead-stop end-point.

Suprun[119] described a determination of methylene blue with excess of 0.1 M iodine monochloride in hydrochloric acid, which gave a precipitate within 15–20 min. This was filtered and unused iodine monochloride in the filtrate was determined iodometrically. This method is probably also based on a substitution reaction.

E. Reactions of Other Groups in the Molecule

Many examples could, of course, be given under this heading. Detection and determination of amines and amino acids through the well known reaction with ninhydrin applies to S-methylmethioninesulphonium salts. Ogawa et al.[40] purified methylmethioninesulphonium derivatives from drugs by precipitation as tungstophosphate and then determined amines and amino acids colorimetrically through reaction with 1,2-naphthoquinone-4-sulphonate.

Salts of stronger bases with weaker acids can be titrated in glacial acetic acid with perchloric acid, generally in the same solvent.

The end-points can be detected potentiometrically (glass electrode) or with indicators, usually triphenylmethane dyes such as crystal or methyl violet. For halides, probably the most commonly encountered salts, mercury(II) acetate is added. This yields the scarcely dissociated mercury(II) halide and acetate ion, which is titratable as above. This is a standard method for many quaternary ammonium salts and some authors quote also examples of sulphonium salts. Thus, Pifer and Wollish[120] mention sulphonium salts, sulphonium sulphides, $(R_3S)_2S^{2-}$, and protonated sulphoxides, $(R_2SOH)^+NO_3^-$. They titrated with perchloric acid in dioxan and indicated the end-point with crystal violet or potentiometrically. Lane[121] found that dioxan was not necessary but confirmed that phosphonium and sulphonium salts could be titrated. He mentioned the example of tri (o-hydroxyphenyl) sulphonium chloride and determined the titration end-point by the 'high frequency' method. Veibel and Nielsen[60] titrated sulphonium picrates with crystal violet. This was the final stage in the identification of alkyl sulphides (Section IIIA7). Similarly, Bock and Hummel[53] checked the purity of triphenylsulphonium chloride by passing a solution through an ion-exchanger resin in the OH⁻ form and titrated the eluate with 0.1 M hydrochloric acid.

IV. PHYSICAL METHODS

A. Spectroscopic Methods

The spectroscopic properties (u.v., i.r., m.s., and n.m.r.) of sulphonium compounds have been studied extensively but general analytical applications are rare. Ultraviolet measurements were used for determination of S-adenosylmethionine. Cantoni[122] reported a molar absorptivity of 16,000 at 260 nm and values near these were used by subsequent workers[62,123–125]. Zappia et al.[44] determined a number of S-adenosylsulphonium compounds at 260 nm (neutral solution, molar absorptivity 15,400) or 257 nm (strongly acid solution, molar absorptivity of 14,700). Some analogous S-inosyl compounds were determined at

249 nm (near neutral solution, molar absorptivity 12,200) and 250 nm (acid, molar absorptivity 10,900).

Alkyl sulphides in petroleum fractions have been determined by treatment with iodine in a solvent and measurement of the light absorbance of the 'complexes' formed. If these are of the form $(R_2S—I)^+I^-$, this is tantamount to determination of sulphonium iodides. Various solvents have been used and the absorption maxima vary slightly. Hastings and co-workers[126] used isooctane (308 nm). Drushel and Miller[127], Hastings et al.[128], and Snyder[129] used carbon tetrachloride (310 nm). Rosenthal et al.[130] dichloromethane (308 nm), and Prokshin et al.[131] light petroleum (315 nm) for dimethyl sulphide in air.

Isotope dilution was used by Baldessarini and Kogin[132] (see also Kogin and Baldessarini[133]) in the assay of S-adenosylmethionine in tissue, using the [^{14}C]methyl compound.

B. Separation Methods

Chromatographic and allied methods are considered under this heading. Most of the procedures use paper (PC) and thin-layer chromatography (TLC) and concern either mixtures of thiazine dyes or of dyes among which at least one thiazine dye is present.

1. Paper chromatography

Ciglar et al [134] carried out the PC of basic dyes, including methylene blue, on Ederol paper No. 202 but obtained satisfactory separation only in reversed-phase PC after impregnation with 5% cetyl alcohol in 95% ethanol. Lindner[135] studied wool dyes, again including methylene blue; for samples including this dye he used a mixture of 95% methanol containing 3% formic acid. Kiel and Kuypers[136] used methanol–5 N ammonia solution to chromatograph basic dyes. Özsös[137] chromatographed a mixture of gentian violet and methylene blue on Whatman No. 1 paper with butanol–ethanol–water. Zoch[138] studied the PC of aerosol components, using model mixtures of methylene blue, fuchsin and pyrocatechol violet and a mobile phase of n-butanol–acetic acid–water. Cosmetic dyes, including methylene blue, were subjected by Legatova[139] to PC with butanone–acetone–water–concentrated ammonia solution. Přistoupil and Kramlová[140] carried out ascending 'membrane chromatography' of mixtures of six or seven dyes, including methylene blue, on nitrocellulose filters. Their mobile phase was ethanol–chloroform, ethanol–chloroform–acetic acid, or ethanol–25% ammonia solution and the best results were obtained with pore sizes of 0.2–0.4 μm.

Růžička[117] checked the purities of a number of thiazine dyes (thionine, methylene blue, methylene green, Azure A) and their bromination products using butanol–pyridine–water or butanol–acetic acid–water on Whatman No. 1 paper. Other studies include those of Šrámek[141] on water-soluble dyes (methylene blue, Azure C and A and ethyl analogues) on Whatman No. 1 paper using pyridine–isoamyl alcohol–25% ammonia solution; Bellin and Ronayne[142], also on water-soluble acid and basic dyes, including Azure A,B, and C, thionine, methylene blue, methylene blue N, methylene green and toluidine blue O using for these thiazines as the best mobile phases t-butanol–acetone–0.2 M hydrochloric acid and t-butanol–acetone–0.2 M ammonia solution–0.2 M hydrochloric acid.

Paper chromatography is encountered in earlier work on the isolation and purification of S-adenosylmethionine. For instance, Schlenk and DePalma[61] separated it from adenosine mono-, di- and triphosphates, homoserine, and methionine on Whatman No. 1 or 3 paper, using ethanol–water–acetic acid, detecting with ninhydrin. Parks[47] did likewise for S-adenosylethionine. Zappia et al.[44] also employed PC procedures for S-adenosyl- and S-inosylmethionine and similar compounds.

Some comprehensive systematic PC studies of large numbers of compounds, often of biological interest, have been performed. These occasionally include S-methylmethionine- and/or S-adenosylmethioninesulphonium compounds. Thus, Gordon et al.[143] worked on the preliminary PC separation of 97 compounds and were able to separate the S-methylmethionine compound from others by ascending PC on Whatman No. 4 paper using isopropanol–pyridine–water–acetic acid. Fink et al.[144] quoted R_F values for about 400 compounds on Whatman No. 1 paper with ten mobile phases; both S-methylmethionine- and S-adenosylmethioninesulphonium acetates were included and showed noticeably low R_F values. Edwards et al.[145] gave R_F values for about 30 compounds on untreated cellulose paper with methanol–water as mobile phase; S-adenosylmethionine again had a low R_F value, the lowest of all studied. They also chromatographed on paraffin-coated paper, using butanol–propionic acid–water and detecting with ninhydrin.

Leaver and Challenger[146] chromatographed methylsulphonium hydroxides, prepared from the iodides with silver oxide, using a descending procedure on Whatman No. 1 paper. Their mobile phase was n-butanol–acetic acid–water and detection was carried out with bromophenol blue, giving blue zones on yellow. Gasparič[147] chromatographed sulphonium methylsulphates of structures $(R_2S—CH_3)^+$ $CH_3SO_4^-$, prepared from the sulphides, R_2S. Kronrád and Pánek[148] reacted aliphatic sulphides containing 1–6 carbon atoms with methyl iodide to yield the methylsulphonium iodides. These were chromatographed using butanol–acetic acid–water as stationary phase and acetic acid in carbon tetrachloride as mobile phase. Detection was effected with Dragendorff reagent.

Keenan and Lindsay[149] refer to the use of PC for the resolution and identification of methylmethioninesulphonium compounds and homoserine, but unfortunately give no details.

2. Thin-layer chromatography

As with PC, most TLC applications have been concerned with dyes and stains and the presence of thiazines among the examples encourages inclusion here. Logar et al.[150] chromatographed acidic and basic dyes, including methylene blue, on silica gel G with the mobile phase butanol–acetic acid–water and obtained better defined zones than in PC. Puech et al.[151] separated amyleine hydrochloride, naphthazoline nitrate and methylene blue in water on silica gel H, using ethanol–ammonia solution. Pinter et al.[152] worked on cosmetic dyes, including methylene blue, on silica gel using the mobile phases chloroform–acetone–propanol–sulphurous acid (9% SO_2) and the upper phase of butanol–acetic acid–water containing 0.4% potassium chloride.

Samples containing several thiazine dyes have been the subject of TLC studies. Dobres and Moats[153] investigated dyes commonly used in biological staining, including methylene blue and Azure A,B, and C. They found separations faster than with PC and the best results were obtained on silica gel G with n-propanol–n-butanol–concentrated ammonia solution–water, or n-propanol–concentrated ammonia solution–water, and on Adsorbosil with the latter solvent mixture. Patel[154] and Agpar and Patel[155] examined the purity of toluidine blue

O (3-amino-7-dimethylamino-2-methylphenothiazonium chloride) by TLC on silica gel with the mobile phase chloroform–acetone–95% ethanol–butanol. Thionine was then close to the solvent front. Loach[156] separated methylene blue and related thiazine dyes on silica gel, using three developments with 95% ethanol–chloroform–acetic acid. Marshall and Lewis[157] studied the so-called Romanowsky blood stains, also containing methylene blue and Azure A, B, and C, using silica gel layers with the butanol layer from butanol–1% ammonium chloride–2% formic acid.

McKamey and Spitznagle[158] used various techniques (column chromatography, TLC, ms, and absorption spectroscopy) to identify impurities in toluidine blue O. They found N-methyl homologues of 2-methylthionine rather than of thionine, and also small amounts of the 3-amino-2-methyl-7-methylamino- and 2-methyl-3,7-diaminophenothiazonium compounds. Methanol–2 N hydrochloric acid and ethanol–concentrated hydrochloric acid on silica gel were used. Similar work was performed on Azure 1 by Paucescu et al.[159], who separated methylene blue and Azure A and B from one another and from other dyes by TLC on silica gel H using methanol–chloroform–acetic acid.

3. Ion exchange

Ion exchange has been used as a step in some of the isolation and purification procedures already mentioned for S-methylmethioninesulphonium[39,43] and S-adenosylmethionine and related sulphonium compounds[41,43–47, 61,62]. For these procedures strongly cationic types, especially cross-linked sulphonated polystyrenes such as Dowex 50 or Amberlite IRC 20 in the H^+ or Na^+ forms, have been employed. Elution requires fairly concentrated acid (up to 6 N).

In a procedure for determining nucleotide levels in foods, Kieffer and Egli[124] employed Dowex 1-X8 and the procedure was appropriate for S-adenosylmethioninesulphonium salts. Eloranto et al.[125] used a phosphocellulose, Cellex-P, in the H^+ form for assay of tissue S-adenosylmethionine and -homocysteine. Gaitonde and Gaull[160] proposed a series of ion exchangers for the analysis of sulphur-containing amino acids in rat tissues. The usual extracts with perchloric acid were passed successively through the anion exhangers Dowex 2 (Cl^- form) and Dowex 1 (CO_3^{2-} form), then the cationic exchangers Amberlite CG-50 (H^+ form) and Zeokarb 225 (H^+ form). S-Adenosyl- and S-ribosylmethionines traverse the first two and are retained in the Amberlite, from which they can be eluted with 0.05 N hydrochloric acid.

Matoki et al.[161] patented the use of Dialon CR-10, Dialon CR-110 or Dowex A-1 for purifying S-adenosyl-L-methionine.

Comprehensive studies of often automatic separation of compounds of biological interest on ion exchangers have been reported. Many deal with amino acids, among which the S-methyl- and/or S-adenosylmethioninesulphonium compounds occasionally appear. Zacharias and Tallay[162] studied the elution behaviour of about 90 naturally occurring ninhydrin-positive compounds in the automatic amino acid analyser of Spackman et al.[163] S-Methylmethioninesulphonium iodide was eluted from the Amberlite IR-120 column with citrate buffer of pH 4.26. Skodak et al.[164] used the columns of Spackman et al.[163] for amino acids from plant material which included S-methylmethioninesulphonium compounds.

Reports of ion-exchange work on thiazine dyes are rare. Atkin et al.[165] separated methylene blue and NN-diethylglycine on strongly acidic cationic exchangers such as Dowex 50W-X8, in the presence of phosphate buffer of pH 7, followed by elution with 3 N hydrochloric acid.

4. Column and liquid chromatography

Column chromatography for the clean-up of dye mixtures and plant extracts has often been performed. Loehr *et al.*[166] purified the thiazine dye Azure A from commercial samples of polychrome methylene blue on columns of silica gel, eluting with acetic acid–formic acid. McKamey and Spitznagel[158], whose work was mentioned above under thin-layer chromatography, carried out column chromatography of toluidine blue O. They eluted the silica gel column first with ethanol and then with ethanol–concentrated hydrochloric acid.

Hoffman[167] applied liquid chromatography to the determination of *S*-adenosylmethioninesulphonium compounds in tissue. He used a Vydac cation-exchange column and eluted with a linear ammonium formate gradient, monitoring the eluate at 254 nm. The Romanowsky blood stains (including methylene blue and Azure A,B, and C) have been subjected to high-performance liquid chromatography (HPLC) by Dean *et al.*[168] on a silica microparticulate column of Zorbax-Sil at 150°C under a pressure of 150–200 atm. They found the best elution solvent to be a mixture of methanol, water, glycine and acetic acid. They also monitored the eluate at 254 nm.

5. Electrophoresis

Electrophoresis has also been one of several purification stages for *S*-adenosylmethionine. Kieffer and Egli[124] included paper electrophoresis on Whatman No. 1 paper in their scheme for determining nucleotides in foods. They used 0.1 M potassium formate (pH 3.5) and 0.33 M borax (pH 9.2) at 3000 V, with u.v. detection. The analytical procedure of Zappia *et al.*[44] for *S*-adenosyl-L-methionine and similar compounds also included paper ionophoresis at 3000 V on Whatman 3MM paper in formate, acetate, and borate buffers ranging from pH 3 to 8.5. Eloranto *et al.*[125], in their assay of *S*-adenosylmethionine and -homocysteine, used low-voltage paper electrophoresis on Whatman No. 1 paper in citrate buffer of pH 3.5.

6. Gas–liquid chromatography

Direct GLC measurements are, of course, scarcely possible on sulphonium salts but Hancock[169] converted adenosine derivatives, including *S*-adenosylmethionine, into trimethylsilyl derivatives. He then carried out GLC on 110–120-mesh Anakrom A impregnated with SE-30.

V. REFERENCES

1. J. M. Preston, *J. Soc. Dyers Colourists*, **61**, 16 (1945).
2. J. M. Preston and S. R. Epton, *Proc. 1st World Congress on Surface Active Agents, Paris*, **1**, 310 (1954).
3. J. M. Lambert, *J. Colloid Sci.*, **2**, 479 (1947).
4. K. J. Karrman, E. Bladh and P.-O. Gedda, *Mikrochim. Acta*, 779 (1959).
5. R. Wickbold, *Seifen-Öle-Fette-Wachse*, **85**, 415 (1959).
6. E. Ullmann, K. Thoma and W. Dörflinger, *Deut. Apoth.-Ztg.*, **100**, 33 (1960).
7. N. Tanaka and R. Tamamushi, *Nature, Lond.*, **179**, 311 (1957).
8. O. Yoda, *Nippon Kagaku Zasshi*, **77**, 905 (1956).
9. G. S. Hartley and D. F. Runnicles, *Proc. Roy. Soc.*, *A***168**, 420 (1938).
10. M. R. J. Saltor and A. E. Alexander, *Research (London)*, **2**, 247 (1949).
11. M. Doležil and J. Bulandr, *Chem. Listy*, **51**, 255 (1957).
12. S. R. Epton, *Trans. Faraday Soc.*, **44**, 226 (1948); *Nature, Lond.*, **160**, 795 (1947).

13. T. Harada and W. Kimura, *Yukagaku*, **9**, 124 (1960).
14. Antara Chem. Co. New York, from S. Siggia, *Quantitative Organic Analysis via Functional Groups*, Wiley, New York, Chichester, 1963, p. 635.
15. F. J. Cahn, *US Pat.*, 2,471,861 (May 31, 1949).
16. G. R. F. Rose and C. H. Bayley, *Nat. Res. Counc. Can. Bull.*, No. 2875 (1952).
17. H. Nigoro and H. Sakurai, *Ann. Rep. Takamine Lab.*, **6**, 89 (1954).
18. E. D. Carkhuff and W. F. Boyd, *J. Amer. Pharm. Assoc.*, **43**, 240 (1954).
19. A. Dister, *J. Pharm. Belg.*, **23**, 41 (1941).
20. G. F. Davidson, *Shirley Inst. Mem.*, **21**, 29 (1947).
21. H. O. Moraw, *J. Assoc. Off. Agric. Chem.*, **32**, 540 (1949).
22. *Pharmaceutica Helvetica VI*, 1976, p. 891b.
23. *Deutsches Arzneibuch 7*, GDR Edition, Vol. 4, 1973.
24. *Deutsches Arzneibuch 7*, GFR Edition, 1968.
25. G. J. W. Ferrey, *Chemist-Druggist*, **140**, 126 (1943).
26. L. Pelet and V. Garuti, *Bull. Soc. Chim. Fr. [3]*, **31**, 1094 (1904).
27. H. B. Klevens, *Anal. Chem.*, **22**, 1141 (1950).
28. R. Schiffner and B. Borrmeister, *Faserforsch. Textiltech.*, **15**, 211 (1964).
29. M. E. Auerbach, *Ind. Eng. Chem. Anal. Ed.*, **15**, 492 (1943).
30. M. E. Auerbach, *Ind. Eng. Chem. Anal. Ed.*, **16**, 739 (1944).
31. E. L. Colichman, *Ind. Eng. Chem. Anal. Ed.*, **19**, 430 (1947).
32. P. Mukherjee, *Anal. Chem.*, **28**, 870 (1956).
33. R. R. Levine, *Nature, Lond.*, **184**, Suppl. 18, 1412 (1959).
34. J. G. Baldinus, quoted in *Treatise on Analytical Chemistry*, Vol. 15, Wiley, New York, Chichester, p. 73.
35. T. Ogawa, *Denki Kagaku*, **25**, 613 (1957).
36. V. Mareš and Z. Stejskal, *Česk. Farm.*, **11**, 354 (1962).
37. C. Burgess, A. G. Fogg and D. T. Burns, *Lab. Pract.*, **22**, 472 (1973).
38. R. P. Ivanova, P. O. Ripatti and K. S. Bokarev, *Fiziol. Rast. (Moscow)*, **24**, 215 (1977); *Chem. Abstr.*, **86**, 102710.
39. R. A. McRorie, G. L. Sutherland, M. S. Lewis, A. D. Barton, M. R. Glazener and W. Shine, *J. Amer. Chem. Soc.*, **76**, 115 (1954).
40. S. Ogawa, M. Morita, A. Kishimoto and K. Dome, *Bitamin*, **32**, 369 (1965); *J. Vitaminol. (Kyoto)*, **12**, 162 (1966); *Chem. Abstr.*, **63**, 16132 and **65**, 15162.
41. F. Schlenk and R. E. DePalma, *J. Biol. Chem.*, **229**, 1051 (1957).
42. J. A. Stekol, E. I. Anderson and S. Weiss, *J. Biol. Chem.*, **233**, 425 (1958).
43. F. Schlenk, J. L. Dainko and S. M. Stanford, *Arch. Biochem. Biophys.*, **83**, 28 (1959).
44. V. Zappia, C. R. Zydek-Cwick and F. Schlenk, *J. Biol. Chem.*, **244**, 4499 (1968).
45. F. Salvatore, V. Zappia and S. K. Shapiro, *Biochim. Biophys. Acta*, **158**, 461 (1968).
46. F. Salvatore, R. Utili and V. Zappia, *Anal. Biochem.*, **41**, 16 (1971).
47. L. W. Parks, *J. Biol. Chem.*, **232**, 169 (1958).
48. R. Bywood and F. Challenger, *Biochem. J.*, **53**, xxvi (1953).
49. F. Challenger and B. J. Hayward, *Chem. Ind. (London)*, **729** (1954).
50. *Pharmacopoeia of Japan*, 8th Edition, Pt. 1, 1972, p. 102.
51. H. O. Moraw, *J. Assoc. Off. Agric. Chem.*, **28**, 705 (1945); **32**, 540 (1949).
52. F. A. Maurina and Neulon Deahl, *J. Amer. Pharm. Assoc.*, **32**, 301 (1943).
53. R. Bock and C. Hummel, *Z. Anal. Chem.*, **198**, 176 (1963).
54. M. François and L. Séguin, *J. Pharm. Chim. [8]*, **10**, 5 (1929).
55. A. Bollinger, *J. Proc. Roy. Soc. N.S.W.*, **67**, 240 (1934).
56. J. Kračmár, *Česk. Farm.*, **5**, 578 (1956).
57. M. R. F. Ashworth and W. Walisch, *Z. Anal. Chem.*, **181**, 193 (1961).
58. S. Weiner, *Chemist-Analyst*, **42**, 9 (1953).
59. M. Jureček, M. Večeřa and J. Gasparič, *Chem. Listy*, **48**, 542 (1954).
60. S. Viebel and B. J. Nielsen, *Acta Chem. Scand.*, **10**, 1488 (1956).
61. F. Schlenk and R. E. De Palma, *J. Biol. Chem.*, **229**, 1037 (1957).
62. J. Janicki, J. Skupin and J. Kowalczyk, *Chem. Anal. (Warsaw)*, **7**, 1167 (1962).
63. F. Challenger and M. I. Simpson, *J. Chem. Soc.*, 1591 (1948).
64. K. Seno and D. Nagai, *Bitamin*, **31**, 214 (1965); *Chem. Abstr.*, **62**, 13498.

65. B. Buděšinský and J. Körbl, *Chem. Listy*, **52**, 1513 (1958).
66. T. Uno, K. Miyajima and H. Tsukatani, *Yakugaku Zasshi*, **80**, 153 (1960).
67. H. Nezu, *Bunseki Kagaku*, **10**, 575 (1961).
68. W. J. Kirsten, A. Berggren and K. Nilsson, *Anal. Chem.*, **30**, 237 (1958); correction, *Anal. Chem.*, **31**, 376 (1959).
69. L. Pelet and V. Garuti, *Bull. Soc. Vaudoise Sci. Nat.* [*5*], **43**, No. 158, 30 (1907).
70. T. Sabalitschka and W. Erdmann, *Chem.-Ztg.*, **49**, 561 (1925).
71. G. Gurmendi and J. de D. Guevara, *Bol. Soc. Quím. Peru*, **4**, 283 (1938).
72. T. Sabalitschka and W. Erdmann, *Pharm. Helv. Acta*, **15**, 162 (1940).
73. W. C. Holmes, *J. Assoc. Off. Agric. Chem.*, **10**, 505 (1927).
74. A. S. DuBois, *Amer. Dyestuff Rep.*, **34**, 245 (1945).
75. E. L. Colichman and D. L. Love, *J. Org. Chem.*, **18**, 40 (1953).
76. M. Shinagawa, H. Matsuo and N. Maki, *Bunseki Kagaku*, **5**, 80 (1956).
77. K. Fujita, H. Nezu and M. Shinagawa, *Nippon Kagaku Zasshi*, **88**, 1076 (1967).
78. A. Lüttringhaus and H. Machatzka, *Justus Liebigs Ann. Chem.*, **671**, 165 (1964).
79. J. M. Savéant, *Bull. Soc. Chim. Fr.*, 481 (1967).
80. N. P. Shimanskaya, L. D. Gritsaenko, A. P. Klimov and V. D. Bezuglyi, *Zh. Obshch. Khim.*, **42**, 41 (1972).
81. A. A. Pozdeeva, F. V. Sharipova, F. G. Gerchikova, L. N. Karaninskaya, A. G. Petunina and S. I. Zhdanov, *Zh. Anal. Khim.*, **31**, 576 (1976).
82. A. A. Pozdeeva, L. N. Karaninskaya and A. G. Petunina, *USSR Pat.*, 525,015 (Aug. 15, 1976).
83. G. Bozsai-Bolda and L. Sata-Mosonyi, *Gyogyszereszet*, **15**, 17 (1971); *Chem. Abstr.*, **75**, 15906.
84. E. Růžička and M. Kotouček, *Z. Anal. Chem.*, **180**, 429 (1961).
85. W. D. Treadwell, *Helv. Chim. Acta*, **5**, 742 (1922).
86. S. R. Sigi and T. B. Babu, *Talanta*, **23**, 465 (1976).
87. J. P. Tandon and R. C. Mehotra, *Z. Anal. Khim.*, **31**, 576 (1976).
88. A. A. Podolenko, *Izv. Akad. Nauk Moldavsk. SSR*, **63** (1963); *Chem. Abstr.*, **64**, 8350.
89. G. F. Davidson, *Shirley Inst. Mem.*, **21**, 29 (1947).
90. J. Šrámek, *J. Chromatogr.*, **11**, 524 (1962).
91. M. Munemori, *Talanta*, **1**, 110 (1958).
92. F. Hein, O. Schwartzkopff, K. Hoyer, K. Klar, W. Eiszner, W. Clausz and W. Just, *Chem. Ber.*, **62**, 1163 (1929).
93. K. D. Shcherbachev, *Khim. Farm. Prom.*, No. 2, 117 (1935); *Chem. Abstr.*, **30**, 1692.
94. E. Knecht, *J. Soc. Dyers Colourists*, **19**, 9 (1903).
95. E. Knecht and E. Hibbert, *Chem. Ber.*, **38**, 3323 (1905).
96. E. Knecht and E. Hibbert, *J. Chem. Soc.*, 1541 (1924).
97. F. W. Atack, *J. Soc. Dyers Colourists*, **29**, 9 (1913).
98. S. Kikuchi, *J. Chem. Soc. Japan*, **43**, 544 (1922); *Sci. Rep. Tohoku Imp. Univ.*, **16**, 707 (1927); *Chem. Abstr.*, **22**, 36.
99. Koo Wha Lee, *J. Chem. Soc. Japan*, **53**, 25 (1932).
100. E. Knecht and E. Hibbert, *J. Soc. Dyers Colourists*, **43**, 94 (1925).
101. I. Gál, *Nature, Lond.*, **138**, 799 (1936).
102. R. Pummerer, F. Eckart and B. Gassner, *Chem. Ber.*, **47**, 1505 (1914).
103. C. Burgess, A. G. Fogg and D. T. Burns, *Lab. Pract.*, **22**, 472 (1973).
104. *British Pharmacopoeia*, 1973, p. 304.
105. P. C. Banerjee, *J. Indian Chem. Soc.*, **19**, 35 (1942).
106. M. Matrka and Z. Ságner, *Chem. Prům.*, **9**, 526 (1959).
107. T. Kiribuchi and T. Yamanishi, *Agric. Biol. Chem. (Tokyo)*, **27**, 56 (1963).
108. F. F. Wong and J. F. Carson, *J. Agric. Food Chem.*, **14**, 247 (1966).
109. R. G. Ackman and J. Dale, *J. Fish Res. Bd. Can.*, **22**, 875 (1965).
110. A. A. Bezzubov and N. N. Gessler, *Prikl. Biokhim. Mikrobiol.*, **13**, 301 (1977); *Chem. Abstr.*, **86**, 167282.
111. M. Wronski and B. Mazurkiewicz, *Chem. Anal. (Warsaw)*, **21**, 949 (1976).
112. M. S. Chernavina, *Tr. Sverdlovsk. Sel'skokhoz. Inst.*, **7**, 363 (1960).
113. R. Wojtas, *Chem. Anal. (Warsaw)*, **7**, 1177 (1962).

114. U. Muralikrishna and G. V. Ramanadham, *J. Indian Chem. Soc.*, **53**, 95 (1976).
115. U. Muralikrishna, G. V. Ramanadham and N. S. Rao, *Z. Anal. Chem.*, **279**, 208 (1976).
116. L. M. Atherden, *Pharm. J.*, **195**, 115 (1965).
117. E. Růžička, *Monatsh. Chem.*, **93**, 1262 (1962).
118. K. Kalinowski and F. Czlonkowski, *Acta Pol. Pharm.*, **26**, 147 (1969).
119. P. P. Suprun, *Med. Prom. SSSR*, **12**, No. 8, 38 (1958); *Chem. Abstr.*, **58**, 12923.
120. C. W. Pifer and E. G. Wollish, *Anal. Chem.*, **24**, 300 (1952).
121. E. S. Lane, *Analyst (London)*, **80**, 675 (1955).
122. G. L. Cantoni, *J. Biol. Chem.*, **204**, 403 (1953).
123. S. K. Shapiro and D. J. Ehninger, *Anal. Biochem.*, **15**, 323 (1966).
124. F. Kieffer and R. H. Egli, *Z. Anal. Chem.*, **221**, 416 (1966).
125. T. O. Eloranto, E. O. Kajander and A. M. Raina, *Biochem. J.*, **160**, 287 (1976).
126. S. H. Hastings, *Anal. Chem.*, **25**, 420 (1953); S. H. Hastings and B. H. Johnson, *Anal. Chem.*, **27**, 564 (1955).
127. H. V. Drushel and J. F. Miller, *Anal. Chem.*, **27**, 495 (1955).
128. S. H. Hastings, B. H. Johnson and H. E. Lumpkin, *Anal. Chem.*, **28**, 1243 (1956).
129. L. R. Snyder, *Anal. Chem.*, **33**, 1538 (1961).
130. I. Rosenthal, G. J. Frisone and J. K. Coberg, *Anal. Chem.*, **32**, 1713 (1960).
131. G. F. Prokshin, L. M. Sofrygina and Yu. G. Khabarov, *Bumazh. Prom.*, 23 (1973); *Anal. Abstr.*, **28**, 4H7.
132. R. J. Baldessarini and I. J. Kopin, *Anal. Biochem.*, **6**, 289 (1963).
133. I. J. Kopin and R. J. Baldessarini, *J. Neurochem.*, **13**, 769 (1966).
134. J. Ciglar, J. Kolšek and M. Perpar, *Chem.-Ztg.*, **86**, 41 (1962).
135. W. F. Lindner, *Chem.-Ztg.*, **86**, 103 (1962).
136. E. G. Kiel and G. H. A. Kuypers, *Tex.*, **22**, 533 (1963).
137. B. Özsös, *Türk Hij. Tecr. Biyol. Derg.*, **24**, 119 (1964); *Chem. Abstr.*, **62**, 2241.
138. E. Zoch, *J. Chromatogr.*, **18**, 198 (1965).
139. B. Legatova, *Rocz. Panstw. Zakl. Hig.*, **17**, 21 (1966); *Chem. Abstr.*, **65**, 5297.
140. T. I. Přistoupil and M. Kramlová, *J. Chromatogr.*, **34**, 23 (1968).
141. J. Šrámek, *J. Chromatogr.*, **15**, 57 (1964).
142. J. S. Bellin and M. E. Ronayne, *J. Chromatogr.*, **24**, 131 (1966).
143. H. T. Gordon, W. W. Thornburg and L. N. Werum, *J. Chromatogr.*, **9**, 44 (1962).
144. K. Fink, R. E. Cline and R. M. Fink, *Anal. Chem.*, **35**, 389 (1963).
145. C. H. Edwards, J. O. Rice, J. Jones, L. Seibles, E. L. Gadsden and G. A. Edwards, *J. Chromatogr.*, **11**, 349 (1963).
146. D. Leaver and F. Challenger, *J. Chem. Soc.*, 39 (1957).
147. J. Gasparič, personal communication, quoted in J. Petránek and M. Večeřa, *Chem. Listy*, **52**, 1279 (1958).
148. L. Kronrád and K. Pánek, *Z. Anal. Chem.*, **191**, 199 (1962).
149. T. W. Keenan and R. C. Lindsay, *J. Chromatogr.*, **30**, 251 (1967).
150. S. Logar, J. Perkavec and M. Perpar, *Mikrochim. Acta*, 712 (1964).
151. A. Puech, G. Kister and J. Chanel, *J. Pharm. Belg.*, **23**, 184 (1968).
152. I. Pinter, M. Kramer and J. Kleeberg, *Elelmiszervizsgalati Kozl.*, **14**, 169 (1968); *Chem. Abstr.*, **70**, 71003.
153. H. L. Dobres and W. A. Moats, *Stain Technol.*, **43**, 27 (1968).
154. J. A. Patel, *Amer. J. Hosp. Pharm.*, **26**, 540 (1969).
155. D. A. Agpar and J. A. Patel, *Amer. J. Hosp. Pharm.*, **26**, 541 (1969).
156. K. W. Loach, *J. Chromatogr.*, **60**, 119 (1971).
157. P. N. Marshall and S. M. Lewis, *Stain Technol.*, **49**, 235 (1974).
158. M. R. McKamey and L. A. Spitznagle, *J. Pharm. Sci.*, **64**, 1456 (1975).
159. S. D. Paucescu, M. D. Rotaru, H. Caldararu and V. Vornov, *Rev. Chim. (Bucharest)*, **26**, 433 (1975).
160. M. K. Gaitonde and G. E. Gaull, *Biochem. J.*, **102**, 959 (1967).
161. G. Motoki, T. Kawahara, K. Uchida and H. Yoshino, *Japan Kokai*, 75.160294 (25 Dec., 1975); 76.06990 (20 Jan., 1976); 76.06989 (20 Jan., 1976); *Chem. Abstr.*, **84**, 150988 (1976); **85**, 21768, 21769 (1976).
162. R. N. Zacharias and E. A. Talley, *Anal. Chem.*, **34**, 1551 (1962).

163. D. H. Spackman, W. H. Stein and S. Moore, *Anal. Chem.*, **30**, 1190 (1958).
164. F. I. Skodak, F. F. Wong and L. M. White, *Anal. Biochem.*, **13**, 568 (1965).
165. C. L. Atkin, E. P. Parry and D. H. Hern, *Anal. Chem.*, **39**, 672 (1967).
166. W. Loehr, I. Sohmer and D. Wittekind, *Stain Technol.*, **49**, 359 (1974).
167. J. Hoffman, *Anal. Biochem.*, **68**, 522 (1975).
168. W. W. Dean, G. J. Lubrano, H. G. Heinsohn and M. Stastny, *J. Chromatogr.*, **124**, 287 (1976).
169. R. L. Hancock, *J. Gas Chromatogr.*, **4**, 363 (1966).

The Chemistry of the Sulphonium Group
Edited by C. J. M. Stirling and S. Patai
© 1981 John Wiley & Sons Ltd.

CHAPTER **4**

Thermochemistry of the sulphonium group

ROBERT SHAW

*Chemistry Department, Lockheed Palo Alto Research Laboratory,
Lockheed Missiles & Space Company, Inc., 3251 Hanover Street, Palo
Alto, California 94304, USA*

I.	INTRODUCTION	101
II.	SULPHONIUM SALTS	102
III.	SULPHONIUM YLIDES	103
IV.	REFERENCES	104

> If in trouble—delegate
> If in doubt—advise
>
> *Anon*

I. INTRODUCTION

The following publications were searched for data: the annual IUPAC *Bulletin of Thermochemistry and Thermodynamics*[1] from 1977 back to 1970, when three comprehensive reviews of thermochemistry were published by Cox and Pilcher[2], Stull *et al.*[3], and Benson *et al.*[4]. Also studied were Domalski's CHNOPS review[5], a list of references kindly provided by Dr. I. C. Paul of the University of Illinois at Urbana-Champaign, two monographs on sulphonium chemistry, one by Stirling[6] and one by Johnson[7], and a recent review by Benson[8] on the thermochemistry of sulphur-containing molecules. Donald Scott of the Bartlesville Energy Research Center, Randolph Wilhoit of Texas A and M's Thermochemical Research Center, and G. T. Armstrong of the National Bureau of Standards were also privately consulted. These searches revealed that there are no data on the thermochemical quantities of direct relevance to this review, namely, the absolute standard molar heat of formation, entropy, and heat capacity at 298.15 K (25°C) of pure sulphonium salts and sulphonium ylides. Accordingly, estimating techniques are outlined below and some candidate compounds for experimental thermochemical evaluation are suggested.

SI units are used, except for calories. The conversion factor is 1 cal = 4.18 J.

II. SULPHONIUM SALTS

Sulphonium salts are analogous in structure and many properties to quaternary ammonium salts[9]. No heats of formation of quaternary ammonium salts have been reported, but the heat of formation of methylammonium chloride (crystal)[10] is -71.2 kcal/mol. The heat of formation of ammonium chloride (crystal)[10] is -75.2 kcal/mol. Hence, the effect of substituting one methyl group is 4.0 kcal/mol. The heat of formation of tetramethylammonium chloride (crystal) may be estimated to be -59.2 kcal/mol, assuming the effects of the methyl group to be additive. Further, from the known[2] heat of formation of trimethylamine (liquid) of -12.0 kcal/mol, the difference in heats of formation of tetramethylammonium chloride (crystal) and trimethylamine (liquid) is -47.2 kcal/mol. Therefore, assuming

$$Me_4NCl_{(c)} - Me_3N_{(l)} = Me_3SCl_{(c)} - Me_2S_{(l)} \tag{1}$$

then, from the known heat of formation of dimethyl sulphide (liquid)[2] of -15.6 kcal/mol, the heat of formation of trimethylsulphonium chloride is calculated to be -62.8 kcal/mol.

The heats of formation of several ammonium salts have been measured[10] (see Table 1).

Assuming that

$$NH_4Cl_{(c)} - NH_4X_{(c)} = Me_3SCl_{(c)} - Me_3SX_{(c)} \tag{2}$$

where X is another anion, the heats of formation of several trimethylsulphonium salts were estimated (see Table 2).

One of the salts, namely trimethylsulphonium iodide, has been synthesized and its structure determined[11]. This is a good candidate for measurement. The heat of formation of dimethylphenylsulphonium perchlorate may be estimated from the value for trimethylsulphonium perchlorate in Table 2, assuming that

$$Me_3SClO_{4(c)} - Me_2PhSClO_{4(c)} = Me_2S_{(l)} - MePhS_{(l)} \tag{3}$$

From the known[2] heat of formation of methyl phenyl sulphide (11.5 kcal/mol) and that of dimethyl sulphide (-15.6 kcal/mol), the heat of formation of dimethylphenylsulphonium perchlorate is -31.1 kcal/mol. This salt has been

TABLE 1. Measured[10] heats of formation of ammonium salts

Ammonium salt, NH_4X	Heat of formation (kcal/mol)	Difference (X − chloride) (kcal/mol)
Fluoride	-110.9	35.7
Chloride	-75.2	0
Bromide	-64.7	-10.5
Iodide	-48.1	-27.1
Perchlorate	-70.6	4.6

TABLE 2. Estimated heats of formation of
trimethylsulphonium salts

Trimethyl sulphonium salt	Heat of formation (kcal/mol)
Fluoride	−98.5
Chloride	−62.8
Bromide	−50.3
Iodide	−35.7
Perchlorate	−58.2

synthesized, its structure is known[12], and it is also a good candidate for measurement.

Finally, the heat of formation of S-methylthianium perchlorate, $C_{10}H_{20}SClO_4$:

may be estimated assuming

$$C_{10}H_{21}SClO_{4(c)} - Me_3SClO_{4(c)} = C_9H_{18}S_{(g)} - Me_2S_{(g)} \qquad (4)$$

The heat of formation of 4-t-butyl-1-thiacyclohexane ($C_9H_{18}S$) has not been measured but it may be estimated in the ideal gas state to within about 1 kcal/mol by group additivity[4,13] (see Table 3).

The heat of formation of dimethyl sulphide in the ideal gas state is −8.9 kcal/mol. Then, from equation (4), the heat of formation of S-methylthianium perchlorate is −87.7 kcal/mol. This salt has been synthesized[14]. The equatorial form[14] is more stable by 0.28 kcal/mol at 373 K. S-Methylthianium perchlorate is an excellent candidate for calorimetry.

III. SULPHONIUM YLIDES

Sulphonium ylides[6] have the general structure $R^1R^2\overset{+}{S}\overset{-}{C}R^3R^4$. One of the simplest structures is the ylide $Me_2\overset{+}{S}-\overset{-}{C}H_2$, which has the canonical form $Me_2S=CH_2$. The heat of formation of this ylide may be estimated from the heat of formation of

TABLE 3. Estimation of the heat of formation of 4-t-butyl-1-thiacyclohexane ($C_9H_{18}S$) using group additivity[4,13] (for the ideal gas phase)

Group	Value (kcal/mol)
3 C—C, H₃	−30.6
C—C₄	0.5
C—C₃, H	−1.9
2 C—C₂, H₂	−9.8
2 C—C, S, H₂	−11.3
S—C₂	11.5
4 gauche	3.2
	−38.4

TABLE 4. Comparison of the measured[2] heats of formation (kcal/mol) in the condensed state of sulphoxides and the corresponding ketones

R	R_2SO^a	R_2CO^a	S compound minus C compound
Me	−52.1 (l)	−59.3 (l)	7.2
Et	−64.0 (l)	−70.9 (l)	6.9
i-Pr	−78.6 (l)	−84.3 (l)	5.7
Ph	2.4 (c)	−8.1 (c)	10.5
			Average 8 ± 3

^{a}(l) = liquid; (c) = crystal.

trimethylsulphonium chloride estimated in the last section by assuming that

$$Me_2SCH_{2(c)} - Me_3SCl_{(c)} = Me_2CCH_{2(l)} - Me_3CCl_{(l)} \qquad (5)$$

From the heats of formation[2] of $Me_2CCH_{2(l)}$ (−9.2 kcal/mol) and of $Me_3CCl_{(l)}$ (−50.6 kcal/mol), together with the estimated value for $Me_3SCl_{(c)}$ (−62.8 kcal/mol), the heat of formation of $Me_2SCH_{2(c)}$ is −21.4 kcal/mol.

The structure of the ylide $Me_2\overset{+}{S}\overset{-}{C}(CN)_2$ has been determined[15]. Its heat of formation may be estimated assuming that

$$Me_2SC(CN)_{2(c)} - Me_2SCH_{2(c)} = 2[CH_2CHCN_{(g)} - CH_2CH_{2(g)}] \qquad (6)$$

From available data[2], the heat of formation of acrylonitrile$_{(g)}$ is 44.1 kcal/mol and of ethylene$_{(g)}$ 12.4 kcal/mol. The estimated value for $Me_2SCH_{2(c)}$ is −21.4 kcal/mol and the heat of formation of $Me_2SC(CN)_{2(c)}$ is estimated to be 42.0 kcal/mol. This compound is an excellent candidate for experimental measurement of heat of formation.

Finally, an approximate but simple method of estimating heats of formation in the condensed state of ylides having the structure R_2SCH_2 utilizes the expression

$$\Delta H_f^0 (R_2SCH_2) = \Delta H_f^0 (R_2CCH_2) + 8 \text{ kcal/mol} \qquad (7)$$

This method is based on the data shown in Table 4.

The heats of formation listed in Table 4 show that substitution of a sulphur atom in a sulphoxide for a carbon atom results in an increase of 8 ± 3 kcal/mol in the heat of formation for the condensed state (neglecting heats of fusion). For example, from equation (7) and the known[2] heat of formation of $Me_2CCH_{2(l)}$ of −9.2 kcal/mol, the heat of formation of $Me_2SCH_{2(l)}$ may be estimated to be −1 kcal/mol, compared with the value of −21 kcal/mol estimated earlier for the solid state. Experimental data are needed to resolve this discrepancy.

IV. REFERENCES

1. R. D. Freeman (Editor), *Bulletin of Thermodynamics*, Oklahoma State University, Stillwater, Oklahoma, 1977; E. F. Westrum (Editor), *Bulletin of Thermochemistry and Thermodynamics*, IUPAC, University of Michigan, Ann Arbor, Michigan, 1970–1976.
2. J. D. Cox and G. Pilcher, *Thermochemistry of Organic and Organometallic Compounds*, Academic Press, New York, 1970.
3. D. R. Stull, E. F. Westrum and G. C. Sinke, *The Chemical Thermodynamics of Organic Compounds*, Wiley, New York, 1969.
4. S. W. Benson, F. R. Cruickshank, D. M. Golden, G. R. Haugen, H. E. O'Neal, A. S. Rodgers, R. Shaw and R. Walsh, *Chem. Rev.*, **69**, 279 (1969).

5. E. S. Domalski, *J. Phys. Chem. Ref. Data*, **1**, 221 (1972).
6. C. J. M. Stirling in *Sulfonium Salts in Organic Chemistry of Sulfur* (Ed. S. Oae), Plenum Press, New York, 1977, p. 473.
7. A. W. Johnson, *Ylid Chemistry*, Academic Press, New York, 1966.
8. S. W. Benson, submitted for publication.
9. J. D. Roberts and M. C. Caserio, *Basic Principles of Organic Chemistry*, Benjamin, New York, 1965, p. 756.
10. *Selected Values of Chemical Thermodynamic Properties*, NBS Technical Note 270–3, US Department of Commerce, Washington, DC.
11. D. E. Zuccaro and J. D. McCullough, *Z. Kristallogr.*, **112**, 401 (1959).
12. A. Lopez-Castro and M. R. Truter, *Acta Crystallogr.*, **17**, 465 (1964).
13. S. W. Benson, *Thermochemical Kinetics*, 2nd ed., Wiley, New York, 1976.
14. E. L. Eliel, R. L. Willer, A. T. McPhail and K. D. Anan, *J. Amer. Chem. Soc.*, **96**, 3021 (1974).
15. A. T. Christensen and W. G. Witmore, *Acta Crystallogr.*, **B25**, 73 (1969).

The Chemistry of the Sulphonium Group
Edited by C. J. M. Stirling and S. Patai
© 1981 John Wiley & Sons Ltd.

CHAPTER **5**

Photochemistry of sulphonium compounds

J. D. COYLE

Chemistry Department, The Open University, Milton Keynes MK7 6AA, England

I.	INTRODUCTION	107
II.	SULPHONIUM SALTS	107
III.	SULPHONIUM YLIDES	112
IV.	SULPHIMIDES	117
V.	SULPHONIUM SPECIES AS PHOTOCHEMICAL INTERMEDIATES	120
VI.	REFERENCES	121

I. INTRODUCTION

The photochemistry of certain classes of organic sulphur compounds, such as thioketones, sulphones and five-membered sulphur heterocycles, has been widely investigated. In comparison, sulphonium compounds have undergone less intensive study, but the results which have been reported allow general features to be seen. The pattern that emerges is one in which the typical photochemical reactions involve homolytic or heterolytic cleavage of the bond between the sulphonium sulphur and one of the adjacent atoms. Needless to say, there are exceptions to this generalization, but they are few in number. In this review the major groups of compounds considered are sulphonium salts, sulphonium ylides and sulphimides. Sulphoxides ($R_2\overset{+}{S}{-}\bar{O}$) are formally analogous to sulphimides ($R_2\overset{+}{S}{-}\bar{N}R$), but their photochemistry is rather different—it will be dealt with in another volume, and here only a few reactions of such related species will be mentioned where they provide a useful analogy.

II. SULPHONIUM SALTS

The irradiation of triarylsulphonium salts in hydroxylic solvents produces a mixture of products[1]. Triphenylsulphonium chloride in methanol gives rise to benzene, chlorobenzene, anisole, biphenyl, diphenyl sulphide, diphenyl sulphoxide, and HCl [equation (1)]. The corresponding iodide gives a high yield of iodobenzene, and less HI [equation (2)]. The difference in product ratios suggested that two competing primary processes are occurring. In one of them [equation (3)] electron

$$\overset{+}{Ph_3S}\overset{-}{X} \xrightarrow[\text{MeOH}]{h\nu\ (254\ \text{nm})} PhH + PhX + PhOMe + Ph_2 + Ph_2S + Ph_2SO + HX$$

X = Cl:	10%	0.5%	11%	2%	30%	2%	49%	(1)
X = I :	10%	47%	6%	1%	43%	0.3%	4%	(2)

$$\overset{+}{Ph_3S}\overset{-}{X} \xrightarrow{h\nu} X\cdot + Ph_3S\cdot \longrightarrow Ph_2S + Ph\cdot \tag{3}$$

transfer from the halide ion to the cation is followed by cleavage of the resulting sulphur radical to give diphenyl sulphide and phenyl radical. The phenyl radical is the source of benzene, biphenyl, and halobenzene products, and because iodide is a much better electron donor than chloride, the iodide gives a very much higher yield of halobenzene (and a higher yield of benzene when the two are compared in ethanol as solvent). The other primary process [equation (4)] involves homolytic

$$\overset{+}{Ph_3S} \xrightarrow{h\nu} Ph\cdot + \overset{\cdot+}{Ph_2S} \xrightarrow{\text{MeOH}} PhS\cdot + PhOMe + H^+ \tag{4}$$

cleavage of a carbon–sulphur bond to give phenyl radical and diphenylsulphinium radical cation. The latter is attacked by the solvent, in part to give anisole and a proton. Other possible reactions of this radical cation also lead to the production of proton acid, and because this is the minor reaction pathway when iodide is the counter ion, little HI is produced. The sulphonium nitrate or tetrafluoroborate gives similar results to the chloride (except that no PhX is formed), and this is expected if homolytic cleavage predominates for these salts. Supporting evidence for the radical route to phenyl ethers comes from the fact that radical scavengers inhibit their formation, whereas an alternative route to the ethers via phenyl cation should not be affected by such reagents.

Numerous examples of β-ketosulphonium salts have been reported to undergo photochemical reaction. The majority are phenacyl or related compounds, and aliphatic sulphonium salts lead largely[2] to polymer [equation (5)] with only very low yields of monomeric product [equation (6)].

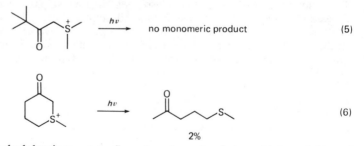

$$\text{(5)} \qquad h\nu \longrightarrow \text{no monomeric product}$$

$$\text{(6)} \qquad h\nu \longrightarrow \qquad 2\%$$

Dimethylphenacylsulphonium tetrafluoroborate gives[2] a high yield of acetophenone [equation (7)] on irradiation in methanol, whilst in *t*-butanol (a much poorer source of abstractable hydrogen atoms) some 1,2-dibenzoylethane is formed, as well as a small amount of a γ-ketosulphide [equation (8)]. The major products can be accounted for in terms of homolytic carbon–sulphur bond cleavage in the initial stages of reaction to give phenacyl radical and dialkylsulphinium radical cation.

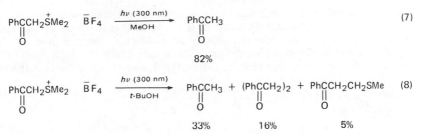

$$PhCCH_3 \qquad (7)$$

82%

$$PhCCH_3 + (PhCCH_2)_2 + PhCCH_2CH_2SMe \qquad (8)$$

33% 16% 5%

Dimethylphenacylsulphonium nitrate on irradiation (Pyrex) in water leads in a similar way to acetophenone (10%), dibenzoylethane (44.5%), and an estimated 78% yield of dimethyl sulphide[3]. Dibutylphenacylsulphonium tetrafluoroborate leads to dibutyl sulphide [equation (9)][2], and the production of dialkyl sulphide

$$PhCCH_2\overset{+}{S}Bu_2 \quad \overset{-}{B}F_4 \quad \xrightarrow[\text{MeOH}]{h\nu\ (300\ nm)} \quad PhCCH_3 + Bu_2S \qquad (9)$$

48% 29%

points to the intermediacy of dialkylsulphinium radical cation $(R_2\overset{+}{S})$, which abstracts hydrogen and then loses a proton.

A number of ring-substituted phenacylsulphonium salts [equation (10)], the naphthyl analogues, and related compounds [equation (11)] undergo the same type

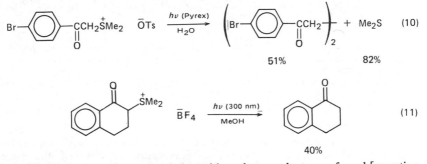

51% 82%

of reaction[2,3]. When the counter ion is bromide, other products are found [equation (12)] that incorporate the bromine, and this is attributed[3,4] to the relative ease of oxidation of bromide ion to bromine atom. The bromine atom combines with the

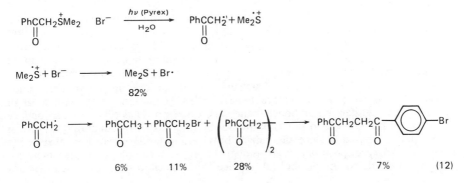

phenacyl radical to give phenacyl bromide, or it attacks the dibenzoylethane to give 1-benzoyl-2-(*p*-bromobenzoyl)ethane. It is established that phenacyl bromide is not on the main reaction pathway to the 'normal' products of the reaction.

p-Bromophenacyldimethylsulphonium bromide gives a high yield of the corresponding phenacyl bromide [equation (13)], and one possibility is that a direct

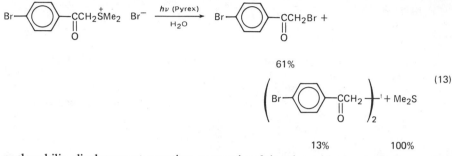

nucleophilic displacement reaction occurs involving bromide ion and the excited state of the sulphonium cation. The effect of substituents[3] is not readily understood; the *p*-methoxyphenacyl compound gives largely radical-derived products, but there is an appreciable amount of *p*-methoxyphenacyl alcohol and *p*-methoxylphenylacetic acid [equation (14)], both of which are best explained on

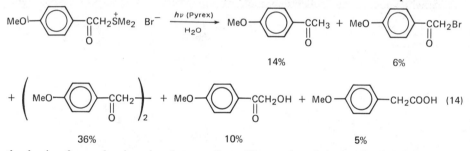

the basis of a carbonium ion intermediate. The *p*-phenylphenacyl sulphonium salt gives *p*-phenylacetophenone, perhaps unexpectedly in view of the stabilizing influence of the phenyl group on a carbonium ion, and dimethyl(1- or 2-naphthoylmethyl)sulphonium bromide reacts only slowly. It may be that differences in the electronic character (n,π^* or π,π^*) of the excited state play some part in the different pathways.

A study has been made[5] of an α-bromophenacylsulphonium bromide, and the products in this reaction [equation (15)] arise either by thermal reaction (dimethylphenacylsulphonium bromide, benzoic acid, dibromomethyldimethyl-sulphonium bromide), by photochemical carbon–sulphur bond cleavage (phenacyl bromide, dimethyl sulphide), or by secondary reaction of phenacyl bromide.

Benzyldimethylsulphonium salts behave rather differently[2], and the major products in an alcohol solvent are benzyl ethers [equation (16)], with a small amount of radical-derived diphenylethane. The other products arise from secondary photolysis of the ether, and this is borne out by their absence [equation (17)] when the alcohol is *t*-butanol, since the bulky *t*-butyl group hinders hydrogen abstraction from the benzylic position. It is thought that a nucleophilic displacement of dimethyl sulphide by the alcohol takes place in the excited state, although it is not

$$\underset{\underset{O}{\overset{\overset{Br}{|}}{\overset{|}{\underset{||}{PhCCH}}}}}{}\!\!-\overset{+}{S}Me_2 \;\; Br^- \quad \xrightarrow[H_2O]{h\nu \text{ (Pyrex)}} \quad \underset{\underset{O}{\overset{||}{PhCCH_2}}}{}\overset{+}{S}Me_2 \;\; Br^- \; + \; PhCOOH \; + \; Br_2CH\overset{+}{S}Me_2 \;\; Br^-$$

30%   20%   13%

$$+ \; \underset{\underset{O}{\overset{||}{PhCCH_2Br}}}{} \; + \; Me_2S \; + \; \underset{\underset{O}{\overset{||}{PhCCH_3}}}{} \; + \; \left(\underset{\underset{O}{\overset{||}{PhCCH_2}}}{} \!-\!\! \right)_{\!2} \tag{15}$$

13%   31%   2%   5%

$$PhCH_2\overset{+}{S}Me_2 \;\; \overset{-}{B}F_4 \quad \xrightarrow[MeOH]{h\nu \text{ (254 nm)}} \quad PhCH_2OMe \; + \; (PhCH_2)_2 \; + \; PhCHO$$

25%   3%   9%

$$+ \; \underset{\underset{OMe}{|}}{PhCH}\!-\!CH_2Ph \; + \; \left(\underset{\underset{OMe}{|}}{Ph}\!-\!CH\!-\!\!\right)_{\!2} \tag{16}$$

8%   7%

$$PhCH_2\overset{+}{S}Me_2 \;\Big|\; \overset{-}{B}F_4 \quad \xrightarrow[t\text{-BuOH}]{h\nu \text{ (254 nm)}} \quad PhCH_2OBu\text{-}t \; + \; (PhCH_2)_2 \tag{17}$$

33%   4%

clear whether heterolytic cleavage of the carbon–sulphur bond occurs before attack by the solvent or in a concerted process. In acetonitrile the major product arises by the action of the nitrile as a (very weak) nucleophile [equation (18)]; the minor

$$PhCH_2\overset{+}{S}Me_2 \;\; \overset{-}{B}F_4 \quad \xrightarrow[MeCN]{h\nu \text{ (254 nm)}} \quad [PhCH_2\overset{+}{N}\!\!\equiv\!\!CMe] \;\longrightarrow$$

$$PhCH_2NH\!-\!\underset{\underset{O}{\overset{||}{C}}}{}CH_3 \; + \; PhCH_2\!\!-\!\!\left\langle\bigcirc\right\rangle\!\!-\!Me \tag{18}$$

70%   7%

product in this reaction is not expected on the basis of either radical or ionic intermediates. With dibenzylmethylsulphonium tetrafluoroborate, benzyl methyl sulphide (PhCH₂SMe) is isolated, and with two cyclic benzylsulphonium salts the primary solvolysis product contains the sulphide group [equation (19)], although secondary photolysis of the ethers leads to the products lacking sulphur.

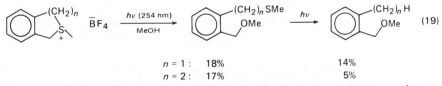

$$\tag{19}$$

n = 1 :   18%   14%
n = 2 :   17%   5%

A compound that incorporates both a benzyl and a phenacyl system gives a benzyl solvolysis product in highest yield [equation (20)], although some of the minor products probably arise via phenacyl–sulphur homolytic cleavage.

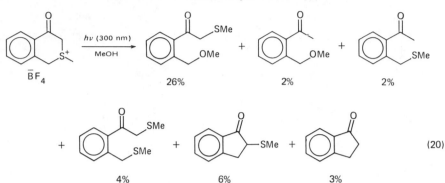

26%   2%   2%

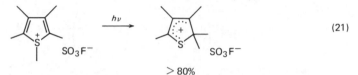

(20)

4%   6%   3%

A different type of sulphonium salt, 1,2,3,4,5-pentamethylthiophenium fluorosulphonate, undergoes an efficiency photochemical 1,5-shift of methyl[6] to give the isomeric 2,2,3,4,5-pentamethyl cation [equation (21)]. If the shift is

(21)

> 80%

concerted it should occur antarafacially according to the basic Woodward–Hoffman rules, and this is not impossible if the excited state of the thiophenium system has a distorted geometry along the lines of that proposed for several ring-transposition reactions of thiophenes and thiazoles. A reasonable alternative mechanism involves homolytic cleavage of the sulphur–methyl bond and recombination of the resulting radicals to give the more stable rearranged cation.

III. SULPHONIUM YLIDES

The photochemistry of compounds of the type $R_2\overset{+}{S}-\overset{-}{C}R'R''$ is influenced by the relationship between the ylides and the sulphide and carbene fragments [equation (22)], although other processes are also important. The ylides can be formed by the

$$R_2\overset{+}{S}-\overset{-}{C}R'R'' \rightleftharpoons R_2S + :CR'R'' \qquad (22)$$

generation of a singlet carbene in the presence of a dialkyl sulphide, for example[7] by irradiation [or Cu(II)-catalysed thermal decomposition] of dimethyl diazomalonate with dimethyl sulphide [equation (23)]. Benzophenone-

$$N_2C(COOMe)_2 + Me_2S \xrightarrow{h\nu \text{ (Pyrex)}} Me_2\overset{+}{S}-\overset{-}{C}(COOMe)_2 \qquad (23)$$

photosensitized reaction does not lead to the ylide because sulphides are efficient traps for singlet, but not triplet, carbenes. This affinity is seen[8] in the reaction at sulphur rather than at the C=C bond in dihydrothiapyran [equation (24)].

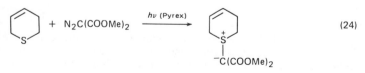

(24)

With Pyrex-filtered (>300 nm) radiation, the ylides can be generated photochemically from diazo precursors, but with shorter wavelengths the same ylides decompose readily [equation (25)][7].

$$Me_2\overset{+}{S}-\overset{-}{C}(COOMe)_2 \xrightarrow[\text{MeOH}]{h\nu\ (254\ nm)} MeOCH(COOMe)_2 + CH_2(COOMe)_2 \qquad (25)$$

Most of the sulphonium ylides whose photochemistry has been studied have ketone or ester groups stabilizing the negative charge on carbon. One exception is diphenylsulphoniopropenylide, which undergoes two competing primary photoreactions[9]. When irradiated at −78°C in the presence of a conjugated diene, a Diels–Alder adduct is formed [equation (26)], which points to the initial production of cyclopropene, probably via vinylcarbene. In addition, two sulphides are formed by formal 1,2- and 1,4-phenyl shifts, which may be concerted or may involve initial phenyl–sulphur cleavage, and a third sulphide probably arises from one of the other two by a base-induced shift of the double bond [equation (27)]. In this system it seems that no cyclopropene is formed when the ylide undergoes thermal decomposition.

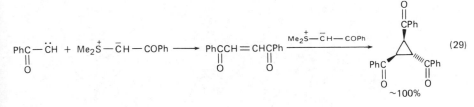

Phenacylides also undergo photochemical cleavage of the ylide carbon–sulphur bond on irradiation. These ylides probably exist largely as a mixture of enolate stereoisomers, and the primary products of photoreaction are dialkyl sulphide and benzoylcarbene [equation (28)]. In the absence of other species[10] the carbene

$$Me_2\overset{+}{S}-CH=C\overset{\displaystyle Ph}{\underset{\displaystyle O^-}{}} \xrightarrow{h\nu\ (Pyrex)} Me_2S + PhC\overset{\displaystyle \cdot\cdot}{-}\overset{\displaystyle}{C}H \qquad (28)$$

attacks a second molecule of ylide to give 1,2-dibenzoylethene, and this in turn reacts with a third molecule of ylide to produce the 1,2,3-tribenzoylcyclopropane isolated in nearly quantitative yield [equation (29)]. In the presence of

$$PhC-\overset{\cdot\cdot}{C}H + Me_2\overset{+}{S}-\overset{-}{C}H-COPh \longrightarrow PhCCH=CHCPh \xrightarrow{Me_2\overset{+}{S}-\overset{-}{C}H-COPh} \qquad (29)$$

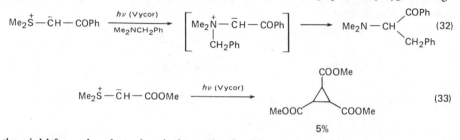

(30)

cyclohexene, some of the carbene is trapped as a bicyclo[4.1.0]heptane [equation (30)]; with a good hydrogen donor such as propan-2-ol, some acetophenone is formed by reduction of the carbene; with ethanol, some ethyl phenylacetate is produced, and this must arise by reaction with the ketone that is the product of carbene (or ylide) rearrangement [equation (31)].

$$PhC - \ddot{C}H \text{ (or ylide)} \longrightarrow PhCH=C=O \xrightarrow{EtOH} PhCH_2COOEt \quad (31)$$

Another group has reported[11] that N,N-dimethylbenzylamine traps benzoylcarbene to give a product via a nitrogen ylide [equation (32)]. This report also provides another example of cyclopropane formation [equation (33)], although

$$Me_2\overset{+}{S} - \overset{-}{C}H - COPh \xrightarrow[Me_2NCH_2Ph]{h\nu \text{ (Vycor)}} \left[Me_2\overset{+}{N} - \overset{-}{C}H - COPh \atop CH_2Ph \right] \longrightarrow Me_2N - CH \overset{COPh}{\underset{CH_2Ph}{<}} \quad (32)$$

$$Me_2\overset{+}{S} - \overset{-}{C}H - COOMe \xrightarrow{h\nu \text{ (Vycor)}} \quad (33)$$

5%

the yield from the photochemical reaction in this case is much lower than that from the Cu(II)-catalysed thermal decomposition. The use of shorter wavelength radiation in this study may be one factor contributing to the low yields of products isolated.

In some phenacylide systems, cleavage of a sulphur–carbon bond other than the ylide linkage occurs. A 2-methylisothiachromanone ylide gives rise[12] to indan-1-one as the sole volatile product on irradiation [equation (34)], and the suggested mechanism involves a 1,2-shift of the benzylic carbon to give 2-(methylthio)indan-1-one, which undergoes an efficient photochemical elimination of thioformaldehyde in a secondary reaction that has been independently demonstrated to take place in nearly quantitative yield (this second photoreaction is

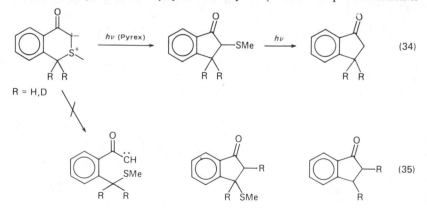

R = H,D

a typical Norrish type 2 process for ketones). An alternative mechanism by way of an acylcarbene and 3-(methylthio)indan-1-one [equation (35)] was ruled out on the basis of results of deuterium-labelling studies.

The 1,2-shift resembles the thermally induced Stevens rearrangement, which may occur by a homolytic cleavage and recombination mechanism[13]. The reaction also occurs on photolysis of an *o*-methylphenacyclide, which produces *o*-methyl-α-(methylthio)propiophenone and its photochemical decomposition product *o*-methylpropiophenone [equation (36)]. Amongst the other products of

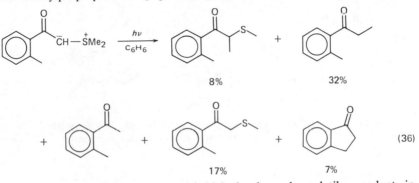

8%   32%

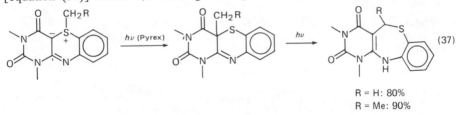

17%   7%   (36)

this reaction is *o*-methylacetophenone (which is the sole volatile product in chloroform solution), and this may arise via *o*-methylbenzoyl carbene or via the other isolated sulphide product. Indan-1-one is also formed, and this is likely to be produced from the carbene by intramolecular insertion.

Other systems give rise to 1,2-photorearrangement products, including pyrimido-1,4-benzothiazine sulphonium ylides[14], which lead to thiazepines on irradiation [equation (37)]. Initial 1,2-rearrangement gives a sulphide which then undergoes

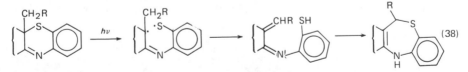

R = H: 80%
R = Me: 90%

further photochemical reaction by way of carbon–sulphur bond cleavage, hydrogen atom transfer and cyclization by thiol addition to alkene [equation (38)]. The

CH₂R equation image (38)

1,2-rearranged products can be isolated after a short irradiation time, although in the case of the *S*-isopropyl compound ($\geq$SCHMe₂) this is the final product (90%) and no thiazepine is formed. The intermediate products undergo independent photochemical rearrangement to the thiazepines. In the thermal reactions of these ylides, 1,2-rearranged products are accompanied by 1,4-shifted products and others, but with no thiazepines.

Stabilized sulphoxonium ylides also undergo cleavage of the ylide carbon–sulphur bond on irradiation[15], giving rise to a ketene that can be trapped by alcohol [equation (39)]. With these compounds, the sulphur-containing photoproduct is a sulphoxide. A cyclic sulphoxonium ylide leads to a single ring-opened photoproduct [equation (40)].

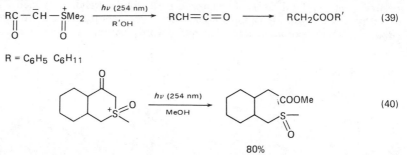

$$R = C_6H_5 \quad C_6H_{11}$$

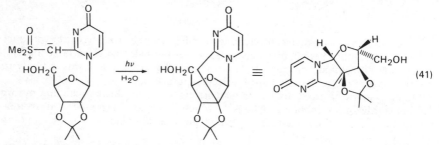

A sulphoxonium ylide incorporated into a nucleoside has been used[16] to generate a carbene that inserts into a C—H bond of the pentose fragment [equation (41)].

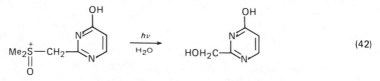

A related sulphoxonium salt undergoes photochemical heterolytic cleavage of a carbon–sulphur bond, and the carbonium ion leads to the isolated hydroxymethyl product [equation (42)].

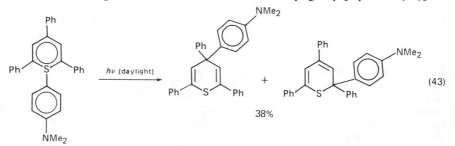

Benzenoid thia-aromatics are formally sulphonium ylides, and a substituted thiabenzene[17] undergoes a 1,2- and a 1,4-shift of an aryl group [equation (43)] on exposure to daylight (or on heating). The shift may involve cleavage of the non-ylide carbon–sulphur bond and recombination of the resulting radicals. A similar 1,2-shift occurs[18] with 1,2,4-triphenyl-1-thianaphthalene [equation (44)], and once again thermal or photochemical reaction leads to the same product.

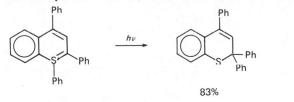

(44)

83%

IV. SULPHIMIDES[19]

In the same way that sulphonium ylides can be generated from diazo compounds and sulphides, so can sulphimides be prepared from azides and sulphides[20]. Ethyl azidoformate on irradiation in the presence of dimethyl sulphide gives[21] N-ethoxycarbonyl-S,S-dimethylsulphimide [equation (45)], and benzoyl azide gives the N-benzoyl analogue[22]. The sulphide intercepts singlet nitrenes with high efficiency.

$$N_3COOEt + Me_2S \xrightarrow{h\nu} Me_2\overset{+}{S}-\overset{-}{N}COOEt \qquad (45)$$

71%

Irradiation of sulphimides with short-wavelength (usually 254 nm) radiation results in fragmentation to give nitrene and sulphide, and the sulphimides offer an alternative source of nitrenes to the more conventional azides. The unstabilized N-methyl-S,S-diphenylsulphimide on irradiation (or heating to 175°C) gives diphenyl sulphide [equation (46)] and a polymer which is presumed to be derived from methylnitrene[23].

$$Ph_2\overset{+}{S}-\overset{-}{N}Me \xrightarrow{h\nu \ (254 \ nm)} Ph_2S \qquad (46)$$

N-Benzoyl-SS-dimethylsulphimide gives rise to benzamide, N-methoxybenzamide, and methyl phenylcarbamate on irradiation in methanol [equation (47)][24]. The

$$PhC-\overset{-}{N}-\overset{+}{S}Me_2 \xrightarrow[\text{MeOH}]{h\nu \ (254 \ nm)} PhCNH_2 + PhCNHOMe + PhNH-COOMe \qquad (47)$$

47%   7%   33%

carbamate probably arises by attack of methanol on the first-formed phenyl isocyanate, and the other products are derived from benzoylnitrene by hydrogen abstraction or by insertion into the O—H bond of methanol. The same sulphimide irradiated in cyclohexane gives benzamide and phenyl isocyanate, and in the presence of 4-methylpent-2-ene it gives the isocyanate and an N-benzoylaziridine [equation (48)][22]. The aziridine is formed with a high degree of stereoselectivity, and it arises by reaction of the alkene with singlet benzoylnitrene. The yield of phenyl isocyanate is more or less independent of the amount of alkene present, and this indicates that the nitrene is *not* a precursor of the isocyanate, which is probably formed in a concerted reaction from the sulphimide.

Irradiation of N-benzoyl-S,S-diphenylsulphimide has been shown[25] to give phenyl isocyanate and diphenyl sulphide [equation (49)] as well as biphenyl and diphenyl disulphide arising from secondary photolysis of the sulphide. In addition, a urea is

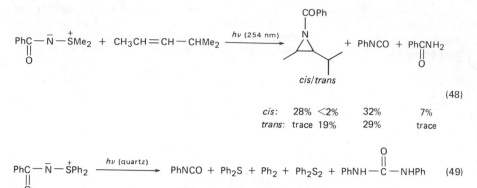

$$(48)$$

cis:	28%	<2%	32%	7%
trans:	trace	19%	29%	trace

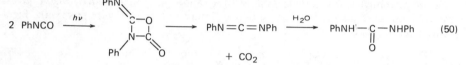

PhC—N̄—⁺SPh₂ $\xrightarrow{h\nu \text{ (quartz)}}$ PhNCO + Ph₂S + Ph₂ + Ph₂S₂ + PhNH—C(=O)—NHPh $\quad(49)$

$\qquad$ 66% $\quad$ 49% $\quad$ 11% $\quad$ trace $\quad$ 48%

formed, and the suggested reaction sequence for its production [equation (50)] involves photochemical dimerisation of the isocyanate followed by elimination of

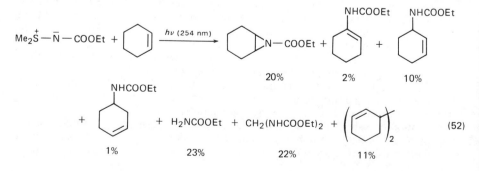

carbon dioxide to give a carbodiimide (detected by its characteristic infrared absorption band). The carbodiimide hydrolyses to the urea on silica in the chromatographic work-up. The urea can be prepared independently by photolysis of phenyl isocyanate.

A sulphimide with a chiral group attached to the carbonyl unit gives rise[25] to a chiral isocyanate [equation (51)] and this is in keeping with a concerted intramolecular reaction for this process.

$$Ph—\overset{*}{C}H(Me)—C(=O)—\overset{-}{N}—\overset{+}{S}Ph_2 \xrightarrow{h\nu} Ph—\overset{*}{C}H(Me)—NCO \qquad (51)$$

N-Ethoxycarbonyl-*S*,*S*-dimethylsulphimide behaves in a similar way to the *N*-benzoyl compound, giving nitrene-derived products in cyclohexane, cyclohexene [equation (52)], or 4-methylpent-2-ene[21]. For this compound the aziridine is formed from the alkene with only partial stereoselectivity, and it seems that for ethoxycarbonylnitrene, singlet → triplet intersystem crossing competes efficiently with reactions of the singlet nitrene, so that a higher proportion of products arise from the triplet nitrene. Products arising from insertion into C—H or O—H bonds, however, are derived solely from the singlet nitrene. Irradiation of this sulphimide in the presence of excess of diethyl sulphide provides an example of a useful exchange process [equation (53)] in which the added sulphide intercepts the singlet nitrene.

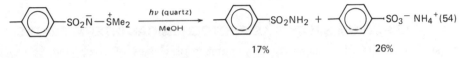

$$Me_2\overset{+}{S}-\overset{-}{N}-COOEt + Et_2S \xrightarrow{h\nu\ (254\ nm)} Et_2\overset{+}{S}-\overset{-}{N}-COOEt \qquad (53)$$

45%

An *N*-sulphonyl sulphimide gives only the sulphonamide and the ammonium sulphonate [equation (54)] on irradiation[26], without the variety of other products

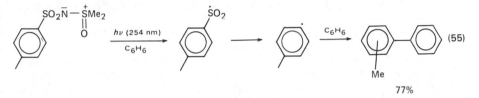

17%   26%

associated with *N*-acyl compounds. A series of related *N*-sulphonyl sulphoximides gives products derived from aryl radicals [equation (55)], and in this system it is likely that homolytic sulphonamide sulphur–nitrogen bond cleavage occurs[27].

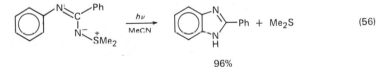

77%

In the photochemical reactions of sulphimides with an *N*-arylimine group[28], the first-formed nitrene undergoes intramolecular insertion into the *ortho* C—H bond to give a benzimidazole [equation (56)]. For a related *N*-(2-pyridyl)imine there is a

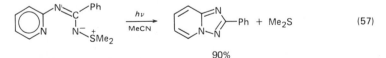

96%

distinct preference for attack at the pyridine nitrogen rather than at a C—H bond, and a good yield of a triazole is obtained [equation (57)]. This preference is

understandable on the basis of the electrophilic nature of nitrenes. If the *ortho* positions are blocked, two products are formed [equation (58)][29], a carbodiimide by phenyl migration and a pyrimidine by a suggested mechanism that involves three 1,5-shifts [equation (59)]. In all of these examples, thermal decomposition of the sulphimide does not give products via the nitrene.

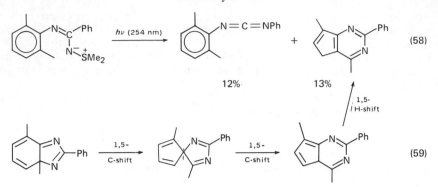

V. SULPHONIUM SPECIES AS PHOTOCHEMICAL INTERMEDIATES

In studies of the photochemistry of a number of classes of compound, it has become clear that the presence of a sulphur atom in the molecule can substantially alter the course of reaction even when it is not part of the chromophore. In the mechanisms proposed for the involvement of sulphur, intermediates that are sulphonium ylides or similar species are often proposed. Whilst not strictly within the scope of this review, it seems worth mentioning two such reactions to illustrate the point.

A cyclic β-ketosulphide on irradiation gives rise[30] to a cyclic thiocarboxylate and a ring-opened product that has incorporated a molecule of methanol solvent [equation (60)]. These products could be explained on the basis of a standard Norrish type I homolytic cleavage reaction of a cyclic ketone, but the position of incorporation of deuterium from O-deuteriomethanol is inconsistent with this mechanism and points to the operation of a route involving sulphur participation and sulphonium intermediates [equation (61)].

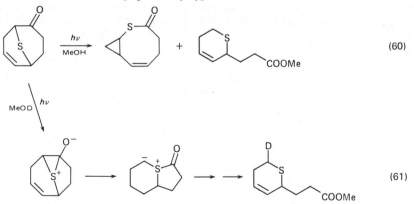

For the ring-scrambling photoreactions of five-membered sulphur heterocycles a variety of mechanistic pathways have been proposed to explain the range of observed transformations. For example, the transposition of ring atoms 2 and 3 in the photolysis of 2-arylthiophens [equation (62)] can be envisaged as proceeding by way of a sulphonium ylide intermediate[31]. However, the weight of evidence suggests that for both thiophenes and thiazoles non-zwitterionic valence isomers

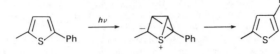

(62)

such as

are better representations of the intermediates[32,33]. There has been a suggestion that the pattern of deuterium incorporation on irradiation of a thiazole in MeOD was more in accord with an intermediate containing carbanionic carbon[34], but the original proposer of zwitterionic intermediates recognized[31] that the zwitterionic and uncharged structures are closely related.

VI. REFERENCES

1. J. W. Knapczyk and W. E. McEwan, *J. Amer. Chem. Soc.*, **91**, 145 (1969); *J. Org. Chem.*, **35**, 2539 (1970); Benzhydryldiethylsulphonium tetrafluoroborate gives products derived from the benzhydryl carbocation. S. Kando, M. Tsumadori and K. Tsuda. *Chem. Abstr.*, **90**, 151724 (1979).
2. A. L. Maycock and G. A. Berchtold, *J. Org. Chem.*, **35**, 2532 (1970).
3. T. Laird and H. Williams, *J. Chem. Soc. C*, 1863 (1971).
4. T. Laird and H. Williams, *J. Chem. Soc. Chem Commun*, 561 (1969).
5. T. Laird and H. Williams, *J. Chem. Soc. C*, 3467 (1971).
6. H. Hogeveen, R. M. Kellogg and K. A. Kuindersma, *Tetrahedron Lett.*, 3929 (1973).
7. W. Ando, T. Yagihara, S. Tozune and T. Migita, *J. Amer. Chem. Soc.*, **91**, 2786 (1969).
8. W. Ando, S. Kondo, K. Nakayama, K. Ichibori, H. Kohoda, H. Yamato, I. Imai, S. Nakaido and T. Migita, *J. Amer. Chem. Soc.*, **94**, 3870 (1972).
9. B. M. Trost and R. W. LaRochelle, *J. Amer. Chem. Soc.*, **92**, 5804 (1970).
10. B. M. Trost, *J. Amer. Chem. Soc.*, **88**, 1587 (1966); **89**, 138 (1967).
11. A. W. Johnson and R. T. Amel, *J. Org. Chem.*, **34**, 1240 (1969).
12. R. H. Fish, L. C. Chow and M. C. Caserio, *Tetrahedron Lett.*, 1259 (1969).
13. S. Oae (Editor), *Organic Chemistry of Sulfur*, Plenum Press, New York, 1977, p. 517.
14. Y. Maki and T. Hiramitsu, *Chem. Pharm. Bull.*, **25**, 292 (1977).
15. E. J. Corey and M. Chaykovsky, *J. Amer. Chem. Soc.*, **86**, 1640 (1964).
16. T. Kuneida and B. Witkop, *J. Amer. Chem. Soc.*, **91**, 7752 (1969); **93**, 3487 (1971).
17. C. C. Price and H. Pirelahi, *J. Org. Chem.*, **37**, 1718 (1972).
18. M. Hori, T. Kataoka and H. Shimizu, *Chem. Pharm. Bull.*, **22**, 2485 (1974).
19. For a general review of sulphimides (sulfilimines), see T. L. Gilchrist and C. J. Moody, *Chem. Rev.*, **77**, 409 (1977).
20. W. Ando, *Accounts Chem. Res.*, **10**, 179 (1977).
21. Y. Hayashi and D. Swern, *Tetrahedron Lett.*, 1921 (1972). The formation of products by insertion has recently been investigated more extensively, N. Torimoto, T. Shingaki and T. Nagai, *Bull. Chem. Soc. Japan*, **51**, 1200 (1978).
22. Y. Hayashi and D. Swern, *J. Amer. Chem. Soc.*, **95**, 5205 (1973).
23. J. A. Franz and J. C. Martin, *J. Amer. Chem. Soc.*, **97**, 583 (1975).
24. U. Lerch and J. G. Moffatt, *J. Org. Chem.*, **36**, 3391 (1971).
25. N. Furukawa, M. Fukumura, T. Nishio and S. Oae, *J. Chem. Soc. Perkin Trans. I*, 96 (1977); Phosphorus and Sulphur **5**, 231 (1978); N. Furukawa, T. Nishio, M. Fukumura and S. Oae, *Chem. Lett.*, 209 (1978).
26. U. Lerch and J. G. Moffatt, *J. Org. Chem.*, **36**, 3686 (1971).
27. R. A. Abramovitch and T. Takaya, *J. Chem. Soc. Chem. Commun.*, 1369 (1969); *J. Chem. Soc. Perkins Trans. I*, 1806 (1975).

28. T. L. Gilchrist, C. J. Moody and C. W. Rees, *J. Chem. Soc. Perkin Trans. I*, 1964 (1975).
29. T. L. Gilchrist, C. J. Moody and C. W. Rees, *J. Chem. Soc. Chem. Commun.*, 414 (1976); *J. Chem. Soc. Perkin Trans. I*, 1871 (1979).
30. A. Padwa and A. Battisti, *J. Amer. Chem. Soc.*, **94**, 521 (1972); A similar sulphonium species is suggested as an intermediate in the photochemistry of 9-thiabicyclo[3.3.1]nonane-2,6-dione: P. H. McCabe and C. R. Nelson, *Tetrahedron Lett.*, 2819 (1978).
31. H. Wynberg, *Accounts Chem. Res.*, **4**, 65 (1971).
32. A. Couture, A. Delevallee, A. Lablache-Combier and C. Párkányl, *Tetrahedron*, **31**, 785 (1975).
33. C. Riou, G. Vernin, H. J. M. Dou and J. Metzger, *Bull. Soc. Chim. Fr.*, 2673 (1972).
34. M. Maeda and M. Kojima, *Tetrahedron Lett.*, 3523 (1973); *J. Chem. Soc. Perkin Trans. I*, 685 (1978).

The Chemistry of the Sulphonium Group
Edited by C. J. M. Stirling and S. Patai
© 1981 John Wiley & Sons Ltd.

CHAPTER **6**

Electronic spectra

G. C. BARRETT

Oxford Polytechnic, Headington, Oxford OX3 0BP, England

I.	INTRODUCTION	123
II.	ABSORPTION CHARACTERISTICS OF THE SULPHONIUM CHROMOPHORE	124
III.	TRIALKYLSULPHONIUM SALTS; THE SULPHONIUM SULPHUR ATOM AS A CHROMOPHORE	125
IV.	ARYLSULPHONIUM SALTS AND ALKENYLSULPHONIUM SALTS: INFLUENCE OF THE SULPHONIUM GROUP ON THE U.V. ABSORPTION BEHAVIOUR OF AN ADJACENT CHROMOPHORE	129
V.	ENVIRONMENTAL EFFECTS: INFLUENCE OF SOLVENT, TEMPERATURE, AND COUNTER ION ON THE U.V. SPECTRA OF SULPHONIUM SALTS	135
VI.	U.V. SPECTRA OF SULPHONIUM SALTS: COMPARISONS WITH U.V. SPECTRA OF ANALOGOUS ORGANIC CATIONS	136
VII.	U.V. SPECTRA OF S-HETEROATOM-SUBSTITUTED SULPHONIUM SALTS	137
VIII.	MISCELLANEOUS U.V. STUDIES INVOLVING SULPHONIUM SALTS .	137
IX.	ACKNOWLEDGEMENT	138
X.	REFERENCES	138

I. INTRODUCTION

This Chapter is arranged in several sections, the first dealing with studies which reveal the absorption characteristics of the sulphonium chromophore; later sections are concerned with chromophoric systems carrying a sulphonium substituent, influence of solvent, temperature, and anion on the u.v. absorption of a sulphonium cation, and comparisons of the u.v. spectra of sulphonium salts with those of analogous organic cations (particularly carbonium and phosphonium salts).

II. ABSORPTION CHARACTERISTICS OF THE SULPHONIUM CHROMOPHORE

Trialkylsulphonium salts ($R_3\overset{+}{S}\ X^-$) in which the anion X^- is devoid of absorption features in the 180–600-nm wavelength range (X = Cl, OH, BF_4, ClO_4, etc.) have been reported to show strong absorption (log $\varepsilon \approx 4.0$) near 200 nm, and weaker absorption (log $\varepsilon \approx 2.8$) near 230 nm[1]. There are some similarities between these features and those for simple alkanethiols ($\lambda\lambda_{max}$ 198, 228 nm; $\varepsilon\varepsilon_{max}$ 2000, 150) and dialkyl sulphides ($\lambda\lambda_{max}$ near 195, 200, 220, 240 nm; log ε 3.45, 3.2, 2.1, $\geqslant 1.9$)[2]. An earlier study[3] reports the wavelength range 208–214 nm for the absorption maximum of the sulphonium chromophore, with much lower intensity (log ε 1.73–2.76, varying with concentration), and the longer wavelength maximum reported later by Ohkubo and Yamabe[1] was not located.

The results of these two studies[1,3] are discussed more fully in the following section. The lability of simple sulphonium salts in solution is well known[4], and is favoured by an anion of high nucleophilicity:

$$Me_3\overset{+}{S}Br^- \;\; \xrightleftharpoons{\qquad} \;\; Me_2S + MeBr \tag{1}$$

This is one factor which must be considered when the reliability of published data for the ultraviolet spectra of sulphonium salts is judged. The u.v. spectrum of trimethylsulphonium bromide (Figure 1) shows minor differences from the spectra of dimethyl-*n*-butylsulphonium tetrafluoroborate and tri-*n*-butylsulphonium tetrafluoroborate (Figure 1) in spite of the possible establishment of the equilibrium shown in equation (1), and the fact that the bromide ion shows absorption maxima at 190 and 199.5 nm ($\varepsilon\varepsilon_{max}$ 12,000 and 11,000, respectively)[5], while the tetrafluoroborate ion is transparent in this wavelength region.

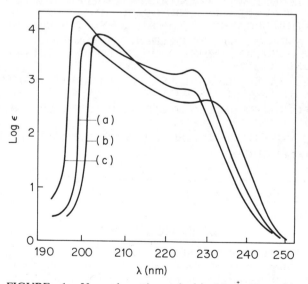

FIGURE 1. U.v. absorption of (a) $Me_3\overset{+}{S}$ Br$^-$, (b) *n*-Bu$\overset{+}{S}$Me$_2$ $\bar{B}F_4$, and (c) *n*-Bu$_3\overset{+}{S}$ $\bar{B}F_4$ in ethanol at 25°C. [Adapted from K. Okhubo and T. Yamabe, *J. Org. Chem.*, **36**, 3149 (1971).]

Charge-transfer interaction between ions or within ion pairs may also need to be considered when interpreting u.v. absorption spectra of salts; the absorption maxima of monatomic (Br^-, Cl^-) or quasi-monatomic anions (OH$^-$, λ_{max} 187 nm, ε 5000) are known[6] to be due to anion–solvent charge-transfer transitions.

Recent work[7] exploring the near-u.v. magnetic circular dichroism (c.d.) behaviour of dialkyl sulphides[7] and trialkylsulphonium, selenonium, and telluronium salts[8] has provided indirect support for the location of an absorption maximum, due to transitions involving the chalcogen atom, near 200 nm for sulphonium salts. The first-reported optical rotatory dispersion (o.r.d.) and circular dichroism study of sulphonium salts[9] deals with the chirospectroscopic behaviour of dialkyl p-tolylsulphonium tetrafluoroborates through the wavelength range ca. 210–300 nm, and is therefore a study of the influence of the chiral sulphonium centre on the o.r.d. generated in the ^1L_a and ^1L_b bands of the p-tolyl chromophore. The reported spectra (isotropic $\lambda\lambda_{max}$ 230, 255, 266.5, 274.5 nm; $\varepsilon\varepsilon_{max}$ 34,000, 1720, 1900, 1800, respectively) include an intense negative c.d. maximum near 220 nm for (R)-(−) ethyl methyl p-tolylsulphonium tetrafluoroborate, which was assigned to the ^1L_a transition rather than to a sulphonium chromophore transition centred near 230 nm, located by Ohkubo and Yamabe[1].

III. TRIALKYLSULPHONIUM SALTS; THE SULPHONIUM SULPHUR ATOM AS A CHROMOPHORE

U.v. absorption data have been reported for trimethylsulphonium bromide, trimethylsulphonium iodide, trimethylsulphonium tetrafluoroborate, n-butyldimethylsulphonium tetrafluoroborate, and tri-n-butylsulphonium tetrafluoroborate[1], and therefore provide examples of the u.v. absorption behaviour of the simplest representative sulphonium salts. The presence of an absorption maximum near 200 nm ($\varepsilon \approx 8000$) is established in this work[1], together with a weaker absorption maximum at ca. 230 nm. The group of spectra shown in Figure 1 indicate the insensitivity of these features to variations in the structure of the alkyl substituent. However, additional examples given in the original paper include the spectrum of tri-n-butylsulphonium tetrafluoroborate in ethanol at $-40°C$, which shows absorption maxima at 237 and 250 nm (in addition to the 200 nm maximum) superimposed on a broad absorption shelf extending to wavelengths beyond 280 mm (Figure 2).

Earlier work[3] had established the existence of an absorption maximum in the wavelength range 208–210 nm for 1,3-bis(methyl-n-butylsulphonio)propane dichloride, n-BuMeSCH$_2$CH$_2$CH$_2$SMeBu-n 2Cl$^-$, and within the wavelength range 208–214 nm for the unsaturated analogue n-BuMeSCH$_2$CH=CHSMeBu-n 2Cl$^-$. No mention was made in that paper[3] of absorption features at longer wavelengths for these compounds.

An interesting conclusion from the earlier study[3], that these compounds show very substantial deviations from Beer's law, was not tested for the simpler aliphatic sulphonium salts studied by Ohkubo and Yamabe[1]. For the former bissulphonium salt, log ε_{max} values vary between 1.73 and 2.76 through the concentration range 126×10^{-4} to 0.9×10^{-4} M, and for the unsaturated analogue the figures are log ε_{max} 2.05 to 2.50 through the concentration range 182×10^{-4} to 7.3×10^{-4} M. Deviations on this scale from Beer's law usually imply aggregation of solute molecules, a conclusion reached by Rothstein and co-workers and open to scrutiny by other physical techniques. Ion-pair phenomena may account for these results.

Ohkubo and Yamabe's study[1] was undertaken in an attempt to clarify the

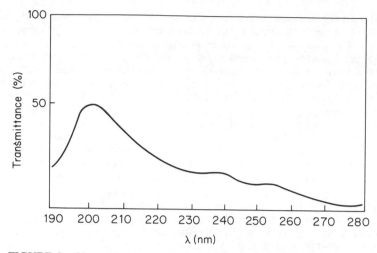

FIGURE 2. U.v. spectrum of tri-*n*-butylsulphonium tetrafluoroborate in ethanol (2.02×10^{-5}M) at $-40°$C. (Adapted from reference 1.)

catalytic role of simple sulphonium salts in the liquid-phase oxidation of hydrocarbons by atmospheric oxygen[10]. These workers stated that, by 1971, the u.v. absorption behaviour of sulphonium salts accompanying electronic excitation of the sulphonium sulphur atom had not been satisfactorily established. This was due to inadequate formulation at that time of the role of 3d orbitals on sulphur in the bonding patterns of ground and excited states of sulphonium salts. Details of the long-wavelength absorption maximum were tabulated by Ohkubo and Yamabe[1] under the heading [1]W (*sic*), but this term was neither defined nor mentioned elsewhere in the text, and there is no discussion of the electronic transition or transitions responsible for the absorption maximum. There is little evidence based on reported solvent and temperature effects from which deductions can be made about the origins of the long-wavelength absorption maximum. However, it is possible to interpret the data as excluding common transitions involving excitation of unshared electrons, since the lack of any influence of solvent polarity on λ_{max} values is not in keeping with familiar generalisations for $n \rightarrow \sigma^*$ and $n \rightarrow \pi^*$ transitions. Ohkubo and Yamabe[1] assigned the absorption maximum near 200 nm to a $3p \rightarrow 3d$ transition, or to the equivalent $(3p_z)^2 \rightarrow (3p_z)(3d)$ transition, and formulated the orbital overlap scheme shown in Figure 3 to depict the interaction between oxygen and a trialkylsulphonium ion. This appeared to be implied by a red-shifted 200-nm peak when O_2 was bubbled through the ethanolic solution of a trialkylsulphonium salt (Figure 4). Relatively unsophisticated AMSO–SCF calculations give general support to the assignment of the 200-nm peak to an electronic transition of the $3p \rightarrow 3d$ type, and the small shifts accompanying change of solvent polarity are also consistent with this assignment. The 200-nm absorption maximum (log $\varepsilon \approx 3.45$) seen in the u.v. spectra of dialkyl sulphides has also been assigned to an 'atomic-like' $3p \rightarrow 3d$ transition, based on molecular orbital calculations[2], and this provides independent support for the assignment made by Ohkubo and Yamabe for sulphonium salts. However, van Tilborg[11] has shown that the catalysis of the autoxidation of cumene by triphenylsulphonium salts does not involve the activation of the oxygen molecule by Ph_3S—O_2 complex formation, but

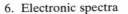

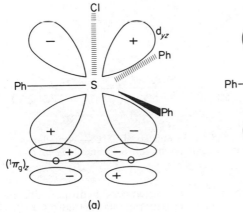

(a)

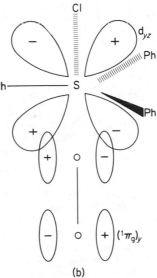

(b)

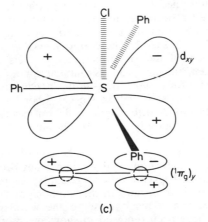

(c)

FIGURE 3. Plausible overlap schemes for $Ph_3\overset{+}{S}$ Cl^-–O_2 complex formation: (a) d_p–σ overlap between $(^1\pi_g)_z$ O_2 orbitals and d_{yz} sulphur orbitals; (b) D_p–σ overlap between $(^1\pi_g)_y$ O_2 orbitals and d_{yz} sulphur orbitals; (c) d_p–π overlap between $(^1\pi_g)_y$ O_2 orbitals and d_{xy} sulphur orbitals. (After reference 1.)

128   G. C. Barrett

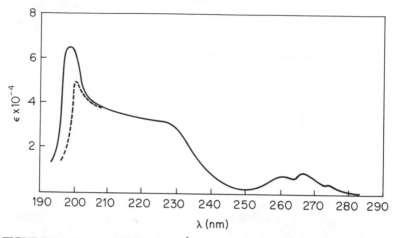

FIGURE 4. U.v. absorption of Ph₃S̈ Cl⁻ in ethanol at 25°C:———, under N$_2$; — — — —, under O$_2$. [Adapted from K. Ohkubo and H. Sakamoto, *Chem. Lett.*, 209 (1973).]

rather through the catalysis of the homolysis of the cumene hydroperoxide by the sulphonium salt. This criticism undermines the experimental support for Ohkubo and Yamabe's 3p → 3d assignment for the 200-nm transition of sulphonium salts, since sulphur orbitals were invoked as the site of overlap with oxygen ($^1\pi_g$) orbitals in complex formation. The 'red shift' found in the presence of oxygen is due to an artifact, an earlier solvent cut-off already observed by Mulliken[12] as a result of charge-transfer complex formation between oxygen and aliphatic alcohols. The 'red shift' does not occur in aqueous solutions, and only the short-wavelength absorption region of the spectrum is affected by oxygenation of the solutions[11]. Further, the 'red shift' was not accompanied by an increase in extinction coefficient (actually, the solvent cut-off caused a decrease in integrated intensity), and the conclusion that charge-transfer complexation does not occur between sulphonium ions and oxygen seems to be well established by these arguments.

Until further exploration of the electronic energy levels in simple sulphonium ions has been undertaken, the 3p → 3d assignment for an absorption maximum near 200 nm remains uncertain since transitions in composite orbitals involving adjacent C—H and C—C bonds, as well as the C—S bond and the S orbitals, may provide an equally satisfactory picture. Circular dichroism measurements can reveal 'hidden' transitions in electronically complex chromophores, and can help the interpretation of isotropic u.v./visible absorption spectra of compounds containing several chromophores[13]; an attempt to deduce the contribution of the sulphonium chromophore to the c.d. of the camphor derivative **2** has been made[8]. Subtraction of the c.d. of **2** from that of **1** gives a difference c.d. spectrum, suggesting that the perturbation of the electronic transitions in the sulphonium chromophore generates c.d. centred near 220–230 nm. This is close to the wavelength region in which simple sulphonium salts have been reported to show an absorption maximum[1] (see Figure 1), but the c.d. behaviour can be interpreted in another way. The ester chromophore in a chiral molecule generates c.d. reaching a maximum near its isotropic u.v. absorption maximum (*ca.* 210 nm) due to perturbation by the chiral environment, and in conformationally mobile systems (such as **1** and **2**) the sulphonium group is likely to influence the conformational relationship between the ester chromophore and the chiral camphor moiety so that it differs from the conformational relationship existing between these two parts of the molecule in the bromo compound **1**. This would result in different c.d. parameters for the compounds **1** and **2**, and therefore the observed c.d. difference spectrum can be accounted for without invoking a contribution from the sulphonium chromophore. However, the c.d. of chiral sulphonium salts of simpler structures, e.g. the adamantyl sulphonium salts employed in recent stereochemical studies[14], could provide useful confirmation of the results described above for the u.v. absorption if simple aliphatic sulphonium salts. Kelstrup *et al.*[15] have recently assigned the absolute configuration shown as **3** to (+)-isopropylmethylethylsulphonium

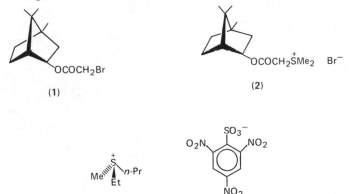

(1)   (2)

(3)

trinitrobenzenesulphonate, and ion-exchange conversion of this salt into an analogue with an anion which is transparent through the near-u.v. and much of the far-u.v. wavelength ranges will provide a suitable sulphonium salt for careful u.v. and c.d. spectroscopic studies (currently available c.d. spectrometers operate routinely through the u.v. region to wavelengths around 185 nm).

IV. ARYLSULPHONIUM SALTS AND ALKENYLSULPHONIUM SALTS: INFLUENCE OF THE SULPHONIUM GROUP ON THE U.V. ABSORPTION BEHAVIOUR OF AN ADJACENT CHROMOPHORE

Early work[16] revealed the substantial bathochromic effect of the sulphonium group on the absorption maxima of the benzene chromophore. These results and their interpretation have been reviewed by Price and Oae[17] and by Stirling[18]. Data on phenyl and substituted phenylsulphonium salts are given in Table 1.

The short-wavelength ^1L_a band, arising at 203.5 nm in benzene itself, appears at 220 nm in dimethylphenylsulphonium perchlorate[16]. Surprisingly, a quaternary ammonium substituent has no such bathochromic effect (PhNMe$_3$CL$^-$:λ_{max} 203 nm, log ε 3.87; λ_{max} 254 nm, log ε 2.31). The Me$_2$S substituent is very similar to a methanesulphonyl group in its auxochromic effect on the benzene chromophore. When other conjugative substituents are present on the benzene ring, further bathochromic shifts are seen which can be attributed to the ability of the sulphonium group to conjugate with the ring and with electron-releasing substituents in the excited state. These shifts are also seen in *m*-substituted

TABLE 1. Absorption features for arylsulphonium salts: λ_{max} and corresponding log ε_{max} (in parentheses)

Compound	Solvent and temperature (°C)[a]	Wavelength (nm)												
		190	200	210	220	230	240	250	260	270	280	290	300	310
Ph_3S^+ Cl^-	EtOH, 25°		204 (4.76)			227 (4.65)				269 (3.59)	276 (3.40)			
Ph_3S^+ Cl^-	EtOH, -40°		202.5 (4.71)			232 (4.27)				270 (3.66)	276 (3.58)			
Ph_3S^+ Cl^-	H_2O, 25°	194 (4.85)				232 (4.25)				268 (3.56)	276 (3.33)			
Ph_3S^+ Cl^-	H_2O, 0°		197.5 (4.85)			232.5 (4.29)				269 (3.56)	277 (3.56)			
Ph_3S^+ $\bar{B}F_4$	EtOH, 25°		204 (4.76)			232 (4.28)				268.5 (3.56)	276 (3.40)			
Ph_3S^+ $\bar{B}F_4$	EtOH, 4°			205 (5.15)		232.5 (4.68)				269.5 (3.55)	277 (3.88)			
Ph_3S^+ $\bar{B}F_4$	EtOH, -40°		201 (4.62)			233 (4.30)				268 (3.71)	276 (4.03)			
$(Ph_3\overset{+}{S})_3PO_4^{3-}$	EtOH, 4°		204 (4.68)			235 (4.00)				270 (3.80)	277.5 (3.74)			
$Ph\overset{+}{S}Me_2ClO_4^-$	H_2O				220 (3.94)				265 (3.04)					
Benzene	EtOH		203.5 (3.87)					254 (2.31)						
$p\text{-}NO_2C_6H_4\overset{+}{S}Me_2$ $MeSO_4^-$	EtOH							252 (4.28)						
Nitrobenzene	EtOH							252 (3.92)				285sh (3.26)		
$p\text{-}HOC_6H_4\overset{+}{S}Me_2$ $MeSO_4^-$	EtOH						242 (4.03)		264 (3.06)					
Phenol	EtOH			210.5 (3.79)						270 (3.26)				
$p\text{-}\bar{O}C_6H_4\overset{+}{S}Me_2$	EtOH									269 (4.29)				

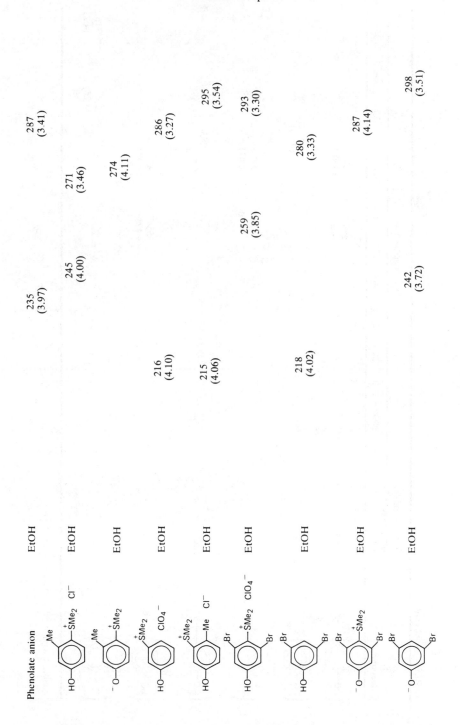

TABLE 1. *continued.*

Compound	Solvent and temperature (°C)[a]	Wavelength (nm)											
		190	200	210	220	230	240	250	260	270	280	290	300	310
	EtOH								258.5 (4.59)	277 (4.48)			
	EtOH										283 (4.56)		
	EtOH								260 (4.50)	277 (4.41)			
	EtOH											283--------300 (4.61)	
	EtOH								258 (4.47)	277 (4.40)			
	EtOH											295 (4.26)	316.5 (4.21)

[a]Ambient temperature unless stated otherwise.

phenylsulphonium salts, even though direct conjugation is not possible in this case. The relatively few examples available (Table 1) suggest that the effect appears to be smaller than in *o*- or *p*-substituted isomers. Oae and Price[19] have suggested that electron-accepting resonance interaction between the *m*-substituent and the Me_2S group in the excited state can be represented by structures such as **4–6**, to account for the observed bathochromic shifts.

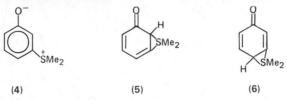

| (4) | (5) | (6) |

The remaining data in Table 1 bring out the point that the u.v. absorption behaviour of aromatic sulphonium salts carrying a bulky substituent *ortho* to the sulphonium group also reveals conjugative interaction. Steric inhibition of p–π resonance by bulky *ortho*-substituents has been observed in dialkylaminobenzenes and it may be that overlap involves the π-orbital of the benzene moiety and d-orbitals of the sulphonium sulphur atom, geometrical restrictions due to a bulky *ortho*-substituent being insignificant. Data for trialkylsulphonium salts (Table 1) show how little the *o*-methyl group affects the conjugative characteristics of the sulphonium sulphur atom in the photoexcited state, and the evidence derived from the u.v. spectra of *o*-substituted phenylsulphonium salts gives consistent support to 2p–3d overlap, an arrangement involving minimum directional requirements as the conjugative interaction.

U.v. spectra of aromatic sulphonium ylides have been reported by research groups headed by Yoshida *et al.*[20] and by Lloyd[21a,b]. Although stable sulphonium fluorenylides have been known for many years[22], the first theoretical study of the u.v. absorption of the simpler analogue, dimethylsulphonium cyclopentadienylide **7**, dates from 1971[23]. The spectrum of the latter compound (Figure 5) in *n*-hexane

(7)

was resolved into three gaussian peaks using a curve resolver; a substantial shift of the long-wavelength maximum to shorter wavelengths with increase in solvent polarity (to 268 nm in methanol) is found. This is consistent with that of other zwitterionic systems (e.g. mesoionic heterocycles[24]) and accounted for by the dependence of the distribution of different charge-separated forms on solvent polarity. The three gaussian peaks have been assigned[23] to an ylide ground state → ylene **7** transition (i.e. to an intramolecular electron-transfer band) for the longest wavelength band (*ca.* 292 nm), and to the E_2 (D_{5h}) and E_1 (D_{5h}) transitions of the cyclopentadienyl anion (*ca.* 248 and 198 nm, respectively). The analogous diphenylsulphonium tetraphenylcyclopentadienylide absorbs at λ_{max} 292 nm (ε 26,000), and a shoulder is seen at 328 nm (ε not given)[21a].

U.v. spectral data have been used[25] to determine which of two isomeric structures **8** or **9** is formed from the corresponding tertiary alcohol as a reaction intermediate. Carbonium ions show absorption features at considerably longer

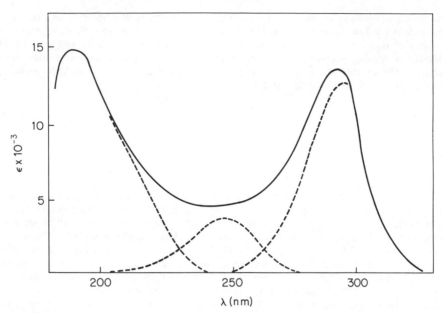

FIGURE 5. U.v. absorption of dimethylsulphoniumcyclopentadienylide (**7**) (——) in *n*hexane, resolved into three gaussian peaks (-----). (Adapted from reference 23.)

wavelengths than their hydrocarbon analogues (see later section) and the data $\lambda\lambda_{max}$ 260, 329, 387, and 448 nm, $\varepsilon\varepsilon_{max}$ 31,000, 7300, 8300, and 6400, respectively, have been taken to suggest the carbonium ion structure **8** rather than **9**.

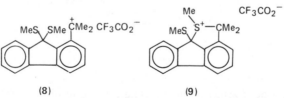

(8)   (9)

A similar use of u.v. spectral data in assigning the butatrienyl structure **10** implies an assumption that the sulphonium group exerts a bathochromic effect on the butatriene chromophore. Braun *et al*.[26] report that the sulphonium ion shows λ_{max} (EtOH) 258 nm, $\varepsilon \geqslant 10^4$, considered to compare better with reported data[27] for butatriene [$\lambda\lambda_{max}$ (MeCN) 228, 438 nm, $\varepsilon\varepsilon$ 1.25×10^4, 13.7, respectively] than with data for alternative butenyne chromophores. N.m.r. data give strong support to the assignment of structure **10**, and alternative isomeric forms were excluded. It can be concluded that the high-intensity absorption maximum of the butatriene chromophore is shifted about 30 nm to longer wavelengths by the attached sulphonium grouping. Cycloaddition of **10** to cyclopentadiene gives the allene **11**, showing λ_{max} 200, $\lambda_{shoulder}$ 250 nm, ε 3.7×10^4, 3×10^3, respectively[28]; the original paper states that the u.v. absorption of **11** is continuous between 200 and 285 nm, which presumably means that the spectrum shows some absorption between the

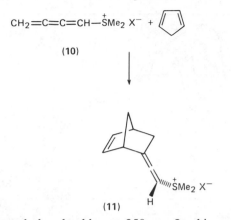

$$CH_2=C=C=CH-\overset{+}{S}Me_2 \ X^- \ +$$

(10)

(11)

maximum at 200 nm and the shoulder at 250 nm. In this case, the sulphonium substituent exerts a bathochromic effect on the absorption maximum due to the allene chromophore, since a simple representative allene, penta-1,2,3-triene ($EtCH=C=CH_2$) shows absorption maxima at 181 nm (log ε 4.3) (vapour), with a shoulder at 188 nm (log ε 3.5)[30], and a maximum at 225 nm (log ε 2.7) in n-hexane[29]. In this case too, the sulphonium substituent enhances the absorption intensity, but the isolated ethene chromophore in compound 11 also contributes to the far-u.v. absorption of this compound (1,2-dialkylethylenes, λ_{max} 160–200 nm log $\varepsilon \approx 4.0$)[6], and may also modify the absorption characteristics of the allene chromophore by through-space interaction which has been demonstrated for compound 13 by comparison of u.v. data with those of compound 12.

(12)   (13) *

$\lambda_{max.}$ (heptane) 182 (ε 6500) $\lambda_{max.}$ (ethanol)   205 (ε 2100)

214 (ε 1480)

220 (ε 870)

230 (ε 200)

V. ENVIRONMENTAL EFFECTS: INFLUENCE OF SOLVENT, TEMPERATURE, AND COUNTERION ON THE U.V. SPECTRA OF SULPHONIUM SALTS

The u.v. spectrum of triphenylsulphonium chloride is solvent dependent (Figure 6)[1].

Reference to Table 1 shows that the nature of the counter ion appears to have little effect on the position or intensity of the 200-nm absorption maximum of sulphonium salts, and this has been taken[1] to show that charge-transfer interaction

*C. F. Wilcox, S. Winstein and W. G. McMillan, *J. Amer. Chem. Soc.*, 82, 5450 (1960).

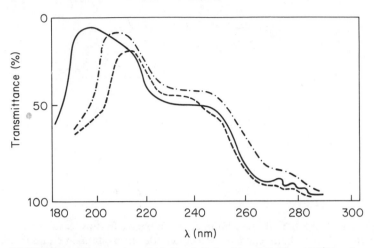

FIGURE 6.　U.v. spectra of triphenylsulphonium chloride at 25°C under nitrogen in water (——), ethanol (– · – · – · –·), and isopropanol (– – – –). (Adapted from reference 1.)

between the sulphonium cation and its accompanying anion is not involved. However, quoting from reference 6, p. 354: 'electronic transfer between the components of an ion pair, e.g. a halide ion and either a metallic or organic cation, gives an excited state less polar than the ground state, and large blue shifts of the charge-transfer absorption result on changing from a less to a more polar environment'. Therefore, interpretation of the solvent shifts seen in Figure 7 may require the consideration of charge-transfer complex formation, and this appears preferable to Ohkubo and Yamabe's statement that 'a polar solvent stabilises . . . through orientation of the solvent', which has no obvious connection with the explanation of the observed solvent shifts. The conclusion drawn by Ohkubo and Yamabe[1] is that, based on the assignment of the 200-nm absorption maximum to a 3p → 3d transition, solvent probably affects the energy of the 3d level but not that of the ground-state 3p level.

VI. U.V. SPECTRA OF SULPHONIUM SALTS: COMPARISON WITH U.V. SPECTRA OF ANALOGOUS ORGANIC CATIONS

Taking the u.v. spectrum of triphenylsulphonium chloride as a basis for comparison, the 200-nm absorption maximum, considered to be characteristic of the sulphonium sulphur atom[1], is also seen in the u.v. spectra of corresponding selenonium and telluronium cations[1]. The spectrum of tetraphenylphosphonium bromide is also reported to be very similar to that of triphenylsulphonium chloride[31], although it should be pointed out that the absorption maximum of the bromide ion at the longest wavelength is located at 199.5 nm (ε 11,000)[5]. More data are required so that the contribution of the organic cation to the u.v. absorption of tetraphenylphosphonium salts can be established, but there is a clear indication that the chalcogen and phosphorus onium salts show electronic transitions of similar energy involving the heteratom. Alkyl carbonium ions, previously thought to be characterized by an intense absorption maximum at 295 nm, in fact are transparent at wavelengths longer than 210 nm[32].

Hydroxycarbonium ions (protonated aliphatic ketones) have been reported to show absorption at shorter wavelengths than corresponding trialkylcarbonium ions[33], but this comparison needs confirmation in view of Olah *et al's*. later study[32], before heteroatom-substituted carbonium ions are compared with *S*-heteroatom-substituted sulphonium ions (next section). Allylic carbonium ions are well known to show substantial absorption at longer wavelengths than their saturated analogues (e.g. $Me_2C\cdots\overset{+}{C}H\cdots CMe_2$, λ_{max} 305 nm; dienylic cations, λ_{max} 397 nm)[33] and although electron delocalization in analogous unsaturated sulphonium ions must be less complete, there is an interesting parallel in their u.v. absorption behaviour, with the maximum shifted some 30 nm to longer wavelengths in comparison with analogous saturated sulphonium salts (see previous section).

Comparisons have been made between the u.v. spectra of sulphonium ylides and their phosphorus and arsenic analogues[34]. Here, the chromophore responsible for the longest wavelength absorption is the ylide $(R_2\overset{+}{S}—\overset{-}{C}=) \leftrightarrow$ ylene $(R_2S=C=)$ group, and no assignment of absorption maxima in the far-u.v. wavelength range to transitions involving only the heteroatom have been made. Triphenylphosphonium fluorenylide shows absorption maxima at 250, 258, and 284 nm (log $\varepsilon\varepsilon_{max}$ 4.6, 4.6, and 4.3, respectively), which are characteristic of the fluorene moiety, and a maximum at 382 nm (log ε 3.6) assigned to the ylide p–d$_\pi$ transition[35]. Dimethylsulphonium fluorenylide shows very similar u.v. absorption characteristics to those of analogous phosphorus and arsenic fluorenylides[34], although the long-wavelength maximum appears at 368 nm.

VII. U.V. SPECTRA OF *S*-HETEROATOM-SUBSTITUTED SULPHONIUM SALTS

Heterosulphonium salts include alkoxysulphonium salts, $RO\overset{+}{S}R_2\ X^-$, involved in the conversion of aryl alkyl sulphoxides into aryl dialkylsulphonium salts[36], and halogenosulphonium salts, e.g. $Cl\overset{+}{S}R_2\ Cl^-$, formed from a sulphide and a halogen[37]. Although no u.v. studies of these species have been reported, protonation of sulphoxides in strongly acidic media has been followed by u.v.[38] and c.d. methods[39]. An absorption band ascribed to the hydroxysulphonium salt formed by protonation appears near 210.nm. This is seen as a shoulder on a shorter wavelength absorption peak in isotropic absorption spectra[38] but is seen clearly in c.d. spectra [as a negative Cotton effect for e.g. (+)-butyl phenyl sulphoxide].

Alkylthiosulphonium salts, $RS\overset{+}{S}R_2\ X^-$, have recently gained importance as powerful alkylthiolating agents but have been created from disulphides or from sulphides in media which are not suitable for exploration of far-u.v. absorption characteristics[40]. No u.v. data are included in reported preparations of protonated disulphides (e.g. $Me\overset{+}{S}HSMe\ X^-$) or bis(alkylthio)sulphonium salts, $RSSRSR\ X^{41}$.

VIII. MISCELLANEOUS U.V. STUDIES INVOLVING SULPHONIUM SALTS

Laird and Williams[42] refer to an increase in u.v. absorbance during the u.v. irradiation of α-bromophenacyldimethylsulphonium bromides, but the experimental section of their paper includes no details of the u.v. characteristics of starting materials or products:

$$PhCOCHBr\overset{+}{S}Me_2\ Br^- \xrightarrow[H_2O]{h\nu} PhCOCH_2\overset{+}{S}Me_2\ Br^- + Me_2S$$

The reaction mixture is the result of more complex consecutive reactions, since prolonged photolysis leads to a slow decrease in absorbance (over a period of 6 h), probably due to partial bromination of the phenyl group.

IX. ACKNOWLEDGEMENT

The author thanks Professor P. H. Laur, Aachen, for providing unpublished data.

X. REFERENCES

1. K. Ohkubo and T. Yamabe, J. Org. Chem., 36, 3149 (1971).
2. J. S. Rosenfield and A. Moscowitz, J. Amer. Chem. Soc., 94, 4797 (1972); see also P. Salvadori, J. Chem. Soc. Chem. Commun., 1203 (1968).
3. D. C. Nicholson, E. Rothstein, R. W. Saville and R. Whiteley, J. Chem. Soc., 4019 (1953).
4. G. C. Barrett in Comprehensive Organic Chemistry (Eds. D. H. R. Barton and W. D. Ollis), Vol. 3, Pergamon Press, Oxford, 1979, p. 105.
5. T. Schiebe, Z. Phys. Chem., B, 5, 355 (1929).
6. S. F. Mason, Quart. Rev., 15, 287 (1961).
7. P. H. Laur and co-workers, Poster Communication at NATO Study Institute, Optical Activity and Chiral Discrimination, University of Sussex, 10–22 September, 1978.
8. P. H. Laur, personal communication.
9. K. K. Andersen, R. L. Caret and D. L. Ladd, J. Org. Chem., 41, 3096 (1976).
10. K. Ohkubo, Tetrahedron Lett., 2897 (1971), and earlier papers by K. Ohkubo and co-workers cited therein.
11. W. J. M. van Tilborg, Tetrahedron, 31, 2841 (1975).
12. R. S. Mulliken, J. Amer. Chem. Soc., 74, 811 (1952).
13. G. C. Barrett, in Elucidation of Organic Structures by Physical and Chemical Methods, Vol. 4, Part 1 of the series Technique of Chemistry (Ed. A. Weissberger; Volume Editors K. W. Bentley and G. W. Kirby), Interscience, New York, 1972, p. 515.
14. B. M. Trost and R. F. Hammen, J. Amer. Chem. Soc., 95, 962 (1973).
15. E. Kelstrup, A. Kjaer, S. Abrahamsson and B. Dahlen, J. Chem. Soc. Chem. Commun., 629 (1975).
16. F. G. Bordwell and P. J. Boutan, J. Amer. Chem. Soc., 78, 87 (1956).
17. C. C. Price and S. Oae, Sulfur Bonding, Ronald Press, New York, 1962, p. 162.
18. C. J. M. Stirling in Organic Chemistry of Sulfur (Ed. S. Oae), Plenum Press, New York and London, 1977, p. 485.
19. S. Oae and C. C. Price, J. Amer. Chem. Soc., 80, 4938 (1958), and references cited therein.
20. Z. Yoshida, S. Yoneda and M. Hazama, J. Chem. Soc. Chem. Commun., 716 (1971); Z. Yoshida, K. Iwata and S. Yoneda, Tetrahedron Lett., 1519 (1971).
21. (a) D. Lloyd and M. I. C. Singer, Chem. Ind. (London), 118 (1967).
    (b) E. E. Ernstbrunner and D. Lloyd, Chem. Ind. (London), 1332 (1971).
22. C. K. Ingold and J. A. Jessop, J. Chem. Soc., 713 (1930); E. D. Hughes and K. I. Kuriyan, J. Chem. Soc., 1609 (1935).
23. K. Iwata, S. Yoneda and Z. Yoshida, J. Amer. Chem. Soc., 93, 6745 (1971).
24. G. C. Barrett and A. R. Khokhar, J. Chem. Soc. C, 1117 (1969).
25. M. Hojo, T. Ichi, T. Nakanishi and N. Takaba, Tetrahedron Lett., 2159 (1977).
26. H. Braun, G. Strobl and H. Gotzler, Angew. Chem. Int. Ed. Engl., 13, 469 (1974).
27. W. M. Schubert, T. H. Liddicoet and W. A. Lanka, J. Amer. Chem. Soc., 76, 1929 (1954).
28. H. Braun and G. Strobl, Angew. Chem. Int. Ed. Engl., 13, 470 (1974).
29. L. C. Jones and L. W. Taylor, Anal. Chem., 27, 228 (1955).
30. G. D. Burr and E. S. Miller, Chem. Rev., 29, 419 (1941).
31. K. Ohkubo and H. Sakamoto, Chem. Lett., 209 (1973).

32. G. Olah, C. U. Pitman, R. Waack and M. Doran, *J. Amer. Chem. Soc.*, **88**, 1488 (1966).
33. N. C. Deno, J. Bollinger, N. Freedman, K. Hafer, J. D. Hodge and J. J. Houser, *J. Amer. Chem. Soc.*, **85**, 2998 (1963).
34. A. W. Johnson and R. B. La Count, *J. Amer. Chem. Soc.*, **83**, 417 (1961).
35. A. W. Johnson, *Ylid Chemistry*, Academic Press, New York and London, 1966; see also B. M. Trost and L. S. Melvin, *Sulfur Ylides*, Academic Press, New York and London, 1975.
36. K. K. Andersen, *J. Chem. Soc. Chem. Commun.*, 1051 (1971).
37. L. A. Paquette, R. E. Wingard, J. C. Philips, G. L. Thompson, L. K. Read and J. Clardy, *J. Amer. Chem. Soc.*, **93**, 4508 (1971).
38. D. Landini, G. Modena, G. Scorrano and F. Taddei, *J. Amer. Chem. Soc.*, **91**, 6703 (1969).
39. U. Quintily and G. Scorrano, *J. Chem. Soc. Chem. Commun.*, 260 (1971).
40. G. Capozzi, O. De Lucchi, V. Lucchini and G. Modena, *Synthesis*, 677 (1976), and references cited therein.
41. G. Capozzi, V. Lucchini, G. Modena and F. Rivetti, *J. Chem. Soc. Perkin Trans. II*, 900 (1975).
42. T. Laird and H. Williams, *J. Chem. Soc. C*, 3467 (1971).

The Chemistry of the Sulphonium Group
Edited by C. J. M. Stirling and S. Patai
© 1981 John Wiley & Sons Ltd.

CHAPTER **7**

Electrochemistry of the sulphonium group

J. GRIMSHAW

Department of Chemistry, David Keir Building, Queen's University of Belfast, Belfast BT9 5AG, N. Ireland

I.	REDUCTION OF SULPHONIUM SALTS	145
II.	FORMATION OF RADICAL CATIONS AND SULPHONIUM SALTS BY	
	OXIDATION	145
	A. Sulphides	145
	B. Six-ring Sulphur Heterocycles	147
	C. Thianthene	150
	D. Phenothiazine	151
	E. Phenoxathin	153
III.	REFERENCE ELECTRODE POTENTIALS	154
IV.	REFERENCES	154

I. REDUCTION OF SULPHONIUM SALTS

Sulphonium salts undergo electrochemical reduction. The resulting polarographic waves at a dropping mercury electrode in aqueous solution have been extensively studied and a list of half-wave potentials are given in Table 1. These waves are irreversible and the half-wave potentials are independent of pH over the range where the salts themselves exhibit no acid–base behaviour. Sulphonium salts with other electroactive groups may show further waves at more negative potentials. Polarographic waves are also to be expected from sulphonium salts in aprotic solvents but there has been no systematic investigation under these conditions.

The half-wave potential for the trimethylsulphonium ion occurs at relatively negative potentials and lies close to the limiting potential due to decomposition of the solvent and supporting electrolyte[1]. Attachment of alkyl groups bearing electronegative substituents to the sulphur atom causes the half-wave potential to assume less negative values[2]. The rate-determining step in the reaction which gives

141

TABLE 1. Polarographic half-wave potentials of sulphonium ions in aqueous solution, pH 7.4. Except where stated, these values are independent of pH

Sulphonium ion	$-E_{1/2}$ (V vs. S.C.E.)	Ref.	Sulphonium ion	$-E_{1/2}$ (V vs. S.C.E.)	Ref.
$Me_3\overset{+}{S}$	1.85	1	$Et_2\overset{+}{S}CH_2Ph$	1.30	2
$Et_2\overset{+}{S}CH_2CH_2Cl$	1.60	2	$(PhCH_2)_3\overset{+}{S}$	0.65	5
$Et_2\overset{+}{S}CH_2CH{=}CH_2$	1.46	2	$Et_2\overset{+}{S}CH_2{-}{-}CH_2\overset{+}{S}Et_2$	1.04	2
$Et\overset{+}{S}CH{=}CH_2$	1.69	2	$PhCOCH_2\overset{+}{S}Et_2$	0.63^a	6
$Et\overset{+}{S}(CH{=}CH_2)_2$	1.51	2	$PhCOCH_2\overset{+}{S}MePh$	0.41^b	6
$Me_2\overset{+}{S}CH{=}C{=}CH_2$	1.15	3	$Me_2\overset{+}{S}CH_2COCO_2Et$	0.31^c	7
$Me_2\overset{+}{S}{-}{-}Me$	1.55	1			
$Ph_3\overset{+}{S}$	1.09	4			

^{a}Independent of pH over the range 2.0–7.1.
^{b}Independent of pH over the range 1.2–6.4.
^{c}Independent of pH below 4.2.

rise to these polarographic waves is the addition of an electron to the sulphonium salt which formally gives rise to a sulphur radical. It is likely that electron addition and carbon–sulphur bond cleavage are concerted processes and give rise to an alkyl radical which is reduced and protonated in a second more facile step. Vinyl and

$$\overset{+}{Me_3S} + e \longrightarrow Me_2S + CH_3^{\cdot} \xrightarrow[+ H^+]{+ e} CH_4$$

phenyl substituents on the sulphur atom lower the half-wave potential for the first reduction step because in the transition state some of the additional electron density is located on the sp^2 hybridized carbon atom and the increased s character from sp^3 to sp^2 hybridized carbon makes the electronegativity of the carbon nucleus more apparent and lowers the half-wave potential[1,2]. Substituents such as benzyl[5] and allyl[2] are particularly effective in lowering the half-wave potential of sulphonium ions because the radical character which results from carbon sulphur bond cleavage can be delocalized by the substituent. Reductive cleavage of benzyldialkylsulphonium salts gives exclusively dialkyl sulphide and benzyl radicals with products derived by further reactions of the latter.

Pyruvyl-[7] and phenacylsulphonium salts[6,8] such as **1** undergo electrochemical carbon–sulphur bond cleavage at relatively low half-wave potentials. The overall reaction for phenacyl derivatives is

$$PhCOCH_2\overset{+}{S}Et_2 + 2e + H^+ \longrightarrow PhCOCH_3 + Et_2S$$

(1)

and pyruvyl derivatives follow a similar pathway. The half-wave potentials for these processes are independent of pH over a wide range. At the acidic end of the range, protonation of the carbonyl group becomes important by giving rise to a species which is more easily reduced than the parent so that the half-wave potential falls with decreasing pH in this region. In strongly basic solution, the equilibrium between the substrate and the ylide becomes important. The ylide is not reducible. This equilibrium lowers the concentration of the parent sulphonium ion at the electrode surface and the electrode potential must be made more negative in order to maintain a given rate of electron transfer. In this alkaline region, the half-wave potential rises with increasing pH.

These cleavage reactions are related to the electrochemical cleavage of the carbon–nitrogen bond in phenacyltrialkylammonium compounds. In both reactions the carbonyl π system is the probable first site of electron addition and bond cleavage and electron addition are thought to be concerted.

Preparative-scale reduction of trialkylsulphonium salts in aqueous solution in the presence of an inert conducting salt leads to cleavage of one alkyl–sulphur bond. There has been no systematic investigation of the relative ease of removal of alkyl groups. However, since reduction[9] of the salt **2** derived from methionine gives

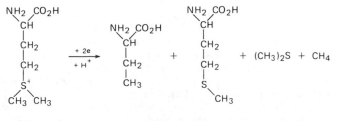

(2)

2-aminobutyric acid in 76% yield, the other product being methionine in 11% yield, it can be concluded that the more substituted alkyl group will be cleaved preferentially. This suggests that the first reaction step involves addition of one electron to the sulphonium ion and then cleavage to give an alkyl radical which is subsequently reduced to the carbanion and protonated rather than addition of two electrons and cleavage to a carbanion directly.

Reduction of simple trialkylsulphonium salts at a mercury cathode in an aprotic solvent such as dimethylformamide or acetonitrile gives rise to trialkylsulphonium amalgams which are stable at temperatures below 0°C[10]. Decomposition of the amalgam occurs at higher temperatures and basic products are formed. If the reduction of trimethylsulphonium salts is conducted in dimethyl sulphoxide at room temperature with a carbon cathode, part of the unreduced sulphonium ion is converted into its ylide by reaction with these basic materials formed by reductive cleavage of the sulphonium ion. Ylides formed in this manner undergo their characteristic reaction with carbonyl compounds to form oxiranes[11,12]. Electrolysis of a solution of a trimethylsulphonium salt and benzaldehyde gave the oxirane 3 in 28% yield[11].

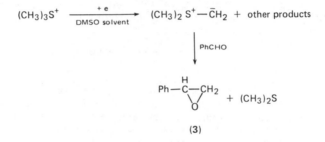

(3)

The ylides 4 and 5 have been prepared[11,13] by this electrochemical process which uses the decomposition of part of the sulphonium salt to prepare the necessary base. Reduction of the sulphonium ion 6 in dimethyl sulphoxide generates a base which causes decomposition of the ion to dimethyl sulphide and acrylonitrile[13].

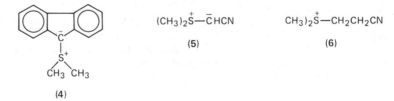

(4)

The reduction products from triphenyl- and tribenzylsulphonium ions depend on the electrode material. Both of these ions, and also their reduction products, are strongly adsorbed at a mercury cathode from aqueous solution. The initial reduction step at a mercury cathode leads to radical products, but since these are generated from species adsorbed at the mercury surface, the radicals attack mercury and the isolated products are diphenylmercury[4,14] or dibenzylmercury[5] together with diphenyl sulphide or dibenzyl sulphide. Reduction of triphenylsulphonium salts in aqueous solution at an aluminium cathode leads to diphenyl sulphide and benzene, the latter in poor yield. The yield of benzene is much improved in the presence of dimethylformamide as a co-solvent, so it appears likely that this product results from the further reaction of the first-formed phenyl radicals by abstraction of a hydrogen atom from any available source[14]. Dimethylformamide serves as a good hydrogen donor towards phenyl radicals.

4-Nitrobenzyldimethylsulphonium salts 7 give good yields of the corresponding 4,4′-dinitrobibenzyl on reduction in aqueous solution at cathodes of mercury,

$$O_2N\text{—}\underset{}{\bigcirc}\text{—}CH_2\overset{+}{S}(CH_3)_2 \xrightarrow[\text{H}_2\text{O solvent}]{+\ e} O_2N\text{—}\underset{}{\bigcirc}\text{—}CH_2CH_2\text{—}\underset{}{\bigcirc}\text{—}NO_2 + (CH_3)_2S$$

(7)

copper, or platinum[15]. This reaction is general for substituted 4-nitrobenzyl salts but 3-nitrobenzyl salts do not undergo the electrochemical carbon–sulphur bond cleavage. Probably in the latter case electrochemical reduction of the nitro group is the faster reaction in aqueous solution. There has been no comparison of the adsorption of benzyl- and 4-nitrobenzylsulphonium salts at the mercury–water interface, which may indicate the reason why these salts give different classes of product on reduction. The disulphonium salts 8 derived from p-xylene are converted into a hydrocarbon polymer on reduction in aqueous solution at cathodes of mercury, aluminium, tin, or lead[16].

$$(CH_3)_2\overset{+}{S}CH_2\text{—}\underset{}{\bigcirc}\text{—}CH_2\overset{+}{S}(CH_3)_2 \xrightarrow[\text{H}_2\text{O solvent}]{+\ e} \left[\text{—}CH_2\text{—}\underset{}{\bigcirc}\text{—}CH_2\text{—}\right]_n + (CH_3)_2S$$

(8)

There is considerable evidence detailed above to indicate that addition of the first electron to a sulphonium ion causes carbon–sulphur bond cleavage to yield the most stable carbon radical. When the electrolysis solvent is acrylonitrile, this can trap the initially formed radical intermediates and lead to useful products[17]. Under these conditions, cyanomethyldimethylsulphonium salts give glutaronitrile in 20% yield and benzyldimethylsulphonium salts give γ-phenylbutyronitrile 9 in 29%

$$PhCH_2\overset{+}{S}(CH_3)_2 \xrightarrow[\text{CH}_2=\text{CHCN}]{+\ e} PhCH_2CH_2CH_2CN + (CH_3)_2S$$

(9)

yield. In a related reaction, reduction of dimethyl(3-cyanopropyl)sulphonium salts in dimethyl sulphoxide containing styrene at a mercury cathode leads to di(3-cyanopropyl)mercury and 6-phenylhexanonitrile by reactions of the first-formed 3-cyanopropyl radical[16].

II. FORMATION OF RADICAL CATIONS AND SULPHONIUM SALTS BY OXIDATION

A. Sulphides

The oxidation of sulphides at a platinum anode in aprotic solvents (acetonitrile is usually employed) leads to cation radicals. Cation radicals derived from simple dialkyl, diaryl, or alkyl aryl sulphides are not stable in solution and so reversible electron transfer between these species and the substrate cannot be demonstrated. Their presence is inferred on the basis of the reaction products which have been isolated. Half-wave potentials for the oxidation of some sulphides are given in Table 2.

The cation radicals from dialkyl sulphide have a positive charge and radical centre on the sulphur atom. An aryl substituent on the sulphur atom can assist the

TABLE 2. Half-wave potentials for the electrochemical oxidation of sulphides. Products formed by oxidation at the potential of this first wave give rise to further waves at more positive potentials

Compound	Acetonitrile		Water	
	$E_{1/2}$ (V vs. S.C.E.)	Ref.	$E_{1/2}$ (V vs. S.C.E.)	Ref.
Ph_2S	1.56	18	1.24	22
	1.45	19		
Dibenzothiophene	1.20	20	1.39	23
$PhSCH_3$	1.47	19		
α-$C_{10}H_7SCH_3$	1.32	18		
$(CH_3)_2S$	1.70	21		

oxidation process by delocalizing the radical cation and in these cases the electrochemical half-wave potential can be correlated with the energy of the highest filled molecular orbital calculated by the Hückel method[18,21].

Electrochemical oxidation of dimethyl sulphide and further reactions of the radical cation formed leads to a mixture of products, including methanesulphonic acid and dimethyl sulphone[21]. Decomposition of the radical cation is thought to occur by loss of a proton and further oxidation to give a species which alkylates unreacted dimethyl sulphide:

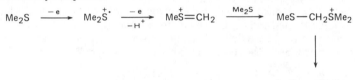

When the dialkyl sulphide to be oxidized possesses a nucleophilic centre in a suitable stereochemical situation, then reaction between this centre and the sulphur radical cation may be sufficiently rapid to lead to reaction products. Methionine is oxidized at a platinum anode in aqueous solution to the dehydro product **10**[24].

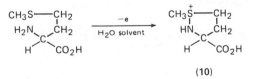

(10)

The radical cation formed by oxidation of diphenyl sulphide in acetonitrile behaves as an electrophile. In the absence of any other reagent, the radical cation attacks unoxidized diphenyl sulphide and the triphenylsulphonium salt **11** can be obtained in good yield[19,25]. Dibenzothiophene shows an analogous reaction on electrochemical oxidation[26]. Added phenol will undergo electrophilic substitution

$$Ph_2S \xrightarrow{-e} Ph_2\overset{+}{S}\cdot \xrightarrow[-e,-H^+]{Ph_2S} Ph_2\overset{+}{S}-\!\!\!\bigcirc\!\!\!-SPh$$

(11)

by the radical cation from diphenyl sulphide[22]. The radical cation reacts with water to yield diphenyl sulphoxide[19]. Oxidation at a platinum anode in aqueous acetic acid is a useful electrochemical process for the conversion of diphenyl sulphides[22] and including dibenzothiophene[23] and thianthrene[27] into the corresponding sulphoxide.

Electrochemical oxidation of methyl phenyl sulphide in acetonitrile gives the cation radical which reacts immediately with traces of water to give the sulphoxide[25]. When the alkyl group can form a stable cation, such as benzyl or *t*-butyl, then oxidation of the alkyl phenyl sulphide results in cleavage of the alkyl sulphur bond:

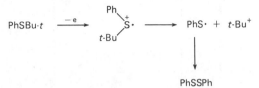

Tetrathioethylenes (12) are able to delocalize the radical cation centre formed by electrochemical oxidation. This group of compounds shows two reversible

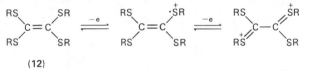

(12)

one-electron oxidation steps at a platinum anode in acetonitrile[28]. The half-wave potentials for some of these processes are given in Table 3. The intermediate radical cations have been characterized by e.s.r. spectroscopy[28–30].

B. Six-ring Sulphur Heterocycles

Sulphur heterocycles of the type 13 (X = S, O, NH, and NCH$_3$) are oxidized at a platinum anode to the corresponding radical anion and these species are stable in acetonitrile. The interconversion of 13 to 14 is a reversible redox equilibrium. Further oxidation of the radical cation gives the species 15 and in suitable solvents this is also a reversible process. Some half wave potentials are given in Table 4.

The oxidation products 14 possess an unpaired electron and thus give rise to e.s.r. spectra. These spectra indicate that the unpaired electron is extensively delocalized over the three rings so that structures such as 16 where the radical is located on a heteroatom are not an adequate representation of the radical cations.

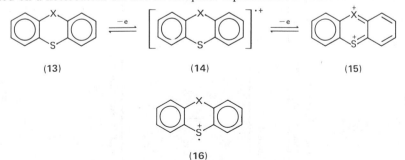

(13)   (14)   (15)

(16)

TABLE 3. Half-wave potentials for the electrochemical oxidation of symmetrical tetrathioethylenes in acetonitrile. Both oxidation steps are reversible unless otherwise stated

Compound	E_1 (V vs. S.C.E.)	E_2 (V vs. S.C.E.)	Ref.
MeS–, MeS– / –SMe, –SMe	0.89	—	29
(bis-cyclopentenedithiole tetrathiafulvalene)	0.68	1.12	30
(bis-dithiane ethylene)	0.70	0.84	30
(bis-tetrahydrobenzodithiole tetrathiafulvalene)	0.33	0.70	28
(dibenzotetrathiafulvalene)	0.50	0.94	28
PhS–, PhS– / –SPh, –SPh	1.18	1.4^a	29

^{a}Irreversible oxidation step.

TABLE 4. Half-wave potentials for the electrochemical oxidation of sulphur heterocycles in acetonitrile. Unless otherwise stated the oxidation steps are reversible

Compound	E_1 (V vs. S.C.E.)	E_2 (V vs. S.C.E.)	Ref.
Thianthrene	1.23	1.73	31, 32
Phenothiazine	0.64	1.1[a]	33
Phenoxathiin	1.19	1.69[b]	32

[a]Reversible but pH dependent.
[b]Irreversible.

Hückel molecular orbital calculations, which make the tacit assumption that the heterocyclic ring is planar, have been carried out in order to predict the e.s.r. spectra of these radical cations. Fair agreement is found between the predicted and the observed e.s.r. spectra of the radical cations from thioanthracene[34], phenoxathiin[34], and phenothiazine[35]. A linear combination of the p-orbitals on C, S, and X atoms of the structure 14 formed the basis for calculation and it was not necessary to employ sulphur d-orbitals.

Phenothiazine forms a crystalline molecular complex with cis-bis(trifluoromethyl-ethylene-1,2-dithiolate)nickel, which in the ground state can be shown from spectroscopic evidence to be composed of a pair of radical ions. Thus, in this material, for which an X-ray crystal structure has been obtained[36], phenothiazine is present as the radical cation. The heterocyclic ring of the phenothiazine radical cation 14 (X = NH) deviates from planarity so that the dihedral angle between the two benzene ring planes is 172°. Many years previously, Malrieu and Pullman[37] predicted on theoretical grounds that the phenazine radical cation would be non-planar.

This deviation from planarity of the phenothiazine radical cation is relatively small and it is accommodated in the molecular orbital calculations by optimizing the resonance integral between the hetero- and adjacent carbon atoms. The e.s.r. spectra of radical cations derived from a series of N-phenylphenothiazines have been studied[35]. Where the phenyl group bears substituents with no strong mesomeric effect, the radical cations are heterocycle based and the plane of the phenyl substituent is twisted at an angle of 65° with respect to the mean plane of the heterocycle. With a substituent such as p-dimethylamino, which exerts a strong mesomeric effect, the radical is based on this p-dimethylaminophenyl substituent and the heteroring can no longer be regarded as planar for the purposes of molecular orbital calculations.

When solutions of one of the radical cations (14, X = S or NH) and a radical anion derived from another aromatic system are mixed, intermolecular electron transfer occurs to annihilate the radicals and the energy which is released causes one of the neutral products to be generated as an excited state. Luminescence occurs as this excited state falls to the ground state. Such chemiluminescence can be electrogenerated at a platinum electrode placed in a solution of thianthrene 13, (X = S) and benzophenone in acetonitrile[38]. The electrode is pulsed successively at a positive potential to generate thianthrene radical cation and at a negative potential to generate benzophenone radical anion. These species mix in the diffusion layer around the electrode and generate the chemiluminescence, which here is emitted from triplet benzophenone. Chemiluminescence has been generated from thianthrene radical cation and the radical anions from 2,5-diphenyl-1,3-oxadiazole[39] and 9,10-diphenylanthracene[40]. Other systems which give rise to chemiluminescence are based on the radical cation from 10-methylphenazine **14** (X = NMe) and the radical anion from fluoranthene[41] or tetraphenylporphine[42].

C. Thianthrene

Thianthrene **13** (X = S) is converted into its radical cation and dication by electrochemical oxidation in a variety of solvents. Both oxidation steps are reversible and both oxidation products are stable in acetonitrile which is kept in contact with alumina to ensure it remains anhydrous[31]. The radical cation is more stable in trifluoroacetic acid than non-rigorously dried acetonitrile and the perchlorate is obtained by oxidation of thianthrene in trifluoroacetic acid containing perchloric acid followed by evaporation of the solvent[43]. Chemical oxidation processes have been used to prepare salts of thianthrene radical cation with other anions, viz. $Cl_3I_2^-$[44], ClO_4^-[44–46], SbF_6^-[45,47], and BF_4^-[46]. Thianthrene radical cation perchlorate undergoes reversible association equilibria in solution[48]. 2,3,7,8-Tetramethoxythianthrene is oxidized at lower potentials and crystalline salts of both the radical cation[49,50] and the dication[50] have been obtained. Spectroscopic evidence indicates tetramethoxythianthrene dication to be a triplet in the ground state[51].

Thianthrene radical cation shows electrophilic behaviour on the sulphur atoms. Thus, when solutions of the radical cation are diluted with water, equimolar amounts of thianthrene and its monoxide **17** are formed[52]. When the

(17)

electrochemical oxidation of thianthrene is carried out in aqueous acetic acid, the monoxide can be obtained in quantitative yield[53].

Electrophilic substitution reactions occur between the radical cation and activated benzene compounds in aprotic solvents. Anisole[54] gives the product **18** with R = Me and phenol[55] gives **18** with R = H. Reaction with ketones (MeCOR; R = Me, Ph, M3₃C) gives the corresponding oxoalkylsulphonium salts **19**[56]. Primary amines (RNH₂) give the sulphimine derivatives **20**[57].

Shine and Murata[52,58] have suggested that these reactions of the thianthrene radical cation are preceded by disproportionation to the dication and thianthrene

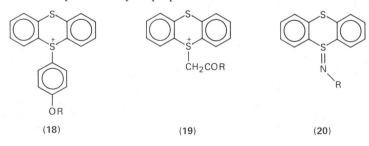

(18)   (19)   (20)

and that the dication undergoes the electrophilic reactions. Demonstration[31,59] of reversibility of the two one-electron redox reactions in suitable solvents allowed the standard redox potentials for these two processes to be measured and so the equilibrium constant for the disproportionation reaction could be calculated. It was concluded that this equilibrium constant is so small that the disproportionation mechanism is not possible for any but the fastest of these electrophilic reactions of the radical cation.

The kinetics of the reaction between thianthrene radical cation and anisole[54] or phenol[55] have been examined by mixing a solution of the preformed radical cation with the aromatic substrate and following the fall in radical cation concentration by an electrochemical technique. At concentrations greater than 10^{-5} M the following kinetic scheme was proposed for the reaction with anisole in acetonitrile:

$$Th^{\cdot +} + C_6H_5OMe \rightleftharpoons [Th\,;\,C_6H_5OMe]^{\cdot +}$$
$$\text{Complex}$$

$$[Complex]^{\cdot +} + Th^{\cdot +} \rightleftharpoons [Complex]^{2+} + Th$$

$$[Complex]^{2+} \longrightarrow 18\ (R = Me) + H^+$$

An identical reaction scheme was proposed for the reaction between the radical cation and phenol in trifluoroacetic acid. However, in acetonitrile the initially formed complex was thought to loose a proton to give the radical **21**, which is oxidized to the product by a second molecule of radical cation.

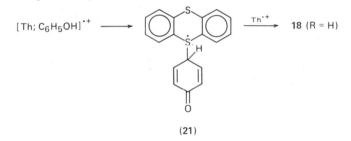

(21)

D. Phenothiazine

Phenothiazine and its derivatives have long been known to undergo oxidation in single electron steps. Salts of phenothiazine radical cation have been described in the early literature[60] but the work here is confused because the interplay of both redox and proton equilibria was not appreciated. Michaelis and Granick[61] investigated the oxidation of phenothiazine derivatives by means of potentiometric titration and demonstrated the importance of these two equilibria in explaining the reactions of these compounds. More recent investigations have used oxidation at a rotating platinum anode or cyclic voltammetry as electrochemical techniques[33,62].

Oxidation of phenothiazine **22** in acetonitrile, preferably containing a small amount of perchloric acid, gives the radical cation **23** in a reversible electrochemical process[33]. The radical cation is stable in solution and the e.s.r. spectrum indicates that the NH proton is present. In acetonitrile containing an acetate buffer this oxidation step gives rise directly to the neutral radical **24** and the half-wave potential for this process differs from that for the process **22** → **23**[62]. This neutral

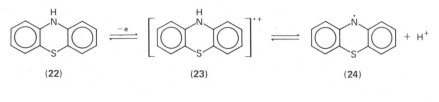

$$(22) \qquad\qquad (23) \qquad\qquad (24)$$

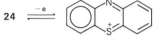

$$(25)$$

radical has only a short lifetime in solution. Both the radical cation and the neutral radical from 3,7-di-*t*-butylphenothiazine have been characterized by e.s.r. spectroscopy. The neutral radical from dibenzol[*c*,*h*]phenothiazine **26** can be obtained as a stable crystalline solid[62]. Dimerization is the principle reaction of the neutral radical from phenothiazine to form 4% of **27** and 45% of **28**. When this

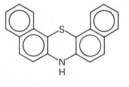

$$(26)$$

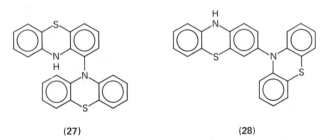

$$(27) \qquad\qquad\qquad (28)$$

radical is generated by oxidation of phenothiazine in buffered solution, the primary coupled product **28** undergoes further oxidation to a green quinonoid cation[62,63]. Structures for **27** and **28** were proposed from spectroscopic evidence[64] and the dimer **28** has been synthesized by a rational procedure[63].

Phenothiazine shows a second oxidation wave at more positive potentials. This wave is reversible in acetonitrile containing an acetate buffer and at this pH the oxidation process is conversion of the neutral radical **24** into the monocation **25**. This monocation is not protonated in acidic media so that electrochemical oxidation of the radical cation **23** gives **25** in a process whose half-wave potential is dependent on the acidity of the solution. The monocation **25** has been characterized as the perchlorate[33].

The related reversible and pH-dependent redox behaviour of phenothiazine derivatives has been used by Michaelis and Granick[61] to determine the acidity of strongly acidic solutions in an alternative procedure to that of Hammett[65]. The latter uses dyes as acid–base indicators. Within certain limits of pH, 3-hydroxyphenothiazine undergoes the following two redox reactions which obey the Nernst equation:

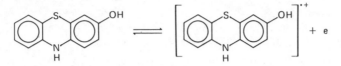

for which $E = E_1^0 + RT/F \ln [\text{Ox}]/[\text{Red}]$, and

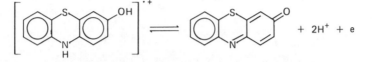

for which $E = E_2^0 + RT/F \ln [\text{H}^+]^2 + RT/F \ln [\text{Ox}]/[\text{Red}]$. By potentiometric titration of 3-hydroxyphenothiazine in solutions of different pH values, Michaelis and Granick obtained the values of E_1^0 and $E_2^0 + RT/F \ln[\text{H}^+]^2$. These qualities could now be more easily obtained by cyclic voltammetry or by determining the half-wave potential for oxidation of the dyestuff at a rotating platinum anode. The difference, ΔE^0, between these values depends on the hydrogen ion concentration and is independent of the reference electrode used:

$$\Delta E^0 = \frac{2RT}{F} \ln [\text{H}^+] + \text{constant}$$

Measurements of ΔE^0 can be used to relate the acidity of an unknown solution to that of a known solution.

Over a wider range of acidity, the redox behaviour of 3-hydroxyphenothiazine becomes complex because further acid–base equilibria become important. Michaelis and Granick extended the range over which pH can be compared by using other phenothiazines such as 3-amino- and 3,7-diaminophenothiazine.

E. Phenoxathiin

Two oxidation waves have been recorded for phenoxathiin in acetonitrile at a platinum anode, and the radical cation which is formed at the potential of the first wave has been characterized by e.s.r. spectroscopy[32]. Phenoxathiin radical cation perchlorate (29) has also been prepared by chemical oxidation[66]. It reacts with primary amines to form sulphilimides such as 30.

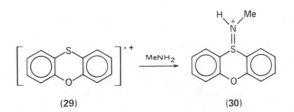

(29)   (30)

III. REFERENCE ELECTRODE POTENTIALS

In this chapter all electrode potentials have been calculated with respect to the aqueous saturated calomel electrode. The following values were used:

Potentials *versus* S.C.E.:

$Ag/AgNO_3$ 0.1 M, CH_3CN	+0.337 V	Ref. 67
$Ag/AgCl$, sat. KCl, H_2O	−0.0455 V	Ref. 68

The reference electrode of Billon[33] and Barry *et al.*[32]:

Ag/Ag 0.01 M	+0.36 V	Refs. 31, 32

Potential *versus* Ag/AgCl, sat. KCl, H_2O:

The reference electrode used by Hünig and co-workers:

$Ag/AgCl$, CH_3CN	−0.070 V	Refs. 28, 69

IV. REFERENCES

1. E. L. Colichman and D. L. Love, *J. Org. Chem.*, **18**, 40 (1953).
2. A. Luettringhaus and H. Machatzke, *Ann. Chem.*, **671**, 165 (1964).
3. R. W. Howsam and C. J. M. Stirling, *J. Chem. Soc. Perkin Trans. II*, 847 (1972).
4. P. S. McKinney and S. Rosenthal, *J. Electroanal. Chem.*, **16**, 261 (1968).
5. H. J. Baer, *Z. Phys. Chem. (Leipzig)*, **243**, 398 (1970).
6. J. M. Saveant, *C. R. Hebd. Seances. Acad. Sci. Ser. C*, **258**, 585 (1964).
7. J. Moiroux and M. B. Fleury, *Electrochim. Acta*, **18**, 691 (1973).
8. P. Zuman and S.-Y. Tang, *Collect. Czech. Chem. Commun.*, **28**, 829 (1963); J. M. Saveant, *C. R. Hebd. Seances. Acad. Sci. Ser. C*, **257**, 448 (1963); J. M. Saveant, *Bull. Soc. Chim. Fr.*, 481 (1967).
9. T. Iwasaki, M. Miyoshi, M. Matsuoka and K. Matsumoto, *Chem. Ind. (London)*, 1163 (1973).
10. J. D. Littlehailes and B. J. Woodhall, *J. Chem. Soc. Chem. Commun.*, 665 (1967); Imperial Chemical Industries Ltd., *Fr. Pat.*, 1,565,481 (1969); *Chem. Abstr.*, **72**, 74129 (1970).
11. T. Shono and M. Mitani, *Tetrahedron Lett.*, 687 (1969).
12. T. Shono, T. Akazawa and M. Mitani, *Tetrahedron*, **29**, 817 (1973).
13. J. H. Wagenknecht and M. M. Baizer, *J. Electrochem. Soc.*, **114**, 1095 (1967).
14. M. Finkelstein, R. C. Petersen and S. D. Ross, *J. Electrochem. Soc.*, **110**, 422 (1963).
15. R. A. Wessling and W. J. Settineri, *U.S. Pat.*, 3,480,527 (1969); *Chem. Abstr.*, **72**, 66593 (1970).
16. R. A. Wessling and W. J. Settineri, *U.S. Pat.*, 3,480,525 (1969); *Chem. Abstr.*, **72**, 32677 (1970).
17. M. M. Baizer, *J. Org. Chem.*, **31**, 3847 (1966).
18. A. Zweig and J. E. Lehnsen, *J. Amer. Chem. Soc.*, **87**, 2647 (1965); A. Zweig, A. H. Maurer and B. G. Roberts, *J. Org. Chem.*, **32**, 1322 (1967).
19. S. Torii and K. Uneyama, *Tetrahedron Lett.*, 4513 (1972); K. Uneyama and S. Torii, *J. Org. Chem.*, **37**, 367 (1972).
20. G. Bontempelli, F. Magno, G.-A. Mazzocchin and S. Zecchin, *J. Electroanal. Chem.*, **43**, 377 (1973).
21. P. T. Cottrell and C. K. Mann, *J. Electrochem. Soc.*, **116**, 1499 (1969).
22. D. S. Houghton and A. A. Humffray, *Electrochim. Acta*, **17**, 1421 (1972).
23. D. S. Houghton and A. A. Humffray, *Electrochim. Acta*, **17**, 2145 (1972).
24. S. Mann, *Z. Anal. Chem.*, **173**, 112 (1960); D. Kyriacou, *Nature, Lond.*, **211**, 519 (1966).
25. F. Mango and G. Bontempelli, *J. Electroanal. Chem.*, **36**, 389 (1972).
26. U. Schmidt, K. Kabitzke and K. Markau, *Angew. Chem.*, **72**, 708 (1960).
27. D. S. Houghton and A. A. Humffray, *Electrochim. Acta*, **18**, 373 (1973).

28. S. Hünig, G. Kiesslich, H. Quast and D. Scheutzow, *Ann. Chem.*, 310 (1973).
29. D. H. Geske and M. V. Merritt, *J. Amer. Chem. Soc.*, **91**, 6921 (1969).
30. D. J. Coffen, J. Q. Chambers, D. R. Williams, P. E. Garrett and N. D. Canfield, *J. Amer. Chem. Soc.*, **93**, 2258 (1971).
31. O. Hammerich and V. D. Parker, *Electrochim. Acta*, **18**, 537 (1973).
32. G. Barry, G. Cauquis and M. Maurey, *Bull. Soc. Chim. Fr.*, 2510 (1966).
33. J. P. Billon, *Bull. Soc. Chim. Fr.*, 1923 (1961).
34. P. D. Sulivan, *J. Amer. Chem. Soc.*, **90**, 3618 (1968).
35. D. Clarke, B. C. Gilbert and P. Hanson, *J. Chem. Soc. Perkin Trans. II*, 114 (1976).
36. A. Singhabhandu, P. D. Robinson, J. H. Fang and W. E. Geiger, *Inorg. Chem.*, **14**, 318 (1975).
37. J. P. Malrieu and B.. Pullman, *Theor. Chim. Acta*, **2**, 293 (1964).
38. S. M. Park and A. J. Bard, *Chem. Phys. Lett.*, **38**, 257 (1976).
39. C. P. Keszthelyi, P. Csaba, H. Tachikawa and A. J. Bard, *J. Amer. Chem. Soc.*, **94**, 1522 (1972).
40. C. P. Keszthelyi, N. E. Tokel-Takvoryan and A. J. Bard, *Anal. Chem.*, **47**, 249 (1975).
41. R. Bezman and L. R. Faulkner, *J. Amer. Chem. Soc.*, **94**, 6331 (1972); D. J. Freed and L. R. Faulkner, *J. Amer. Chem. Soc.*, **94**, 4790 (1972).
42. N. E. Tokel, C. P. Keszthelyi and A. J. Bard, *J. Amer. Chem. Soc.*, **94**, 4872 (1972).
43. O. Hammerich, N. S. Moe and V. D. Parker, *J. Chem. Soc. Chem. Commun.*, 156 (1972).
44. Y. Murata, L. Hughes and H. J. Shine, *Inorg. Nucl. Chem. Lett.*, **4**, 573 (1968); Y. Murata and H. J. Shine, *J. Org. Chem.*, **34**, 3368 (1969).
45. E. A. C. Lucken, *J. Chem. Soc.*, 4963 (1962).
46. W. Rundel and K. Schaffler, *Tetrahedron Lett.*, 993 (1963).
47. M. Kinoshita, *Bull. Chem. Soc. Jap.*, **35**, 1137 (1962).
48. M. de Sorgo, B. Wasserman and M. Szware, *J. Phys. Chem.*, **76**, 3468 (1972).
49. K. Fries, H. Koch and H. Stukenbrock, *Ann. Chem.*, **468**, 162 (1929).
50. R. S. Glass, W. J. Britt, W. N. Miller and G. S. Wilson, *J. Amer. Chem. Soc.*, **95**, 2375 (1973).
51. I. B. Goldberg, H. R. Crowe, G. S. Wilson and R. S. Glass, *J. Phys. Chem.*, **80**, 988 (1976).
52. Y. Murata and H. J. Shine, *J. Org. Chem.*, **34**, 3368 (1969).
53. K. Uneyama and S. Torii, *Tetrahedron Lett.*, 329 (1971).
54. U. Svanholm, O. Hammerich and V. D. Parker, *J. Amer. Chem. Soc.*, **97**, 101 (1975).
55. U. Svanholm and V. D. Parker, *J. Amer. Chem. Soc.*, **98**, 997 (1976).
56. K. Kim and H. J. Shine, *Tetrahedron Lett.*, 4413 (1974).
57. H. J. Shine and K. Kim, *Tetrahedron Lett.*, 99 (1974; K. Kim and H. J. Shine, *J. Org. Chem.*, **39**, 2537 (1974).
58. H. J. Shine and Y. Murata, *J. Amer. Chem. Soc.*, **91**, 1872 (1969).
59. V. D. Parker and L. Eberson, *J. Amer. Chem. Soc.*, **92**, 7488 (1970); O. Hammerich and V. D. Parker, *J. Electroanal. Chem.*, **36**, App. 11 (1972).
60. F. Kermann and L. Diserens, *Chem. Ber.*, **48**, 318 (1915).
61. L. Michaelis and S. Granick, *J. Amer. Chem. Soc.*, **64**, 1861 (1942).
62. G. Cauquis, A. Deronzier and D. Serve, *J. Electroanal. Chem.*, **47**, 193 (1973).
63. P. Hanson and R. O. C. Norman, *J. Chem. Soc. Perkin Trans. II*, 264 (1973).
64. Y. Tsujino, *Tetrahedron Lett.*, 2545 (1968), 4111 (1968), 763 (1969).
65. L. P. Hammett, *Physical Organic Chemistry*, 2nd ed., McGraw-Hill, New York, 1970, p. 263.
66. R. Serugudi and H. J. Shine, *J. Org. Chem.*, **40**, 2756 (1975).
67. R. C. Larson, R. T. Iwamoto and R. N. Adams, *Anal. Chim. Acta*, **25**, 371 (1961).
68. R. N. Gerke, *J. Amer. Chem. Soc.*, **44**, 1684 (1922); N. Randall and L. E. Lowry, *J. Amer. Chem. Soc.*, **50**, 989 (1928).
69. S. Hünig, G. Kiesslich, F. Linhart and H. Schlaf, *Ann. Chem.*, **752**, 182 (1971).

The Chemistry of the Sulphonium Group
Edited by C. J. M. Stirling and S. Patai
© 1981 John Wiley & Sons Ltd.

CHAPTER **8**

Isotopically labelled sulphonium salts

LEONARD F. BLACKWELL

*Department of Chemistry, Biochemistry and Biophysics, Massey University,
Palmerston North, New Zealand*

I.	INTRODUCTION		158
II.	HYDROGEN EXCHANGE		158
	A. The Acid Strengthening Effect of Positive Sulphur . . .		158
	B. Structural Effects on Rates of Exchange		160
III.	SULPHONIUM YLIDES		162
	A. Introduction		162
	B. Ylide Reactions Involving Labelled Sulphonium Salts . . .		163
	1. Formation of carbonyl compounds		163
	2. Reaction of sulphonium ylides with $NaBH_4$ . . .		164
	3. Epoxidation and cyclopropanation		165
	4. Reaction of diazomethane and diallyl sulphide . .		167
	C. Elimination via Ylide Intermediates (α'–β Elimination) . . .		167
IV.	ISOTOPE EFFECTS IN ELIMINATION REACTIONS . . .		170
	A. Introduction		170
	B. Theoretical Considerations		170
	1. Primary deuterium kinetic isotope effects . . .		170
	2. Heavy atom isotope effects		172
	3. Secondary isotope effects		173
	4. Variation of transition state structure in E2 eliminations .		173
	C. Experimental Evidence Relating to E2 Transition State Structure .		174
	1. Hammett ρ values		174
	2. Primary deuterium kinetic isotope effects in Me_2SO—H_2O mixtures .		176
	3. Heavy atom isotope effects in Me_2SO—H_2O mixtures . .		178
	4. Isotope effects in systems other than dimethyl-2-phenethylsulphonium ion .		179
	D. Transition State Structure for Elimination From Dimethyl-2-phenethyl-sulphonium Ion		180
V.	SUBSTITUTION REACTIONS		182
VI.	REFERENCES		184

I. INTRODUCTION

The main objective of this chapter will be to consider the various roles which isotopically labelled sulphonium salts have played in the understanding of the general chemistry of such salts and in particular the reactivity special to the functional group. A sulphonium salt will be understood to mean a structure of the type $R^1R^2R^3S^+$, where the R groups may be alkyl, aryl, or a mixture of both. Salts in which one of the R groups may be alkoxy, aryloxy, or $=O$ will be included in the definition. In general, the counter ion of the salt will be ignored, on the basis that there seems to be little evidence of it influencing chemical behaviour (at least for isotopically labelled sulphonium salts).

No attempt will be made to discuss or assess the various methods of preparation of labelled sulphonium salts since in the main these are straightforward and will be discussed elsewhere in this volume. In any case, the only difference in the synthesis is usually the employment of an appropriately labelled reagent, such as $^{14}CH_3I$ or CD_3I instead of CH_3I, for example. In the case of heavy atom isotope effects, natural abundances are sufficient and no special preparation is necessary.

The major section of this chapter relates to the investigation of transition state structure for E2 elimination reactions using isotope effects as probes for the extent of bond breaking of the isotopically substituted bond. Of necessity, therefore, a brief non-mathematical discussion of the background theory of isotope effects has been included but the interested reader will need to consult more detailed reference works for a greater understanding. This section will be discussed from the point of view of the transition state structure for the sulphonium salts and the conclusions reached cannot, therefore, necessarily be generalized. When discussions are related to more general mechanistic areas, attention will be focused on conclusions which can be drawn from data from *labelled* sulphonium salts, although in some instances data from *non-labelled* sulphonium salts, and other compounds will be included for comparison and completion.

The material is arranged in terms of reactivity, rather than in terms of the isotopic label, since this allows a more integrated and coherent account.

II. HYDROGEN EXCHANGE

A. The Acid Strengthening Effect of Positive Sulphur

The acidic nature of protons alpha to the positive sulphur atom in sulphonium salts was first clearly demonstrated by Doering and Hoffman in 1955[1], in a comprehensive study using trimethylsulphonium (1), tetramethylammonium ion (2), and tetramethylphosphonium ion (3). When 1 was dissolved in a solution of sodium

$$(CH_3)_2\overset{+}{S}CH_3 \qquad (CH_3)_3\overset{+}{N}CH_3 \qquad (CH_3)_3\overset{+}{P}CH_3$$

$$(1) \qquad\qquad (2) \qquad\qquad (3)$$

deuteroxide in D_2O, 98% of the C—H bonds were replaced by C—D bonds (in 3 h) whereas under identical conditions 2 showed only 1% exchange after 15 days. Although 1 and 2 are not strictly comparable, since sulphur and nitrogen belong to different groups of the Periodic Table, the acid strengthening effect of S^+ is not likely to be due solely to an electrostatic interaction between the positive charge and the adjacent negative charge, which is assumed to be developed in the conjugate base 4, as this interaction should be similar in 5. In agreement with this argument, when 2 was compared with 3 it was found that 74% exchange occurred

$$(CH_3)_2\overset{+}{S}\overset{-}{CH_2} \qquad (CH_3)_3\overset{+}{N}\overset{-}{CH_2}$$

$$\text{(4)} \qquad\qquad \text{(5)}$$

$$1 \xrightarrow{\quad^-OD\quad} [(CH_3)_2\overset{+}{S}-\overset{-}{CH_2} \longleftrightarrow (CH_3)_2S=CH_2] \xrightarrow{\quad D_2O\quad} (CH_3)_2\overset{+}{S}CH_2D$$

$$\text{(4)} \qquad\qquad\qquad \text{(6)}$$

SCHEME 1

for **3** within 3 h, even though the C—P bond is longer than the C—N bond[2]. Since both sulphur and phosphorus are second-row elements, explanations of the increased acidity and rates of exchange for **1** and **3**, compared with **2**, in terms of d-orbital interactions have commonly been invoked. The net result of such an interaction is an extra stabilization of the negative charge, as shown by the resonance structures **4** and **6** for the ylide intermediate derived from **1**, compared with the ylide **5** for which such d-orbital interactions are not possible[3].

The importance of the resonance form written with a sulphur–carbon double bond **6** has been questioned, since dipole moment studies suggest a large amount of charge separation[4,5], more in agreement with **4**. However, the temperature dependences of the rate constants for exchange of **1** and **2** have been determined and show that the large difference in rates of exchange are almost equally due to a lower heat of activation and a more positive entropy of activation for **1**. This would be expected if d-orbital participation were important (as represented by **6**) since delocalization of the negative charge in the transition state should result in a lower enthalpy of activation and at the same time there should be less ordering of the solvent molecules in the transition state than in the ground state, and thus an increase in entropy relative to **2**.

That the activating effect of S[+] is several orders of magnitude greater than that of N[+] was confirmed by the observation that vinyldimethylsulphonium ion **7** reacted rapidly with a nucleophilic reagent in a base-catalysed addition reaction (Scheme 2)[6], whereas there was no evidence that the corresponding vinyltrimethylammonium ion **8** underwent any base-catalysed addition. In agreement with these findings is the observation that 2-bromoethyldimethylsulphonium ion **9** experienced a rapid elimination of HBr to form **7** in the presence of hydroxide ion but 2-bromoethyltrimethylammonium ion **10** did not produce **8** but instead underwent

$$CH_2=\overset{+}{C}HS(CH_3)_2 \qquad CH_2=\overset{+}{C}HN(CH_3)_3 \qquad BrCH_2CH_2\overset{+}{S}(CH_3)_2 \qquad BrCH_2CH_2\overset{+}{N}(CH_3)_3$$

$$\text{(7)} \qquad\qquad \text{(8)} \qquad\qquad \text{(9)} \qquad\qquad \text{(10)}$$

the much slower displacement of the bromide ion. These observations can all be rationalized by proposing the formation of ylide intermediates with d-orbital interactions (which are not possible for the ammonium ion analogues) leading, as in the deuterium exchange of **1**, to a lower enthalpy and a more positive entropy of activation. For example, the base-catalysed addition reaction of **7** can be expressed as shown in Scheme 2 and the elimination from **9** can be envisaged by reading Scheme 2 in the opposite direction.

$$Nu\!:\ + 7 \longrightarrow [NuCH_2\overset{-}{C}H\overset{+}{S}(CH_3)_2 \longleftrightarrow NuCH_2CH=S(CH_3)_2] \xrightarrow{\quad H_2O\quad} NuCH_2CH_2\overset{+}{S}(CH_3)_2$$

SCHEME 2

If d-orbitals are involved in the enhanced acidity of sulphonium salts there appears to be little geometric constraint, since bicyclo[2.2.1]heptane-1-sulphonium ion exchanges deuterium at the alpha positions, despite the fact that a planar arrangement around the sulphur cannot easily be achieved. This conclusion is somewhat limited by the fact that exchange could be carried out for only a limited period of time since decomposition occurred under the alkaline conditions employed.

B. Structural Effects on Rates of Exchange

Fava and co-workers studied the effect of structural changes on the kinetic acidity of alpha protons in sulphonium ions of the type $R^1CH_2\overset{+}{S}R^2CH_3$ and in nearly all cases studied, the S-methyl groups exchanged their protons about 300 times more rapidly than the S-methylene or S-methine protons[7]. There was little effect of R^1 and R^2 on the S—CH$_3$ exchange rate constants whether R^1 and R^2 were alkyl groups, or whether the sulphur atom was included in a 5-, 6- or 7-membered ring system. The differences in rate constants between S—CH$_3$ exchange on the one hand and S—CH$_2$R^1 exchange on the other may be considered as resulting from the relative substituent effects of hydrogen and R^1. If the transition state for the exchange has carbanionic character (as seems intuitively reasonable), then the effect of a methyl group, for example, will be to destabilize it, relative to hydrogen, because of the electron-releasing tendency of methyl. Since the effect of alkyl substituents on the exchange is very large ($k_{CH3}/k_{CH2R'} \approx 300$), it may be indicative of a very carbanionic transition state (**4**, for example). The ρ value of about −4 which may be estimated from the data is similar to that reported by Crosby and Stirling[8] in the ElcB elimination of phenoxide from phenoxyethyldimethyl-sulphonium ion which proceeds via an ylide intermediate.

Despite the general insensitivity of the S—CH$_3$ exchange rate constants to structural effects, the thietanium salt (**11**) shows a rate acceleration of over 100 for its S—CH$_3$ group. This remarkable increase in rate constant is undoubtedly due to the steric strain imposed by the four-membered ring[9]. For example, the ring hydrogens of cyclobutane undergo deuterium exchange 28 times faster than the corresponding hydrogens in cyclohexane[10].

(11)

Ring size has little significant effect on the exchange rates of the α-methylene protons of **12**, **13** and **14**[7], but some interesting stereoselectivity is observed. In **12**, **13** and **14** the α-methylene protons may be divided into two sets, those *cis* to the S-methyl group and those *trans* to it. For **13** and **14** all four hydrogens are

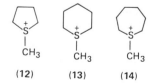

(12)   (13)   (14)

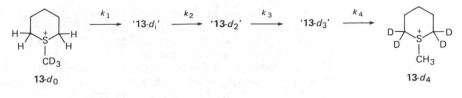

SCHEME 3

exchanged and for **13** very little preference (if any) is shown for one diastereotopic pair of α-methylene protons over the other (and presumably in **14**)[11]. What is measured in practice is the total incorporation of deuterium, irrespective of its stereochemical position, as a function of time, that is, whether a total of zero, one, two, three, or four deuterium atoms have been incorporated at any given time and the relative percentages of each. These composition *versus* time data can be fitted by assuming four consecutive processes, governed by rate constants k_1, k_2, k_3, and k_4, for interconversion of the **13**-d_0, '**13**-d_1, '**13**-d_2', **13**-d_3', and **13**-d_4 species respectively. The *S*-methyl group is shown as deuterated since the methyl protons exchange more rapidly than the methylene protons under the reaction conditions[11]. In the case of '**13**-d_2', for example, we do not know the relative amounts of the *cis*-dideuterio, *trans*-dideuterio and *cis*, *trans*-deuterio forms. Nevertheless, the model presented above accounts for the observed *total* composition *versus* time data, with rate constant ratios k_1:k_2:k_3:k_4 of 4.4:3.0:1.83:0.84. If it is assumed that there is no stereochemical preference during the exchange, then the statistically determined ratio would be 4:3:2:1, which is very close to the observed values, thus leading to the conclusion that the *cis* and *trans* protons of **13** exchange at the same rate. A similar analysis has not been carried out for **14** but n.m.r. studies show that the peaks for the two diastereotopic protons disappear at the same rate[7].

On the basis of n.m.r. studies, Fava and co-workers[7,12] originally claimed that **12** exchanged its *cis*-α-methylene protons at a rate 400 times faster than the *trans* pair, although more recently they have amended this figure to a ratio of 28:1 when decomposition reactions were taken into account[13]. The latter figure agrees reasonably well with the value of about 14 which may be estimated from the observed ratio of rate constants k_1:k_2:k_3:k_4 of 5.66:3.0:0.48:0.22 quoted by Hofer and Eliel on the basis of their mass spectral measurements[11].

The difference between **12** and **13** has generated considerable interest since, according to a theoretical idea based on quantum mechanical calculations[14], a carbanion intermediate should be most stable when the orbital containing the lone pair of electrons on the carbon atom is oriented *gauche* to the sulphur lone pair orbital. The so called *gauche* effect explains the results for **12**, since the carbanion derived from it by removal of a *cis*-α-methylene proton has its orbital more or less *gauche* to the sulphur atom lone pair. The results for **13**, however, are inconsistent with this idea. It has been argued that the lack of stereochemical preference shown by **13** could be the result of a highly selective abstraction of the *cis*-proton[7], as required by the theory, followed by a re-protonation with inversion so that the proton which was originally *trans* now becomes *cis*. If such an inversion were sufficiently fast, it would effectively scramble the α-methylene protons, masking any inherent reactivity differences. Such an explanation, although plausible, violates the principle of microscopic reversibility. If true reactivity differences exist, then fast exchange with retention *must* be the favoured pathway, since this transition state *must* be of lowest free energy[15], protonation being the microscopic reverse of deprotonation. Thus, it appears that there is truly a difference in the rates of exchange between the diastereotopic pairs of α-methylene protons in **12**, but no detectable differences for **13**, irrespective of the predictions of the *gauche* effect. It has been pointed out that the *gauche* effect is based on predictions about thermodynamic acidity, whereas what is measured in base-catalysed hydrogen exchange experiments is in fact kinetic acidity[11,13]. There is no guarantee that thermodynamic and kinetic acidity will run in parallel, especially if extensive solvent and structural reorganization occurs on deprotonation. Thus, the explanation of the different behaviour of **12** and **13** must await further work. A comparison of the sulphur inversion rates of **12** and **13** might prove interesting since a rapid inversion rate for **13** would appear to offer a pathway for stereochemical scrambling of the α-methylene protons.

The relative ease of base-catalysed deuterium exchange of protons in positions alpha to positive sulphur is now commonly exploited in synthetic organic chemistry as a method of preparing deuterium-labelled sulphonium salts. It also has other consequences which will be discussed in the following sections.

III. SULPHONIUM YLIDES

A. Introduction

The species formed when a sulphonium salt such as **1** is treated with a base is known as an ylide and is usually written as a resonance hybrid of two canonical forms such as **4** and **6**. As discussed in section IIA, there is some argument over the importance of the form drawn with a sulphur–carbon double bond, but some stabilization over and above electrostatic interactions seems necessary to account for the acidity of sulphonium salts compared with ammonium salts. A range of stabilities have been observed for sulphonium ylides, ranging from those which can be stored easily to those which can only be generated *in situ*, depending on the structure of the ylide. The general chemistry of sulphonium ylides and their general

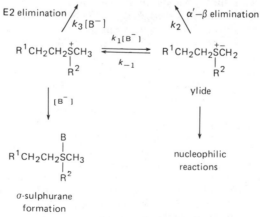

SCHEME 4

properties have been comprehensively reviewed in this volume (Chapter 16) and elsewhere[16].

The fate of a given ylide species depends on the nature of the alkyl groups and the base and several pathways are possible as shown in Scheme 4. If β-protons are available, an α'–β, or E2, elimination reaction may take place and in hydroxylic solvents re-protonation may occur to such an extent that the equilibrium concentration of the ylide is very small. In some cases the base can react with the corresponding sulphonium salt to form a σ-sulphurane. However, if suitable choices of base and sulphur substituents are made, the ylide can exist in sufficient concentrations to undergo a variety of synthetic reactions[15]. In these synthetic reactions the ylide usually functions as a nucleophile and can be used to form new carbon to carbon bonds.

B. Ylide Reactions Involving Labelled Sulphonium Salts

1. Formation of carbonyl compounds

Alkoxysulphonium salts such as dimethylmethoxysulphonium ion (15) undergo a number of interesting reactions with nucleophiles and in particular with alkoxides in alcoholic solvents. For example, if 15 is labelled with carbon-14 in the O-methyl group and treated with sodium methoxide in methanol, methanal is produced which

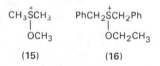

(15) (16)

contains less than 1% of the initial radioactivity, whereas if sodium hydride in dimethyl sulphoxide is used the methanal is radioactive[17]. The difference in behaviour between the two nucleophilic systems is though to be due to alkoxide exchange, which is possible with NaOMe/MeOH but not for NaH/Me₂SO (Scheme 5). If ¹⁴C-labelled 15 is partially reacted with NaOMe/MeOH, the recovered 15 shows a considerable reduction in radioactivity. Other evidence for alkoxide exchange includes the fact that when other alkoxide base systems are employed, different carbonyl compounds are formed. Treatment of 15 with NaOEt/EtOH, for example, yields ethanal whereas reaction with *i*-PrO⁻/*i*-PrOH yields propanone. When dibenzylethoxysulphonium ion 16 is dissolved in radioactive ethanol for 1 h at 70°C, the recovered salt shows a small amount of radioactivity exchange even in the absence of alkoxides.

The most obvious mechanism for production of methanal from 15 involves a base-catalysed elimination of dimethyl sulphide, as shown in Scheme 6. When diphenylmethoxysulphonium ion 17 is treated with sodium methoxide, however, the main product is dimethyl ether instead of methanal. This is inconsistent with the mechanism proposed in Scheme 6. The importance of alpha protons is demonstrated by the fact that when methylphenylmethoxysulphonium ion 18 is treated with NaOMe/MeOH, methanal and methyl phenyl sulphide 19 are readily

$$CH_3\overset{+}{S}CH_3 + CH_3O^- \rightleftharpoons CH_3\overset{+}{S}CH_3 + {}^{14}CH_3O^-$$
$$\underset{O^{14}CH_3}{|} \qquad\qquad \underset{OCH_3}{|}$$

SCHEME 5

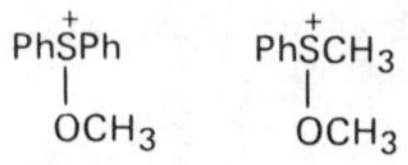

$$\overset{+}{PhSPh}$$
$$|$$
$$OCH_3$$
$$(17)$$

$$\overset{+}{PhSCH_3}$$
$$|$$
$$OCH_3$$
$$(18)$$

produced. Thus, production of an ylide seems to be an essential requirement for production of aldehydes or ketones. When this cannot be achieved, an S_N2 displacement of sulphoxide and formation of ethers becomes the favoured reaction. Deuterium labelling experiments have enabled the existence of ylide intermediates to be clearly demonstrated, since when **18** was trideuteriated in the O-methyl group and reacted with NaH/THF, the resulting **19** had one deuterium atom in the S-methyl group per molecule, whereas when **18** was trideuteriated in the S-methyl group and subjected to the same treatment, unlabelled methanal and dideuteriated **19** were obtained. These results are best explained by invoking a methoxyphenylsulphonium methylide intermediate, formed by abstraction of the S-methyl proton, which can function as an *internal* base to form the carbonyl compound more rapidly than it is re-protonated (Scheme 7). Consistent with this mechanism is the observation that reaction of **18** with sodium methoxide in methanol-O-d produced virtually no incorporation of deuterium into the resulting **19**. Unfortunately, this reaction is of limited synthetic importance since a large excess of alcohol is required to produce acceptable yields of carbonyl compounds.

2. Reaction of sulphonium ylides with NaBH₄

When **18** is treated with NaBH₄ in alcohols or tetrahydrofuran as solvent, **19** is produced in good yields[18]. For example, in methanol 78% of the sulphide is obtained but no methanal is produced. A number of possible mechanistic schemes have been proposed[18], the majority involving an ylide intermediate formed by abstraction of a proton from **18** by $\bar{B}H_4$, followed either by a reaction similar to Scheme 7, or an internal displacement of methoxide concurrent with, or followed by, hydride transfer to form **19**.

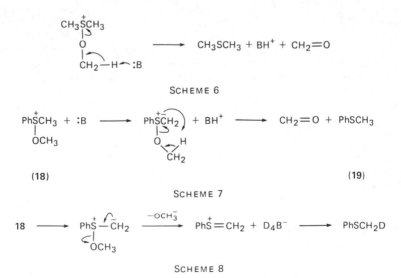

SCHEME 6

SCHEME 7

SCHEME 8

$$18 + \bar{B}H_4 \longrightarrow \overset{+}{PhSCH_3} + \bar{O}CH_3 \xrightarrow{-H^+} PhSCH_3$$
$$\underset{H}{|}$$

SCHEME 9

Deuterium-labelling experiments eliminate reaction via ylide intermediates, since all of these mechanisms should produce labelled **19** when either deuterated *O*-methyl-**18** or NaBD$_4$ is used in the reaction. However, only trace amounts of deuterium were in fact incorporated, as shown by mass spectral analysis. In agreement with the conclusion that ylides are not involved in this reaction is the finding that other sulphonium salts also undergo this reaction, among them **17**, which in methanol gives **19** in 78% yield despite the fact that it cannot form an ylide intermediate. The remaining most probable mechanism is that involving displacement of the alkoxy group by a hydride ion (Scheme 9)[18].

3. Epoxidation and cyclopropanation

Sulphonium ylides can be used in reactions involving nucleophilic addition to carbonyl compounds, or in Michael-type addition reactions with α–β unsaturated carbonyl compounds. Addition of an ylide to a carbonyl compound results in epoxide formation, often in high yield, and consequently this reaction has found much use in synthetic organic chemistry[16]. Although the mechanism of carbonyl addition has not been extensively studied it is presumed to proceed as follows. The first step involves nucleophilic attack at the carbonyl group by the ylide carbanion to form a tetrahedral intermediate (Scheme 10), which takes advantage of the leaving group propensity of the dialkyl sulphide group to form an epoxide ring in an internal nucleophilic substitution reaction. If such a reaction is to proceed in good yields, it is obviously necessary to use a sulphonium salt which does not undergo ready α'–β elimination and employ conditions which do not favour reprotonation of the ylide, or σ-sulphurane formation.

An interesting use of a labelled sulphonium salt in epoxide formation has been found in studies relating to sterol biosynthesis, which is presumed to involve cyclization of 2,3-oxidosqualene **20** by an enzyme known as 2,3-oxidosqualene-sterol cyclase[19,20]. The enzyme has been isolated in a partially

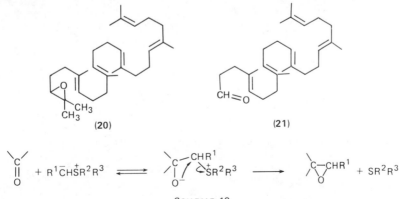

(20)　　　　　　　　　　(21)

SCHEME 10

purified form and has been found to catalyse the cyclization of structural analogues to **20** also[21], and there has therefore been interest in preparing such analogues for mechanistic studies. In biosynthetic experiments it is usual to prepare radioactively labelled substrates to allow relatively easily monitoring of metabolites. A ^{14}C-labelled 2,3-oxidosqualene has been conveniently prepared by reaction of the aldehyde **21** with ^{14}C-labelled diphenylsulphonium isopropylide **22**[22]. The isopropylide **22** can be easily prepared *in situ* from diphenylsulphonium ethylide **23** and ^{14}C-labelled methyl iodide. In both cases, use is made of the acidic nature

$$Ph_2\overset{+}{S}\overset{-}{C}{}^{14}CH_3 \qquad Ph_2\overset{+}{S}\overset{-}{C}HCH_3$$
$$\underset{|}{\overset{|}{CH_3}}$$

$$(22) \qquad\qquad (23)$$

of the protons alpha to the positive sulphur atom. Other structural analogues of **20** can be prepared by this reaction merely by varying the nature of the *S*-alkyl group and hence various stereochemical aspects of the cyclization reaction have been studied in this way[23].

Non-labelled **22** has been found to react with conjugated carbonyl compounds, such as 2-cyclohexenone **24**, to give a cyclopropane derivative instead of an epoxide derivative (Scheme 11). This is an α,β-addition reaction rather than a carbonyl addition reaction and contrasts with the behaviour of **4** which yields only the epoxide with **24**. Whether epoxide or cyclopropane ring formation predominates depends on the structural features of the enone system[24]. For example, 3-methyl-2-cyclohexenone reacts with **22** to give mainly the epoxide whereas 2-methyl-2-cyclohexenone gives 86% of the corresponding cyclopropane derivative[24]. Structural features in the ylide itself are also obviously important, as noted with the different behaviour of **22** and **4** with **24**.

Dimethyloxosulphonium ylides can also be used to prepare cyclopropane derivatives by reacting with α,β-unsaturated ketones. For example, 1-phenyl-2-benzoyl-3,3-dideuteriocyclopropane **25** has been synthesized by reaction of 1-phenyl-2-benzoylethene with dimethyloxosulphonium dideuteriomethylide (Scheme 12)[25]. The configuration of the cyclopropane ring has been shown to be *trans* by n.m.r. spectroscopy[25].

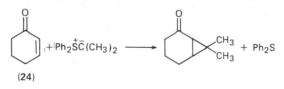

(24)

SCHEME 11

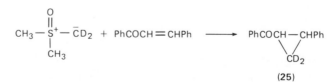

(25)

SCHEME 12

4. *Reaction of diazomethane and diallyl sulphide*

An ylide intermediate has been postulated to explain the results of a reaction between diazomethane and diallyl sulphide catalysed by copper(I) chloride[26]. It is known, for example, that allyl halides react with diazomethane and copper(I) halide to give a halomethylcyclopropane and a rearranged 4-halogenoalkene[27], presumably via an ylide intermediate (Scheme 13). When deuterium-labelled diallyl sulphide **26** is reacted under similar conditions, the expected cyclopropane derivative is formed in 17% yield but the major product is the rearranged derivative **27**, which is isolated in 83% yield (Scheme 14). Some sort of rearrangement is obviously indicated and, considering the known alkylating properties of diazomethane, the most likely mechanism for the rearrangement involves the intermediate formation of a sulphonium ylide which can rearrange as a result of the presence of the allyl group (Scheme 15).

C. Elimination via Ylide Intermediates (α′–β Elimination)

Although the elimination reaction of a sulphonium salt with a base is normally considered to proceed via an E2 mechanism, the acid strengthening effect of positive sulphur on the α-hydrogens makes elimination via an ylide intermediate as outlined in Scheme 4 a possibility which must be considered. Which elimination route actually predominates in a particular case will depend on the relative

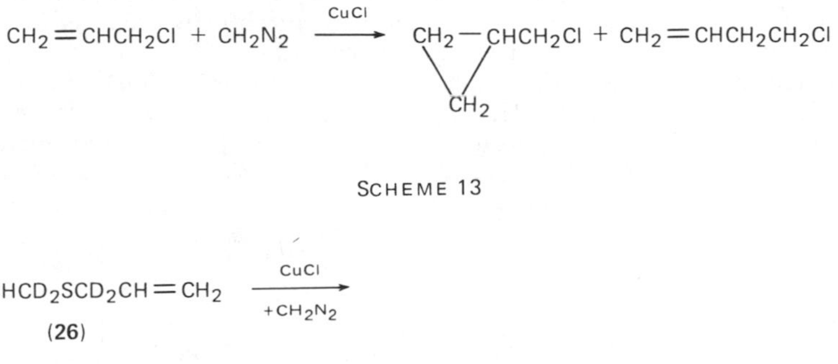

CuCl
CH2=CHCH2Cl + CH2N2
CH2—CHCH2Cl + CH2=CHCH2CH2Cl
CH2

SCHEME 13

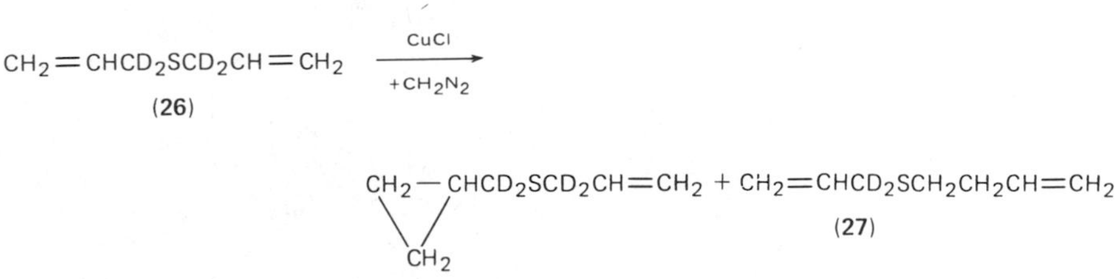

CH2=CHCD2SCD2CH=CH2
(26)
CuCl
+CH2N2
CH2—CHCD2SCD2CH=CH2 + CH2=CHCD2SCH2CH2CH=CH2
(27)
CH2

SCHEME 14

RSCD2CH=CH2 + CH2N2
CuCl
R
CH2—S+
CD2
CH2=CH
RSCH2CH2CH=CD2

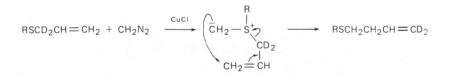

SCHEME 15

magnitudes of k_1, k_{-1}, k_2, and k_3. Since the kinetics of the α'–β reaction are the same as for an E2 or E1cB reaction, depending on whether formation of the ylide ($k_1 < k_2$) or its breakdown ($k_2 < k_1$) is rate determining, non-kinetic methods must be used to demonstrate its presence. The usual approach to detect the intermediacy of an ylide has been to introduce an isotopic label (usually deuterium or tritium) into the β position, since during an α'–β reaction this will be transferred to the departing sulphide instead of to the attacking base, as in an E2 process. Thus, provided that exchange between the solvent and the sulphide is not possible, the existence of the α'–β pathway will be revealed by the presence of the isotopic label.

Saunders and Pavlović[28] first attempted to demonstrate the existence of the α'–β pathway for the hydroxide ion-catalysed elimination of dimethyl sulphide from dimethyl-2-phenethylsulphonium ion **28** at 96°C. When the 2-position was

$$ZC_6H_4CH_2CH_2\overset{+}{S}(CH_3)_2$$

Z = H, unless otherwise stated

(28)

dideuterated, no significant incorporation of the deuterium label into Me_2S could be detected and thus the conclusion was drawn that the reaction was pure E2. This appears to be a general conclusion for the elimination of sulphonium salts catalysed by hydroxide ion in water, but when **28** in which *all* of the alpha-protons were replaced with deuterium was reacted with trityl sodium in diethyl ether[29], the triphenylmethane product was nearly 80% deuterated. This suggests that nearly 80% of the elimination occurred via the α'–β pathway, although since the actual alkene yield was not determined, this figure is not as convincing as it could be. Some of the factors which favour the α'–β pathway over the E2 pathway were recently elucidated by Saunders and co-workers for **28**, dimethyl-2-propyl **29**, dimethyl-3-pentyl **30**, dimethylcyclopentyl **31** and dimethylcyclohexyl **32** sulphonium ions. For each substrate the amount of α'–β elimination increased in the order $HO^-/H_2O < n\text{-}BuO^-/n\text{-}BuOH < t\text{-}BuO^-/t\text{-}BuOH$ and in $t\text{-}BuO^-/t\text{-}BuOH$ the amount of α'–β elimination increased in the order **32** (4.3% at 40°C) <**29** (22.2%

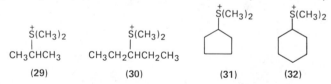

(29)   (30)   (31)   (32)

at 35°C) < **30** (60% at 35°C) < **31** (82.4% at 35°C). The rate of ylide reaction is rapid compared with the rate of E2 elimination, even for DO^-/D_2O when α'–β elimination is expected to be virtually insignificant, and therefore ylide formation is likely to be a side-reaction for all sulphonium salts[7,30]. For a given base–solvent system, since Fava and co-workers[7] have shown that the rate constant for ionization (k_1) of the S-methyl protons is approximately independent of substrate structure, whether or not the α'–β pathway is important, will be determined mainly by the ratio of $k_2:k_3$ in Scheme 4. If $k_3 \gg k_2$, E2 elimination will predominate, whereas if $k_2 \gg k_3$, α'–β elimination will predominate.

If the simple alkyl sulphonium salts **29** and **30** are compared, it can be seen that there is more α'–β elimination from **30** than from **29** in $t\text{-}BuO^-/t\text{-}BuOH$. It is likely that the rate constant (k_2) for elimination from the corresponding ylide **33** is greater than that from **34**, since in elimination from the ylide **35**, in which a similar choice of β-protons is possible in the same molecule, elimination to form

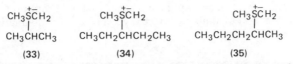

$$\underset{(33)}{\underset{\underset{CH_3CHCH_3}{|}}{CH_3\overset{+}{\underset{..}{S}}CH_2}} \qquad \underset{(34)}{\underset{\underset{CH_3CH_2CHCH_2CH_3}{|}}{CH_3\overset{+}{\underset{..}{S}}CH_2}} \qquad \underset{(35)}{\underset{\underset{CH_3CH_2CH_2CHCH_3}{|}}{CH_3\overset{+}{\underset{..}{S}}CH_2}}$$

pent-1-ene is the predominant reaction. Thus, on this basis there should be more α'–β elimination from **29**. The fact that the reverse is actually the case must mean that the decreased rate constant for E2 elimination from **30** (k_3) is the controlling factor[30]. When **31** and **32** are compared, also in t-BuO$^-$/t-BuOH, there is considerably more α'–β elimination for **31**, despite the fact that the E2 elimination rate constants are probably faster for **31**[30]; thus, in this case, the increased amount of α'–β elimination must be due to an even greater increase in k_2 for **31**. Such an increase in k_2 (for α'–β elimination from **31**) is not unreasonable when the different conformational effects in the ylide intermediates derived from **31** and **32** are considered[30].

Conformational effects have also been invoked to explain the fact that dimethyl-1-phenethylsulphonium ion (**36**), trideuterated in the 2-position, eliminates

$$\underset{(36)}{\underset{\underset{PhCHCD_3}{|}}{\overset{+}{S}(CH_3)_2}}$$

with ethanolic ethoxide at 35°C via the α'–β pathway to an extent of 11%[31], whereas **29**, **30**, **31**, and **32** eliminate with n-BuOK/n-BuOH to give between 0 and 3% α'–β elimination. The overall elimination rates (and presumably the rate constants) are faster for **36**, and therefore again it must be a faster elimination from the ylide derived from **36** than from **29**, **30**, **31**, and **32** which accounts for the increase in contribution from the α'–β pathway. It has been argued that the aryl group will favour an ylide structure in which the negatively charged methylene carbon atom is oriented towards the β-methyl group, resulting in an increased rate constant for elimination, whereas no such effects are expected for the ylides derived from **29**, **30**, **31**, and **32**[31].

The observed increase in the amount of α'–β elimination as the base–solvent system is changed along the series HO$^-$, n-BuO$^-$, t-BuO$^-$ is harder to rationalize, since k_1 and k_{-1} will presumably change as well as k_2 and k_3 and hence the equilibrium concentration of ylide will vary along the series. An increase in k_1 (or a decrease in k_{-1}) or k_2 relative to k_3 will result in an increased proportion of α'–β elimination. Both of these changes may be anticipated as the base strength increases and the solvent polarity decreases.

An interesting feature of the results obtained by Saunders and co-workers is the finding that additions of Me$_2$SO increase the percentage of elimination by the α'–β pathway[30]. When Me$_2$SO (95 mol%) is added to t-BuO$^-$/t-BuOH, the amount of α'–β elimination from **30** increases from 60% to 73% of the total and the addition of 84 mol% of Me$_2$SO to the hydroxide ion-catalysed elimination from **28** in water results in an increase in the proportion of α'–β elimination from 0 to 12%. Most of the effect occurs, in fact, over the range 61–84 mol% Me$_2$SO–H$_2$O. Again the main effect is probably on k_1 or k_2, since rate constants for total elimination from **28** in Me$_2$SO–H$_2$O mixtures, catalysed by hydroxide ion, show very large increases (90,000). Although the amount of α'–β elimination is not large, it may be a significant factor when interpreting ρ values and isotope effects for E2 reactions of **28**. It should also be noted that no attempt has been made to correct the reported percentages of α'–β elimination for isotope effects, even though there is some evidence that this factor will favour the α'–β mechanism to some extent. However, the correction is unlikely to change any of the qualitative conclusions which have been reached.

IV. ISOTOPE EFFECTS IN ELIMINATION REACTIONS

A. Introduction

Elimination from 2-phenethyl compounds is one of the most extensively studied reaction systems[32-34], since the elimination is normally quantitative, gives rise to an easily monitored alkene product, is amenable to structural variation by way of meta- or para-phenyl substitution, base–solvent composition, or changes in the nature of the leaving group, and is susceptible to isotopic substitution in a number of positions. The two most studied systems are for elimination from sulphonium salts and ammonium salts. More systematic data are available for the sulphonium salts, making them good experimental candidates for testing theories relating to E2 transition state structures.

The importance of isotopically labelled compounds in the study of elimination reactions resides in the fact that a rate retardation on the introduction of a heavier isotope at a specific position is believed to yield information on the extent of bond breaking (or bond making) of the isotopically substituted bond in the transition state. Hence information about transition state structure may, at least in principle, be obtained by a systematic study of the various appropriate isotope effects. Whilst the *observation* of an isotope effect may reasonably be interpreted as indicating which atoms are undergoing bonding changes during the rate-determining processes, the interpretation of the *magnitude* of the isotope effect in terms of the *extent* of bond breaking obviously depends on the underlying assumptions of the theoretical treatments. Kinetic isotope effects are usually considered within the framework of the absolute reaction rate theory, and models for the transition state are postulated in which assumptions are necessarily made about its geometry and force constants. It is not normally possible to determine which detailed model is most appropriate for a given reaction. However, with modern computing facilities, the approach is to calculate isotope effects for a wide range of models, covering all forseeable possibilities. For details of the theoretical basis and calculations of isotope effects, the interested reader is referred to recent articles and reviews[32,33,35-40] and only a brief outline of the essential features of the calculations is given here.

B. Theoretical Considerations

1. Primary deuterium kinetic isotope effects

For reactions in which a carbon to hydrogen bond is broken in the rate-determining step, as for base-catalysed E2 elimination from **28**, a retardation in the rate of reaction is observed if deuterium is substituted for hydrogen ($k^H/k^D = 3$–10). In simple terms, this occurs because the mass difference between hydrogen and deuterium atoms gives rise to a zero point energy difference, which leads to a larger energy of activation for the breaking of the carbon–deuterium bond. The isotope effect calculated on this basis is approximately 7 at 25°C and many reactions proceed with isotope effects of around this value. However, there are also many reactions which proceed with values both greater than and less than this theoretical value[41,42]. Westheimer[43] has proposed a three-centre model for the transition state in a proton transfer reaction in order to explain the low isotope effect values obtained in some reactions, and this treatment has led to the prediction that the isotope effect will pass through a maximum value when the proton is equally bonded to the abstracting base and the substrate in the transition state (the proton is half transferred). Since lower values are expected when the proton is either close to the base (more than half transferred) or to the substrate (less than half transferred), an ambiguity is introduced into the interpretation of the *magnitude* of the isotope effect. In Westheimer's terms, the lower values predicted for non-symmetrical transition states are the result of partial cancelling of zero point energy differences as a result of a real vibration of the transition state which does not exist in either reactant. When the proton is half transferred it does not move in this vibrational mode and hence there is no mass-dependent zero point energy difference in the transition state and the maximum value is expected.

The three-centre model is obviously a gross oversimplification for concerted elimination reactions in which more than three atoms are involved, and Katz and Saunders have calculated the isotope effect for the E2 reaction of hydroxide with ethyldimethylsulphonium ion (37) on the assumption that the transition state

$$CH_3CH_2\overset{+}{S}(CH_3)_2$$

(37)

geometry is essentially the same as the ground state for 37 except that the force constants for the reacting bonds are less than in the ground state. Changes in the force constants of the reacting bonds were simulated by varying these force constants from 0 to 100% of their normal values. For this model also, k^H/k^D was predicted to pass through a maximum when the β-proton was half transferred to the abstracting base for all the variations in force constants which were considered and was thus in agreement with the simpler three-centre model[42,43]. As well as the ambiguity imposed by the so called Westheimer maximum, these calculations indicated that the magnitude for k^H/k^D to be expected for half transfer of the β-proton depends on the extent of coupling with other atomic motions in the transition state. For instance, a lower value of k^H/k^D would be expected for a half-transferred proton when its motion is extensively coupled with motion of the departing sulphur atom and the carbon atoms of the developing double bond, than when proton transfer is the main component of the reaction coordinate motion. All k^H/k^D values are predicted to be lower than those calculated from the three-centre model.

In these calculations, Katz and Saunders ignored the effect of a proton tunnelling correction since they rightly considered that it complicated even further an already complex situation. However, the tunnel correction has the same logical status as zero point energy[39], on which isotope effect theories are based, since both concepts follow from the uncertainty principle. Therefore, it seems unreasonable to neglect it in theoretical calculations. Inclusion of a tunnel correction does add a further complication since now assumptions must be made relating to the curvature and width of the potential energy barrier in the region of the transition state.

More recently, Saunders has carried out detailed model calculations investigating the effects of transition state geometry, a tunnel correction, and barrier curvature on the value of k^H/k^D for a model similar to the previous one, except that the reactant did not include the attacking hydroxide ion in the ground state as previously[37,38]. Again, a Westheimer type maximum is predicted which is determined mainly by the extent of C—H bond rupture and is independent of the extent of C—S bond rupture. However, to obtain agreement with the experimental magnitudes a significant barrier curvature had to be assumed and, perhaps more significantly, the degree of coupling of the proton transfer to other atomic motions had to be set low. A substantial tunnelling correction also had to be included. Some years ago Willi and Wolfsberg[35] demonstrated, on the basis of a three-centre model (to which the more recent Saunders model approximates), that the *shape* of the Westheimer maximum depended very much on the amount of barrier curvature, becoming very broad when the imaginary frequency corresponding to barrier curvature, becomes large. In the absence of a tunnelling correction the maximum k^H/k^D value did not depend on the amount of barrier curvature. However, the limited calculations which did include a tunnelling correction showed a dependence on the barrier curvature, the value for k_H/k_D at the maximum increasing with an increasing tunnelling correction. The imaginary frequencies used by Willi and Wolfsberg ($>1756i \, cm^{-1}$) were all larger than the value of $800i \, cm^{-1}$ used by Saunders; this probably accounts for the more truncated curve predicted in the former case.

Bell et al.[44] used an electrostatic model for calculating the deuterium isotope effect and concluded that a maximum in the isotope effect should occur, but that it arises primarily from changes in the tunnelling correction rather than from symmetry related changes in the zero point energy of the transition state. In the absence of the tunnelling correction the calculated isotope effects were much smaller than the experimental values. Bell et al. applied their model to concerted reactions and again concluded that coupling of the proton transfer motion to other motions should result in low values for the isotope effect. Despite the different models used it has been argued that both explanations of the occurrence of a maximum in k^H/k^D can be related to the position of the proton in the transition state[39].

None of the theoretical treatments has so far considered the solvent, even though it is known to exert an important influence on the isotope effect in some proton transfer reactions[45]. Thus, there is a considerable degree of uncertainty as to the significance to be attached to the magnitude of k^H/k^D values and to the changes observed as structural and environmental factors are varied. As we have just seen, recourse to theoretical treatments is of no avail since these predict an increase, a decrease, or no change in k^H/k^D as the extent of proton transfer to the abstracting base increases, depending on the underlying assumptions regarding coupling, tunnelling, and barrier curvature. Equally serious is the possibility that the isotope effect may change, although the position of the proton in the transition state may not. If an unambiguous Westheimer maximum can be demonstrated for an elimination reaction, then it is likely that some of the difficulties mentioned have been overemphasized and that k^H/k^D values can be interpreted in terms of the proton position in the transition state. The ambiguity imposed by the Westheimer maximum will still present some difficulties in interpretation, however. As a working hypothesis, it will be assumed while discussing the experimental data for elimination from sulphonium salts that primary deuterium kinetic isotope effects *do* give a measure of the extent of C—H bond rupture in the transition state.

2. Heavy atom isotope effects

The sulphur isotope effect, which is expressed as the percentage by which k^{32}/k^{34} differs from unity, has been calculated by Saunders and co-workers and does not appear to be subject to the range of assumptions discussed in Section IVB1[32,36–38]. Using hydroxide ion-catalysed E2 elimination from **37** as a model, the calculations show a more or less linear increase in the sulphur isotope effect with the increase in extent of C—S bond rupture, up to a maximum value of about 1.1–1.3%. The values show little or no dependence either on the extent of proton transfer or in the assumed transition-state geometry; a tunnelling contribution does not appear to be important since the magnitude of the calculated and experimental sulphur isotope effects are in reasonable agreement without it. In any case, a tunnelling contribution is not normally as large for the heavier elements.

Of the other isotopic substitutions which are possible in **37**, the only one which has so far been extensively investigated is the β-^{12}C/^{13}C isotope effect. The calculations carried out by Saunders predict an S-shaped dependence on the extent of C—H bond rupture[37,38]. As the extent of C—H bond rupture increases from 0 to 100% the carbon isotope effect first passes through a minimum and then through a maximum. In all cases considered, a small β-carbon isotope effect indicates a low extent of proton transfer in the transition state, while a large value (3–5%) indicates a high degree of C—H bond rupture[37]. Thus, in principle, the β-carbon isotope effect, in conjunction with the primary deuterium isotope effects, may allow removal of the ambiguity imposed by the Westheimer maximum. No special preparative methods are required in order to measure these heavy atom isotope effects, since natural abundances are sufficient to allow accurate determination of the isotope ratios by mass spectrometry.

3. Secondary isotope effects

Isotopic substitution in non-reacting bonds can also give information about transition-state structure. For example, the rate of elimination catalysed by hydroxide ion in water is less than the rate for deuteroxide ion in deuterium oxide. The isotope effect (k^{DO^-}/k^{HO^-}) is expected to increase in a more or less linear fashion with the extent of proton transfer to the base. The value for complete transfer depends on the temperature of the measurements, since the isotope effect for complete proton transfer should be equal to the ratio of equilibrium basicities of DO^- and HO^-[46]. Thus, at 80°C a value of 2.0 is expected[46]. If a Brønsted relationship is assumed to hold for k^{DO^-} and K^{DO^-} (or k^{HO^-} and K^{HO^-}), then for half transfer of the proton in the transition state for a base-catalysed reaction k^{DO^-}/k^{HO^-} should be $2.0^{0.5}$, or 1.4. The exponent 0.5 is the value for the slope of a Brønsted plot when the proton is half transferred. Values greater than 1.4 are taken to indicate a more than half transferred proton, whereas values less than 1.4 are supposed to indicate less than half transfer of the proton in the transition state. Thus a comparison of k^{DO^-}/k^{HO^-} and k^H/k^D should reveal whether a low value for k^H/k^D signifies extensive or very little C—H bond rupture in the transition state.

4. Variation of transition state structure in E2 eliminations

The possible mechanisms for base-catalysed E2 elimination and the experimental methods for distinguishing between them have been very comprehensively reviewed in recent years[32-34] and thus will not be discussed here. Of particular relevance to the present discussion, however, are the theories which have been proposed to predict the effect of a structural or environmental change on the transition-state structure[47,48]. Winey and Thornton[49] have proposed a comprehensive model which purports to account for all known experimental data for elimination reactions. It is claimed that the known data are accommodated by this model if it is assumed that compounds with good leaving groups such as 2-phenethyl halides react via central

TABLE 1. Winey–Thorton predictions for change in transition-state structure as a result of structural and environmental changes[49]

	Transition state type			
	Central		ElcB-like	
Structural or environmental change	C—H bond rupture	C—X bond rupture	C—H bond rupture	C—X bond rupture
Poorer leaving group	Increase	Little change	Little change	Decrease
Stronger base	Little change	Decrease	Decrease	Little change
Electron-withdrawing group at C_β	Increase	Decrease	Decrease	Decrease
Electron-donating group at C_α	Decrease	Increase	Little change	Increase

transition states while 2-phenethyl onium compounds react via ElcB-like transition states. The precise details of the model need not concern us and in any case application of the theory is not easy; however, predictions have been made expressly for E2 elimination reactions and are summarized in Table 1. These predictions will serve as a basis for discussion of the isotope effects.

C. Experimental Evidence Relating to E2 Transition State Structure

1. Hammett ρ values

Although Hammett correlations do not strictly involve isotopically labelled sulphonium salts, the results which have been obtained are relevant to the interpretation of the isotope effects and have been widely used as experimental tests of the theory of variable E2 transition states. Most of the reported Hammett ρ values for sulphonium salts have been determined for base-catalysed E2 eliminations from dimethyl-2-phenethylsulphonium ions **28** and are collected in Table 2. Unfortunately, the data were obtained in different solvent systems and at different temperatures, which makes a direct comparison difficult; however, one can say that the values are very similar in magnitude and all are significantly positive. In any discussion of the ρ values, it is important to remember that the range of substituents used to measure ρ is limited, usually ranging only from

TABLE 2. Hammett ρ values for E2 elimination reactions of dimethyl-2-phenethylsulphonium ions (**28**)

Base–solvent system	Temperature (°C)	ρ	Reference
HO^-/H_2O		2.11	50
$EtO^-/EtOH$	30	2.64 ± 0.16	51
$i\text{-}PrO^-/i\text{-}PrOH$	30	3.47 ± 0.19	52
$HO^-/H_2O/20$ mol% Me_2SO	50	2.54 ± 0.05	50
$HO^-/H_2O/38$ mol% Me_2SO	50	2.56 ± 0.04	50
$HO^-/H_2O/50$ mol% Me_2SO	50	2.61 ± 0.1	50
$HO^-/H_2O/60$ mol% Me_2SO	50	2.59 ± 0.05	50
$HO^-/H_2O/50$ mol% Me_2SO	20	2.13 ± 0.04	53

$Z = p\text{-OMe}$ to $Z = p\text{-Cl}$, and that ρ values of around 0.2 are not normally considered to be significantly different from zero.

It is obvious from the data in $Me_2SO\text{-}H_2O$ mixtures that there is no significant effect on the ρ values as the Me_2SO concentration increases, even though the basicity of the medium is increasing and the dielectric constant is decreasing. Since ρ values normally show an inverse relationship with the dielectric constant of the medium, this factor alone should result in an *increase* in the ρ value. Between 7 and 70 mol-% $Me_2SO\text{-}H_2O$ there is a rate acceleration of about 90,000 for **38** $(Z{=}H)$ and **28** $(Z{=}H$ and $p\text{-COCH}_3)^{50,54,55}$, which is well accommodated by plots of $\log k$

$$ZC_6H_4CH_2CH_2Br$$

Z = H, unless otherwise stated

(38)

against $H_- + \log C_{H_2O}$, where H_- represents the thermodynamic basicity of the solvent[56]. The slopes of the plots are all near unity, as expected for a rate-determining proton transfer (consistent with either an E2 or ElcB mechanism), showing that the increase in rate constant is due primarily to those factors which lead to an enhanced solvent basicity and is not mainly a dielectric constant effect. The Winey–Thornton prediction (Table 2) for the effect of such an increase in basicity on an ElcB-like transition state is a decrease in ρ, since the extent of C—H bond rupture is expected to decrease, while little change is predicted for C—S bond rupture. As Table 2 shows, this is clearly not the case, however, it is possible that the predicted decrease in ρ occurs but is offset by the effect of the decrease in dielectric constant.

As a result of the dielectric constant effect the differences between ρ in H_2O and EtOH on the one hand and ρ in H_2O and $Me_2SO\text{-}H_2O$ on the other are probably not significant either. However, since the dielectric constants for EtOH and i-PrOH are not very different, the difference in ρ for these two solvents may reflect changes in transition-state structure. If the increase in ρ is due to a change in transition-state structure, it must indicate a more carbanionic transition state for the i-PrO$^-/i$-PrOH system, a medium which is less polar and more basic than ethanolic sodium ethoxide. In this case there is no doubt that the observed increase in ρ is contrary to the Winey–Thornton predictions for an ElcB-like transition state. The increase would be consistent with the predictions for a central transition state but a truly central transition state is inconsistent with the magnitude and sign of the ρ values for the sulphonium salts, since if C—H and C—S bond rupture have made equal progress very little substituent effect would be expected. Central transition states are unlikely even from the bromides **38** since a ρ value of 2.03 ± 0.12 in 50 mol% $Me_2SO\text{-}H_2O$ has been obtained at 20°C and in order to correlate the data σ^- values were required[53]. This must imply considerable advance of C—H bond rupture at the transition state as for the sulphonium salts **28**.

The ρ value for elimination from dimethyl-1-phenethylsulphonium ions (**39**) in

$$\overset{+}{S}(CH_3)_2$$
$$|$$
$$ZC_6H_4CHCH_3$$

(39)

ethanolic sodium ethoxide has been determined at 35°C as 0.95 ± 0.2^{31}. This value is significantly less than for elimination from **28** in the same solvent, but the positive sign clearly shows that C—H bond rupture is still advanced over C—S bond rupture despite the absence of a β-phenyl substituent, and clearly indicates that **39** does not utilize an El-like transition state (see Section IVC4).

2. Primary deuterium kinetic isotope effects in $Me_2SO–H_2O$ mixtures

As for the ρ values, most of the reported data on isotope effects have been obtained for elimination from the sulphonium ions **28** and most pertain to $Me_2SO–H_2O$ mixtures, although a few other systems have been investigated. The data for $Me_2SO–H_2O$ mixtures are collected in Tables 3 and 4. The most notable feature of the results is that there is little variation in the mean k^H/k^D values with increasing Me_2SO concentration for both $Z = H$ and $Z = p\text{-}COCH_3$ (Table 4), even though the rate increases by over 90,000[50,55]. β-Phenyl substitution, however, produces a more or less linear increase with the electron withdrawing power of Z (Table 3) for a more modest rate increase of only 2300[55]. When the data are

TABLE 3. Substituent effect on elimination from substituted dimethyl-2-phenethylsulphonium ions (**28**) in $HO^-/H_2O/50.2$ mol% Me_2SO[53]

Substituent Z	k^H/k^D	Temperature (°C)	k^{DO^-}/k^{HO^-}	Temperature (°C)
p-OMe	6.8 ± 0.1	20		
H	7.4 ± 0.1[a]	20	1.57 ± 0.04[c]	80.45
p-Cl	8.1 ± 0.2	20		
m-Br	8.4 ± 0.1	20		
p-COCH₃	9.6 ± 0.5[b]	20		
p-NMe₃	—	—	1.54 ± 0.02[c]	60

[a]For **38** $k^H/k^D = 8.5 ± 0.2$.
[b]For **38** $k^H/k^D = 9.7 ± 0.3$.
[c]k^{DO^-} determined in D_2O and k^{HO^-} determined in H_2O (reference 49).

TABLE 4. Isotope effects for the hydroxide ion-catalysed E2 elimination from **28** in mixtures of water and dimethyl sulphoxide. Temperatures (°C) are given in parentheses

Concentration of Me₂SO (mol%)	k^H/k^D		$^{32}S/^{34}S$, $Z = H^{[a,57]}$	$^{12}C/^{13}C$, $Z = H^{[a,58]}$
	$Z = H^{[50]}$	$Z = p\text{-}COCH_3^{[55]}$		
0	5.03(50°)		0.74(40°)	2.17 ± 0.18(30°)
2			0.65 ± 0.03(40°)	
5				3.15 ± 0.08(30°)[b]
6		7.9 ± 0.2(30°)		
10				2.32 ± 0.09(30°)
19.4	5.29 ± 0.09(50°)		0.11 ± 0.06(40°)	
22.2		8.6 ± 0.4(30°)		
26.3	5.55 ± 0.12(50°)			
32.6		9.2 ± 0.6(30°)		
40.0				2.01 ± 0.09(30°)
50.2	5.73 ± 0.03(50°)	9.4 ± 0.5(30°)		
59.5		9.5 ± 0.5(30°)		
60.3	5.61 ± 0.03(50°)			1.83 ± 0.11(30°)
63.5		9.2 ± 0.6(30°)		
69.6		8.9 ± 0.3(30°)		
71.8	4.79 ± 0.09(50°)			

[a]Isotope effects expressed as $(k^{light}/k^{heavy} - 1) \times 100$.
[b]A second determination gave 3.40 ± 0.14.

corrected for temperature differences by adding about 1.5 to the data for Z = H in Table 4[53,55], the entire range of results for Z = H lies between the values for Z = p-OMe and Z = H listed in Table 3. The data for Z = p-COCH$_3$ in Table 4 lie approximately between the values for Z = p-COCH$_3$ and Z = m-Br from Table 3 when also corrected for temperature differences; thus, it is clear that β-phenyl substitution is the more important determinant of the isotope effect.

The two values for Z = H in 50 mol% Me$_2$SO determined in different laboratories (Tables 3 and 4) agree reasonably well. Any differences in the values are probably due to the fact that the measured k^D values will be too large if the data points are collected too early in the reaction for deuteriated **28**, since deuteriation in the β-position is not complete. In the first half-life of this reaction there is a significant contribution from non-deuterated **28** which must be allowed for. This effect has been allowed for in the data in Table 3 but not in the data for Z = H in Table 4.

The k^H/k^D values for β-phenyl-substituted compounds in Table 3 and for Z = p-COCH$_3$ in Table 4 are larger than those calculated from simple treatments of concerted E2 reactions and are sufficiently different to give cause for concern[36,59]. For **28** (Z = p-OMe), it has been shown that the Arrhenius parameters (A) are abnormal even though the k^H/k^D value for this substituent is the *lowest* of the series (Table 3). No temperature dependence studies have yet been carried out for the remainder of the series since the differences in k^H/k^D caused by a 10°C rise in temperature are of the same order of magnitude as the uncertainty in the k^H/k^D values themselves, making extrapolation to obtain the A_H/A_D ratio imprecise[55]. Secondary isotope effects at C$_\beta$ are believed to be too small to account for the magnitude of the k^H/k^D values[38,60]. Therefore, in the absence of compelling experimental evidence for a tunnelling correction, it has been assumed that elimination from **28** (and **38**) is not very concerted, especially for Z = p-COCH$_3$[55]. This conclusion is supported on theoretical grounds by Saunders[38], who had to assume both a tunnelling correction and *minimal* coupling of the proton transfer motion with other atomic motions in the reaction coordinate motion of the transition state in order to force agreement between the theoretical calculations and the experimental values (Section IVA1).

The increase in k^H/k^D for **28** with an increase in the electron-withdrawing power of Z in 50 mol% Me$_2$SO–H$_2$O could indicate either a decreasing, or an increasing, extent of proton transfer to the hydroxide ion at the transition state depending on its initial structure. The so called Westheimer effect does not allow a decision between these two possibilities to be made on the basis of the k^H/k^D values alone. However, the secondary isotope effects (k^{DO^-}/k^{HO^-}) determined in aqueous sodium hydroxide (or deuteroxide) suggest that the change to a more electron-withdrawing Z group results in a *decrease* in the extent of proton transfer, since the secondary value for Z = p-NMe$_3$ is probably less than for Z = H when the temperature difference between the two measurements in Table 3 is taken into account. The magnitude of the secondary isotope effects probably represents about half transfer of the proton to the hydroxide ion in H$_2$O for Z = p-NMe$_3$; thus for Z = H the proton will be more than half transferred in H$_2$O. The fact that k^H/k^D increases for Z = H as a change is made from H$_2$O to 50 mol% Me$_2$SO–H$_2$O must therefore mean that the proton is less transferred to the base in the latter solvent than in pure H$_2$O. It must, however, still be more than half transferred, since the change to Z = p-COCH$_3$ causes a decrease in the extent of proton transfer according to the secondary isotope effects but the k^H/k^D value continues to rise. The value for Z = p-COCH$_3$ of nearly 10.0 is one of the largest values reported for an elimination reaction and presumably reflects an almost half transferred proton at the transition state for this substituent.

The conclusion that in 50 mol% Me_2SO the proton is more than half transferred to the base for Z = H is also supported by a comparison of the effect on k^H/k^D of changing from the sulphonium ions **28** to the bromides **38**. Table 3 shows that this change results in an increase in k^H/k^D, which again must indicate a decrease in the extent of proton transfer. When a change from **28** to **38** is made for Z = p-COCH$_3$ there is no significant change in k^H/k^D, both values suggesting a highly symmetrical proton position in the transition state.

When the data for the effect of adding increasing amounts of Me_2SO to a solution of **28** in HO^-/H_2O are examined, it can be seen that although there is an increase in going from H_2O to 50 mol% Me_2SO-H_2O, most of the increase has taken place by about 20 mol% Me_2SO-H_2O. Beyond this there is no further significant change which cannot be accounted for by experimental error even though the thermodynamic basicity of the solution increases by a factor of about 10^5. These results have been fairly widely claimed as evidence for a Westheimer maximum with the proton half transferred to the hydroxide ion at about 50 mol% $Me_2SO-H_2O^{50}$, but such a conclusion is inconsistent with the substituent effects on k^H/k^D and the secondary isotope effect data. The maximum rests mainly on the value for 7.19 mol% Me_2SO, which is likely to be imprecise since the reported half-life is at the limits of time resolution by conventional techniques (about 5 s) and the solution is prone to autoxidation[50]. Also at this concentration there will be a small contribution from the α'-β mechanism (see Section IIIC), which would be expected to lower the observed k^H/k^D value since the transition state for this mechanism is probably non-linear. The incursion of the α'-β pathway is probably not sufficient to lower the value as far as is observed; however, in view of the other factors mentioned, it must at present be viewed with suspicion.

The results for Z = p-COCH$_3$ provide even less convincing evidence for a Westheimer maximum, even though this compound must have a more symmetrical transition state, with respect to the position of the proton, than the unsubstituted compound and should therefore be more susceptible to changes in Me_2SO concentration. Apart from a modest increase from 6 to 32.6 mol% Me_2SO-H_2O, k^H/k^D shows no significant change, even at 70 mol% Me_2SO. If one nevertheless wishes to regard the maximum for Z = H as real, then it must be admitted that it is a very broad one and some problems of interpretation arise. Most seriously, it is difficult to explain why the value for Z = p-COCH$_3$ is so much larger than the largest value for Z = H at 50 mol% Me_2SO-H_2O. Even if it is argued that the higher value of k^H/k^D for Z = p-COCH$_3$ is a result of a less synchronous reaction coordinate motion, one is still left with the problem of the absence of a maximum as the Me_2SO concentration varies in this case. Also, **38** has a higher isotope effect than **28** even though the coupling of the various atomic motions should be *greater* for **38**.

3. Heavy atom isotope effects in Me_2SO-H_2O mixtures

Some work has been reported for sulphur isotope effects in elimination from **28** which sheds some light on changes in the extent of C—S bond rupture as the concentration of Me_2SO increases. Cockerill and Saunders[58] measured the sulphur isotope effect in the hydroxide ion catalysed elimination of Me_2S from **28** in Me_2SO-H_2O mixtures at 40°C. Unfortunately, the temperature is different yet again, but the temperature effect can be estimated since the sulphur isotope effect has been measured in water at two different temperatures (0.74-% at 40°C and 0.64-% at 59°C). As observed for the primary deuterium isotope effects, there is a decrease in the sulphur isotope effect between 0 and about 20 mol% Me_2SO-H_2O, but unfortunately no data are available at greater Me_2SO concentrations, since presumably the rates become too fast for the accurate determination of isotope ratios after partial reaction. According to the theoretical calculations these isotope effects represent a decrease in the extent of C—S bond rupture from about 50% in H_2O to less than 10% in 20 mol% Me_2SO-H_2O[37].

The β-carbon isotope effect has also been measured as a function of the Me_2SO concentration for elimination from **28** and the results are given in Table 4[57]. Despite some scatter at low Me_2SO concentrations, there is a general decrease in the values as the Me_2SO concentration is increased, steeply at first and then more or less levelling out at about 60 mol% Me_2SO-H_2O. It is interesting to note that again the main change in the isotope effect occurs at low Me_2SO concentration, between 0 and 10 mol% Me_2SO-H_2O in this case. Although it is not possible *a priori* to say how much proton transfer and how much double bond formation that a value of say 2% indicates, the calculations do suggest that small β-carbon isotope effects are associated with low extents of C—H bond rupture and larger values (up to 3–5%) indicate larger extents of C—H bond rupture. Thus, increasing amounts of Me_2SO are shown to indicate a decreasing extent of C—H bond rupture up to about 10 mol% Me_2SO, and beyond that no further significant change. It is an interesting observation that k^H/k^D, $^{32}S/^{34}S$, and $^{12}C/^{13}C$ ratios all show a decrease in the extent of both C—H and C—S bond rupture for Z = H from pure H_2O to around 20 mol% Me_2SO-H_2O, but no significant changes thereafter. Similar results are obtained for k^H/k^D when p-$COCH_3$ is the substituent.

4. Isotope effects in systems other than dimethyl-2-phenethylsulphonium ion

While the majority of the published work has focused on elimination from isotopically labelled **28**, some work has been carried out on simpler sulphonium salts. The sulphur isotope effect has been determined for E2 elimination reactions of t-butyldimethylsulphonium ion (**40**) in $EtO^-/EtOH$[61]. The value (0.72 ± 0.4% at

$$(CH_3)_3C\overset{+}{S}(CH_3)_2$$

(40)

24°C) indicates a reasonable amount of C—S bond rupture at the transition state, similar in fact to that indicated in HO^-/H_2O for elimination from **28** (0.74% at 40°C). This is surprising since **40** might be expected to eliminate via an E1-like transition state, when one considers the tertiary α-carbon and lack of a β-activating group, and thus have a largely ruptured C—S bond. The E1 reaction of **40** with HO^-/H_2O, for example, *does* have a large sulphur isotope effect of about 1.8% consistent with a carbonium ion-like transition state[62]. The value for E2 elimination from **40** indicates only about 50% rupture of the C—S bond, which would be consistent with some carbanion character in the transition state as required by the observed adherence of dimethyl-t-alkylsulphonium salts to the Hofmann rule[63].

A value of 5.6 for the primary deuterium isotope effect has been obtained for the elimination from **39** in $EtO^-/EtOH$ at 35°C[31]. This value has been calculated from the observed value of 5.9 by subtracting the contribution from the small amount of α'-β elimination, for which an isotope effect of 3.5 has been estimated. The lower value for the α'–β elimination is reasonable since the transition state for this mechanism must be non-linear and model calculations give smaller isotope effects for non-linear than for linear proton transfers[37,64]. The estimated isotope effect for the E2 reaction of **39** is very similar to that found for elimination from **28** and suggests an appreciable amount of C—H bond rupture in the transition state for **39**, despite the lack of an activating group in the β-position.

These observations lead one to raise the question of whether an E1-like E2 elimination is in fact attainable in practice. For example, the Hammett ρ value for E2 elimination from benzyldimethylcarbinyl chlorides **41** in methanol at 66°C to form β,β-dimethylstyrene **42** is +1.0, which again suggests a degree of carbanion

$$ZC_6H_4CH_2C(CH_3)_2Cl \qquad ZC_6H_4CH{=}C(CH_3)_2$$

$$\text{(41)} \qquad\qquad\qquad \text{(42)}$$

character despite the tertiary nature of the α-carbon atom and a good ionizing solvent[65]. An interesting feature of this reaction is the observation of a very low (2.5) primary deuterium isotope effect for Z = H, which originally led Bennett *et al.*[66] to assign this reaction to the nearly E1 side of the variable E2 transition state spectrum. In the light of the ρ value, however, one must assume either a carbanionic, reactant-like transition state or a very product-like transition state with extensive C—H and C—Cl bond rupture, although C—H bond rupture is of course still greater than C—Cl bond rupture.

In all three cases, **39**, **40**, and **41**, for which an E1-like transition state would have been predicted, the available evidence suggests that the transition states have carbanion character.

D. Transition State Structure for Elimination From Dimethyl-2-phenethyl sulphonium Ion

We are now in a position to compare the picture of the transition state for elimination from **28**, built up by assuming that isotope effects reflect the extent of rupture of the bond to the isotope, with the theoretical predictions provided by Winey and Thornton[49] (Table 1). If we take **28** in water as the reference point, the combined isotope effect data indicate that the C—H bond is more than half ruptured and the C—S bond is about half ruptured; thus there is carbanion character at C_β and a considerable degree of double bond character. This picture of the transition state is essentially that assumed by Winey and Thornton for an ElcB-like transition state. For the addition of a limited amount of Me_2SO (up to about 20 mol%) the extents of both C—H and C—S bond rupture decrease, but the extent of C—S bond rupture probably decreases more than does C—H bond rupture; thus the double bond character decreases and the extent of carbanion character increases. The transition state in effect becomes more reactant-like.

The environmental factor leading to these changes is believed to be the increase in thermodynamic basicity of the solutions (see Section IVC2). If increases in thermodynamic basicity are paralleled by increases in kinetic basicity, as seems reasonable on the basis of good linear plots of log k against $H_- + \log C_{H_2O}$ (a sort of Brønsted plot), then the Winey–Thornton predictions for an ElcB-like transition state suggest a decrease in the extent of C—H bond rupture (in agreement with the observed k^H/k^D values) but little change in the extent of C—S bond rupture (contrary to experiment). The ρ values should therefore decrease in more basic solutions below the value determined in water, since a decrease in carbanion character at C_β is expected. An increase in ρ is observed over this range of Me_2SO

concentrations, but the increase is modest and may or may not be indicative of increasing carbanion character. The one ρ value which is significantly different from the value in water is larger ($\rho_{i\text{-PrOH}} > \rho_{H_2O}$), clearly in disagreement with the Winey–Thornton predictions.

As the concentration of Me_2SO is further increased, and the thermodynamic basicity increases, little further change is detected by the primary deuterium or β-carbon isotope effects or Hammett ρ values. As previously mentioned, the sulphur isotope effect was not measured above 20 mol% Me_2SO, but it is unlikely to become much lower than the reported value. This lack of further variation is hard to understand and is certainly not predicted by theory.

When a change is made from **28** to the phenethyl bromide series **38**, the results indicate a further decrease in the extent of C—H bond rupture with what is assumed to be a change to a better leaving group. This is contrary to prediction but for Z = p-$COCH_3$ there is no change in the isotope effect, which is consistent with prediction. Again, the reason for the difference in behaviour is not obvious.

Finally, a change to more electron-withdrawing β-phenyl substituents causes a decrease in the extent of proton transfer in 50 mol% Me_2SO—H_2O until a fairly centrally situated proton is indicated in the transition state for Z = p-$COCH_3$. The substituent effect on the extent of C—H bond rupture is consistent with predictions but unfortunately no sulphur isotope effect data are available to test the prediction that the extent of C—S bond rupture should also decrease.

Thus, for sulphonium salts on the evidence available the combined results of both isotope effects and Hammett ρ values are not well accommodated by the Winey–Thornton predictions for the effect of structural and environmental changes on an ElcB-like transition state. The data do, however, support the assumption of an ElcB-like transition state at least in so far as C—H bond rupture is advanced over C—S bond rupture, that the C—H bond is more than half ruptured in all situations studied, and that the proton motion is not very highly coupled with other transition-state atomic motions. We are therefore forced to the conclusion that either the Winey–Thornton predictions are inappropriate, or that our original assumption that isotope effects give information about the extent of bond rupture in the transition state is invalid. The former conclusion seems more likely at present since some results by Saunders and Cockerill for trimethyl-2-phenethylammonium ions[32] appear to constitute a Westheimer maximum which cannot be accounted for by the experimental errors, and thus may provide evidence that isotope effects are indeed related to the extent of bond rupture. Unfortunately, there were not yet sufficient systematic data available for ammonium ions to allow the Winey–Thornton predictions to be tested adequately. Some suspicions about the interpretation of the isotope effects in Me_2SO–H_2O mixtures might be raised, however, by the pattern of most changes occurring in the region of 0–20 mol%, where solvation changes are likely to be most important, and little further change in regions where the thermodynamic basicity is increasing markedly.

While this is not the place to attempt a detailed criticism of the variable transition-state theory predictions, some comments can be made. Most of the theoretical treatments at some stage invoke the Hammond postulate[3], but as has been correctly pointed out by Fărcasiu[67], the Hammond postulate does not apply to E2 elimination reactions, since the potential energy of the transition state lies considerably above the potential energies of both reactants and products and no high energy intermediates are involved along the reaction coordinate. The transition state may therefore be influenced by factors which are not present in either reactants or products. A criticism can also be made of the use of so called potential energy surfaces to estimate the effects of structural changes on bonds remote from the site of substitution. Carbanion **43** and carbonium ion **44** species

$$BH^+ \quad \bar{C}-CX \qquad B \; HC-\overset{+}{C} \; X^-$$

$$(43) \qquad\qquad (44)$$

are represented as stable species on such surfaces[33], although such species do not actually exist for a normal E2 process. While it is conceivable that **43** might not be too different in potential energy from the E2 transition state for **28** (or **38**), it is unlikely that **44** is even remotely similar, in view of the marked reluctance of elimination reactions to populate the E1-like end of the E2 transition state spectrum. Thus, comparing the E2 transition state with such hypothetical intermediates is equivalent to comparing it with itself[67]

V. SUBSTITUTION REACTIONS

Sulphonium salts undergo substitution reactions with bases as well as elimination reactions and there has also been considerable interest in determining the effect of changes in substrate structure and reaction conditions on the transition state for substitution. Isotopically labelled sulphonium salts have been used as probes to decide whether a given structural or environmental change results in a shift to a more reactant-like or more product-like transition state. For an S_N2 process, a reactant-like transition state is one in which the bond between the nucleophile and the α-carbon atom is not formed to any great extent and the carbon-leaving group bond is not extensively ruptured. A product-like transition state results, on the other hand, if the new bond is extensively formed and the carbon-leaving group bond is extensively ruptured. Between these extremes we have the situation normally depicted in text books, in which both the new bond and the old bond have made equal progress constituting a 'central' transition state.

For an S_N1 reaction, which has the formation of a carbonium ion intermediate as the rate-determining step, we can refer to the transition state as being close to, or remote from, the carbonium ion in structure with extensive rupture of the carbon-leaving group bond, on the basis of the Hammond postulate[3]. Some variation in the degree of resemblance to the carbonium ion must be possible since in the electrophilic substitution reactions of benzene, for example, a range of ρ values is observed which suggests varying degrees of resemblance of the transition state, for these reactions, to the intermediate σ-complex.

S_N2 processes will also presumably utilize a range of transition-state structures, from reactant-like to product-like. The Hammond postulate is only of predicitive value for very exothermic reactions (reactant-like transition state) and for highly endothermic reactions (product-like transition state), since otherwise there are no high-energy intermediates whose structures may be investigated and hence serve as models for the transition state. To cope with reactions for which the Hammond postulate is inapplicable, Thornton[48] has proposed what has become known as the Swain–Thornton rule. This rule is based on predictions of the effects of perturbations on the vibrational potentials for the normal coordinate motions parallel and perpendicular to the reaction coordinate motion. For example, the prediction is made that change to a more reactive nucleophile will give a more reactant-like S_N2 transition state.

Most of the reported isotope effects relate to sulphur isotope effects, although some work has been carried out on the secondary isotope effects which result on α-deuteration. Saunders and co-workers have measured the sulphur isotope effect for the S_N1–El reaction of **40** in 97% ethanol at 40°C and in water at 59°C[61,62]. In these two solvents, **40** undergoes an S_N1 reaction via the rate-determining formation of the 2-methypropyl carbonium ion and thus the sulphur isotope effect should give an indication of the degree of resemblance of the transition state to the carbonium ion. In 97% ethanol, a value of $1.03 \pm 0.11\%$ has been obtained, whereas in water the value increases to $1.8 \pm 0.14\%$, leading to the conclusion that the C—S bond is more ruptured at the transition state in water than in ethanol. Although the absolute percentage of C—S bond rupture cannot be estimated from the results, both values confirm that the transition states are carbonium ion-like. Why the transition state should be more like the carbonium ion in H_2O is not obvious and is inconsistent with the predictions of the Swain–Thornton rule and the Hammond postulate.

In the S_N2 displacement of dimethyl sulphide by HO^-/H_2O from benzyldimethylsulphonium ions (**45**) at 60°C, the sulphur isotope effect decreases

$$ZC_6H_4CH_2\overset{+}{S}(CH_3)_2$$

(**45**)

as the electron-withdrawing power of the phenyl substituent increases, from $Z = p\text{-}CH_3$ ($0.96 \pm 0.03\%$) to $Z = m\text{-}Cl$ ($0.82 \pm 0.07\%$)[68]. The highest yield of alcohol product, obtained for the $Z = m$-chloro salt, was only 89%; thus the possibility of some E2 reaction cannot be excluded. Although the change in the sulphur isotope effect is very small, it would indicate a more product-like transition state for the $Z = p$-methyl salt, apparently as predicted by Thornton[48]. One could also take the view, however, that the isotope effects were not significantly different from each other and this conclusion would also be consistent with predictions for an unsymmetrical transition state[48]. Despite the uncertainty of interpretation with respect to changes in the transition state structure, the results do indicate an appreciable amount of C—S bond rupture. The extent of C—S bond rupture is almost as great as for the S_N1 solvolysis reactions of **45** which have also been measured in H_2O at 69°C[69]. In this case the sulphur isotope effects again show that the amount of C—S bond breaking at the transition state becomes less as the substituent changes from $Z = p\text{-}OMe$ ($1.08 \pm 0.02\%$) to the more electron-withdrawing $Z = m\text{-}Br$ ($0.97 \pm 0.04\%$), in agreement with the Swain–Thornton rule[48]. As judged from the magnitude of the isotope effects, the extent of C—S bond breaking at the transition state is almost the same in both the S_N1 and S_N2 reactions of **45** and thus the sulphur isotope effect is not a good indicator of mechanistic type.

Saunders and co-workers[70] studied the S_N2 reactions in ethanol at 60°C of trimethylsulphonium ion **1** with a variety of nucleophiles and found that the sulphur isotope effect decreased along the series Br^- ($1.36 \pm 0.03\%$), PhS^- ($1.20 \pm 0.01\%$), PhO^- ($1.04 \pm 0.03\%$), EtO^- ($0.96 \pm 0.2\%$), more or less in parallel with increasing basicity. Thus, decreasing basicity leads to more C—S bond rupture and hence a more product-like transition state, in agreement with prediction[48]. Since factors such as the strength of the forming carbon to nucleophile bond, or the mass of the nucleophile, which are unrelated to the extent of C—S bond rupture, could also influence the magnitude of the sulphur isotope effect, Saunders and co-workers[70] varied the basicity of the $EtO^-/EtOH$ nucleophile while keeping the nucleophilic atom constant, by the now traditional measure of adding varying amounts of Me_2SO. Addition of Me_2SO causes a very modest increase in the rate of reaction by a factor of about 10 over the range 0–77 mol% Me_2SO (compared with 90,000 for HO^-/H_2O catalysed elimination from **28**) and an increase in the reactant-like nature of the transition state, as shown by an almost linear *decrease* in the sulphur isotope effect from $0.95 \pm 0.03\%$ for EtO$^-$/EtOH to $0.35 \pm 0.04\%$ for 65 mol% Me$_2$SO–EtOH/EtO$^-$. This result is in agreement with the Swain–Thornton rule, since the carbon–nucleophile bond is likely to be much the same in all of these solutions, as is the mass of the nucleophile, and thus the most important variable is the increasing basicity of the solution.

Very little work has been reported on secondary α-deuterium isotope effects for sulphonium salts, partly because of the difficulty on measuring them accurately and interpreting them, but Wu and Robertson[71] have measured some k^H/k^D values for the S$_N$2 reactions of **1** in ethanol with a range of nucleophiles. The results for PhO$^-$ (1.211 at 76°C), EtO$^-$ (1.074 at 76°C) and PhS$^-$ (0.907 at 51°C) demonstrate a tendency to give inverse effects with stronger nucleophiles, since the expected temperature effect for PhS$^-$ is small. However, the values are difficult to interpret, especially as the departing SMe$_2$ group is also deuteriated and contributes a secondary isotope effect as well.

Some interesting results have been obtained for the substitution reactions of **39** with EtO$^-$/EtOH at 35°C. Careful analysis of the products of this reaction showed the formation of three substitution products, **46**, **47**, and **48**, and Hammett ρ values

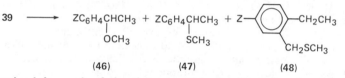

were determined for each of the three reactions by dissecting the measured rate constants on the basis of the product analysis[33]. Although the data are poor, there is a suggestion of curvature in the data for the formation of **46** since the rate constants for Z = *p*-Me and Z = *p*-Br are both faster than for Z = H. Such curved Hammett plots are usually explained by assuming changes in the ratio of bond formation to bond rupture in the transition state as the phenyl substituents are varied and are not uncommon for 1-phenethyl derivatives. The isotope effect, which is slightly inverse ($k^H/k^D = 0.95$) for **39** (Z = H) is consistent with a merged S$_N$1–S$_N$2 mechanism.

The second possible S$_N$2 reaction presumably involves an S$_N$2 attack by ethoxide ion on one of the *S*-methyl carbon atoms to give **47** and has a ρ value of about $+1.0$, as expected for development of negative charge on the leaving group, relative to the ground state, during the substitution. The rearrangement to give **48**, presumably via a Sommelet–Hauser rearrangement, has a large positive ρ value (4.8 ± 0.4) which is consistent with an aromatic nucleophilic substitution, as the rate-determining step, presumably by the *S*-methylene carbanion of the ylide resulting from **39**.

VI. REFERENCES

1. W. Von E. Doering and A. K. Hoffman, *J. Amer. Chem. Soc.*, **77**, 521 (1955).
2. M. L. Huggins, *J. Amer. Chem. Soc.*, **75**, 4126 (1953).
3. G. S. Hammond, *J. Amer. Chem. Soc.*, **77**, 334 (1955).
4. A. F. Cook and J. G. Moffatt, *J. Amer. Chem. Soc.*, **90**, 740 (1968).
5. H. Behringer and F. Scheidl, *Tetrahedron Lett.*, 1757 (1965).
6. W. Von E. Doering and K. C. Schreiber, *J. Amer. Chem. Soc.*, **77**, 514 (1955).
7. G. Barbarella, A. Garbesi and A. Fava, *Helv. Chim. Acta*, **54**, 2297 (1971).
8. J. Crosby and C. J. M. Stirling, *J. Chem. Soc. B*, 679 (1970).

9. D. J. Cram, *Fundamentals of Carbanion Chemistry*, Academic Press, New York, 1965, p. 48.
10. A. Streitwieser, Jr., R. A. Caldwell and W. R. Young, *J. Amer. Chem. Soc.*, **91**, 529 (1969).
11. O. Hofer and E. L. Eliel, *J. Amer. Chem. Soc.*, **95**, 8045 (1973).
12. A. Garbesi, G. Barbarella and A. Fava, *J. Chem. Soc. Chem. Commun.*, 155 (1973).
13. G. Barbarella, A. Garbesi, A. Boicelli and A. Fava, *J. Amer. Chem. Soc.*, **95**, 8051 (1973).
14. S. Wolfe, A. Rauk, L. M. Tel and I. G. Czismadia, *J. Chem. Soc. B*, 136 (1971).
15. R. R. Fraser, F. J. Schuber and Y. Y. Wigfield, *J. Amer. Chem. Soc.*, **94**, 8795 (1972).
16. B. M. Trost and L. S. Melvin, Jr., in *Sulphur Ylides*, Academic Press, London, 1975.
17. C. R. Johnson and W. G. Phillips, *J. Org. Chem.*, **32**, 1926 (1967).
18. C. R. Johnson and W. G. Phillips, *J. Org. Chem.*, **32**, 3233 (1967).
19. E. J. Corey and W. E. Russey, *J. Amer. Chem. Soc.*, **88**, 4751 (1966).
20. E. J. Corey, W. E. Russey and P. R. Ortiz de Montellano, *J. Amer. Chem. Soc.*, **88**, 4750 (1966).
21. P. D. G. Dean, P. R. Ortiz de Montellano, K. Bloch and E. J. Corey, *J. Biol. Chem.*, **242**, 3014 (1967).
22. E. J. Corey, K. Lin and M. Jautelat, *J. Amer. Chem. Soc.*, **90**, 2724 (1968).
23. E. J. Corey, P. R. Ortiz de Montellano and H. Yamamoto, *J. Amer. Chem. Soc.*, **90**, 6254 (1968).
24. E. J. Corey and M. Jautelat, *J. Amer. Chem. Soc.*, **89**, 3912 (1967).
25. D. N. Boykin, A. B. Turner and R. E. Lutz, *Tetrahedron Lett.*, 817 (1967).
26. W. Kirmse and M. Kapps, *Chem. Ber.*, **101**, 994 (1968).
27. W. Kirmse, M. Kapps and R. B. Hager, *Chem. Ber.*, **99**, 2855 (1966).
28. W. H. Saunders, Jr., and D. Pavlović, *Chem. Ind. London*, 180 (1961).
29. V. Franzen and M. J. Schmit, *Chem. Ber.*, **94**, 2937 (1961).
30. W. H. Saunders, Jr., S. D. Bonadies, M. Braunstein, J. K. Borchardt and R. T. Hargreaves, *Tetrahedron*, **33**, 1577 (1977).
31. F. L. Roe, Jr., and W. H. Saunders, Jr., *Tetrahedron*, **33**, 1581 (1977).
32. W. H. Saunders, Jr., and A. F. Cockerill, in *Mechanisms of Elimination Reactions*, Wiley–Interscience, New York, 1973, p. 79.
33. R. A. More O'Ferrall, in *The Chemistry of the Carbon Halogen Bond*, (Ed. S. Patai), Wiley–Interscience, London, 1973, p. 609.
34. A. F. Cockerill, in *Comprehensive Chemical Kinetics*, Vol. 9, (Ed. C. H. Bamford and C. F. H. Tipper), Elsevier, Amsterdam, 1973, p. 163.
35. A. V. Willi and M. Wolfsberg, *Chem. Ind. London*, 2097 (1964).
36. A. M. Katz and W. H. Saunders, Jr., *J. Amer. Chem. Soc.*, **91**, 4469 (1969).
37. W. H. Saunders, Jr., *Chem. Scr.*, **8**, 27 (1975).
38. W. H. Saunders, Jr., *Chem. Scr.*, **10**, 82 (1976).
39. R. P. Bell, *Quart. Rev.*, 513 (1974).
40. J. R. Jones, in *The Ionisation of Carbon Acids*, Academic Press, London, 1973, p. 150.
41. J. Bigeleisen, *J. Chem. Phys.*, **17**, 675 (1949).
42. L. Melander, in *Isotope Effects in Reaction Rates*, Ronald Press, New York, 1960.
43. F. H. Westheimer, *Chem. Rev.*, **61**, 265 (1961).
44. R. P. Bell, W. H. Sachs and R. L. Tranter, *Trans. Faraday Soc.*, **67**, 1995 (1971).
45. E. F. Caldin and S. Mateo, *J. Chem. Soc. Chem. Commun.*, 854 (1973).
46. R. L. Schowen, *Progr. Phys. Org. Chem.*, **9**, 275 (1972).
47. J. F. Bunnett, *Surveys Progr. Chem.*, **5**, 53 (1969).
48. E. R. Thornton, *J. Amer. Chem. Soc.*, **89**, 2915 (1967).
49. D. A. Winey and E. R. Thornton, *J. Amer. Chem. Soc.*, **97**, 3102 (1975).
50. A. F. Cockerill, *J. Chem. Soc. B*, 964 (1967).
51. W. H. Saunders, Jr., and R. A. Williams, *J. Amer. Chem. Soc.*, **79**, 3712 (1957).
52. A. F. Cockerill and W. J. Kendall, *J. Chem. Soc. Perkin Trans. II*, 1352 (1973).
53. L. F. Blackwell and J. L. Woodhead, *J. Chem. Soc. Perkin Trans. II*, 234 (1975).
54. L. F. Blackwell and J. L. Woodhead, *J. Chem. Soc. Perkin Trans. II*, 1218 (1975).
55. L. F. Blackwell, *J. Chem. Soc. Perkin Trans. II*, 488 (1976).

56. M. Anbar, M. Bobtelsky, D. Samuel, B. Silver and G. Yagil, *J. Amer. Chem. Soc.*, **85**, 2380 (1963).
57. J. Banger, A. Jaffe, A.-C. Lin and W. H. Saunders, Jr., *Faraday Symp. Chem. Soc.*, **10**, 113 (1975).
58. A. F. Cockerill and W. H. Saunders, Jr., *J. Amer. Chem. Soc.*, **89**, 4985 (1967).
59. R. P. Bell, *Discuss. Faraday Soc.*, **39**, 16 (1965).
60. H. Simon and G. Mullhofer, *Chem. Ber.*, **97**, 2202 (1964).
61. W. H. Saunders, Jr., and S. E. Zimmerman, *J. Amer. Chem. Soc.*, **86**, 3789 (1964).
62. W. H. Saunders, Jr., and S. Asperger, *J. Amer. Chem. Soc.*, **79**, 1612 (1957).
63. E. D. Hughes, C. K. Ingold and L. I. Woolf, *J. Chem. Soc.*, 7084 (1948).
64. R. A. More O'Ferrall, *J. Chem. Soc. B*, 785 (1970).
65. L. F. Blackwell, A. Fischer and J. Vaughan, *J. Chem. Soc. B*, 1084 (1967).
66. J. F. Bunnett, G. T. Davis and H. Tanida, *J. Amer. Chem. Soc.*, **84**, 1606 (1962).
67. D. Fǎrcasiu, *J. Chem. Educ.*, **52**, 76 (1975).
68. C. G. Swain and E. R. Thornton, *J. Org. Chem.*, **26**, 4808 (1961).
69. M. P. Friedberger and E. R. Thornton, *Diss. Abstr. Int. B*, **36**, 2222 (1976).
70. R. T. Hargreaves, A. M. Katz and W. H. Saunders, Jr., *J. Amer. Chem. Soc.*, **98**, 2614 (1976).
71. C.-Y. Wu and R. E. Robertson, *Chem. Ind. London*, 1803 (1964).

The Chemistry of the Sulphonium Group
Edited by C. J. M. Stirling and S. Patai
© 1981 John Wiley & Sons Ltd.

CHAPTER **9**

Electronic effects of the sulphonium group

JOHN SHORTER

Department of Chemistry, The University, Hull HU6 7RX, England

I.	INTRODUCTION	188
	A. Sulphur Bonding	188
	B. Electronic Effects and π-Bonding	189
	C. The Scope of this Chapter	191
II.	THE HAMMETT EQUATION	191
	A. Introduction	191
	B. Multiparameter Extensions	194
	C. The Behaviour of Unipolar Substituents	195
III.	SIGMA VALUES IN AROMATIC SYSTEMS	197
	A. Sigma Values from the Ionization of Benzoic Acids or Phenols	197
	B. Sigma Values from Spectroscopic Measurements	198
IV.	ELECTRONIC EFFECTS IN ELECTROPHILIC SUBSTITUTION	201
	A. Early Studies of the Nitration of Sulphonium Compounds	201
	B. Modern Studies of Electrophilic Substitution in Sulphonium Compounds	201
V.	ELECTRONIC EFFECTS IN NUCLEOPHILIC SUBSTITUTION	204
	A. The Sulphonium Group in the Substrate	204
	B. Aspects of Attack at the Sulphonium Group	205
	C. The Sulphonium Group in the Nucleophile	206
VI.	STABILIZATION OF CARBANIONIC CENTRES BY THE SULPHONIUM GROUP	207
	A. Introduction	207
	B. pK_a Values for C—H Bond Ionization	208
	C. Kinetics of H–D Exchange Reactions	212
	D. Promotion of Reactions via a Carbanionic Intermediate	215
VII.	ELECTRONIC EFFECTS IN ALIPHATIC SYSTEMS	217
	A. Introduction	217
	B. Ester Hydrolysis	217
	C. Charged and Uncharged Substituents: a Common Scale for σ^*	219
VIII.	THE *ORTHO*-EFFECT OF THE SULPHONIUM GROUP	220

IX.   THE IRRELEVANCE OF *d*-ORBITAL CONJUGATION(?)   .   .   .   222
X.   REFERENCES AND NOTES   224

I. INTRODUCTION

A. Sulphur Bonding[1,2]

The electronic structure of the sulphur atom in its ground state may be written as $1s^2\,2s^2\,2p^6\,3s^2\,3p_x^2\,3p_y^1\,3p_z^1$. When sulphur forms two single bonds to carbon, as in dimethyl sulphide, this may be pictured in terms of the overlap of singly occupied $3sp^3$ hybridized orbitals on sulphur with singly occupied $2sp^3$ orbitals on carbon. Two doubly occupied, localized molecular orbitals of the σ-type are thereby formed. Two unshared pairs of electrons in the valence shell of the sulphur atom are left in the remaining $3sp^3$ orbitals. In accord with this simple picture, the bond angle ∠CSC in a dialkyl sulphide is approximately tetrahedral (109° 28′; see **1**), e.g. in dimethyl sulphide it is *ca*. 105°.

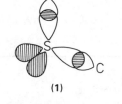

(1)

The formation of a further single bond between sulphur and carbon, as in the trimethylsulphonium cation, may be pictured as involving a $3sp^3$ unshared pair orbital on sulphur and an empty $2sp^3$ orbital on carbon in a methyl cation. Thus the three σ bonds and the remaining unshared pair (in a $3sp^3$ orbital) in a trialkylsulphonium ion are distributed approximately tetrahedrally, i.e. the ion is pyramidal, with the sulphur atom at the apex (**2**).

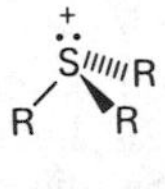

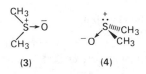

(2)

When the third bond is formed to oxygen rather than carbon, the situation is not so simple. Firstly, the formation of dimethyl sulphoxide can be pictured in a way analogous to the above, an empty $2sp^3$ orbital on oxygen being involved. The bond between sulphur and oxygen is then a coordinate bond and the structure is appropriately written as **3** or **4**. At this point, however, the possible contribution of

CH₃
  \\
   S⁺→O̅
  /
CH₃

(3)

        ⁺
    ..S⁗CH₃
   /   \\
⁻O   CH₃

(4)

a 3d orbital on sulphur must be considered. One of the two electrons in an unshared-pair $3sp^3$ orbital of dimethyl sulphide may be pictured as transferred to an appropriate 3d orbital, e.g. $3d_{xy}$. The oxygen atom is considered to be in a $2sp^2$

hybridized state, with two unpaired electrons, one in one of the $2sp^2$ orbitals and the other in the unhybridized $2p_y$ orbital. A σ bond is now formed by the end-on overlap of the singly occupied $3sp^3$ orbital of sulphur, with the singly occupied $2sp^2$ orbital of oxygen, while a π bond is formed by the sideways overlap of the $3d_{xy}$ orbital with the unhybridized $2p_y$ orbital (see **5**).

Considered in this way, the structure of dimethyl sulphoxide involves a double bond and a valence shell of ten electrons for sulphur (**6** and **7**)[3,4].

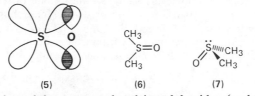

(5)   (6)   (7)

The nature of the sulphur–oxygen bond in sulphoxides (and also in sulphones) has been much discussed. The actual structure must be intermediate between the canonical structures **3** and **6**, and the discussion concerns the relative contributions of these. We need not go into this here.

While the expansion of the valence shell beyond the electron octet and the participation of 3d orbitals are not issues for trialkylsulphonium ions, the matter is of considerable importance for dialkylsulphonium groups attached to unsaturated systems, as, for example, in $PhSMe_2$ or $CH_2=CHSMe_2$, or in the ylide $RCHSMe_2$, each of which may be regarded as a resonance hybrid of two or more canonical structures (**8, 9, 10**).

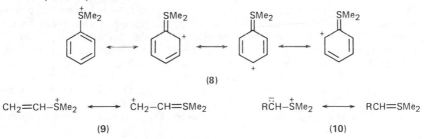

(8)

(9)   (10)

B. Electronic Effects and π-Bonding

The importance of $\pi(pd)$ bonding in the ground-state structures of molecules such as those portrayed in **8–10** and in governing the electronic effects of the sulphonium group (and other sulphur substituents) on chemical reactivity and physical properties has long been recognized[1-4]. However, the most important feature of the sulphonium group in determining its electronic effects is its unit positive charge. Whether the group is attached to a saturated or an unsaturated system, it exerts an electron-withdrawing inductive effect. While such an effect is often regarded as relayed to a distant point in the molecule via polarization of successive bonds, there is now much evidence that the so-called inductive effect is often largely a field effect, transmitted partly through the molecular cavity and partly through the surrounding medium[5]. Such an interpretation seems particularly appropriate for unipolar substituents such as the sulphonium group.

A sulphonium substituent such as the dimethylsulphonium group would be expected to show considerable similarity in its electronic effects to other unipolar substituents, e.g. the trimethylammonium group, $\overset{+}{N}Me_3$. Thus the substitution of NMe_3 or $\overset{+}{S}Me_2$ for H in the methyl group of acetic acid increases greatly the acid strength, the relevant pK_a values in water being 1.83 (25°C)[6] and 1.15 ± 0.1 (30°C; ionic strength 0.2)[7], respectively, compared with 4.76 (30°C)[8] for acetic acid. Further, it has long been known that the trimethylphenylammonium[9] and dimethylphenylsulphonium[10] ions are nitrated almost entirely in the *meta*-position in preference to the *ortho*- or *para*-position.

Certain other processes, however, show an important difference between the groups: that valence shell expansion can occur with sulphur but not with nitrogen. Thus, Doering and Schreiber[11] found in 1955 that the 2-bromoethyldimethyl-sulphonium ion rapidly reacted with hydroxide ion to give the dimethylvinyl-sulphonium ion (equation 1). In contrast, the 2-bromoethyltrimethylammonium

$$BrCH_2CH_2\overset{+}{S}Me_2 + OH^- \longrightarrow CH_2{=}CH\overset{+}{S}Me_2 + Br^- + H_2O \qquad (1)$$

ion, $Br(CH_2)_2\overset{+}{N}Me_3$ reacted with hydroxide ion much more slowly and substitution, not elimination, occurred. This difference was attributed to the stabilization of the dimethylvinylsulphonium ion by resonance (see **9** above). The same authors[11] found that a variety of bases would add rapidly to the dimethylvinylsulphonium ion to give 2-substituted ethyldimethylsulphonium ions (equation 2), but similar additions

$$CH_2{=}CH\overset{+}{S}Me_2 + B^- + H_2O \longrightarrow BCH_2CH_2\overset{+}{S}Me_2 + OH^- \qquad (2)$$

to the trimethylvinylammonium ion did not occur. The difference in behaviour was attributed to the stabilization of the relevant transition state by resonance involving valence shell expansion in the case of the sulphonium ion.

The important role of the 3d orbitals in the reactions of sulphonium compounds was confirmed by Doering and Hoffmann[12] in the discovery that these sometimes resembled those of the analogous compounds of the neighbouring second short period element, phosphorus. Thus, $\overset{+}{S}Me_3$ and $\overset{+}{P}Me_4$ were found readily to exchange H for D in $OD^-{-}D_2O$ solutions, while $\overset{+}{N}Me_4$ reacted much more slowly.

In principle there are other kinds of orbital interaction involving sulphur and unsaturated carbon which would produce π bonding. Thus, in the dimethyl-phenylsulphonium ion the unshared pair of electrons in the valence shell of sulphur could be delocalized into the aromatic system, **11**. It is well known that an SMe

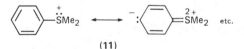

(11)

group may engage in such π(pp) bonding[13]. However, in sulphonium compounds the large charge separation involved makes such interactions less likely and there is little experimental evidence to support the idea. A further possibility is π(pp) bonding involving the movement of electrons from the aromatic ring to the side-chain. For this to occur the unshared pair on S would have to be in a 3d orbital.

There is evidence against any important role for π(pp) bonding in the case of a sulphonium group attached to a benzene ring. When π(pp) bonding is known to occur between the benzene ring and a side-chain, it is always subject to rather stringent stereochemical requirements for the p–p overlap to be effective, i.e. the side-chain must adopt a certain conformation with respect to the ring so that 'sideways' overlap of the p orbitals can occur. The introduction of bulky groups into the positions *ortho* to the side-chain may prevent the necessary conformation being attained, and there is then 'steric inhibition of resonance'; that is the partial or complete elimination of π bonding between the benzene ring and the side-chain, with observable consequences in chemical reactivity or spectra[14]. There is little evidence for steric inhibition of the π bonding between sulphur and aromatic carbon in sulphonium compounds[15]. (The same is true for aromatic sulphoxides and sulphones[16].) This is reasonable for π(pd) bonding, which should be less sensitive to conformation than π(pp) bonding, because of the geometry and number of 3d orbitals available on sulphur. Rotation about the ring–sulphur bond by 90° brings into line another 3d orbital of sulphur which can be used for π(pd) bonding, cf. the situation for p orbitals and π(pp) bonding[17].

In the later sections of this chapter, the important role of the 3d orbitals of sulphur will be emphasized. It is, however, necessary to enter a *caveat*. For many years it has been maintained that it is not necessary to invoke a bonding role for the 3d orbitals of sulphur (or in analogous situations for various other elements)[18]. Such views do not seem to have made much impact on organosulphur chemistry but recently there has been a move towards taking them more seriously[18,19]. In this chapter it seems best to adhere to tradition and make frequent use of the symbol π(pd) and related terminology, but it must be recognized that within a few years such an approach may be regarded as outmoded. The recent developments are referred to briefly in the last section of this chapter.

There is one other *caveat*. In discussing the electronic effect of the sulphonium group it is easy to lose sight of the fact that sulphonium compounds are made up of sulphonium ions and *counter ions*, such as chloride, bromide, tosylate, and fluorosulphonate. In many cases the observed properties are those of the sulphonium ion and are independent of the nature of the counter ion, particularly when the experiments involve very dilute solutions in highly polar solvents. However, a possible role of the counter ion must not be forgotten altogether, particularly when it is known to be prone to enter into ion pairing with cations and when relatively concentrated solutions are involved (as they may be, for instance, in n.m.r. studies). For this reason, the nature of the counter ion will frequently be mentioned in the later sections.

C. The Scope of this Chapter

The next section contains a summary of the salient features of the Hammett equation and its extensions, with particular reference to the behaviour of unipolar substituents in such linear free-energy relationships. This prepares the ground for a discussion of the electronic substituent parameters (sigma values) of the sulphonium group attached to aromatic systems. Later sections deal with substituent effects in electrophilic substitution, nucleophilic substitution, the stabilization of carbanionic centres, and aliphatic systems. A section on the *ortho*-effect of the sulphonium group is followed by a section on the current dispute about the role of d-orbitals.

II. THE HAMMETT EQUATION[20]

A. Introduction

The Hammett equation is the best known example of a linear free-energy relationship (LFER); that is an equation which implies a linear relationship between free energies of reaction or activation for two related processes[21]. It describes the influence of polar *meta-* or *para*-substituents on reactivity for side-chain reactions of benzene derivatives.

The Hammett equation (1937)[22,23] takes the form

$$\log k = \log k^0 + \rho\sigma \tag{3}$$

or

$$\log K = \log K^0 + \rho\sigma \tag{4}$$

The symbol k or K is the rate or equilibrium constant, respectively, for a side-chain reaction of a *meta-* or *para*-substituted benzene derivative, and k^0 or K^0 denotes the statistical quantity approximating to k or K for the 'unsubstituted' or 'parent' compound. The *substituent constant*, σ, measures the polar effect (relative to hydrogen) of a substituent (in the *meta-* or *para*-position) and is, in principle, independent of the nature of the reaction. The *reaction constant*, ρ, depends on the nature of the reaction (including conditions such as solvent and temperature) and measures the susceptibility of the reaction to polar effects. The ionization of benzoic acids in water at 25°C is the standard process for which ρ is defined as 1.000. The value of σ for a given substituent is thus $\log(K_a/K_a^0)$, where K_a is the ionization constant of the substituted benzoic acid, and K_a^0 is that of benzoic acid itself. Selected values of σ for well known substituents are given in Table 1. They are readily interpreted qualitatively in simple electronic terms, i.e. through the inductive (I) effect and the resonance or conjugative (K) effect.

Jaffé (1953)[24] showed that while many rate or equilibrium data conform well to the Hammett equation (as indicated by the correlation coefficient), many such data are outside the scope of the equation in its original form and mode of application.

Deviations are commonly shown by *para*-substituents with considerable $+K$ or $-K$ effect[25]. Hammett himself found that p-NO$_2$ showed deviations in the correlation of reactions of anilines or phenols. The deviations were systematic in that a σ value of *ca.* 1.27 seemed to apply, compared with 0.78 based on the ionisation of p-nitrobenzoic acid. Other examples were soon discovered and it became conventional to treat them similarly in terms of a 'duality of substituent constants'.

TABLE 1. Selected values[a] of σ, σ^+, and σ^- constants

Substituent	σ_m	σ_p	σ_p^+	σ_p^-
Me	−0.07	−0.17	−0.31	—
OMe	0.12	−0.27	−0.78	—
SMe	0.15	0.00	−0.60	0.21
OH	0.12	−0.37	−0.92	—
SH	0.25	0.15	—	—
NMe$_2$	−0.15	−0.63	−1.7	—
F	0.34	0.06	−0.07	—
Cl	0.37	0.23	0.11	—
CF$_3$	0.43	0.54	—	0.65
CN	0.56	0.66	—	0.88
NO$_2$	0.71	0.78	—	1.24
SO$_2$Me	0.64	0.72	—	0.98

[a]These values, drawn from various sources, are presented solely for illustration. The table should not itself be used uncritically as a source of σ values for correlations. See rather references 23 and 60.

When σ values based on the ionization of benzoic acid are used, deviations may occur with $-K$ *para*-substituents for reactions involving $+K$ electron-rich reaction centres, and with $+K$ *para*-substituents for reactions involving $-K$ electron-poor reaction centres. The explanation of these deviations is in terms of 'cross-conjugation', i.e. conjugation involving substituent and reaction centre.

In the ionization of the *p*-nitroanilinium ion, the free base is stabilized by delocalization of electrons involving the canonical structure **12**. An analogous

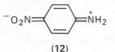

(12)

structure is not possible for the *p*-nitroanilinium ion. In the ionization of *p*-nitrophenol, analogous delocalization is possible in both phenol and phenate species, but is more marked in the ion. Thus, in both the aniline and the phenol system p-NO$_2$ is effectively more electron-attracting than in the ionization of benzoic acid, where the reaction centre is incapable of a $+K$ effect, and indeed shows a small $-K$ effect (**13**).

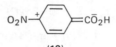

(13)

An example of a reaction series in which large deviations are shown by $+K$ *para*-substituents is provided by the rate constants for the solvolysis of substituted *t*-cumyl chlorides, ArCMe$_2$Cl[26]. This reaction follows an S$_N$1 mechanism, with intermediate formation of the cation ArCMe$_2^+$. A $+K$ *para*-substituent such as OMe may stabilize the activated complex, which resembles the carbocation–chloride ion pair, through delocalization involving structure **14**. Such delocalization will clearly be more pronounced than in the species involved in the ionization of *p*-methoxybenzoic acid, which has a reaction centre of feeble $-K$ type (**15**). The

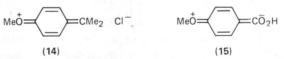

(14)   (15)

effective σ value for *p*-OMe in the solvolysis of *t*-cumyl chloride is thus -0.78, compared with the value of -0.27 based on the ionization of benzoic acids.

The special substituent constants for $-K$ *para*-substituents are denoted by σ⁻, and those for $+K$ *para*-substituents are denoted by σ⁺[26]. They are based respectively on the reaction series discussed above. Selected values are given in Table 1. The range of applicability of the Hammett equation is greatly extended by means of σ⁻ and σ⁺, notably to nucleophilic (by σ⁻) and to electrophilic (by σ⁺) aromatic substitution.

However, the 'duality of substituent constants' and the attempt to deal with cross-conjugation by selecting σ⁺, σ, or σ⁻ in any given case is somewhat artificial. The contribution of the resonance effect of a substituent relative to its inductive effect must in principle vary continuously as the electron-demanding quality of the reaction centre is varied, i.e. whether it is electron rich or electron poor. A sliding scale of substituent constants would be expected for each substituent having a resonance effect and not just a pair of discrete values: σ⁺ and σ for $+K$, or σ⁻ and σ for $-K$ substituents[27].

B. Multiparameter Extensions[28]

Two main types of treatment have emerged, both involving multiparameter extensions of the Hammett equation.

In the Yukawa–Tsuno equation (1959)[29] (equation 5), the 'sliding scale' is provided by multiple regression on σ, and $(\sigma^+ - \sigma)$ or $(\sigma^- - \sigma)$ depending on whether the reaction is more or less electron demanding than the ionization of benzoic acid.

$$\log k = \log k^0 + \rho[\sigma + r^{\pm}(\sigma^{\pm} - \sigma)] \tag{5}$$

(There is a corresponding form of the equation for equilibria.) The quantity $r^{\pm}$ gives the contribution of the enhanced $\pm K$ effect in a given reaction. (The equation was modified in 1966[30] to use σ^0 instead of σ values, see below, but the essential principles are unaltered.)

In the form of treatment developed by Taft and his colleagues since 1956[31-33], the Hammett constants are analysed into inductive and resonance parameters, and the sliding scale is then provided by multiple regression on these. Equations 6 and 7 show the basic relationships.

$$\sigma_m = \sigma_I + 0.33\sigma_R(\text{BA}) \tag{6}$$

$$\sigma_p = \sigma_I + \sigma_R(\text{BA}) \tag{7}$$

The σ_I scale is based on alicyclic and aliphatic reactivities (see below), and the factor 0.33 in equation 6 is the value of a 'relay coefficient', α, giving the indirect contribution of the resonance effect to σ_m. However, the ionization of benzoic acids is not regarded as an entirely satisfactory standard process, since it is subject to some slight effect of cross-conjugation (see structure 15 above). Consideration of 'insulated series', not subject to this effect, e.g. the ionization of phenylacetic acids, is used as the basis of a σ^0 scale, which can be analysed by equations 8 and 9[34,35]. (Note the different value of α.)

$$\sigma_m^0 = \sigma_I + 0.5\sigma_R^0 \tag{8}$$

$$\sigma_p^0 = \sigma_I + \sigma_R^0 \tag{9}$$

Analysis of σ^+ and σ^- constants correspondingly involves σ_R^+ and σ_R^-.

Multiple regression on σ_I- and σ_R-type parameters employs the 'dual substituent-parameter' equation, which may be written as in the equation[36]

$$\log(k/k^0) = \rho_I\sigma_I + \rho_R\sigma_R \tag{10}$$

For any given reaction series the equation is applied to *meta*- and *para*-substituents separately, and so values of ρ_I and ρ_R characteristic both of reaction and of substituent position are obtained. The various σ_R-type scales are linearly related to each other only approximately. In any given application the scale which gives the best correlations must be found[37].

Values of σ^0, σ_I, and σ_R-type parameters for certain substituents are given in Table 2.

The correlation analysis of spectroscopic properties in terms of σ_I- and σ_R-type parameters has been very important. Substituent effects on ^{19}F n.m.r. shielding in fluorobenzenes have been studied in great detail by Taft and his colleagues[34,38,39]. For δ_m^F linear regression on σ_I is on the whole satisfactory, but a term in σ_R^0 with a small coefficient is sometimes introduced. The correlation analysis of δ_p^F, however,

TABLE 2. Selected values[a] of σ^o, σ_I. and σ_R-type constants

Substituent	σ^o_m	σ^o_p	σ_I	$\sigma_R(BA)$	σ^0_R	σ^+_R	σ^-_R
Me	−0.07	−0.15	−0.05	−0.12	−0.10	−0.25	—
OMe	0.06	−0.16	0.26	−0.53	−0.41	−1.02	—
NO$_2$	0.70	0.82	0.63	0.15	0.19	—	0.61
F	0.35	0.17	0.52	−0.46	−0.35	−0.57	—
Cl	0.37	0.27	0.47	−0.24	−0.20	−0.36	—

[a]See footnote to Table 1.

requires terms in both σ_I and σ_R-type parameters, with σ^0_R being widely applicable. Many new values of these parameters have been assigned from fluorine chemical shifts.

The correlation analysis of i.r. data has been much examined by Katritzky, Topsom, and their colleagues[40,41]. Thus, the intensities of the ν_{16} ring-stretching bands of mono- and disubstituted benzenes may be correlated with the σ^0_R values of the substituents and these correlations may be used to find new σ^0_R values. Also, Schmid[42] has shown that ν_{CH} intensities in substituted benzenes may be analysed in terms of σ_I.

Finally, in this account of multiparameter extensions of the Hammett equation, we comment briefly on the origins of the σ_I scale. This had its origins around 1956[33] in the σ' scale of Roberts and Moreland[43] for substituents X in the reactions of 4-substituted bicyclo[2.2.2]octane-1 derivatives. However, at that time few values of σ' were available. A more practical basis for a scale of inductive substituent constants lay in the σ^* values for XCH$_2$ groups derived from Taft's analysis of aliphatic ester reactions[33,34]. For the few σ' values available it was shown that σ' for X was related to σ^* for CH$_2$X by the equation $\sigma' = 0.45\sigma^*$. Thereafter, the factor 0.45 was used to calculate σ_I values of X from σ^* values of CH$_2$X[45].

C. The Behaviour of Unipolar Substituents

In the original presentation of the Hammett equation (1937)[46], substituent constants were not tabulated for any unipolar substituent. By the time Jaffé (1953)[24] reviewed the state of the Hammett equation, σ values were available for several 'cationic' or 'anionic' substituents, e.g. m- and p-NMe$_3$, m-NH$_3$, m- and pCO$_2^-$, p-SO$_3^-$, and others. Jaffé commented[47], 'Since the available data indicate no greater uncertainty for substituent constants of ionic substituents than for those of neutral groups, the Hammett equation also appears to be applicable to substituents which carry an integral charge'. He also pointed out, however, that ionic substituents will interact strongly with polar solvents, and that their substituent constants might be expected to be particularly solvent dependent. (Solvent dependence was already indicated for certain dipolar substituents, e.g. m- or p-OH.) Further, a given reaction of a substrate containing an ionic substituent is of a charge type different from that of the same reaction of a substrate carrying a dipolar substituent. This difference would have consequences for entropy of activation or reaction. Thus, a reaction series involving both ionic and dipolar substituents would be unlikely to meet the requirement of constant entropy change or a linear relationship between entropy and enthalpy believed necessary for conformity to the Hammett equation.

Almost 20 years later, Exner considered that the peculiarities of unipolar substituents were such that systems involving these substituents should not be regarded as conforming to the Hammett equation. He wrote[48], 'A special problem is posed by unipolar substituents, which should be excluded from the validity range owing to the dependence of their effects on solvent and ionic strength. A number of formal σ constants of cationic and anionic substituents have been evaluated. These should be discussed separately and not included in ordinary correlations'. In support of this contention, Exner cited several papers and their salient features are summarized here.

Zollinger et al.[49] determined substituent constants for m- and p-SO_3^- from the ionization of the appropriate benzoic acids, phenols, and anilinium ions in water. The values depended markedly on ionic strength, particularly those based on the ionization of benzoic acids, e.g. for p-SO_3^-, σ = 0.09 when μ = 0.001, and σ = 0.32 when μ = 0.1. Zollinger[50] also measured rate constants for the coupling reactions of substituted benzenediazonium ions with the anion of 2,6-naphthylaminesulphonic acid. For dipolar substituents there was a good correlation of log k with ordinary Hammett σ values, but p-SO_3^- could in no way be made to conform: it behaved as if it had a σ value of about 0.4.

In one of their early papers on the separation of inductive and resonance effects, Taft and Lewis[31] commented that while the data for unipolar substituents were somewhat limited, the points for such substituents frequently deviated seriously from correlation lines established by means of dipolar substituents.

Okamato and Brown[51] studied the influence of m- or p-CO_2^-, and of m- or p-NMe_3 on the rate of solvolysis of t-cumyl chloride. CO_2^- in either position exerted a small accelerating effect and this was readily explicable. NMe_3 exerted a strong retarding effect, corresponding to σ+ values of ca. 0.4, i.e. very much lower than $σ_m$ = 1.02 and $σ_p$ = 0.88, based on the ionization of benzoic acid. The authors found this situation surprising, since for other strongly meta-directing groups, such as the nitro-group, σ and σ+ agree closely. They comment that 'the position of groups carrying a charge in both the usual Hammett treatment and in our proposed extension to electrophilic reactions is quite uncertain, and great caution should be exercised in utilizing these treatments to correlate rates and equilibria involving such groups'. They also refer to the entropy question mentioned above.

Willi[52] studied the influence of m- or p-NMe_3 on the ionization of benzoic acid, phenol, anilinium ion, and dimethylanilinium ion in water. The benzoic acids were also studied in 20% and 40% aqueous dioxan. Values of σ were extrapolated to zero ionic strength. In water, depending on the system studied, $σ_m$ for NMe_3 varied between 0.82 and 0.99, and $σ_p$ between 0.80 and 0.96. The addition of dioxan, to decrease the dielectric constant, increased the difference between $σ_m$ and $σ_p$, while it seemed possible that in solvents of dielectric constant >100, the order might be reversed, i.e. $σ_p > σ_m$.

The purpose of the above discussion is to emphasize that the substituent constants for sulphonium groups to be presented in the following section must be discussed and applied cautiously. It is most unwise to analyse or interpret the substituent constants of these groups, to compare values with those for dipolar or other unipolar groups, or to include data for sulphonio-substituted compounds in correlations without regard to the limitations which these substituent constants inevitably have.

III. SIGMA VALUES IN AROMATIC SYSTEMS

A. Sigma Values from the Ionization of Benzoic Acids or Phenols

Bordwell and Boutan (1956)[53] were the first to obtain σ values for m- and p-$\overset{+}{S}Me_2$. Dissociation constants of m- or p-dimethylsulphoniobenzoic acid and of m- or p-dimethylsulphoniophenol were measured by potentiometric titration at 25°C, with water as solvent. The counter ion was p-tosylate for the benzoic acid system and the same or p-brosylate for the phenol system. The concentrations varied but were up to *ca.* 0.025 M. Ionic strength effects were not considered.

In both systems the polar substituent constant for m-$\overset{+}{S}Me_2$ was found to be 1.00. This may be regarded as an ordinary Hammett σ value. For p-$\overset{+}{S}Me_2$ the benzoic acid system gave 0.90, while the phenol system gave 1.16. The former is an ordinary σ value, but the latter is a σ^- value, corresponding to the (relative) stabilization of the phenate ion by resonance involving π(pd) bonding (**16**). (Tosylate or brosylate as counter ion gave the same result.)

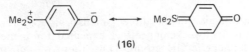

(16)

The authors contrasted this behaviour with results for the $\overset{+}{N}Me_3$ group. The substituent constants based on benzoic acid ionization in 50% ethanol were 1.02 for m-$\overset{+}{N}Me_3$ and 0.88 for p-$\overset{+}{N}Me_3$, while values from phenol ionization in water were 0.83 and 0.70, respectively, i.e. $\sigma_m > \sigma_p$ in both systems, because N cannot expand its valence shell. [For $\overset{+}{N}Me_3$ the authors quote five pairs of values (theirs and other workers'), and in each case $\sigma_m > \sigma_p$, but σ_m ranges from 0.67 to 1.02, and σ_p from 0.65 to 0.88, with ($\sigma_m - \sigma_p$) ranging from +0.02 to +0.14.]

Bordwell and Boutan[53] confirmed the electron-accepting conjugative properties of $\overset{+}{S}Me_2$ by studies of u.v. spectra: the primary band of benzene was markedly shifted to longer wavelengths by $\overset{+}{S}Me_2$, whereas it was scarcely affected by $\overset{+}{N}Me_3$. The effect of $\overset{+}{S}Me_2$ in u.v. spectra was found to be very similar to that of $MeSO_2$, suggesting that these substituents have similar possibilities of conjugation.

The effects of $\overset{+}{S}Me_2$ and $\overset{+}{N}Me_3$ in phenol ionization were also examined by Oae and Price (1958)[54]. The counter ion was I^- but otherwise the conditions were very similar to those used by Bordwell and Boutan[53]. The values of the substituent constants for $\overset{+}{S}Me_2$ agreed almost exactly with those determined by Bordwell and Boutan, while the values for m-$\overset{+}{N}Me_3$ and p-$\overset{+}{N}Me_3$ were 0.84 and 0.76, respectively (cf. 0.83 and 0.70). These authors also examined the influence of a methyl group adjacent to $\overset{+}{N}Me_3$ or $\overset{+}{S}Me_2$, as in structures **17** and **18**. The effective σ values for

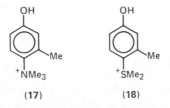

(17)   (18)

the combined influence of m-Me and the unipole in each case were 0.70 and 1.03, respectively. If the effect of m-Me is allowed for as $\sigma_m = -0.07$, the apparent sigma values of $\overset{+}{N}Me_3$ and $\overset{+}{S}Me_2$ in the presence of the methyl group are 0.77 and 1.10 respectively (cf. 0.76 and 1.16 in the absence of the methyl group). Thus,

p-$\overset{+}{\text{N}}$Me$_3$ is unaffected by the adjacent methyl while the effect of p-$\overset{+}{\text{S}}$Me$_2$ seems to be slightly decreased. Oae and Price considered this to be a slight indication of steric inhibition of resonance by the adjacent methyl group.

However, strong evidence against steric inhibition of the conjugative effect of p-$\overset{+}{\text{S}}$Me$_2$ was obtained later by the same authors[55], through examining the effect of placing substituents in *both* positions adjacent to p-$\overset{+}{\text{S}}$Me$_2$. In the ionisation of 3,5-dichloro- or 3,5-dibromo-4-dimethylsulphoniophenol a strictly additive effect of substituents was demonstrated. Further, Oae and Zalut[56] measured the (first) pK_a values in 39.9% aqueous ethanol of the tris(4-hydroxyphenyl)sulphonium ion (19) and its tris-2-methyl derivative (20). Species 20 is an extremely hindered molecule.

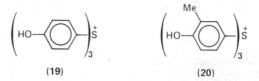

(19)   (20)

Nevertheless, the introduction of the three 2-methyl groups produces no influence on pK_a but the slight acid-weakening effect expected for a methyl group in the position *meta* to the phenolic OH group. Thus, there was no indication of steric inhibition of the conjugative effect of the sulphonium group. The σ value for the effect of the p-bis(4-hydroxyphenyl)sulphonium group on the acidity of the remaining phenolic group was given as 1.48, compared with 1.19 for p-$\overset{+}{\text{S}}$Me$_2$ under the same conditions. This agrees with the electron-attracting properties of the aryl compared with the methyl group. It seems to the present author, however, that a statistical correction should have been applied to the pK_a value of tris(4-hydroxyphenyl)sulphonium ion before the σ value was calculated. If this is done, σ⁻ for the p-bis(4-hydroxyphenol)sulphonium group becomes 1.28, a value which is still slightly greater than that of p-$\overset{+}{\text{S}}$Me$_2$.

The ionization of m- and p-dimethylsulphoniobenzoic acids has been re-examined by Hojo *et al*. (1971)[57]. Thermodynamic pK_a values were determined very carefully for solutions in water, 65% v/v aqueous dimethyl sulphoxide (DMSO), and 50% v/v aqueous ethanol, with p-tosylate as the counter ion. The derived σ values for $\overset{+}{\text{S}}$Me$_2$ were as follows: in water $\sigma_m = 1.10$, $\sigma_p = 1.13$; in 65% DMSO, $\sigma_m = 1.07$, $\sigma_p = 0.99$; in 50% ethanol, $\sigma_m = 1.23$, $\sigma_p = 1.16$. The values in water show poor agreement with those of Bordwell and Boutan[53]: $\sigma_m = 1.00$, $\sigma_p = .090$. Indeed, the values of Hojo *et al*.[57] are in the order $\sigma_m < \sigma_p$, while the earlier values show $\sigma_m > \sigma_p$. There is a considerable effect on adding an organic component to the solvent, and here $\sigma_m > \sigma_p$. The trimethylammonio-substituted benzoic acids were studied for comparison, and it was found that in the three solvents, equations 11 and 12 applied.

$$\sigma_{m\text{-}\overset{+}{\text{S}}\text{Me}_2} = \sigma_{m\text{-}\overset{+}{\text{N}}\text{Me}_3} + 0.06 \tag{11}$$

$$\sigma_{p\text{-}\overset{+}{\text{S}}\text{Me}_2} = \sigma_{p\text{-}\overset{+}{\text{N}}\text{Me}_3} + 0.13 \tag{12}$$

These relationships were held to indicate the importance of π(pd) bonding for $\overset{+}{\text{S}}$Me$_2$ as a factor even in the ionization of benzoic acids. Other aspects of this work will be discussed later in connection with the *ortho*-effect.

B. Sigma Values from Spectroscopic Measurements

Substituent constants of the $\overset{+}{\text{S}}$Me$_2$ group have been obtained by studies of ¹⁹F and ¹H n.m.r. and infrared spectra.

Sheppard and Taft[58] included $\overset{+}{S}Me_2$ in studies of the effect of sulphur substituents on the ^{19}F n.m.r. of fluorobenzene. This was part of Taft's very extensive work on the correlation analysis of ^{19}F chemical shifts through σ_I and σ_R-type constants and the dual substituent-parameter equation[34,38,39]. Substituents in the *meta*-position influence the ^{19}F chemical shift (relative to H as substituent) according to equation 13. Correlation equations for the difference between the *meta* and the *para* shift are equations 14 and 15, applying to $+K$ and $-K$ substituents, respectively.

$$\int_{H}^{m-X} = -7.1\,\sigma_I + 0.60 \tag{13}$$

$$(+K) \qquad \int_{m-X}^{p-X} = -29.5\sigma_R^0 \tag{14}$$

$$(-K) \qquad \int_{m-X}^{p-X} = -29.5\bar{\sigma}_R \tag{15}$$

The reason for distinguishing between $+K$ and $-K$ substituents is that for the latter there is cross-conjugation in the molecule of a *para*-substituted fluorobenzene (structures **21**). This means that very precise correlation with σ_R^0 is not to be

(21)

expected, so the σ_R-type value is denoted as $\bar{\sigma}_R$, the resonance parameter which is effective in the particular situation. For many $-K$ substituents $\bar{\sigma}_R$ is more positive than σ_R^0 by a few units in the second place of decimals.

From equations 13 and 15 and the observed ^{19}F shifts in acetonitrile as solvent (perchlorate as the counter ion) Sheppard and Taft determined $\sigma_I = 0.89$ and $\bar{\sigma}_R = 0.17$ for $\overset{+}{S}Me_2$. The σ_I value is fairly close to that of $\overset{+}{N}Me_3$; this is *ca.* 0.9 as measured by ^{19}F n.m.r. in protic solvents[38]. The value of $\bar{\sigma}_R$ is slightly less than the value 0.25 calculated for a σ_R-type parameter by English *et al.*[59] from the results of Oae and Price[54] on the ionization of phenols. This seems reasonable insofar as the latter should be a $\sigma_{\bar{R}}$ value. However, the calculation used the relation $\sigma_R = 1.5\,(\sigma_p - \sigma_m)$, based on equations 6 and 7, and it is by no means certain that the relay coefficient $\alpha = 0.33$ applies to unipolar substituents.

Exner[60] has applied Taft's equations 8 and 9 to the above σ_I and σ_R values for $\overset{+}{S}Me_2$ to calculate $\sigma_m^0 = 0.97$ and $\sigma_p^0 = 1.06$. These correspond qualitatively to the σ_m and σ_p values of Hojo *et al.*[57] based on the ionization of benzoic acids in water. Because $\bar{\sigma}_R$ cannot be taken as σ_R^0 for this substituent, the status of Exner's calculated values as σ^0 values is doubtful.

Sheppard and Taft[58] also record $\sigma_I = 0.72$ and $\bar{\sigma}_R = 0.10$ for the $\overset{+}{S}MeBCl_3$ group (dichloromethane as solvent).

Adcock *et al.*[61] included $\overset{+}{S}Me_2$ in a study of the ^{19}F n.m.r. of 1- and 2-fluoronaphthalenes, with various relative dispositions of substituent and ^{19}F probe. Dimethylformamide was the solvent, with fluorosulphonate as the counter ion. For comparison, the inductive and resonance parameters of $\overset{+}{S}Me_2$ were determined from ^{19}F shifts of fluorobenzenes dissolved in DMF: $\sigma_I = 0.74$ and $\bar{\sigma}_R = 0.13$ (cf. 0.89 and 0.17, respectively, in acetonitrile). (The authors mistakenly refer to the derived resonance constant for $\overset{+}{S}Me_2$ as σ_R^0; see above.) Values based on correlations involving 6- or 7-substituted 2-fluoronaphthalenes were obtained: $\sigma_I = 0.69$, $\bar{\sigma}_R = 0.08$. The authors claim that these values, when inserted in a correlation expression for 4-substituted 1-fluoronaphthalenes, give good agreement with the observed shift. This is held to indicate that the steric effect of $H_{(5)}$ does not influence the $\pi(pd)$ bonding of 4-$\overset{+}{S}Me_2$.

The small positive values of $\bar{\sigma}_R$ for $\overset{+}{S}Me_2$ obtained by the [19]F n.m.r. method contrast with the value of -0.09 obtained from correlations of i.r. intensities by Angelelli et al.[62]. This is claimed to be a genuine σ_R^0 value, and to indicate that when $\overset{+}{S}Me_2$ is not subject to cross-conjugation with a donor substituent it is a weak donor by the resonance effect. This may involve the participation of the sulphonium unshared pair in $\pi(pp)$ bonding. Adcock et al.[61], however, question the reliability of the i.r. method for weakly interacting substituents, and describe the negative sign of σ_R^0 as given by this method as 'undoubtedly spurious'. If indeed $\pi(pp)$ bonding of $\overset{+}{S}Me_2$ is possible, it is difficult to understand why such interaction is not greatly enhanced in highly electron-demanding reactions, e.g. electrophilic aromatic substitution (cf. the behaviour of SMe)[13].

Adcock et al.[61] showed that a very considerable increase in the electron-attracting power of $\overset{+}{S}Me_2$ occurs when the solvent is changed from DMF to trifluoroacetic acid (TFA). NMe_3 behaves similarly, so no effect on $\pi(pd)$ bonding is involved. It is suggested that the observed increase is due to a reduction in ion-pair interactions as the ionizing power of the solvent is increased. Low nucleophilicity and hydrogen-bonding effects may also be important with TFA.

This mention of ion pairing will serve to remind us that the counter ion is often forgotten in discussing unipolar substituents, and that the solutions generally used in [19]F n.m.r. studies are by no means highly dilute. Thus, Adcock et al.[61] used 15% w/w. Sheppard and Taft[58], however, state that ^{19}F shifts were measured at several concentrations and extrapolated to infinite dilution.

Fujio et al.[63] included $\overset{+}{S}Me_2$ (with iodide as counter ion) in a study of substituent effects on [1]H n.m.r. of the hydroxy proton in phenol dissolved in DMSO. The apparent substituent constants were $\bar{\sigma}_m = 0.721$ and $\bar{\sigma}_p = 0.926$. The $\bar{\sigma}_p$ value was almost unaffected either by buttressing the OH between methyl groups or by introducing a methyl group adjacent to the $\overset{+}{S}Me_2$. These values are much lower than those based on the ionization of phenol in water and this indicates specific interactions between the substituent and DMSO. NMe_3 behaves similarly but many dipolar substituents are only slightly affected. In a further paper, the same group[64] discussed their results in relation to the Yukawa–Tsuno equation[29,30] and their so-called linear substituent free energy (LSFE) relationship, with its distinctive substituent constants σ_i and σ_π[65]. They concluded that in aqueous solutions, σ_p^0 for $\overset{+}{S}Me_2$ should be regarded as 1.00, and σ_p^- as ca. 1.25, while σ_p^0 modified by DMSO is 0.73 and σ_p^- modified by DMSO is 1.01.

We mention finally two relevant studies of [13]C n.m.r. Olah et al.[66] have correlated ^{13}C chemical shifts and CNDO/2 charge distributions in phenylcarbenium ions and related phenyl-substituted onium ions, including $\overset{+}{S}Me_2$ as a substituent. There is a linear relationship between para-carbon shifts and total charge density; $\overset{+}{S}Me_2$ conforms to this. There are more complicated relationships for ortho-, meta-, and ipso-carbon shifts. There is no discussion involving σ values.

Recently, Ricci et al.[67] measured the ^{13}C chemical shifts in an aromatic ring influenced by a positively charged substituent; $\overset{+}{S}Me_2$ was included. A multiple regression of $(\delta C_p - \delta C_m)$ on σ_I and σ_R^0 for dipolar substituents had previously been reported by Syrova et al.[68] (equation 16).

$$\delta C_p - \delta C_m = 3.6\sigma_I + 23.5\sigma_R^0 \tag{16}$$

$\overset{+}{N}Me_3$, $\overset{+}{P}Me_3$, and $CH_2\overset{+}{N}Me_3$ were shown to conform approximately to this equation, although the relevant plot appears to be drawn as a gentle curve rather than a straight line. The authors were apparently not aware of values for inductive and resonance parameters of $\overset{+}{S}Me_2$. If we use Sheppard and Taft's[58] values of σ_I and $\bar{\sigma}_R$, the value of $(\delta C_p - \delta C_m)$ calculated according to equation 16 is 7.2, or with the values of Adcock et al.[61], 5.7, in both cases higher than the observed value of 4.1. A $\bar{\sigma}_R$ value smaller than the values given by ^{19}F n.m.r. seems to be applicable.

IV. ELECTRONIC EFFECTS IN ELECTROPHILIC SUBSTITUTION

A. Early Studies of the Nitration of Sulphonium Compounds

By the late 1920s the strongly *meta*-directing influence of positively charged substituents was well established[9] and was being extensively investigated by Robinson, Ingold, and their respective schools. A variety of substituents was studied. We need not concern ourselves here either with most of the results or with the somewhat contentious discussion which took place. Determination of the proportions of isomers in the nitration products involved either quantitative separation or complicated analytical procedures.

Two papers are of sufficient lasting interest for mention here. Baker and Moffitt[10] (of Ingold's Leeds school) nitrated $Ph\overset{+}{S}Me_2$ and $Ph\overset{+}{S}eMe_2$ as the picrates. The products were found to contain no trace of *o*- or *p*-nitro compounds, and thus $\overset{+}{S}Me_2$ and $\overset{+}{S}eMe_2$ appeared to be exclusively *meta*-directing, like $\overset{+}{N}Me_3$[9]. They concluded that the $+K$ effect (they used the symbol $+T$) was inoperative, 'the mobility of the unshared electron pairs being effectively restricted by the charge on the atom'. They also nitrated $PhCH_2\overset{+}{S}Me_2$ and $PhCH_2\overset{+}{S}eMe_2$, and found that the *m*-nitro compounds were formed only to the extent of 52% and 16%, respectively, compared with 88% for $PhCH_2\overset{+}{N}Me_3$. These results were held to show the damping effect of the methylene group and of additional electron shells in the central atom (N, S, or Se) on the 'electron strain' in the benzene ring caused by the positive charge.

Pollard and Robinson[69] nitrated $PhCH_2\overset{+}{S}Et_2$ as its picrate, and found only 28% of *m*-nitro compound, with most of the product being the *para*-isomer. Their explanation of this poor *meta*-directing effect, compared with that observed in relevant quaternary ammonium compounds, involved a hypothetical role for the picrate ion: '. . . the sulphur atom, softer than nitrogen and with a larger nuclear charge, polarizes the attached anion (or anions) to the greater extent and so allows of a more complete neutralization of its surplus positive charge, which thus develops a field of relatively weaker electric intensity at corresponding points in the surroundings, including the attached aromatic group'. This type of explanation, applied to other results, had already been criticized by Ingold et al.[70] on the grounds that it conflicted with Fajans' rules[71], a criticism which seems justified. However, it is of interest that Robinson did not forget the counter ion, an aspect of compounds involving unipolar substituents which, as we have already noted, is often neglected.

B. Modern Studies of Electrophilic Substitution in Sulphonium Compounds

Further investigation of the influence of positively charged substituents on aromatic nitration began in the early 1960s with the discovery that appreciable *ortho-* and *para*-substitution occurred with certain substituents which were previously believed to be exclusively *meta*-directing. Thus PhṄMe₃, as its nitrate in 98% sulphuric acid, was found to give as much as 11% of *p*-nitro compound[72]. This work and much of the later work was carried out by Ridd and his colleagues, and depended on improved separation techniques (e.g. ion-exchange chromatography) and improved analytical procedures (e.g. u.v. spectroscopy).

The reinvestigation of the nitration of sulphonium compounds was begun by Gilow and his colleagues. PhṠMe₂ and PhṠeMe₂ were nitrated as their methylsulphates with concentrated nitric acid in concentrated sulphuric acid[73]. As shown in Table 3, appreciable amounts of *ortho-* and *para*-isomers were found. The analytical procedure depended on the ready dealkylation of dimethylnitrophenylsulphonium and -selenonium salts with sodium methoxide at low temperatures, and separation of the resulting isomeric methyl nitrophenyl sulphides or selenides by g.l.c. The authors suggested that the analytical procedure used by Baker and Moffitt[10] probably destroyed the small amounts of *ortho-* and *para*-isomers present in the product. Gilow and Walker[73] suggested that the appreciable *ortho*-nitration indicated that resonance interaction of the sulphonium group does occur (cf. NMe₃ which is *meta-* and *para*-directing), but that it was not clear whether π(pp) (donor S) or π(pd) (acceptor S) interaction was involved. That the difference in behaviour between the sulphur and the selenium compound was so small was attributed to the cancelling out of opposed electronic and steric factors.

Gilow *et al.*[74] studied the chlorination or bromination of the same sulphonium or selenonium compounds with halogens activated by silver ion. The results were compared with those of nitration and showed an increased tendency towards *ortho*-substitution, particularly in the case of chlorination. *Ortho/para* and *meta/para* ratios tended to be higher than for nitration. The authors suggested that a cyclic intermediate might be aiding *ortho*-substitution. The uncatalysed bromination of both the sulphur and the selenium compound in acetic acid at 100°C produced mainly *para* derivatives, with no *meta*-substitution taking place. No explanation was offered for this.

Gilow *et al.*[75] extended their study to the nitration of Ph(CH₂)ₙŻ(CH₃)₂, where $n = 0$, 1, or 2 and Z = S or Se. They measured both relative rates and isomer distributions. As the positive pole was removed further from the ring, or S was replaced with Se, the overall rate of nitration increased, the rate of *para*-substitution increased, and the percentage of *meta*-substitution decreased. These results, in association with corresponding results for ṄMe₃, were interpreted in terms of the field/inductive effect of the positive pole and π(pd) interaction for ṠMe₂ or ṠeMe₂ attached directly to the ring. π(pp) interaction was now regarded as unimportant, but an *ortho/para* hyperconjugative effect of CH₂ṠMe₂ was considered to make a contribution.

Marziano *et al.*[76] determined isomer distributions and rates of nitration of PhṠMe₂ in media containing from 84% to 98% sulphuric acid. The mean isomer

TABLE 3. Isomer distribution in nitration[73]

Substrate	Ortho(%)	Meta(%)	Para(%)
PhṠMe₂	3.6 ± 0.2	90.4 ± 0.3	6.0 ± 0.2
PhṠeMe₂	2.6 ± 0.2	91.3 ± 0.3	6.1 ± 0.2

distributions for 85–96% sulphuric acid were: *ortho* 2.18, *meta* 93.97, and *para* 3.95%, and thus differ slightly from those recorded by Gilow and Walker[73]. The authors considered that their results support the importance of the electrostatic effect of the positive charge, increased by participation of 3d orbitals on sulphur.

Gilow and Ridd[77] collaborated in a study of the bromination and the nitration of various disubstituted benzenes, involving either one or two positively charged substituents. $CH_2\overset{+}{S}Me_2$ and $(CH_2)_2\overset{+}{S}Me_2$ were included. Values of log(relative rate) of bromination or nitration were calculated by the additivity principle from results for the relevant monosubstituted compounds and were then compared with the observed values. With strongly deactivating pairs of substituents, such as $1,4\text{-}(CH_2\overset{+}{S}Me_2)_2$, the observed rate tended to be lower than the calculated value (negative deviation) but two weakly deactivating substituents, such as $1,4\text{-}[(CH_2)_2\overset{+}{S}Me_2]_2$ gave a positive deviation. The negative deviations were rationalized in terms of a move towards a 'later' transition state in the disubstituted compound. The positive deviations were attributed to electrostatic interactions between the substituents favouring conformations in which the positive charges are as far as possible apart, and hence less effective in deactivating the ring.

Sulphonium salts have recently been included in studies[78] of nitration and bromination of bridged systems and corresponding open-chain systems (structures **22** and **23**). Bridging causes a retarding effect, which decreases with an increase in

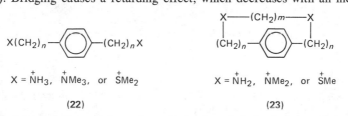

$$X(CH_2)_n\!-\!\!\bigotimes\!\!-\!(CH_2)_nX$$

$$X = \overset{+}{N}H_3, \ \overset{+}{N}Me_3, \ \text{or} \ \overset{+}{S}Me_2$$

(22)

$$X\!-\!\!-\!\!(CH_2)_m\!-\!\!-\!X$$
$$(\overset{|}{C}H_2)_n\!-\!\!\bigotimes\!\!-\!(\overset{|}{C}H_2)_n$$

$$X = \overset{+}{N}H_2, \ \overset{+}{N}Me_2, \ \text{or} \ \overset{+}{S}Me$$

(23)

the degree of methylation of the poles and the length (m) of the methylene chain. Retardation is greater when $n = 2$ than when $n = 1$. The retardation is attributed mainly to the change in the form of the molecular cavity, which transmits the field effect, with a further contribution when $n = 2$ from the change in the distance of the poles from the aromatic ring.

Rees *et al.*[79] recently summarized the salient features of their own and other work on the electrophilic substitution of systems bearing positive poles. Results for the rates of nitration of $Ph(CH_2)_nX$, where $X = \overset{+}{N}R_3, \overset{+}{S}R_2, \overset{+}{P}R_3$, etc., were analysed in terms of non-conjugative and conjugative interaction of the side-chain with the aromatic ring. Logarithms of partial rate factors for *meta*-substitution, f_m, were taken to measure non-conjugative interaction, and were converted into a free-energy function. The latter is related inversely and curvilinearly to the maximum distance of the positive pole from the centre of the benzene ring, in a way that is more consistent with deactivation by a field effect than by a through-bonds inductive effect. There is some scatter about the curve, but $CH_2\overset{+}{S}Me_2$ and $(CH_2)_2\overset{+}{S}Me_2$ conform well. $\overset{+}{S}Me_2$ attached directly to the ring does not appear on the graph, although $\overset{+}{N}Me_3$ does and conforms well. Insofar as f_m for $\overset{+}{S}Me_2$ is smaller than that for $\overset{+}{N}Me_3$ by a factor of *ca.* 8[75], while the 'distance' is slightly larger for $\overset{+}{S}Me_2$ than for $\overset{+}{N}Me_3$, $\overset{+}{S}Me_2$ would not conform well to the curve.

There is, however, no simple relationship between deactivation at the *meta*-position and orientation of substitution, because the latter is also determined by conjugative interaction. An approximate linear free-energy relationship is based on Exner's modification[23,80] of Taft's separation of inductive and resonance effects[31–34,36,37]. Thus, there is a rough linear relationship between $(\log f_p - 1.17 \log f_m)$ and σ_R^0 for six positive substituents (not including $\overset{+}{S}Me_2$), with a regression coefficient of -5.0. A strict treatment would require σ_R^+ values, but these are not available for the substituents in question. The authors suggest, however, that σ_R^0 and σ_R^+ values for positive substituents may well not differ greatly. $\overset{+}{S}Me_2$ would, in fact, fit the LFER at $\sigma_R^0 \approx 0.12$, which is fairly close to the values indicated by ^{19}F studies, but remote from the value of -0.09 given by the i.r. method. For $CH_2\overset{+}{S}Me_2$, a σ_R^0 value of ca. -0.18 is indicated. This perhaps confirms Gilow's suggestion of the importance of C–H hyperconjugation for this group (see p. 202)[75]. For $(CH_2)_2\overset{+}{S}Me_2$ σ_R^0 is indicated as ca. -0.26. If such values do indeed approximate to σ_R^+ values, it appears that as n in $(CH_2)_n\overset{+}{S}Me_2$ is increased, the behaviour of the group rapidly tends to that of a normal alkyl chain, for which σ_R^+ is ca. -0.28.

There is little information on the influence in electrophilic substitution of sulphonium groups other than $(CH_2)_n\overset{+}{S}Me_2$, where $n = 0$, 1, or 2. Kamiyama et al.[81], however, found $\overset{+}{S}OMe_2$ to be 100% meta-directing in nitration, with great deactivation. They derived a σ_m value for this group of ca. 1.38 (cf. $\overset{+}{S}Me_2$ ca. 1.0 and N_2^+ ca. 1.76). It is suggested on spectroscopic grounds that there is $\pi(pp)$ interaction between the benzene ring and the S to O bond (see structures 24). The

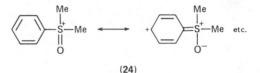

(24)

authors suggest that $\overset{+}{S}OMe_2$ may find considerable application as a highly electron-attracting group in reactivity studies.

V. ELECTRONIC EFFECTS IN NUCLEOPHILIC SUBSTITUTION

A. The Sulphonium Group in the Substrate

The sulphonium group has not been widely used in the study of substituent effects in nucleophilic aromatic substitution. This is doubtless because it is liable to become directly involved in the nucleophilic attack, i.e. R_2S or ArSR acts as the leaving group.

Daly et al.[82] compared the activating power of SMe, $\overset{+}{S}Me_2$, and $\overset{+}{N}Me_3$ in the reactions of 4-substituted 1-chloro-2-nitrobenzenes with methoxide ion in methanol. The counter ion was sulphate for the sulphonium compound and iodide for the quaternary ammonium compound. This reaction series conforms fairly well to the σ^- values of dipolar substituents, with $\rho = 3.90$ at $50°C$. The effective σ^- values for SMe, $\overset{+}{N}Me_3$, and $\overset{+}{S}Me_2$ are 0.34, 1.11, and 1.41, respectively. The σ_p value for SMe is ca. 0, so an appreciable conjugative electron-withdrawing effect in nucleophilic substitution is indicated for uncharged divalent sulphur. This effect is greatly magnified for $\overset{+}{S}Me_2$ (cf. σ_p value, ca. 1.0), and accounts for its activating power being much greater than that of $\overset{+}{N}Me_3$.

The σ_p^- value for $\overset{+}{S}Me_2$ from phenol ionization is 1.16[53,54]. Daly et al.[82] comment that the value of 1.41 for $\overset{+}{S}Me_2$ in nucleophilic substitution includes the 'additional activation consequent on reaction being between an anion and a cation' (cf. the situation in phenol ionization). The authors measured the effect of substitution on the activation parameters by replacing $H_{(4)}$ by various substituents. The value of $\Delta\Delta H^{\ddagger}$ is only -1.45 kcal/mol for $\overset{+}{N}Me_3$, but -5.95 kcal/mol for $\overset{+}{S}Me_2$, while $\Delta\Delta S^{\ddagger}$ is 15.3 cal/K·mol for $\overset{+}{N}Me_3$ and $+6.7$ cal/K·mol for $\overset{+}{S}Me_2$. Thus, activation by $\overset{+}{N}Me_3$ is mainly due to a very favourable entropy term, while enthalpy makes much more contribution in the case of $\overset{+}{S}Me_2$. This difference in behaviour may be ascribed to the importance of the $\pi(pd)$ effect for $\overset{+}{S}Me_2$[83].

The activation of nucleophilic substitution by the diphenylsulphonium group has also been studied. Oae and Khim[84] studied the alkaline hydrolysis of 1-chloro-2-, 3-, or -4-diphenylsulphoniobenzene perchlorate in aqueous ethanol. The *meta*-isomer underwent preferential decomposition by attack at the sulphonium group, and nucleophilic attack at $Cl_{(1)}$ was certainly much slower than for the *ortho*- and *para*-isomers. The *para*-isomer was about three times more reactive towards nucleophilic substitution than *p*-chloronitrobenzene, and for the *ortho* isomers the reactivity ratio was about 5. Competitive attack at the diphenylsulphonium group occurred but at about one-ninth of the rate of attack at $Cl_{(1)}$. The activating effect of *o*- or *p*-$\overset{+}{S}Ph_2$ compared with *m*-$\overset{+}{S}Ph_2$ was attributed to the $\pi(pd)$ conjugative effect. The insertion of a 3-methyl group in the *para*-isomer retarded reaction by a factor slightly less than two, confirming that the $\pi(pd)$ effect is fairly stereo-insensitive. (The corresponding effect for 1-chloro-4-nitrobenzene would involve retardation by a factor of about 4.)

The great activating effect of *p*-$\overset{+}{S}Ph_2$ in nucleophilic aromatic substitution has recently been confirmed by Ignatov *et al.*[85] in a study of the reactions of 1-chloro-2-nitro-4-diphenylsulphoniobenzene tetrafluoroborate with amines. This compound was more reactive than 1-chloro-2,4-dinitrobenzene.

B. Aspects of Attack at the Sulphonium Group

Nucleophilic attack in general on the sulphonium group, or on atoms attached to it, is outside the scope of this chapter, since the sulphonium group furnishes the leaving group. However, some comments are appropriate since the variety of behaviour observed is a manifestation of the electronic effect of the sulphonium group[86].

At first sight the positive charge on the sulphonium group and the possibility of valence shell expansion should encourage direct nucleophilic attack on the sulphur atom. This does indeed sometimes occur but often an easier alternative is provided in the attack on an atom attached to the sulphur, and activated by it towards nucleophilic attack. The polarization produced by the positive charge on the sulphonium group, together with its ability to act as a leaving group, are clearly responsible for this. Thus, the trimethylsulphonium ion is attacked by hydroxide ion at methyl carbon in an S_N2 process leading to dimethyl sulphide and methanol. In the extreme case where one of the groups can form a carbocation, an S_N1 mechanism is possible, e.g. the *t*-butyldimethylsulphonium ion gives the *t*-butyl cation and dimethyl sulphide, the former then undergoing reactions with the hydroxylic solvent. Nucleophilic attack on sp^2 aromatic carbon is much more difficult than attack on sp^3 aliphatic carbon, so methoxide ion attacks a triarylsulphonium ion directly on sulphur, to form a tetracoordinate species, Ar_3SOMe, which then undergoes further reactions via a radical pair to give aryl methyl ether, diaryl sulphide, and some hydrocarbon ArH[87].

Some interesting examples of the various possibilities of nucleophilic attack have recently been found by Matsuyama *et al.*[88]. Ions of the type **25** may undergo competitive nucleophilic attack on Me, R, or carbon attached to X. Also,

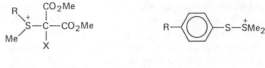

R = Et, *i*-Pr, *n*-Bu, PhCH$_2$, etc.   R = H or Me

X = Br, Cl, PhS, H

(25)   (26)

arylthiodimethylsulphonium ions **26** are attacked by various nucleophiles exclusively at the sulphide S[89]. These substrates thus undergo quite different reactions from ArNHSMe$_2$ or ArOSMe$_2$, where reaction is initiated by removal of a proton from one of the methyl groups to give an ylide.

Some studies of substituent effects in nucleophilic attack at the sulphonium group are of interest as showing the way in which this group responds to electronic effects.

Martin and Basalay[90] have studied degenerate interconversions of sulphonium ions, which involve intramolecular nucleophilic displacement by neighbouring sulphide S. The system shown in **27** was examined, rate of interconversion being

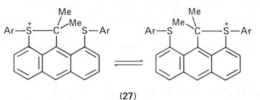

(27)

measured by n.m.r. line-shape analysis. For substituents Y in the 3- or 4-position of Ar, the rate constants are said to be well correlated by σ^+ (rather than σ), with $\rho = 1.04$. The authors suggest that the principal electronic effect is the stabilization of the ground-state sulphonium centre by $\pi(pd)$ conjugation.

These findings contrast with those of Coward and Sweet[91] in the kinetics and mechanism of methyl transfer from sulphonium compounds to various nucleophiles, as in equation 17.

$$4\text{-XC}_6\text{H}_4\overset{+}{\text{S}}\text{Me}_2 + \text{Nuc}^- \longrightarrow 4\text{-XC}_6\text{H}_4\text{SMe} + \text{MeNuc} \qquad (17)$$

Here there is said to be correlation with σ (rather than σ^+), giving positive values of ρ, which vary only slightly with the nucleophile.

It is claimed that both of these systems involve S$_N$2 mechanisms. Possibly the different types of correlation reflect the differing extents to which electron delocalization changes as between initial and transition states. Further investigations with a wider range of substituents might be helpful.

C. The Sulphonium Group in the Nucleophile

There is little information about this topic. Epstein *et al.*[92] measured the rates of reaction, in dilute aqueous solution, of isopropyl methylphosphonofluoridate (nerve gas GB or Sarin, structure **28**) with substituted phenols, displacing fluoride ion. It was established that the nucleophile was the phenate anion. For a wide range of *meta*- and *para*-substituted phenols involving dipolar substituents there was good correlation between log k_2 (second-order rate constants) and the pK_a values of the

(28)

phenols. A large number of positively charged nitrogen substituents and p-$\overset{+}{S}Me_2$ showed positive deviations from this correlation, i.e. the phenols were more reactive than predicted from their pK_a values. The deviation for p-$\overset{+}{S}Me_2$ was 0.23 log unit. Thus, for dipolar substituted phenols the nucleophilicity of the phenate ion towards a neutral substrate is directly related to its basicity towards the oxonium ion, but this is not the case with positively charged phenols, for which the phenate 'ion' is in fact an overall neutral molecule. The authors attribute this difference in behaviour to the dominating influence of electrostatic repulsion between the positive pole and the hydrogen ion for the pK_a value, while such repulsion plays no part in the attack of the nucleophile on the neutral substrate. The reaction of a unipole-substituted compound is of a different charge type from that of a dipole-substituted compound, so the same LFER cannot be expected to hold.

VI. STABILIZATION OF CARBANIONIC CENTRES BY THE SULPHONIUM GROUP

A. Introduction

In 1930 Ingold and Jessop[93] reacted fluorenyl-9-dimethylsulphonium bromide with sodium hydroxide and obtained a yellow crystalline precipitate of dimethylsulphonium-9-fluorenylide (**29**). This compound was sufficiently stable for

(29)

analysis and for the determination of its molecular weight in benzene. It provided the first example of the stabilization of a carbanionic centre by a sulphonium group. The authors had previously[94] found indications of the slight formation of an analogous $\overset{+}{N}Me_3$ species in solution, and had been led to examine sulphonium compounds by recognizing that sulphur has a greater ability to form 'semipolar' linkages to oxygen than nitrogen has, e.g. sulphoxides (R_2SO) are more stable to heat than amine oxides (R_3NO).

The stability of compound **29** is due substantially to the delocalization of the negative charge throughout the fluorene system. However, the peculiar ability of sulphonium groups to stabilize adjacent carbanionic centres is due to valence shell expansion, so that the structure of a molecule $R^1R^2SCR^3R^4$ is a hybrid of two canonical forms, **30a** and **30b**, which are known as the 'ylide' and 'ylene' forms, respectively. This aspect of sulphonium compound chemistry is, of course, part of the extensive subject of ylide chemistry (Chapter 15), in which various heteroatoms

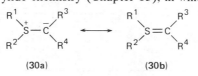

(30a)   (30b)

(notably phosphorus in $R^1R^2R^3P^+$ compounds) can function in a similar way[95,96].

When the negative charge cannot be delocalized away from the $\overset{+}{S}$—$\overset{-}{C}$ system, it seems likely that the ylene structure is very important, even though many of the reactions are best understood in terms of the ylide form. When delocalization of negative charge away from the $\overset{+}{S}$—$\overset{-}{C}$ system can occur, the S—C bond may approximate to a single bond as in the fluorene derivative **29**, and in various other molecules, e.g. $C_5H_4SMe_2$ is said to have predominantly the ylide structure **31**, and in consequence its cyclopentadiene ring shows aromatic behaviour[97,98]. However, for **32**, where delocalization of negative charge into the CN groups is possible, it has nevertheless been concluded that the bond between S and $C(CN)_2$ has considerable double-bond character[99].

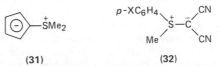

(31)   (32)

We shall discuss first the pK_a values for ionization of C—H bonds adjacent to sulphonium groups, and then the kinetics of H—D exchange reactions involving such bonds. Finally, the role of sulphonium groups in promoting reactions via the formation of carbanionic centres will be discussed.

B. pK_a Values for C—H Bond Ionization

The pK_a values of simple sulphonium compounds in water or aqueous organic solvents are too high for direct measurement. From appropriate data, Crosby and Stirling[100] have calculated the pK_a value of the trimethylsulphonium ion to be ca. 18.9. For direct measurement to be possible there must be present in the molecule another electron-attracting group which can delocalize negative charge through a conjugative effect. Substituents with carbonyl groups are particularly useful in this respect.

Aksnes and Songstad[101] measured the pK_a values, in water or aqueous ethanol, of various compounds of the type XCH_2COY, where $X = R_3N$, R_3P, R_3As, or R_2S (the R groups being Me or Ph in various numbers) and $Y = Ph$, p-BrC_6H_4, Me, or OEt. For one series, with $Y = Ph$, the pK_a values in water are given in Table 4. The inductive/field effect alone would produce the order $N < S < P < As$, so these values show the importance of $\pi(pd)$ bonding in the elements of the second short period, with the S compound being more acidic than the P or As compound because it has the greater inductive or π-bonding effect.

These authors found that, depending on the compound, the value of pK_a either increased or decreased when ethanol was added. They sought to explain this in

TABLE 4. Ionization of
onium salts (XCH_2COPh)
in water at 20°C[101]

X	pK_a
$Me_3\overset{+}{N}$	~10.5
$Me_3\overset{+}{P}$	9.20
$Me_3\overset{+}{As}$	9.15
$Me_2\overset{+}{S}$	8.25

terms of competing factors: decreasing the dielectric constant favours the charge-neutralized ylene structure, and thereby pK_a is decreased, but disfavours charge delocalization in the ylide structure, and thereby pK_a is increased. Finally, replacing Me with Ph attached to S (or the other heteroatoms) increases the acidity in accord with the usual electronic effects of these groups.

Work of this kind was later pursued by Johnson and Amel[102] for a series of P, S, and As phenacyl onium salts, with Me, n-Bu, or Ph attached to the heteroatom. The pK_a values were measured in 95% ethanol and a selection of data is given in Table 5. For given groups attached to the heteroatom, the acidifying effect increases in the order As < P < S, and for a given heteroatom in the order Me < n-Bu < Ph. The position of n-Bu in this series is surprising, since the normal electron-releasing inductive effect of alkyl groups would be expected to produce the order n-Bu < Me[33]. (See, however, a discussion of the electronic effects of alkyl groups in reference 44.) Johnson and Amel[103] later showed that both Ph and n-Bu deshielded the CH_2 protons in $R_2\overset{+}{S}CH_2COPh$ relative to Me, and the order of the chemical shifts agrees with the above acidifying effect of the groups.

The introduction of substituents into the benzene ring of phenacylsulphonium salts affects the acidity. For the system **33**, Ratts and Yao[104] found ρ to be *ca.* 2.6 for pK_a values in *ca.* 90% aqueous methanol. Ratts[105] extended this work to arylmethylphenacylsulphonium salts (**34**), finding ρ to be *ca.* 1.18 for substituents in

$Me_2\overset{+}{S}CH_2COC_6H_4X$-$p$

(33)   (34)

the aryl group and *ca.* 2.65 for substituents in the aroyl group. The author comments that the inductive effect of a substituted aryl ring is thus greater when the ring is attached to a $\pi(pp)$ delocalization system than when it is attached to a $\pi(pd)$ system.

Ratts[105] also studied a series of 4-bromophenacylsulphonium salts with various groups on the sulphur (see Table 6). Replacing Me_2 with Et_2 *increases* the acidity by *ca.* 1 pK unit, which recalls the similar effect of replacing Me with n-Bu as found by Johnson and Amel[102]. Ratts interpreted this in terms of a special role of Me in delocalizing the sulphonium positive charge. (See, however, the alternative view of Lillya, below). Incorporation of the alkyl chains in a ring in $(\overline{CH_2})_4\overset{+}{S}$ increases pK_a, while an enlargement of the ring to $(\overline{CH_2})_5\overset{+}{S}$ decreases pK_a again.

TABLE 5. Ionization of onium salts (XCH_2COPh) in 95% ethanol–water at 25 °C[102]

X	pK_a
$Me_3\overset{+}{As}$	9.80
$Me_3\overset{+}{P}$	8.60
$Me_2\overset{+}{S}$	7.68
n-$Bu_2\overset{+}{S}$	6.65
$Ph_2\overset{+}{S}$	5.36

TABLE 6. Ionization of 4-bromophenacylsulphonium salts
($XCH_2COC_6H_4Br$) in *ca*. 90% methanol–water at ambient
temperature[105]

X	pK_a	X	pK_a
$Me_2\overset{+}{S}$	7.4	$(CH_2)_5\overset{+}{S}$	7.00
$Et_2\overset{+}{S}$	6.46	$S(CH_2CH_2)_2\overset{+}{S}$	6.63
$(CH_2)_4\overset{+}{S}$	7.54	$S(CH_2CH_2CH_2)_2\overset{+}{S}$	8.13

Strain in the five-membered ring is thought to produce increased p character in the ring-C to S bonds, and increased s character in the *exo*-C to S bond, which in turn decreases the acidity of protons attached to the *exo*-C atom. Replacing the middle methylene group in $(CH_2)_5\overset{+}{S}$ with S, i.e. $S(CH_2CH_2)_2\overset{+}{S}$, is slightly acid strengthening, but an enlargement of the ring to $S(CH_2CH_2CH_2)_2\overset{+}{S}$ is acid weakening. The low acidity here is ascribed to a transannular effect involving the interaction of the sulphide lone pairs with the sulphonium centre.

pK_a values of arylmethylphenacylsulphonium salts have also been measured by Kissel *et al.*[106]. For *para*-substituents in the aroyl ring ρ was found to be 2.0. This was for aqueous solution (cf. ρ = 2.65 as found by Ratts[105] for solution in *ca*. 90% methanol). Taft's dual substituent-parameter equation was also applied (see p. 194[31-34,36,37]), but $ρ_I$ was found to be approximately equal to $ρ_R$ and there was no advantage over the simple Hammett equation. By comparing their results with those of other workers for other onium salts, the authors showed that ρ was practically independent of the nature of the onium centre to which *p*-$XC_6H_4COCH_2$ is attached. For *para*-substituents in the aryl ring ρ was found to be 1.4 (cf. 1.18 as found by Ratts). Since conformity to ordinary Hammett σ values was found, there was no evidence for transmission of conjugative effects through the sulphur atom from this ring.

Apart from the above-mentioned studies of substituent effects in an aroyl group attached to the ionizable CH group, there has been little systematic study of the effect on pK_a of varying the groups X and Y attached to CH in R^1R^2SCHXY. There are a few observations worth recording.

9-Dimethylsulphoniofluorene has a pK_a value of 7.3 in 31.7% aqueous dioxan[107], i.e. it is an acid of comparable strength to the dimethylphenacylsulphonium ion. The sulphonium ion (35) has a pK_a value of 10.2 (0.1 M LiCl, in 66.7% ethanol),

$$Et_2\overset{+}{S}CH_2CO_2Me$$

(35)

i.e. CO_2Me is greatly inferior to COPh (pK_a = 7.2) in promoting the acidity of sulphonium salts[108]. Ratts and Yao[104] were unable to titrate Me_2SCH_2COR, where R = OEt or NEt_2.

Lillya *et al.*[109] measured the pK_a values of bis(dialkylsulphonium)methylides (as tetrafluoroborates). These are monobasic acids (equation 18).

TABLE 7. Ionization of bis(dialkylsulphonium)methylides ($R^1R^2\overset{+}{S}CH_2\overset{+}{S}R^1R^2$) in water at ambient temperature[109]

R^1	R^2	pK_a	R^1	R^2	pK_a
Me	Me	9.19	Et	Et	7.71
Me	Et	8.51	Et	n-Pr	7.63
Me	n-Pr	8.52	Et	i-Pr	7.14
Me	i-Pr	7.76			

$$R^1R^2\overset{+}{S}CH_2\overset{+}{S}R^1R^2 \rightleftharpoons R^1R^2\overset{+}{S}\overset{-}{C}H\overset{+}{S}R^1R^2 + H^+ \qquad (18)$$

The pK_a values (aqueous solutions) are given in Table 7. They lie in the range 7–9, i.e. they are comparable to those of normal alkyl sulphonium salts in which ionization is promoted by a benzoyl group. Increasing the chain length and branching of the groups decreases pK_a, contrary to what might have been expected from the ordinary electron-releasing effect of alkyl groups, but fitting in with certain results of Johnson and Amel[102] and Ratts[105] described already. The authors suggest that the trend can be understood in terms of steric inhibition of solvation of the bis-sulphonium cation, which encourages ionization at CH_2. This idea might have application to the other anomalous results.

The pK_a value of an onium salt may often be taken as a guide to the nucleophilic reactivity of the conjugate ylide. Freeman et al.[110] point out that this is not necessarily so. For the tetraphenylcyclopentadienylides (**36**) the basicity decreases in the order Sb > As > P ≫ Se ≈ S, whereas nucleophilicity decreases in the order Sb > As > Se > S > P.

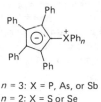

$n = 3$: X = P, As, or Sb
$n = 2$: X = S or Se

(36)

Finally, reference to cases in which a sulphonium centre promotes the ionization of an NH bond is appropriate. Daves et al.[111] found the pK_a value of 3-dimethylsulphonioindole (**37**) to be about 11.2, the ionization being of the N—H

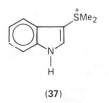

(37)

bond. The ylide, however, is stable, even though it is about 3 pK units more basic than most carbonyl-stabilized sulphonium ylides.

Kise et al.[112] recently measured the pK_a values in water of iminosulphonium ions (chloride as counter ion) (see equation 19).

$$R^1R^2\overset{+}{S}NHCOR^3 \rightleftharpoons R^1R^2\overset{+}{S}\bar{N}COR^3 + H^+$$

(19)

$$R^1R^2S{=}NCOR^3$$

For a series $(\overline{CH_2)_4\overset{+}{S}}{}^+NHCOC_6H_4X\text{-}p$, ρ was found to be *ca.* 0.8 (cf. *ca.* 2.0 for substituents in the aroyl ring of arylmethylphenacylsulphonium salts[106]). For a series $(\overline{CH_2)_4\overset{+}{S}}{}^+NHCOR$ or $Me_2\overset{+}{S}NHCOR$, with various aliphatic groups as R (CHCl$_2$, Et, etc.), there was a curvilinear relationship to σ^* values[33]. This was attributed to the intervention of $\pi(pd)$ bonding between the S and the N in the ylide. Corresponding $(\overline{CH_2)_4\overset{+}{S}}{}^+$ and $Et_2\overset{+}{S}$ compounds had very similar pK_a values (cf. Ratts' observations in CH ionization series[105]). For a series $R_2\overset{+}{S}NHCOMe$, pK_a increased with increasing chain length and branching of the alkyl groups, R. This corresponds to the ordinary electron-releasing inductive effect of alkyl groups, and again contrasts with what has been found in the CH ionization series[102,105,109].

C. Kinetics of H–D Exchange Reactions

Until fairly recently, the kinetic acidity of C—H bonds adjacent to a sulphonium group had been little investigated. As mentioned already (p. 190), Doering and Hoffmann[12], as long ago as 1955, studied the $\overline{O}D$-catalysed exchange of $(CH_3)_4\overset{+}{N}$, $(CH_3)_4\overset{+}{P}$, and $(CH_3)_3\overset{+}{S}$. Measurements of second-order rate constants at two temperatures gave the enthalpies and entropies of activation shown in Table 8. The contrast between S and P on the one hand and N on the other was explained in terms of d-orbital resonance lowering the energy of the S and P transition states, which were assumed to resemble closely the ylides $(CH_3)_{n-1}\overset{+}{X}CH_2$. The 'bicyclo[2.2.1]heptane-1-sulphonium' ion **38** was also shown to undergo exchange

(38)

indicating that 'the special geometrical constraint of the bicyclic sulphonium salt has not drastically inhibited d-orbital resonance'.

A more systematic study began in 1971 with the work of Barbarella *et al.*[113] They measured rate constants for the base-catalysed exchange of α-methyl or α-methylene (or in one case α-methine) hydrogens in eight methylsulphonium compounds. In some cases rate constants were measured at several temperatures

TABLE 8. Activation parameters for $\overline{O}D$-catalysed deuterium exchange[12]

Substrate	$\Delta H_{(expt.)}$ (kcal/mol)	$\Delta S^{\ddagger}$ (cal/K.mol)
$Me_4\overset{+}{N}$	32.2	-15
$Me_4\overset{+}{P}$	25.6	$+4$
$Me_3\overset{+}{S}$	22.4	-1

TABLE 9. Second-order rate constants[a] for base-catalysed H–D exchange in D_2O at 35°C[113]

Sulphonium ion[b]	$10^5 k$, α-CH$_3$ (M^{-1} s^{-1})	$10^7 k$, α-CH$_2$ (M^{-1} s^{-1})
Et$_2$MeṠ	5.0	1.9
n-Pr$_2$MeṠ	4.3	1.4
i-Pr$_2$MeṠ	6.0	<0.8[c]
i-Bu$_2$MeṠ	3.7	~0.7
Me$_2$ ◇ S$^+$Me[d]	770	<2.5
(CH$_2$)$_4$ S$^+$Me[e]	4.6	20
(CH$_2$)$_5$ S$^+$Me	7.3	1.2
(CH$_2$)$_6$ S$^+$Me	3.1	12
Me$_3$Ṡ	3.1[f]	

[a]Statistically corrected.
[b]I$^-$ as counter ion, except as indicated.
[c]Methine H, 55°C.
[d]BF$_4^-$ as counter ion.
[e]Only two of four methylene hydrogens exchange.
[f]Reference 12.

and activation parameters were calculated. Some of their results are given in Table 9. (The rate constants are statistically corrected to a basis of one hydrogen atom.)

In the acyclic compounds the α-methyl hydrogens exchange much more rapidly than the methylene (or methine) hydrogens. In the alkyl series $R_2\dot{S}CH_3$ the methyl rate constants show no clear relationship to the ordinary polar effects of alkyl groups in solution. (This resembles the situation for the pK_a values already discussed.) The structural effect is small and is probably the outcome of combined polar and steric effects.

The most remarkable methyl rate constant in Table 9 is that of the four-membered ring compound. The authors attribute the greatly enhanced kinetic acidity to a high s character in the exocyclic C—S bond, arising from the strain in the four-membered ring (cf. the explanation given by Ratts[105], see p. 210) for *decreased thermodynamic* acidity in a somewhat similar situation.

In the $(RCH_2)_2\dot{S}CH_3$ series the methylene hydrogens exchange more slowly than the methyl hydrogens by a factor of over 200 in each case. This is interpreted in terms of the electron-releasing effect of the alkyl groups destabilizing the transition state, which is held to be well advanced towards a carbanionic species.

In the five- and the seven-membered ring compounds the methylene hydrogens are considerably more reactive than in the six-membered ring, where they have reactivities comparable to those in an aliphatic system. In the five-membered ring the diastereotopic ring protons (see **39**) show high stereoselectivity towards abstraction by a base, those *cis* to the methyl group exchanging much more rapidly

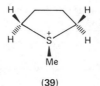

(39)

than those in a *trans* disposition. The authors believed that the rate factor involved was about 400, but later work has shown it to be much smaller than this (see below). There was no stereoselectivity in the six- or the seven-membered rings.

The work on the ring systems has been extended. Garbesi *et al.*[114] have confirmed by nuclear Overhauser experiments that in the five-membered ring in **39** the labile protons are those which are *cis* to the methyl group. Hofer and Eliel[115] have shown, however, that Barbarella *et al.*[113] greatly overestimated the stereoselectivity, which in fact involves a reactivity ratio of *ca.* 12:1. The earlier work on the six-membered ring was confirmed: there is almost no stereoselectivity. Hofer and Eliel[115] discuss the apparent incompatibility of these results with the predictions of the 'gauche effect', and they give several reasons why the incompatibility may be only apparent rather than real.

Barbarella *et al.*[116] have pursued the question of stereoselectivity in a study of *trans*-3-methyl-3-thioniabicyclo[4.3.0]nonane iodide which involves a conformationally rigid thiolanium cation **40a**. Considerable stereospecificity in exchange was found and nuclear Overhauser experiments showed the fastest exchanging proton to be that [$H_{(2)}$] closest to the methyl group (see **40b**), with the

(40a)   (40b)

reactivity ratios $H_{(2)}$:$H_{(4)}$:$H_{(3)}$:$H_{(1)}$ being approximately 200:3:3:1. These authors[117] later also examined the question of stereospecificity in the sulphoxides related to the ring sulphonium cations. In the monocyclic system, there is no stereospecificity in methylene hydrogen exchange in the sulphoxide **41**. In the bicyclic system, however, the exchange is stereospecific in the sulphoxide **42**, but less so than in

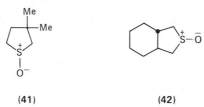

(41)   (42)

the sulphonium cation **40a**. Both of these papers contain much detailed discussion which cannot be summarised adequately here.

Peterson[118] has discussed some of the above-mentioned results on the H–D exchange of cyclic sulphonium salts, together with related data on the reactivity of other heterocyclic derivatives containing d-orbital acceptor atom. He shows that there is evidence for an 'alternation' effect, i.e. dependence on whether the number of ring atoms is odd or even. A molecular orbital treatment is proposed which is an extension of the Walsh formulation of cyclopropane, and a rationalization of odd–even effects is thereby achieved.

Studies of base-catalysed exchange reactions have also been made by Japanese workers. Yano *et al.*[119] have measured methyl hydrogen exchange rates of straight-chain alkyldimethylsulphonium halides, $C_nH_{2n+1}\overset{+}{S}(CH_3)_2X^-$, with X = Cl, Br, or I, and the n = 10 or 12 members being under micellar conditions. The rates for the longer chain compounds were much higher than those for the shorter chains (n = 1, 4, 6, or 8), and showed Michaelis–Menten type saturation kinetics with respect to $[OD^-]$, with the rates increasing in the order $Cl^- < Br^- < I^-$. Okonogi *et al.*[120] have also observed remarkable effects of adding DMSO both in micellar and non-micellar reaction systems.

Finally, the unusually basic sulphonium ylide derived from 3-dimethylsulphonioindole **37**[111] undergoes ready H–D exchange in its methyl groups when dissolved in CD_3OD or $CDCl_3$. It is suggested that the exchange occurs via the sulphonium ion **37** and a methylene ylide **43**.

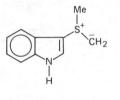

(43)

The varieties of behaviour in H–D exchange that have been discussed above are all basically caused by the unipolar and the d-orbital acceptor character of the sulphonium group, but this is subject to moderation by various other effects, sometimes with very striking results.

D. Promotion of Reactions via a Carbanionic Intermediate

Simple trialkylsulphonium compounds may react with bases and thereby undergo elimination reactions in which the sulphur is in the leaving group (equation 20).

$$R^1R^2\overset{+}{S}CH_2CH_2Ph + OH^- \longrightarrow R^1R^2S + CH_2{=}CHPh + H_2O \qquad (20)$$

If the molecule also contains a rather better leaving group, Y, then the reaction may take a different course (equation 21).

$$R^1R^2\overset{+}{S}CH_2CH_2Y + OH^- \longrightarrow R^1R^2\overset{+}{S}CH{=}CH_2 + Y^- + H_2O \qquad (21)$$

Doering and Schreiber's study[11] (1955) of the formation of the vinyldimethyl-sulphonium ion by elimination of HBr from 2-bromoethyldimethyl-sulphonium ion has already been mentioned (p. 190).

Crosby and Stirling[100] have studied extensively the kinetics of the elimination of phenoxide from β-substituted ethyl phenyl ethers (equation 22).

$$XCH_2CH_2OPh + B^- \longrightarrow XCH{=}CH_2 + BH + \bar{O}Ph \qquad (22)$$

The second-order rate constants can vary with X over eleven powers of 10 for the reaction with ethanolic sodium ethoxide. $Me_2\overset{.}{S}$ is one of the best substituents at promoting this reaction. ($Ph_3\overset{+}{P}$ and NO_2 would be better, but the reactions are inconveniently fast for direct study.) The $Me_3\overset{+}{N}$ compound reacts more slowly by a factor of about 10^{10}. The reaction is believed to occur by the E1cB mechanism, as in Scheme 1, with observed $k = k_1k_2/k_{-1}$.

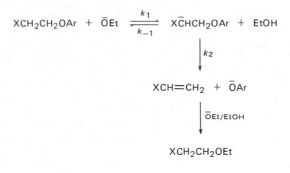

$$XCH_2CH_2OAr + \bar{O}Et \underset{k_{-1}}{\overset{k_1}{\rightleftharpoons}} X\bar{C}HCH_2OAr + EtOH$$

$$\downarrow k_2$$

$$XCH{=}CH_2 + \bar{O}Ar$$

$$\downarrow \bar{O}Et/EtOH$$

$$XCH_2CH_2OEt$$

SCHEME 1

$Me_2\overset{+}{S}$ thus exerts its influence in facilitating the formation of the intermediate carbanion through $\pi(pd)$ bonding.

When $X = Me_2\overset{+}{S}CH_2$ the situation is different and a Hofmann-type reaction occurs. Phenol is not formed, and the products of the reaction with ethoxide ion in ethanol are allyl phenyl ether (45%) and methyl 3-phenoxypropyl sulphide (55%). (The $Me_3\overset{+}{N}CH_2$ compound behaves somewhat analogously.)

For variation of the group X, log k is best correlated with σ_R^- and not with σ^* or σ_p^-, thus confirming that the resonance stabilization of a carbanionic species is very important in this system.

Crosby and Stirling[121] also carried out a more limited study of the elimination reactions with hydroxide ion in water. The rate for the $Me_2\overset{+}{S}$ compound is reduced by a factor of $ca.$ 3000 from that in ethanol. The rates for dipolar substituents are reduced by factors of up to 10. Thus, in water the activating properties of $Me_2\overset{+}{S}$ are comparable to those of CO_2Et. This dramatic effect is ascribed to the much greater solvation of the initial reactant in water in the case of $Me_2\overset{+}{S}$ compared with the situation for dipolar substituents. This solvation greatly inhibits the formation of the ylidic carbanion, which is poorly solvated.

$Me_2\overset{+}{S}$ has also been involved in studies of the carbanion mechanism of ester hydrolysis. Holmquist and Bruice[7] studied the pH–log(rate) profile for the hydrolysis of the ester **44** in water at ionic strength $\mu = 1.0$. At high pH the rate of alkaline hydrolysis becomes independent of pH. This is attributed to specific base-catalysed α-proton abstraction, with the formation of an ylide (**45**), which then breaks down to give the o-nitrophenate anion and a ketene (see equation 23).

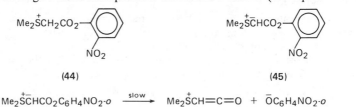

(44)   (45)

$$Me_2\overset{+}{\underset{-}{S}}CHCO_2C_6H_4NO_2\text{-}o \xrightarrow{slow} Me_2\overset{+}{S}CH{=}C{=}O + \bar{O}C_6H_4NO_2\text{-}o \qquad (23)$$

The latter adds on water to give the final product, as in equation 24.

$$Me_2\overset{+}{S}CH{=}C{=}O + H_2O \xrightarrow{fast} Me_2SCH_2CO_2H \qquad (24)$$

This mechanism of ester hydrolysis is essentially E1cb and is rare compared with other mechanisms of ester hydrolysis. It depends on the resonance stabilization of the intermediate carbanionic centre, and $Me_2\overset{+}{S}$ is obviously very suitable for accomplishing this. The cyano group is also able to function in this way[7].

VII. ELECTRONIC EFFECTS IN ALIPHATIC SYSTEMS

A. Introduction

In this section we discuss the effect of the sulphonium group in situations where there can be no delocalization of electrons into or out of the substituent, i.e. the group behaves as a unipole, and its electronic effect should not differ in kind from that of $R_3\overset{+}{N}$. There have been relatively few studies of the influence of the group in aliphatic systems.

We have already noted that the pK_a value of $Me_2\overset{+}{S}CH_2CO_2H$ is 1.15 ± 0.1 ($30°C$; ionic strength, 0.2)[7], compared with 1.83 for $Me_3\overset{+}{N}CH_2CO_2H$ ($25°C$)[6] and 4.76 ($30°C$)[8] for acetic acid. By applying Charton's[122] equation (equation 25) for determining the σ_I value of substituent X from the pK_a value of XCH_2CO_2H in water, the σ_I value for $Me_2\overset{+}{S}$ may be calculated as $ca.$ 0.92 (Holmquist and Bruice[7] calculated 0.86).

$$\sigma_I = -0.257pK_a + 1.204 \tag{25}$$

The σ_I value for $Me_3\overset{+}{N}$ is lower (0.73). Using the relation $\sigma_I = 0.45\sigma^*$ as between the σ_I value of X and the σ^* value of CH_2X (see p. 195), σ^* for $Me_2\overset{+}{S}CH_2$ is $ca.$ 2.0.

The σ_I value of $ca.$ 0.92 based on aliphatic pK_a values is reasonable in relation to a σ_m value of $ca.$ 1.0 based on aromatic pK_a values (see pp. 197 and 198; σ_m is largely governed by σ_I, see p. 194). It also corresponds closely to the σ_I value of 0.89 based on aromatic ^{19}F n.m.r. for solution in acetonitrile[58]. However, this may not be particularly significant as the value obtained by ^{19}F n.m.r. for solution in DMF is lower (0.74) (see p. 199)[61].

Holmquist and Bruice[7] examined the applicability of the σ_I values for $Me_3\overset{+}{N}$ and $Me_2\overset{+}{S}$ (based on acid ionization) to kinetic data for the reactions of o-nitrophenyl esters. They concluded that the effective values of σ_I for these substituents in the kinetic studies were much lower (0.34 and 0.58, respectively). They ascribed this to the different charge types of the processes under consideration.

B. Ester Hydrolysis

Bell and Coller[123,124] studied the kinetics of the alkaline hydrolysis in water at $25°C$ of a considerable number of esters, CH_2XCO_2Et, where X was a unipolar or a dipolar substituent. The most reactive substrate has $Me_2\overset{+}{S}$ as X (bromide as counter ion), the rate constant being about three times that for the $Me_3\overset{+}{N}$ species, the next most reactive (see Table 10). From a plot[123] of $\log k_{OH}$ for some of the esters $versus$ pK_a values for the corresponding acids, the pK_a value of $Me_2\overset{+}{S}CH_2CO_2H$ was read as 1.3[124], which agrees fairly well with the value of 1.15 ± 0.1 obtained by Holmquist and Bruice[7]. The kinetic results were interpreted in detail in terms of the electrostatic effects of substituent charges and dipoles on the free energy of formation of the transition state, i.e. a Kirkwood–Westheimer type of treatment was used[125,126]. The authors considered that their treatment gave an adequate interpretation of the results, while admitting that the work did not constitute a critical test of the importance or otherwise of through-σ-bonds inductive effects.

TABLE 10. Second-order rate constants for the alkaline hydrolysis in aqueous solution of esters (CH$_2$XCO$_2$Et) at 25°C and zero ionic strength[124]

X	k_{OH}(M^{-1} s^{-1})	X	k_{OH}(M^{-1} s^{-1})
S$^-$	0.0064	SO$_2$Me	12.8
CO$_2^-$	0.0145	Me$_3$N	66
H	0.111	Me$_2$S	204
SMe	0.92		

Bell and Coller[124] attached particular importance to the observation that S$^-$ retarded reaction by a factor of only 17 (X = H as standard), compared with an acceleration by a factor of 600 for Me$_3$N. The relatively poor deactivating effect of S$^-$ was attributed to its three lone pairs being directed away from the region of the σ bond, 'so that the centre of charge does not coincide with the sulphur nucleus'. The nitrogen atom in Me$_3$N has no lone pairs but in Me$_2$S the sulphur atom still carries one lone pair, and this situation may be responsible for the latter group producing a larger activating effect than Me$_3$N.

The exact interpretation of the effect of Me$_2$S in this aliphatic ester hydrolysis needs further consideration in the light of work on the hydrolysis of Me$_2$SCH$_2$CO$_2$Me (*p*-tosylate as counter ion) carried out by Casanova and Rutolo[127]. In the pH range 7–8.5 the reaction was found to be first order in each reactant (ester and OH$^-$), as found by Bell and Coller[124] for the ethyl ester in the pH range 8.3–9.3. However, in the course of this work Casanova and Rutolo[128] found that the methyl ester formed a stable ylide (**46**). The pK_a value for the

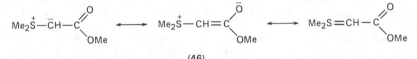

(46)

methyl ester is given as *ca.* 8 (medium not specified [cf. 10.2 for Et$_2$SCH$_2$CO$_2$Me (in 66.7% ethanol, 0.1 M LiCl)[108]]. Thus in the pH range around 8, as used in studying the hydrolysis, a considerable proportion of the ester would be present as ylide.

Casanova and Rutolo[127] further found that the reaction was characterized by a substantial *positive* entropy of activation of 17.7 ± 2.3 cal/K·mol whereas a negative value of *ca.* −20 cal/K·mol might have been expected. The authors suggest that their observations are in accord with reaction via a sulphonium cation–hydroxide anion intimate ion pair. They also discuss the possibility of hydrolysis by way of the ylide, but conclude that this is unlikely.

It must also be noted that Holmquist and Bruice[129] are sceptical about 'electrostatic catalysis', i.e. the interpretation of the influence of unipoles on ester reactions in terms of coulombic interactions between substrate and attacking reagent. Their views are supported by extensive studies of the hydrolysis of a variety of *o*-nitrophenyl esters over a wide range of pH. The LFER (equation 26) is obeyed by *all* substrates, neutral or positively charged (sulphonio-substituted compounds were not studied), where k_{OH} is the rate constant for OH$^-$-catalysed hydrolysis and k_{H_2O} is that for spontaneous general base-catalysed hydrolysis (30°C)[129].

$$\log k_{OH} = 0.84 \log k_{H_2O} + 8.0 \tag{26}$$

The conformity of both neutral and charged esters to the same LFER is held to mean that electrostatic facilitation of the nucleophilic displacement of o-nitrophenate ion by OH^- is unimportant.

It must finally be said that the treatment by Bell and Coller[123,124] does not include the possibility of steric effects, which are commonly believed to play an important role in the alkaline hydrolysis of aliphatic esters[33,44].

C. Charged and Uncharged Substituents: a Common Scale for σ*

Koppel, Palm, and their colleagues have worked on the problem of obtaining a common scale of σ* values for charged and uncharged substituents. The sulphonium group does not appear to be discussed explicitly, presumably for lack of pertinent data, but the general principles of the discussion are important, so we will summarize them here. The problem is similar to the one we have already noted for the Hammett equation (p. 195): that of including both charged and uncharged substituents in the same LFER (cf. Holmquist and Bruice[129], above).

For most readers, the paper by Koppel *et al.* (1974) will be the most helpful, since it is available in a full English version[130], but the basis is laid in two earlier papers in Russian, with English summaries[131,132].

The basic idea is that both in calculating and in applying σ* values of charged substituents the effect of electrostatic interaction between the substituent and the reaction centre must be eliminated. If this be done, σ* values for charged and uncharged substituents are on a common scale of inductive substituent constants. The original papers must be consulted for details of how the correction is made. We give now some examples of the results, with a possible extension to Me_2S.

For calculating a σ* value from the pK_a value of $Me_3NCH_2CO_2H$, the correction was +0.88[133], i.e. the corrected pK_a value was 1.83 + 0.88 = 2.71. The σ* value for Me_3NCH_2 was then read off a plot of pK_a *versus* σ* established by using uncharged substituents. The corrected σ* value was thus 1.15. The correction given is the same for Me_3NCH_2, Bu_3PCH_2, and Ph_3PCH_2 acids[133], so it seems safe to assume that it would be +0.88 for the Me_2SCH_2 acid also. This would give a corrected pK_a of 1.15 + 0.88 = 2.03, which would in turn give a σ* value of 1.5, corresponding to a σ_I value for Me_2S of 0.45 × 1.5 = 0.68.

It must be emphasized that in the subsequent application of such corrected values, the data to be correlated must also be corrected for electrostatic effects when appropriate. Koppel *et al.*[130] illustrate this for several processes including the ionization of alcohols and of thiols. The points for several charged substituents lie convincingly on the same correlation lines as for uncharged groups.

Work related to the above and which concerns sulphonium groups has been carried out by Järv and co-workers[134,135]. A study of the rates of alkaline hydrolysis of the esters (**47**, n = 1–6), together with the pK_a values of the thiols $EtS(CH_2)_nSH$ (n = 1–6), provided σ* values of the $EtS(CH_2)_n$ groups and the ρ* value for the ester hydrolysis was established as 1.90[135]. The work was then extended to the hydrolysis of the sulphonium ions (**48**, m = 2, 3, 4, 6)[134]. The reactions of these

(47) (48)

substrates are subject to a salt effect, and it was possible to obtain two sets of σ^* values for the sulphonium substituents: $\sigma^*_{\mu=0}$ by extrapolation to zero ionic strength and $\sigma^*_{\mu\to\infty}$ by extrapolation to infinite ionic strength. It was considered that the electrostatic effect discussed by Koppel et al.[130] would be eliminated in the $\sigma^*_{\mu\to\infty}$ values. This idea was tested on the ^{31}P chemical shifts of the esters[134]. Good correlation with σ^* for the uncharged esters was established. To place the sulphonium compounds on the same line, the $\sigma^*_{\mu\to\infty}$ values had to be used. The $\sigma^*_{\mu=0}$ values generated a separate line.

Finally, some recent papers on n.m.r. studies of sulphonium compounds shed light on the electronic effects of the sulphonium group in an aliphatic environment.

Barbarella et al.[136] studied the ^{13}C chemical shifts of alkyl chains under the influence of the $Me_2\overset{+}{S}$ group, as well as of other sulphur groups. Strong deshielding effects occur for the α-carbon, and weak for the β-, but weak shielding occurs for γ-carbon. Saturated sulphur heterocycles involving $>\overset{+}{S}Me$ and other sulphur functions have also been studied. An extensive study of [13]C n.m.r. of alkyl-substituted thiacyclohexanes and the corresponding thianium ions has also been carried out[137].

Minato et al.[138] prepared alkoxy-, dialkoxy-, and alkoxyaminosulphonium salts. P.m.r. spectra of ions of the types **49** and **50** were obtained for about 20

$$R^1R^2\overset{+}{S}OCH_AH_BR^3 \qquad\qquad R^1R^2\overset{+}{S}N(R^3)CH_AH_BR^4$$

(49)　　　　　　　　　　　　　　　　(50)

compounds. The relationship between $(\delta_A - \delta_B)$ and the nature of R^1 and R^2 was examined. The difference $(\delta_A - \delta_B)$ is caused by the chiral sulphur atom, but in some cases is zero or effectively so.

VIII. THE *ORTHO*-EFFECT OF THE SULPHONIUM GROUP

The term *ortho*-effect' has long been used to denote the peculiar influence of an *ortho*-substituent on the reactivity of a side-chain[139,140]. Studies of the *ortho*-effect of the sulphonium group are few, and are restricted to $Me_2\overset{+}{S}$.

Casanova et al.[141] studied the alkaline hydrolysis of the *ortho*-substituted methyl phenylacetate **(51)**. For comparison, rate measurements for the p-$\overset{+}{S}Me_2$, o-i-Pr, and

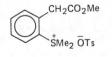

(51)

p-i-Pr compounds were also carried out. The o-$\overset{+}{S}Me_2$ coupound compound was hydrolysed about 5 times faster than its *para*-isomer. The i-Pr compounds showed the reverse rate order by a factor of about 3. The i-Pr group was considered to be an almost non-polar steric model for the $Me_2\overset{+}{S}$ group and on this basis a correction was calculated for steric retardation by o-$\overset{+}{S}Me_2$. With this correction the *ortho*/*para* rate ratio for $Me_2\overset{+}{S}$ was about 14. The authors suggested that this value did not indicate any particular neighbouring group effect involving covalent bonding. Probably a field effect enhanced by proximity to the reaction centre is involved. The activation energies and entropies for the reactions of the o-$\overset{+}{S}Me_2$ and p-$\overset{+}{S}Me_2$ compounds differ only slightly, which is also evidence against any fundamental difference in the two systems. However, the $\Delta S^{\ddagger}$ values are unusual in being slightly positive (*ca*. 7 cal/K·mol). The possibility that this could indicate reaction via a carbanionic intermediate (ylide) is mentioned. It was disposed of in Casanova and Rutolo's later work[128] on the hydrolysis of $Me_2\overset{+}{S}CH_2CO_2Me$, in which it was suggested that both systems react via an intimate ion pair (see p. 218).

There is a more distinctive effect of o-$\overset{+}{S}Me_2$ in the acidity of the corresponding acid. The o-$\overset{+}{S}Me_2$ acid is 100 times stronger (in 45% aqueous dioxan, 29°C) than the corresponding o-SMe acid, while the ratio for the corresponding *para*-acids is *ca*. 10. The ΔpK_a between $\overset{+}{S}Me_2$ acids is 0.75. This may correspond to no more than the enhanced field effect of $\overset{+}{S}Me_2$ from the *ortho*-position, although stabilization of the ionization product by ring formation is conceivable, as in **52**.

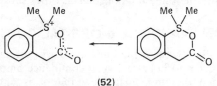

(52)

Hojo *et al*.[142] studied the ionization of o-$\overset{+}{S}Me_2C_6H_4CO_2H$. In one paper the effect of o-SMe, o-SOMe, o-SO$_2$Me, and o-$\overset{+}{S}Me_2$ on the ionization of benzoic acid was measured for DMSO–water mixtures containing 0–90% v/v DMSO. Corresponding *para*-isomers were also studied.

The effective σ_p values for p-SOMe and p-$\overset{+}{S}Me_2$ decrease considerably as the DMSO content of the medium is increased: p-SO$_2$Me shows a smaller decrease. The *ortho*-effects are assessed as $pK_a(p$-X$)$ − $pK_a(o$-X$)$ for the various sulphur substituents at the various solvent compositions and are compared with those of well known substituents. The *ortho*-effect of SMe seems to resemble that of i-Pr and is probably mainly steric in origin, i.e. the secondary steric effect. SO$_2$Me shows a strong *ortho*-effect, which is decreased greatly by the addition of DMSO to a minimum at about 65% DMSO; an analogy with o-NO$_2$ is seen. SOMe has only a weak *ortho*-effect in water (comparable to SMe), but this is greatly increased as DMSO is added. An interpretation in terms of the stabilization of the carboxylate anion by the S–O dipole is offered. The *ortho*-effect of Me$_2$S is strong in water and is enhanced greatly by the addition of DMSO; an analogy to the behaviour of o-OH is seen. It is considered that in water the effect of o-$\overset{+}{S}Me_2$ is little more than that of its secondary steric effect (assessed as that of i-Pr), which enhances acidity by twisting the carboxyl group out of the plane of the benzene ring[14]. A corresponding analysis for Me$_3$N (assumed isosteric with t-Bu) finds a considerable field effect in water. However, the field effect of o-$\overset{+}{S}Me_2$ is enhanced by the addition of DMSO, but stabilization of the carboxylate centre by ring formation (**53**) is not suggested.

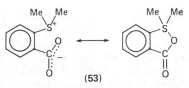

(53)

Hojo *et al*. were led to investigate both Me$_2\overset{+}{S}$ and Me$_3\overset{+}{N}$ further[57]. The pK_a values of o-, m-, and p-$\overset{+}{S}Me_2$ and o-, m-, and p-$\overset{+}{N}Me_3$ substituted acids were carefully measured in water, 65% aqueous DMSO, and 50% ethanol. Part of this work was described earlier in this chapter (p. 198). As far as the *ortho*-effect is concerned, it is held that the conjugative effect of $Me_2\overset{+}{S}$, which is acid strengthening from the *meta-* or the *para*-position, becomes acid weakening in the *ortho*-position. The field effects of $o\text{-}SMe_2$ and $o\text{-}\overset{+}{N}Me_3$ are considered comparable, while much of the difference between the strengths of the $o\text{-}\overset{+}{S}Me_2$ and $o\text{-}\overset{+}{N}Me_3$ acids arises from differences in the secondary steric effects.

In a third paper by the same group[143], values of ΔH^0 and ΔS^0 for the ionization of various *ortho-* and *para*-substituted benzoic acids are presented for solutions in 65 and 95% DMSO and in water. Various relationships between the thermodynamic quantities were found, involving linear plots with some scatter but, in 95% DMSO, groups such as $Me_2\overset{+}{S}$, OH, and SOMe with pronounced *ortho*-effects conform to the same relationships as other substituents.

IX. THE IRRELEVANCE OF d-ORBITAL CONJUGATION (?)

The words in this heading (without the question mark) are taken from the title of a recent paper[19]. They seem to indicate a new confidence among those who have for many years maintained the unpopular view that it is unnecessary to invoke a bonding role for the 3d orbitals of sulphur, or in analogous situations with various other elements. In this chapter we have carried out our declared intention of adhering to tradition (p. 191) and making frequent use of the symbol $\pi(pd)$ and related terminology. Many workers in sulphur chemistry continue to make effective use of this approach (see, e.g., Miller and Wan[144]). We must now, however, turn briefly to the developments which within a few years may make the $\pi(pd)$ approach outmoded.

On the general question of formulating the theory of sulphur chemistry in ways which do not require a bonding role of d orbitals, the reader is referred to a paper by Bent[18]. The main topic is the bonding in SF_6, the molecule that seems to many really to 'require' d orbitals. However, Bent shows that the MO model involving the use of d orbitals is only one of several possibilities. The new confidence in the theoretical treatment of the chemistry of sulphur without d-orbital bonding comes mainly from *ab initio* molecular orbital calculations. Before describing certain results of these, we refer to some experimental evidence that casts doubt on d-orbital participation in bonding.

Glass and Duchek[145] carried out a thorough X-ray crystallographic study of dehydromethionine (54). The most striking feature of their results is that the

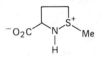

(54)

geometry at the N is pyramidal. This suggests that there is no $\pi(pd)$ bonding between the N and the S. The authors point out that this system is essentially isoelectronic with a sulphonium ylide system, which may therefore behave similarly in this respect, i.e. the usually assumed resonance 55 may not be correct. Pyramidal geometry at an ylide C^- is likely to be difficult to detect, since in many ylides the charge is delocalized into a carbon skeleton, and the ylidic carbon will adopt planar geometry whether or not there is resonance as in 55.

The 1,3-bis-sulphonium ylides present a favourable situation for detecting pyramidal geometry at C^-. These have recently been studied by Wolfe *et al.*[146]. After elaborate n.m.r. and other studies, the authors concluded that 'the static,

(55)

dynamic, and chemical properties of 1,3-bis-sulphonium ylides are consistent with a charge-localized structure in which both sulphur atoms and the ylide carbon have pyramidal configurations'.

Streitwieser and Ewing[147] measured the equilibrium ion-pair acidities of 1,3-dithians **56** towards their caesium salts in cyclohexylamine at 25°C. Varying R

(56)

has an enormous structural effect, equivalent to $\rho^* \approx -20$ in a correlation with σ^*. This large value is held to indicate that the negative charge in the carbanion is essentially localized on $C_{(2)}$ and is not delocalized by $\pi(pd)$ bonding with S. It is suggested that the main mechanism whereby sulphur stabilizes adjacent carbanionic centres involves the polarization of the polarizable sulphur electrons.

Streitwieser and Williams[148] show that this suggestion is in accord with *ab initio* SCF calculations on CH_3SH and $^-CH_2SH$. Two different basis sets were tried: 'split shell' (SS) and 'split shell plus d-orbitals' (SS + d). The marked carbanion-stabilizing effect of sulphur was confirmed but the introduction of d orbitals affected the energies of CH_3SH and $^-CH_2SH$ almost equally. Hence the d orbitals have no effect on the proton affinity of $^-CH_2SH$. Calculations of electron density differences as between CH_3SH and $^-CH_2SH$ showed increased electron density on S in the carbanion, in accord with the polarization mechanism suggested previously.

The paper by Bernardi *et al.*[19], to which reference has already been made, deals with a non-empirical molecular orbital investigation of the reactions

$$^-CH_2XH + H^+ \longrightarrow CH_3XH \longleftarrow H^+ + CH_3X^- \tag{27}$$

for the systems X = O or S. Computations were performed with several basis sets containing either sp or spd functions on the heteroatoms. No evidence was found for the differences in behaviour between the oxygen and sulphur systems being due to $\pi(pd)$ bonding for sulphur. As in Streitwieser's papers[147–148], the importance of the polarizability of sulphur was emphasized. It should be noted, however, that the greater polarizability of sulphur compared with oxygen is attributed, at least in part, to the presence of low lying d orbitals on sulphur. Thus the 'irrelevance' of d orbitals as such is not asserted, only that of their role in bonding.

Bernardi, Wolfe, and their associates have pursued these matters in two recent papers. Thus, Epiotis *et al.*[149] made a theoretical analysis of the factors determining the conformations and stabilities of oxy- and thiocarbanions. A one-electron MO model, supported by *ab initio* calculations, is proposed to rationalize the enhanced stability of a carbanionic centre adjacent to sulphur, as compared with oxygen, without postulating $\pi(pd)$ bonding. In the most recent work, Bernardi *et al.*[150] made a comparative quantum chemical investigation (SCF–MO–4–31G) of the bonding in first- and second-row ylides. The different characteristics of these ylides are rationalized in terms of group orbital interaction diagrams, which incorporate the stabilizing and destabilizing interactions between a carbon lone pair, and π and

$\pi^* XH_n$ group orbitals. Destabilizing interaction dominates when X is a first-row atom and stabilizing interaction dominates when X is a second-row atom.

In a recent monograph on sulphur chemistry, the role of d-orbitals is critically examined[151]. (See also ref. 152.)

X. REFERENCES AND NOTES

1. C. C. Price and S. Oae, *Sulfur Bonding*, Ronald Press, New York, 1962.
2. P. J. Durrant and B. Durrant, *Introduction to Advanced Inorganic Chemistry*, Longmans, London, 1962, pp. 810–861, (This is a detailed account of the chemistry of sulphur, with much reference to hybridization and bonding.)
3. G. Cilento, *Chem. Rev.*, **60**, 147 (1960). (This is a fairly comprehensive review of the literature, up to 1959, on the expansion of the sulphur outer shell.)
4. K. A. R. Mitchell, *Chem. Rev.*, **69**, 157 (1969). (This deals with the use in bonding of outer d orbitals.)
5. See, for example, C. L. Liotta, W. F. Fisher, E. L. Slightom and C. L. Harris, *J. Amer. Chem. Soc.*, **94**, 2129 (1972).
6. O. Weider, *Ber. Deut. Chem. Ges.*, **68**, 263 (1935).
7. B. Holmquist and T. C. Bruice, *J. Amer. Chem. Soc.*, **91**, 3003 (1969).
8. G. Kortüm, W. Vogel and K. Andrussov, *Dissociation Constants of Organic Acids in Aqueous Solution*, Butterworths, London, 1961.
9. D. Vorländer and E. Siebert, *Ber. Deut. Chem. Ges.*, **52**, 283 (1919).
10. J. W. Baker and W. G. Moffitt, *J. Chem. Soc.*, 1722 (1930).
11. W. von E. Doering and K. C. Schreiber, *J. Amer. Chem. Soc.*, **77**, 514 (1955).
12. W. von E. Doering and A. K. Hoffman, *J. Amer. Chem. Soc.*, **77**, 521 (1955).
13. This is indicated, *inter alia*, by the *ortho*/*para*-directing and the strongly activating properties of the group for electrophilic aromatic substitution.
14. R. W. Taft, in *Steric Effects in Organic Chemistry* (Ed. M. S. Newman), Wiley, New York, 1956, p. 580.
15. Ref. 1, pp. 155–158.
16. S. Oae, M. Yoshihara and W. Tagaki, *Bull. Chem. Soc. Jap.*, **40**, 951 (1967).
17. L. Goodman and R. W. Taft, *J. Amer. Chem. Soc.*, **87**, 4385 (1965).
18. H. A. Bent, in *The Organic Chemistry of Sulfur* (Ed. S. Oae), Plenum Press, New York, 1976, Chapter 1. (This article surveys the whole question of the use of 3d orbitals by sulphur.)
19. *E.g.* F. Bernardi, I. G. Csizmadia, A. Mangini, H. B. Schlegel, M.-H. Whangbo and S. Wolfe, *J. Amer. Chem. Soc.*, **97**, 2209 (1975).
20. Certain material in this section is adapted (by kind permission of the Oxford University Press) from J. Shorter, *Correlation Analysis in Organic Chemistry; an Introduction to Linear Free Energy Relationships*, Oxford Chemistry Series, 1973, Chapter 2.
21. Ref. 20, Chapter 1.
22. L. P. Hammett, *Physical Organic Chemistry*, 2nd ed., McGraw-Hill, New York, 1970, Chapter 11.
23. O. Exner, in *Advances in Linear Free Energy Relationships* (Eds. N. B. Chapman and J. Shorter), Plenum Press, London, 1972, Chapter 1. (This is a review of the Hammett equation.)
24. H. H. Jaffé, *Chem. Rev.*, **53**, 191 (1953).
25. The symbol and sign conventions used for substituent effects are those of C. K. Ingold, *Structure and Mechanism in Organic Chemistry*, 2nd ed., Bell, London, 1969. *I* or *K* effects which withdraw electrons from the ring are regarded as negative. The term 'resonance effect' will on occasion be used as synonymous with 'conjugative effect', and, as will appear later, *R* will have its conventional use as a subscript in denoting resonance substituent constants. '*K*' is based on the German for conjugative.
26. H. C. Brown and Y. Okamoto, *J. Amer. Chem. Soc.*, **80**, 4979 (1958).
27. H. van Bekkum, P. E. Verkade and B. M. Wepster, *Rec. Trav. Chim. Pays Bas.*, **78**, 815 (1959).

28. J. Shorter, in *Correlation Analysis in Chemistry: Recent Advances* (Eds. N. B. Chapman and J. Shorter), Plenum Press, New York, 1978, Chapter 4. (This is a general review of multiparameter extensions of the Hammett equation.)
29. Y. Yukawa and Y. Tsuno, *Bull. Chem. Soc. Jap.*, **32**, 971 (1959).
30. Y. Yukawa, Y. Tsuno and M. Sawada, *Bull. Chem. Soc. Jap.*, **39**, 2274 (1966).
31. R. W. Taft and I. C. Lewis, *J. Amer. Chem. Soc.*, **80**, 2436 (1958).
32. R. W. Taft and I. C. Lewis, *J. Amer. Chem. Soc.*, **81**, 5343 (1959).
33. R. W. Taft, in ref. 14, Chapter 13.
34. R. W. Taft, S. Ehrenson, I. C. Lewis and R. E. Glick, *J. Amer. Chem. Soc.*, **81**, 5352 (1959).
35. By a different procedure the authors of ref. 27 devised an analogous σ^n scale (n = normal, *i.e.* free of the effects of cross-conjugation).
36. S. Ehrenson, *Progr. Phys. Org. Chem.*, **2**, 195 (1964).
37. S. Ehrenson, R. T. C. Brownlee and R. W. Taft, *Progr. Phys. Org. Chem.*, **10**, 1 (1973).
38. R. W. Taft, E. Price, I. R. Fox, I. C. Lewis, K. K. Andersen and G. T. Davis, *J. Amer. Chem. Soc.*, **85**, 709 (1963).
39. R. W. Taft, E. Price, I. R. Fox, I. C. Lewis, K. K. Andersen and G. T. Davis, *J. Amer. Chem. Soc.*, **85**, 3146 (1963).
40. A. R. Katritzky and R. D. Topsom, in *Advances in Linear Free Energy Relationships* (Eds. N. B. Chapman and J. Shorter), Plenum Press, London, 1972, Chapter 3, and references cited therein.
41. G. P. Ford, A. R. Katritzky and R. D. Topsom, in *Correlation Analysis in Chemistry: Recent Advances* (Eds. N. B. Chapman and J. Shorter), Plenum Press, New York, 1978, Chapter 6.
42. E. D. Schmid, *Spectrochim. Acta*, **22**, 1659 (1966), and references cited therein.
43. J. D. Roberts and W. T. Moreland, *J. Amer. Chem. Soc.*, **75**, 2167 (1953).
44. J. Shorter, in *Advances in Linear Free Energy Relationships* (Eds. N. B. Chapman and J. Shorter), Plenum Press, London, 1972, Chapter 2. (This is a review of this part of Taft's work and cognate topics.) See also ref. 20, Chapter 3.
45. R. W. Taft, *J. Amer. Chem. Soc.*, **79**, 1045 (1957).
46. L. P. Hammett, *J. Amer. Chem. Soc.*, **59**, 96 (1937).
47. Ref. 24, p. 239.
48. Ref. 23, p. 10.
49. H. Zollinger, W. Büchler and C. Wittwer, *Helv. Chim. Acta*, **36**, 1711 (1953).
50. H. Zollinger, *Helv. Chim. Acta*, **36**, 1730 (1953).
51. Y. Okamoto and H. C. Brown, *J. Amer. Chem. Soc.*, **80**, 4976 (1958).
52. A. V. Willi, *Z. Phys. Chem. (Frankfurt)*, **27**, 233 (1961).
53. F. G. Bordwell and P. J. Boutan, *J. Amer. Chem. Soc.*, **78**, 87 (1956).
54. S. Oae and C. C. Price, *J. Amer. Chem. Soc.*, **80**, 3425 (1958).
55. S. Oae and C. C. Price, *J. Amer. Chem. Soc.*, **80**, 4938 (1958).
56. S. Oae and C. Zalut, *J. Amer. Chem. Soc.*, **82**, 5359 (1960).
57. M. Hojo, M. Utaka and Z. Yoshida, *Tetrahedron*, **27**, 4255 (1971).
58. W. A. Sheppard and R. W. Taft, *J. Amer. Chem. Soc.*, **94**, 1919 (1972).
59. P. J. Q. English, A. R. Katritzky, T. T. Tidwell and R. D. Topsom, *J. Amer. Chem. Soc.*, **90**, 1767 (1968).
60. O. Exner, in *Correlation Analysis in Chemistry: Recent Advances* (Eds. N. B. Chapman and J. Shorter), Plenum Press, New York, 1978, Chapter 10. (This is an extensive compilation of various types of substituent constant.)
61. W. Adcock, J. Alste, S. Q. A. Rizvi and M. Aurangzeb, *J. Amer. Chem. Soc.*, **98**, 1701 (1976).
62. J. M. Angelelli, R. T. C. Brownlee, A. R. Katritzky, R. D. Topsom and L. Yakhontov, *J. Amer. Chem. Soc.*, **91**, 4500 (1969).
63. M. Fujio, M. Mishima, Y. Tsuno, Y. Yukawa and Y. Takai, *Bull. Chem. Soc. Jap.*, **48**, 2127 (1975).
64. Y. Tsuno, M. Fujio and Y. Yukawa, *Bull. Chem. Soc. Jap.*, **48**, 3324 (1975).
65. Y. Yukawa and Y. Tsuno, *Mem. Inst. Sci. Ind. Res. Osaka Univ.*, **23**, 71 (1966).
66. G. A. Olah, P. W. Westerman and D. A. Forsyth, *J. Amer. Chem. Soc.*, **97**, 3419 (1975).

67. A. Ricci, F. Bernardi, R. Danieli, D. Macciantelli and J. H. Ridd, *Tetrahedron*, **34**, 193 (1978).
68. G. P. Syrova, V. F. Bystrov, V. V. Orda and L. M. Yagupol'skii, *Zh. Obsch. Khim.*, **39**, 1395 (1969). (English edition, p. 1364.)
69. A. Pollard and R. Robinson, *J. Chem. Soc.*, 1765 (1930).
70. C. K. Ingold, F. R. Shaw and I. S. Wilson, *J. Chem. Soc.*, 1280 (1928).
71. K. Fajans, *Naturwissenschaften*, **11**, 165 (1923).
72. M. Brickman, J. H. P. Utley and J. H. Ridd, *J. Chem. Soc.*, 6851 (1965).
73. H. M. Gilow and G. L. Walker, *J. Org. Chem.*, **32**, 2580 (1967).
74. H. M. Gilow, R. B. Camp and E. C. Clifton, *J. Org. Chem.*, **33**, 230 (1968).
75. H. M. Gilow, M. De Shazo and W. C. Van Cleave, *J. Org. Chem.*, **36**, 1745 (1971).
76. N. C. Marziano, E. Maccarone and R. C. Passerini, *Tetrahedron Lett.*, 17 (1972).
77. R. Danieli, A. Ricci, H. M. Gilow and J. H. Ridd, *J. Chem. Soc. Perkin Trans. II*, 1477 (1974).
78. R. Danieli, A. Ricci and J. H. Ridd, *J. Chem. Soc. Perkin Trans. II*, 290 (1976).
79. J. H. Rees, J. H. Ridd and A. Ricci, *J. Chem. Soc. Perkin Trans. II*, 294 (1976).
80. O. Exner, *Coll. Czech. Chem. Commun.*, **31**, 65 (1966).
81. K. Kamiyama, H. Minato and M. Kobayashi, *Bull. Chem. Soc. Jap.*, **46**, 2255 (1973).
82. N. J. Daly, G. Kruger and J. Miller, *Aust. J. Chem.*, **11**, 290 (1958).
83. J. Miller, *Aromatic Nucleophilic Substitution*, Elsevier, Amsterdam, 1968, pp. 71–75.
84. S. Oae and Y. H. Khim, *Bull. Chem. Soc. Jap.*, **42**, 1622 (1969).
85. L. Ya. Ignatov, O. A. Ptitsyna and O. A. Reutov, *Dokl. Akad. Nauk. SSSR*, **231**, 874 (1976).
86. C. J. M. Stirling, in *The Organic Chemistry of Sulfur* (Ed. S. Oae), Plenum Press, New York, 1976, Chapter 9. (This is a general review of the chemistry of sulphonium salts. Displacement reactions are discussed on pp. 491–497.)
87. J. Knapczyk and W. E. McEwen, *J. Amer. Chem. Soc.*, **91**, 145 (1969).
88. H. Matsuyama, H. Minato and M. Kobayashi, *Bull. Chem. Soc. Jap.*, **48**, 3287 (1975).
89. H. Minato, T. Miura and M. Kobayashi, *Chem. Lett. (Japan)*, 1055 (1975).
90. J. C. Martin and R. J. Basalay, *J. Amer. Chem. Soc.*, **95**, 2572 (1973).
91. J. K. Coward and W. D. Sweet, *J. Org. Chem.*, **36**, 2337 (1971).
92. J. Epstein, R. E. Plapinger, H. O. Michel, J. R. Cable, R. A. Stephani, R. J. Hester, C. Billington and G. R. List, *J. Amer. Chem. Soc.*, **86**, 3075 (1964).
93. C. K. Ingold and J. A. Jessop, *J. Chem. Soc.*, 713 (1930).
94. C. K. Ingold and J. A. Jessop, *J. Chem. Soc.*, 2357 (1929).
95. A. W. Johnson, *Ylid Chemistry*, Academic Press, New York, 1966.
96. P. A. Lowe, *Chem. Ind. (London)*, 1070 (1970). (This is a brief review of ylide chemistry.)
97. E. E. Ernstbrunner and D. Lloyd, *Chem. Ind. (London)*, 1332 (1971).
98. J. Fabian, in *Organic Compounds of Sulphur, Selenium, and Tellurium*, Vol. 3 (D. H. Reid, Senior Reporter), Chemical Society, London, 1975, p. 740.
99. H. Goetz, B. Klabuhn, F. Marschner, H. Hohberg and W. Skuballa, *Tetrahedron*, **27**, 999 (1971).
100. J. Crosby and C. J. M. Stirling, *J. Chem. Soc. B*, 671 (1970).
101. G. Aksnes and J. Songstad, *Acta Chem. Scand.*, **18**, 655 (1964).
102. A. W. Johnson and R. T. Amel, *Can. J. Chem.*, **46**, 461 (1968).
103. A. W. Johnson and R. T. Amel, *J. Org. Chem.*, **34**, 1240 (1969).
104. K. W. Ratts and A. N. Yao, *J. Org. Chem.*, **31**, 1185 (1966).
105. K. W. Ratts, *J. Org. Chem.*, **37**, 848 (1972).
106. C. Kissel, R. J. Holland and M. C. Caserio, *J. Org. Chem.*, **37**, 2720 (1972).
107. A. W. Johnson and R. B. LaCount, *Tetrahedron*, **9**, 130 (1960).
108. K. Issleib and R. Lindner, *Justus Liebig's Ann. Chem.*, **707**, 120 (1967).
109. C. P. Lillya, E. F. Miller and P. Miller, *Int. J. Sulfur Chem.*, **1A**, 89 (1971).
110. B. H. Freeman, D. Lloyd and M. I. C. Singer, *Tetrahedron*, **28**, 343 (1972).
111. G. D. Daves, W. R. Anderson and M. V. Pickering, *J. Chem. Soc. Chem. Commun.*, 301 (1974).

112. H. Kise, Y. Sugiyama and M. Senō, *J. Chem. Soc. Perkin Trans. II*, 1869 (1976).
113. G. Barbarella, A. Garbesi and A. Fava, *Helv. Chim. Acta*, **54**, 2297 (1971).
114. A. Garbesi, G. Barbarella and A. Fava, *J. Chem. Soc. Chem. Commun.*, 155 (1973).
115. O. Hofer and E. L. Eliel, *J. Amer. Chem. Soc.*, **95**, 8045 (1973).
116. G. Barbarella, A. Garbesi, A. Boicelli and A. Fava, *J. Amer. Chem. Soc.*, **95**, 8051 (1973).
117. G. Barbarella, A. Garbesi and A. Fava, *J. Amer. Chem. Soc.*, **97**, 5883 (1975).
118. P. E. Peterson, *J. Org. Chem.*, **37**, 4180 (1972).
119. Y. Yano, T. Okonogi and W. Tagaki, *J. Org. Chem.*, **38**, 3912 (1973).
120. T. Okonogi, T. Umezawa and W. Tagaki, *J. Chem. Soc. Chem. Commun.*, 363 (1974).
121. J. Crosby and C. J. M. Stirling, *J. Chem. Soc. B*, 679 (1970).
122. M. Charton, *J. Org. Chem.*, **29**, 1222 (1964).
123. R. P. Bell and B. A. W. Coller, *Trans. Faraday Soc.*, **60**, 1087 (1964).
124. R. P. Bell and B. A. W. Coller, *Trans. Faraday Soc.*, **61**, 1445 (1965).
125. J. G. Kirkwood and F. H. Westheimer, *J. Chem. Phys.*, **6**, 506 (1938).
126. F. H. Westheimer and J. G. Kirkwood, *J. Chem. Phys.*, **6**, 513 (1938).
127. J. Casanova and D. A. Rutolo, *J. Amer. Chem. Soc.*, **91**, 2347 (1969).
128. J. Casanova and D. A. Rutolo, *J. Chem. Soc. Chem. Commun.*, 1224 (1967).
129. B. Holmquist and T. C. Bruice, *J. Amer. Chem. Soc.*, **91**, 2982 (1969).
130. I. A. Koppel, M. M. Karelson and V. A. Palm, *Org. React. (USSR)*, **11**, 101 (1974).
131. I. A. Koppel, M. M. Karelson and V. A. Palm, *Reakts. Sposobn. Org. Soedin.*, **10**, 497 (1973).
132. V. A. Palm, V. M. Nummert, T. O. Püssa, M. M. Karelson and I. A. Koppel, *Reakts. Sposobn. Org. Soedin.*, **10**, 223 (1973).
133. Ref. 130, p. 104.
134. J. L. Järv, T. I. Pehk, A. A. Aaviksaar, D. I. Lobanov and N. N. Godovikov, *Izv. Akad. Nauk. SSSR, Ser. Khim.*, 1649 (1976). (English edition, p. 1564.)
135. J. L. Järv, A. A. Aaviksaar, N. N. Godovikov and N. A. Morozova, *Reakts. Sposobn. Org. Soedin.*, **9**, 813 (1972).
136. G. Barbarella, P. Dembech, A. Garbesi and A. Fava, *Org. Magn. Reson.*, **8**, 108 (1976).
137. G. Barbarella, P. Dembech, A. Garbesi and A. Fava, *Org. Magn. Reson.*, **8**, 469 (1976).
138. H. Minato, K. Yamaguchi, K. Okuma and M. Kobayashi, *Bull. Chem. Soc. Jap.*, **49**, 2590 (1976).
139. H. B. Watson, *Modern Theories of Organic Chemistry*, 2nd ed., Clarendon Press, Oxford, 1941, p. 241.
140. Ref. 44, p. 103.
141. J. Casanova, N. D. Werner and H. R. Kiefer, *J. Amer. Chem. Soc.*, **89**, 2411 (1967).
142. M. Hojo, M. Utaka and Z. Yoshida, *Tetrahedron*, **27**, 4031 (1971).
143. M. Hojo, M. Utaka and Z. Yoshida, *Tetrahedron*, **27**, 4255 (1971).
144. J. Miller and K.-Y. Wan, *J. Chem. Soc. Perkin Trans. II*, 1320 (1976).
145. R. S. Glass and J. R. Duchek, *J. Amer. Chem. Soc.*, **98**, 965 (1976).
146. S. Wolfe, P. Chamberlain and T. F. Garrard, *Can. J. Chem.*, **54**, 2847 (1976).
147. A. Streitwieser and S. P. Ewing, *J. Amer. Chem. Soc.*, **97**, 190 (1975).
148. A. Streitwieser and J. E. Williams, *J. Amer. Chem. Soc.*, **97**, 191 (1975).
149. N. D. Epiotis, R. L. Yates, F. Bernardi and S. Wolfe, *J. Amer. Chem. Soc.*, **98**, 5435 (1976).
150. F. Bernardi, H. B. Schlegel, M.-H. Whangbo and S. Wolfe, *J. Amer. Chem. Soc.*, **99**, 5633 (1977).
151. E. Block, *Reactions of Organosulfur Compounds*, Academic Press, New York, 1978. See especially pp. 18–21 and 44–50.
152. H. Kwart and K. King, *The Question of d Orbital Involvement in the Chemistry of Silicon, Phosphorus, and Sulfur*, Verlag Chemie, Weinheim, 1977.

The Chemistry of the Sulphonium Group
Edited by C. J. M. Stirling and S. Patai
© 1981 John Wiley & Sons Ltd.

CHAPTER **10**

Stereochemistry and chiroptical properties of the sulphonium group

K. K. ANDERSEN

Department of Chemistry, University of New Hampshire, Durham, New Hampshire 03824, USA

I.	INTRODUCTION	230
II.	RESOLUTION OF SULPHONIUM SALTS	231
III.	SYNTHESIS OF OPTICALLY ACTIVE SULPHONIUM SALTS	233
IV.	ABSOLUTE CONFIGURATIONS OF SULPHONIUM SALTS	236
V.	CYCLIC SULPHONIUM SALTS	245
	A. Thiiranium and Thiirenium Salts	245
	B. Thietanium Salts	246
	C. Thiolanium Salts	247
	D. Thianium Salts	248
	E. 1,3-Dithianium Salts	253
	F. Thioxanthenium Salts	255
VI.	STEREOMUTATION OF SULPHONIUM SALTS	255
	A. Trialkylsulphonium Salts	255
	B. Dialkylarylsulphonium Salts	258
	C. Alkyldiarylsulphonium Salts	259
	D. Triarysulphonium Salts	259
	E. Sulphonium Ylides	260
VII.	USE OF SULPHONIUM SALTS IN ASYMMETRIC SYNTHESIS	260
VIII.	BASE-CATALYSED H–D EXCHANGE IN SULPHONIUM SALTS	262
IX.	HETEROSULPHONIUM SALTS	263
	A. Oxosulphonium Salts	263
	B. Alkoxysulphonium Salts	263
	C. Aminosulphonium Salts	264
X.	CHIROPTICAL PROPERTIES OF SULPHONIUM SALTS	264
XI.	REFERENCES	264

I. INTRODUCTION

Reports of stereochemical studies on sulphonium salts were published during the years 1899–1905, 1924–1930, and from 1966 to the present. Few papers appeared at other times.

The 1890s were a time of great interest in the stereochemistry of nitrogen. At issue, at least in part, was the resolution of ammonium salts, compounds considered to be five coordinate, for this was before the independent existence of cation and anions was appreciated by most and before the octet theory was formulated. The period is discussed in both Freudenberg's[1] and Wittig's[2] treatise on stereochemistry and also in Mills' obituary of Pope[3]. Not too surprisingly, other heteroatom stereochemical studies accompanied the investigations of nitrogen; these studies included sulphonium salts.

In 1899, Aschan[4] mentioned the possibility of resolving sulphonium salts, examples of 'vierwerthigen Schwefel'. A year later, Strömholm[5] reported his unsuccessful attempts to prepare isomeric sulphonium salts whose existence was predicted by a planar tetracoordinate model with a central sulphur atom surrounded by four ligands. Vanzetti[6] was also unsuccessful in his efforts; he attempted the resolution of sulphonium salts 1 and 2 using a *Penicillium* and a *Mucor* mould and *Beggiatoa* sulphur bacteria.

$$\underset{(1)}{\overset{\overset{\displaystyle Et \quad Me}{|\qquad|}}{Me-S^+-CHCO_2H}} \qquad\qquad \underset{(2)}{\overset{\overset{\displaystyle Et}{|}}{Me-S^+-CH_2CH_2CO_2H}}$$

However, two successful resolutions were reported in 1900. Pope and Peachey[7] resolved ethylmethylthetin 3 via its bromocamphorsulphonate and camphorsulphonate salts. Smiles[8] obtained both enantiomers of methylethylphenacylsulphonium picrate 4 also using bromocamphorsulphonate anion as a resolving agent; his enantiomers were not optically pure, for Pope and Neville[9] later improved the resolution.

$$\underset{(3)}{\overset{\overset{\displaystyle Et}{|}}{Me-S^+-CH_2CO_2H}} \qquad\qquad \underset{(4)}{\overset{\overset{\displaystyle Et}{|}}{Me-S^+-CH_2COPh}}$$

After 1900, sulphonium salts were considered to be tetrahedral with the sulphur atom in the centre of the tetrahedron and the four ligands, one of which was the anion, at the vertices. The anion was still a source of confusion in some instances; e.g. R_3SHgI_3 was considered to be 'sexavalent'[9]. However, Adolph Werner apparently had a more correct picture of the structure of both ammonium and sulphonium salts. His conception corresponded to the present day picture in which the cations and anions exist separately without affecting one another's stereochemistry[1,2,10].

Smiles[11] brought this primal period to a close with a paper on the attempted asymmetric syntheses of a trialkylsulphonium salt (equation 1). No measurable asymmetric induction took place since 3 formed from 5 via 6 had no rotation.

$$\underset{(5)}{MeSEt} \xrightarrow{\ (-)\text{-Menthyl-OCOCH}_2Br\ } \underset{(6)}{(-)\text{-Menthyl-OCOCH}_2\overset{\overset{\displaystyle Me}{|}}{S^+}-Et} \longrightarrow 3 \qquad (1)$$

Werner's ideas and the octet theory were established by 1924; e.g. Meisenheimer *et al.*[10] noted the structural similarity of amines and sulphonium salts. Still, there remained difficulties in resolution. Wedekind[12] was unsuccessful in resolving the bisulphonium salt **7**. He did succeed, however, in separating the diastereomers. Diastereomers of cyclic sulphonium salts **8** and **9** were prepared by Bennett and

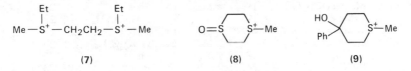

(7)   (8)   (9)

coworkers[13,14]. Ethylmethylphenacylsulphonium salt (**4**) was resolved again, this time by Balfe *et al.*[15], who showed that iodide ions greatly accelerated the racemization of the salt. A reversible dissociation into sulphide and alkyl iodide was postulated as responsible for the loss of optical activity. A period of general quiescence followed this paper, although several publications should be noted.

Firstly, the geometry of sulphonium salts was established by X-ray crystallography (see Chapter 2)[16]. Secondly, two separate calculations of barriers to pyramidal inversion were reported. Kincaid and Henriques[17] calculated an activation energy of 100 kcal/mol, which was later shown to be too high, but they successfully predicted a lowering of the barrier by increasing the steric bulk around sulphur. Weston[18], on the other hand, calculated barriers which were too low to be consistent with the existence of optically active sulphonium salts.

Renewed interest in sulphur stereochemistry in the early 1960s included renewed interest in sulphonium salts. The first paper of this third and present period appeared in 1966. Darwish and Tourigny[19] showed that trialkylsulphonium salts racemized not only by dissociation but also by pyramidal inversion.

II. RESOLUTION OF SULPHONIUM SALTS

Early resolutions of sulphonium salts were accomplished by fractional crystallization of diastereomeric salts formed by using *d*-camphorsulphonic acid (**10**) or *d*-bromocamphorsulphonic acid (**11**) as sources of the anion. Most recent resolutions have employed (−)-(2*R*,3*R*)-dibenzoylhydrogentartrate salts, although (−)-malate[20,21] and (−)-(2*R*,3*R*)-di-*o*-toluylhydrogentartrate[22] salts have also been used. The structure of the parent acids (**12**, **13**, and **14**) are shown in the same

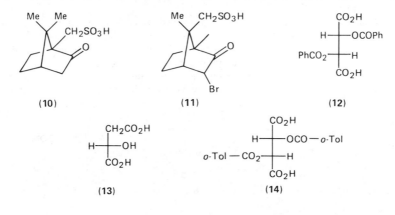

(10)   (11)   (12)

(13)   (14)

TABLE 1. Optically active sulphonium ions, $RR^1R^2S^+$, obtained by resolution

R	R^1	R^2	Ref.	R	R^1	R^2	Ref.
Me	Et	i-Pr	23	Me	Et	$PhCOCH_2$	8, 9, 15, 24, 26
Me	Et	t-Bu	19	Me	Et	$PhMeOC{=}CH$	27
Me	Et	Me_2MeOCH_2C	19	Me	Et	Ph	23
Me	Et	$EtMe_2C$	19	Me	Et	p-$MeOC_6H_4$	23
Me	Et	1-Adamantyl	20, 21	Me	Et	p-HOC_6H_4	23
Me	Allyl	1-Adamantyl	21	Me	Et	p-$NO_2C_6H_4$	23
Me	Et	$PhCH_2$	23, 24	n-Pr	Et	Ph	23
Me	Et	p-$NO_2C_6H_4$	24, 25	n-Bu	Et	Ph	23
Me	Et	p-$MeOC_6H_4$	24	Ph	Et	p-$MeOC_6H_4$	23
Me	Et	p-ClC_6H_4	25	Ph	Me	p-$MeOC_6H_4$	23
Me	Et	3-NO_2-4-$MeOC_6H_3$	24	Me	Et	$HOCOCH_2$	7

order as named. A list of some sulphonium salts resolved by crystallization is given in Table 1.

A convenient resolution procedure is to form the sulphonium hydroxide by passage of a methanolic solution of the sulphonium halide, the usual cation–anion combination obtained when a sulphonium salt is synthesized by alkylation of a sulphide, through an ion-exchange column[27]. The resulting basic eluent is then neutralized by addition of the appropriate acidic resolving agent. After fractional crystallization, the acid-derived anion of the partially or completely resolved salt is replaced with perchlorate ion, e.g. by addition of silver perchlorate or by passage through a hydroxide ion-exchange column into 70% perchloric acid. Naturally, caution should be exercised in handling sulphonium perchlorates. They are potentially hazardous although no explosions of these compounds, to our knowledge, have been reported.

No triarylsulphonium salts have ever been resolved and attempts to do so have failed. A low barrier to pyramidal inversion with consequent racemization is the probable cause of these failures. A 15-min half-life at 25°C was predicted for a cyclic triarylsulphonium salt[23]. If this prediction is correct, as seems likely, any resolution via diastereomeric salt formation would require operation at low temperatures or the use of molecules whose structure is modified so as to inhibit pyramidal inversion. Further discussion of this topic can be found in the section on stereomutation.

Several sulphonium ylides have been prepared in various states of enantiomeric purity by deprotonation of sulphonium salt precursors. Some, such as **15**[23,27], **16**[27], **17**[27], **18**[27], and **19**[21], are relatively stable to decomposition, but in general ylides racemize more readily than their sulphonium salt precursors. Other ylides such as **20**[21], **21**[25], and **22**[25] rearrange to more stable compounds and have not been isolated. Further discussion of ylide stereochemistry is given in the section on sulphonium ylides.

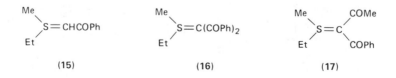

(15)   (16)   (17)

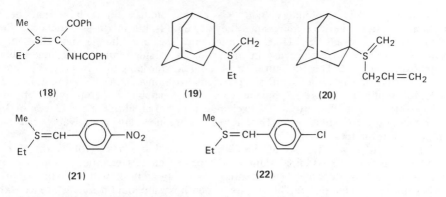

(18)   (19)   (20)

(21)   (22)

III. SYNTHESIS OF OPTICALLY ACTIVE SULPHONIUM SALTS

Optically active sulphonium salts may be synthesized by treating alkoxysulphonium salts, derived from sulphoxides by alkylation, with organocadmium or Grignard reagents[28]. Equations 2–5 show that this process proceeds with a uniform stereochemistry since enantiomeric sulphonium salts are obtained from sulphoxides **23** and **24** and from **24** and **25**[29]. The reactions proceed with inversion as shown[30].

Various enantiomerically pure sulphoxides are available by treatment of diastereomerically pure (–)-methyl sulphinate esters **26** with Grignard reagents (equation 6)[31]. Diastereomerically pure esters **26** are available when R is an aryl group such as p-tolyl, phenyl, p-anisyl, or 1-naphthyl. The epimers of configuration S at sulphur are crystalline and are usually isolated more easily than those of configuration R. When R is benzyl, however, the R epimer is obtained. Simple

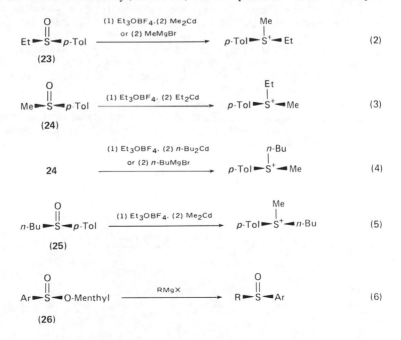

(−)-menthyl alkanesulphinates (e.g. R = methyl and *n*-butyl) have not been obtained diastereomerically pure since they are liquids, but mixtures enriched in the *R* epimer are available. Thus, enantiomerically pure diaryl and alkyl aryl sulphoxides are available for use in the synthesis of sulphonium salts, but dialkyl sulphoxides are not. The configuration of a sulphoxide prepared from a sulphinate ester is known since the synthesis proceeds with inversion of configuration at sulphur. Optically active sulphoxides are also available by conversion of one sulphoxide into another by ligand exchange using organolithium reagents[32] or by chromatographic resolution[33]. This latter technique may conceivably become important in the future.

An alternative to the use of optically active sulphoxides involves the direct synthesis of chiral alkoxysulphonium salts, albeit of modest enantiomeric purity, by treatment of a sulphide, (−)-menthol, and pyridine mixture with *t*-butyl hypochlorite in acetonitrile. The resulting menthoxysulphonium salts **27** react with sodium dimethylmalonate to give optically active ylides[34].

$$\begin{array}{c} \text{O-Menthyl} \\ | \\ \text{Ar} \blacktriangleright \text{S}^+ \blacktriangleleft \text{Ar}^1 \end{array} \xrightarrow{\text{NaCH(CO}_2\text{Me)}_2} \begin{array}{c} \text{C(CO}_2\text{Me)}_2 \\ || \\ \text{Ar}^1 \blacktriangleright \text{S} \blacktriangleleft \text{Ar} \end{array} \qquad (7)$$

(27)

In principle, a given sulphonium salt, $RR'R''S^+$, should be obtainable from an organometallic—sulphoxide (or menthoxysulphonium salt) combination in three ways, i.e. any one of the three R groups could originate from the organometallic reagent. Unfortunately, this optimistic view is not warranted; there are certain limitations to the reaction which proscribe its generality[29].

Firstly, trialkylsulphonium salts have not been obtained using organocadmium or Grignard reagents (equation 8). Secondly, triarylsulphonium salts, although readily available by this route, are racemic; apparently they racemize by pyramidal inversion before they can be isolated (equation 9). Thirdly, ethylphenyl-*p*-tolyl-sulphonium tetrafluoroborate **30** prepared from (+)-phenyl *p*-tolyl sulphoxide **29** was racemic even though the closely related *p*-anisylethylphenylsulphonium salt is fairly stable towards pyramidal inversion with a half-life of over 2 h at room temperature (equation 10). Fourthly, good chemical yields are obtained at the expense of enantiomeric purity.

It therefore seems this synthesis at present is limited to the preparation of dialkylarylsulphonium salts. Further, not many halide-free alkylcadmiums or Grignard reagents are readily available, nor are many functional groups compatible

$$\begin{array}{c} \text{OMe} \\ | \\ \text{Me} \blacktriangleright \text{S}^+ \blacktriangleleft n\text{-Bu} \end{array} \xrightarrow[\text{//}]{\text{Et}_2\text{Cd}} \begin{array}{c} \text{Et} \\ | \\ \text{Me} \blacktriangleright \text{S}^+ \blacktriangleleft n\text{-Bu} \end{array} \qquad (8)$$

(28)

$$\begin{array}{c} \text{O} \\ || \\ \text{Ph} \blacktriangleright \text{S} \blacktriangleleft p\text{-Tol} \end{array} \xrightarrow{\text{(1) EtO}_3\text{BF}_4. \text{ (2) } o\text{-Tol}_2\text{Cd}} \begin{array}{c} o\text{-Tol} \\ | \\ (\pm)\text{-Ph} - \text{S}^+ - p\text{-Tol} \end{array} \qquad (9)$$

(29)

$$\textbf{(29)} \xrightarrow[\text{(2) EtMgBr}]{\text{(1) Et}_3\text{OBF}_4.} \begin{array}{c} \text{Et} \\ | \\ (\pm)\text{-Ph} - \text{S}^+ - p\text{-Tol} \end{array} \qquad (10)$$

(30)

with these reactive reagents. Nevertheless, this method remains the only synthesis of any generality and it does provide dialkylarylsulphonium salts of known absolute configuration even if of variable and unknown enantiomeric purity.

The reason for failure of the reaction to produce trialkylsulphonium salts is not clear. Treatment of (±)-methoxy-*n*-butylmethylsulphonium tetrafluoroborate **28** with ethylcadmium yielded propane, indicating that attack by the organometallic reagent occurred on the methoxy carbon. Yet, when methoxy was replaced with adamantoxy, a group for which such nucleophilic attack is precluded, still no trialkylsulphonium salt was produced. Triethylsulphonium tetrafluoroborate was recovered quantitatively after treatment with methylcadmium or methoxide ion, which indicates that such salts, and presumably the hoped-for products, are stable to the reaction conditions. Neither the organometallic reagent nor the alkoxide ion liberated from the alkoxysulphonium salt destroyed the hoped-for sulphonium salt[29].

The formation of racemic ethylphenyl-*p*-tolylsulphonium tetrafluoroborate **30** from optically active methoxyphenyl-*p*-tolylsulphonium tetrafluoroborate was unexpected, since *p*-anisylethylphenylsulphonium perchlorate has a half-life of 2.3 h in methanol at 25°C (equation 10). However, ylides are known to racemize more readily than their parent sulphonium salts. Perhaps ylide formation occurred in the reaction mixture with consequent racemization prior to isolation of the product[29].

Even though dialkylarylsulphonium salts are stable to pyramidal inversion, they are not enantiomerically pure when prepared from enantionmerically pure alkoxysulphonium salts. This is evident from the variable rotations observed for a given salt obtained in different reactions. When a halide-free alkylcadmium is used, the formation of sulphonium salt takes several hours. During this time, displaced alkoxide ion causes some racemization of alkoxysulphonium salt. Hence an increased reaction time improves the yield of product at the expense of its enantiomeric purity. Grignard reagents give lower chemical and optical yields than do distilled alkylcadmiums. Side reactions caused by bromide ion are believed to be partly responsible for these decreases[29,30].

Alkylation of enantiotopic faces of a sulphide sulphur atom by a chiral alkylating agent or of diastereotopic faces by an achiral or chiral reagent will, in principle, constitute an asymmetric synthesis of the sulphonium unit. An example of the former is the alkylation of ethyl methyl sulphide by (−)-menthyl α-bromoacetate; however, no measurable asymmetric synthesis occurred, for on hydrolysis of the ester racemic sulphonium salt was obtained (see equation 1)[11]. Methylation of diastereotopic faces, however, has proved successful in preparing one of the two possible diastereomers. The penicillanate **31** gave only **32** and not the epimer with methyl fluorosulphonate[35]. Similarly cephalosporin (**33**) gave **34**, but with inversion of configuration of $C_{(6)}$, probably through ring opening and closing via **35**[36]. Generally, alkylation leads to a mixture of diastereomers. For example, (S)-β-thia-δ-caprolactone **37** gave a 1:1 mixture of diastereomers **38**. These were separated by fractional crystallization and then one of the isomers was converted into ethylmethylpropylsulphoniun ion by chemical means not involving C—S

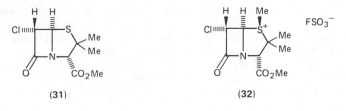

(31)   (32)

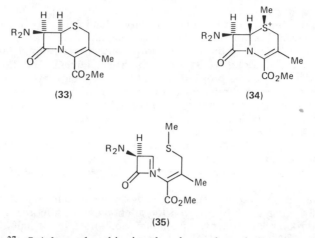

(33)   (34)

(35)

bond cleavage[37]. S-Adenosylmethionine has been degraded to S-carboxymethyl-methionine[38]. These constitute other syntheses of optically active sulphonium salts, but examples of such transformations are rare since sulphonium salts are rather labile. Further discussion of these two examples is given in the section on absolute configuration.

IV. ABSOLUTE CONFIGURATIONS OF SULPHONIUM SALTS

Very few chiral sulphonium ions have been assigned an absolute configuration. The first group described below includes those sulphonium salts whose absolute configurations were assigned by comparison of the stereochemical relationship of the sulphur atom to another chiral centre of known configuration in the molecule. Chemical transformations of these compounds into other sulphonium salts were used to extend these assignments. The second group includes sulphonium salts assigned configurations by chemical processes involving bonds to sulphur. These processes proceed with known stereochemistry.

Ethylmethylpropylsulphonium ion, the simplest trialkylsulphonium cation, had its configuration related to lactic acid by chemical means (Scheme 1)[37]. X-ray analysis of **38**, one of the two epimeric sulphonium salts produced upon ethylation of sulphide **37**, showed that the S-ethyl and C-methyl groups were *trans* to one another. Since the carbon atom's configuration was derived from (S)-ethyl lactate **36**, the configuration at sulphur was established for **38** and for the compounds derived from it, **39–42**.

S-Adenosylmethionine **43** has also had its configuration determined at sulphur by a combination of X-ray and chemical techniques[38]. Rather than establish the configuration of **43** directly by X-ray analysis, which in principle should be possible since **43** contains several centres of known chirality, the molecule was degraded to S-carboxymethyl-(S)-methionine **44**. Crystals of **43** suitable for X-ray analysis were not available. The degradation involved treatment of **43** with cold, dilute sodium hydroxide solution followed by oxidation of the resulting products with periodate ion at pH 4.0. Scheme 2 shows a degradation pathway proposed by the investigators. Radioactive **43**, labelled with $^{14}CH_3$, was used; portions of the resulting labelled methionine derivative **44** were mixed with separate samples of S-carboxymethyl-(S)-methionine epimeric at sulphur. Upon chromatography,

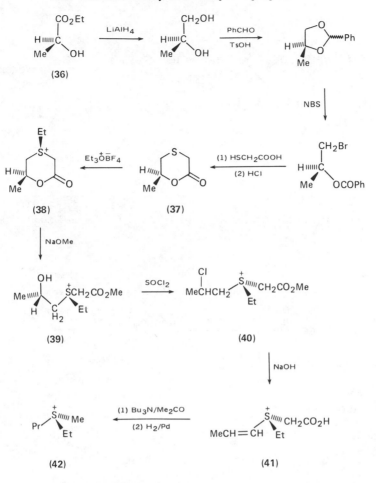

SCHEME 1

labelled **44** co-eluted almost exclusively with the epimer having configuration *S* at sulphur. The epimer of **44**, i.e. **45**, had its structure determined by X-ray analysis. It was of configuration *R* at sulphur, so **44** had to be *S*. The $^{14}CH_3$-labelled **43**, prepared enzymatically, was the natural, biologically important diasteromer.

As mentioned earlier, methylation of the penicillanate **31** and the cephalosporin **33** is stereoselective and gives, in each case, only one of two possible diastereomers, **32** and **34**, respectively[35,36]. Tentative assignments of stereochemistry were made by n.m.r. spectroscopy. A significant nuclear Overhauser effect between the *S*-methyl and $C_{(6)}$ proton was observed for **34**. Therefore, the methyl and $C_{(6)}$ proton are in close proximity, i.e. *cis*. Since the absolute configuration of the starting molecule **33** is known, the configuration of **34** follows from the *cis* stereochemistry.

Optically active dialkylarylsulphonium salts synthesized from alkoxysulphonium salts and organometallic reagents as shown in equations 2–5 are formed with

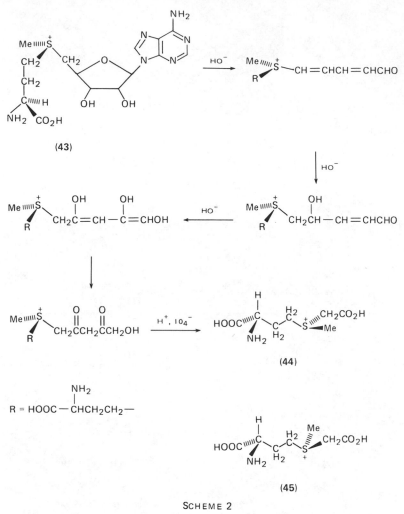

(43)

(44)

R = HOOC—CHCH₂CH₂—
with NH₂ above

(45)

Scheme 2

inversion of configuration at sulphur. This was demonstrated using *cis*- and *trans*-2-methyl-2,3-dihydrobenzothiophene 1-oxides (**46** and **47**) as model compounds for the acyclic alkylarylsulphoxides[30]. The stereochemistry of their conversion to *cis*- and *trans*-1,2-dimethyl-2,3-dihydrobenzothiopenium cations (**48** and **49**) upon methylation followed by treatment with methylcadmium or methylmagnesium bromide was determined. Predominant inversion at sulphur occurred. Generalization of this finding coupled with the known absolute configurations of alkyl aryl sulphoxides allows configurational assignments to be made to the dialkylarylsulphonium salts produced in this way.

Sulphoxides **46** and **47** were prepared as a 2:3 mixture by oxidation of sulphide **50** using hydrogen peroxide in acetic acid and were separated by column chromatography. Configurational assignments were based on chromatographic

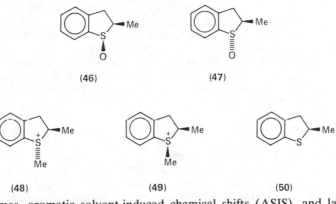

retention times, aromatic solvent-induced chemical shifts (ASIS), and lanthanide-induced chemical shifts (LIS).

Molecular models suggest that the *cis*-sulphoxide would have its sulphinyl oxygen more hindered to binding with a chromatographic substrate than would the *trans* isomer. Therefore, the *cis*-sulphoxide should elute before the *trans* isomer. Consequently, the more rapidly eluting isomer was assigned the *cis* configuration **46** and the more retained sulphoxide the *trans* configuration **47**. Production of a lesser amount (40%) of **46** relative to **47** (60%) in the oxidation of **50** is consistent with this interpretation of the chromatographic results; the hydrogen peroxide is expected to oxidize the least hindered sulphide face most readily. More important, however, the same assignments were made by ASIS and LIS studies.

Benzene is believed to form a weak complex with the positive end of the S–O dipole, thereby shielding neighbouring protons and moving their resonances upfield compared with their signals in an inert solvent such as chloroform. Indeed, as expected, the *C*-methyl protons of **47** underwent a greater upfield shift (0.51 p.p.m.) in benzene *versus* chloroform than did *C*-methyl protons of **46** (0.34 p.p.m.), whereas the α-proton in **46** was shifted upfield more (*ca.* 0.8 p.p.m.) than the corresponding proton in **47** (*ca.* 0.4 p.p.m.).

In LIS studies[30] employing Eu(dpm)$_3$, which complexes with the sulphinyl oxygen, the *C*-methyl protons of **46** were shifted downfield ($\Delta = -8.26$) considerably more than the *C*-methyl protons of **47** ($\Delta = -3.40$). This was as expected; the methyl group closest to the europium underwent the largest shift. The α-protons underwent the appropriate shifts as well ($\Delta = -9.48$ for **46** and -10.2 for **47**), although the shift difference between the two protons is not very large.

Configurational assignments to sulphonium salts **48** and **49** were based on an equilibrium study, a kinetic study, and ^{13}C n.m.r. spectroscopy. Thermal equilibration via pyramidal inversion of a 93:7 mixture of **48** and **49** in acetic acid at 110°C gave a 3:2 mixture of **48** and **49**. Since the sterically less crowded *trans*-sulphonium ion **48** was judged thermodynamically more stable than the sterically more crowded *cis*-isomer **49**, the *cis–trans* assignments were made as indicated by the diagrams. Methylation of sulphide **50**, as with its oxidation, was expected to proceed more rapidly on the less hindered diastereotopic sulphide face, leading to a preponderance of the *trans* salt **48**. Trimethyloxonium tetrafluoroborate in nitromethane or methylene chloride and methyl fluorosulphonate in methylene chloride all gave a 4:1 mixture of **48** and **49**. Demethylation of salts **48** and **49** to sulphide **50** by nucleophilic attack on the *C*-methyl group was predicted to proceed more rapidly for the *trans* isomer **48** than for the *cis* isomer **49**. This prediction was based on an expected larger separation in energy for the two S_N2 diastereomeric transition states lying between sulphide and the two salts (based on the 4:1 ratio) compared with the lesser energy separation between the two salts (based on the 3:2 ratio). As predicted, **48** was demethylated by pyridine twice as fast as **49**. Finally, ^{13}C n.m.r. resonances of the *cis*-*S*-methyl and -*C*-methyl in **49** were shifted upfield compared to the corresponding methyl carbon signals of **48**. Such upfield shifts are expected for non-bonded carbon atoms in close proximity.

Now that the configurations of **46–49** had been established, the stereochemistry of sulphonium salt formation could be elucidated. It was already known that the reactions shown by equations 2–5 gave optically impure products. The sulphonium salts could be crystallized with resulting changes in rotation. Moreover, two supposedly identical preparations yielded salts with different rotations. Such variability was also found when alkoxysulphonium salts **51** and **52** formed upon methylation of sulphoxides **46** and **47** were treated with organometallic reagents. Mixtures of **48** and **49** were formed from either diastereomerically pure alkoxysulphonium salt (**51** or **52**). For example, methylcadmium and **52** at room temperature gave a 45% yield of sulphonium salts consisting of **48** and **49** in the ratio 15:85. Over 30% of sulphoxides **46** and **47** was recovered. Formation of sulphonium salts obviously took place with predominant inversion in this instance. When the reaction time was decreased to 20 min, the yield of salts decreased from 45% to 32%, but the product was diastereomerically pure. The amount of recovered sulphoxide, formed in part by hydrolysis of unreacted alkoxysulphonium salt, was increased. Similarly, **51** gave **48** (inversion) in about 30% yield, but almost no **49**. Methylmagnesium bromide and **51** at $-78°C$ gave a 9:1 mixture of **48** and **49** in 45% yield; the major by-product was sulphide **50** in about 20% yield and not the sulphoxides. The alkoxysulphonium salt **52**, however, was much less stereoselective in its reaction with Grignard reagent, giving only 50–65% inversion. The yields of sulphonium salts were also lower (30–40%) and the yields of sulphide higher (60–70%). There is no question, however, that the sulphonium salts in these reactions are formed by predominant inversion.

Since the sulphonium salts **48** and **49** were stable to methylmagnesium bromide and magnesium methoxide under the reaction conditions, it appeared that loss of stereospecificity involved reactions of the alkoxysulphonium salt and not epimerization of the product. Pseudorotation of a presumed trigonally bipyramidal intermediate was judged to be of no consequence and not the source of epimerization. Ligand permutation should not lead to a decrease in stereospecificity with increasing reaction time as is the actual case; a constant ratio of **48** to **49** should result from each pseudorotating intermediate.

Treatment of *O*-methylated (*R*)-methyl *p*-tolyl sulphoxide with magnesium bromide, one of the constituents of a Grignard reagent, followed by hydrolysis gave sulphoxide (33%) and methyl *p*-tolyl sulphide. Most of the sulphoxide (59%) was of retained configuration rather than completely inverted as is the usual outcome of hydrolysis. When a stream of ethylene was swept through the reaction mixture prior to hydrolysis, 1,2-dibromoethane was formed. These results allowed an explanation for the formation of sulphide in the Grignard synthesis of sulphonium salts. Bromide ion, acting as a nucleophile, displaced methyl from oxygen and methoxy from sulphur; magnesium ion might function as an acidic catalyst. Displacement on carbon gave sulphoxide with retention at sulphur; displacement on sulphur gave a bromosulphonium ion which, in a subsequent displacement on bromine, yielded sulphide.

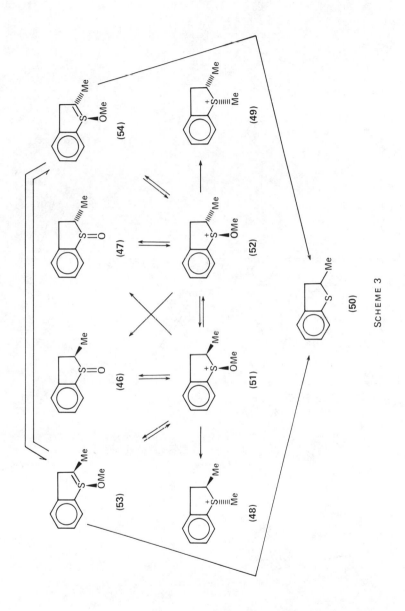

SCHEME 3

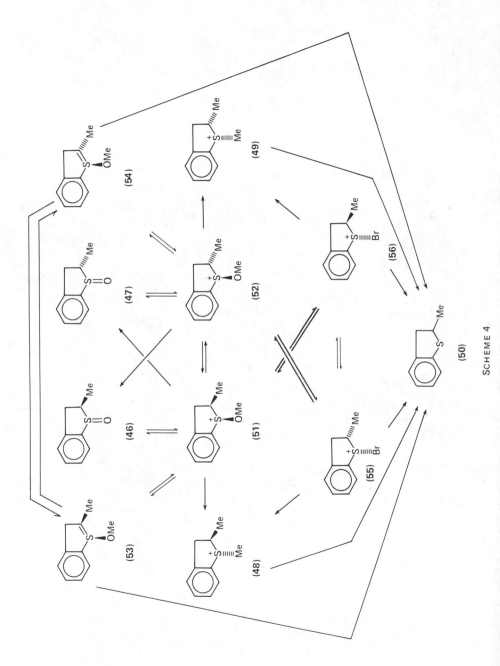

SCHEME 4

Treatment of O-methylated (R)-methyl p-tolyl sulphoxide with magnesium methoxide gave back the sulphoxide upon hydrolysis, but with 59% inversion. Almost complete inversion was expected. The magnesium methoxide must have caused some racemization by attack on the O-methyl group.

With these results at hand, two explanatory reaction schemes were proposed – the one for the methylcadmium reactions, the other for the methylmagnesium bromide reactions (Schemes 3 and 4).

Dimethylcadmium reacted with methoxysulphonium salts **51** and **52** at sulphur to give sulphonium salts **48** and **49**. Since no sulphide **50** was detected, the pathways leading to it must be negligible. Epimerization of **51** and **52** occurred by methoxy

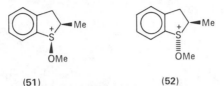

(51)   (52)

group exchange, which then resulted in diastereomerically impure sulphonium salts. This process was manifest only with longer reaction times. Attack by nucleophilies on the O-methyl carbon regenerated retained sulphoxide; hydrolysis generated inverted sulphoxide. Methoxide attack on sulphur may have generated both **46** and **47**.

Methyl Grignard reagent, by virtue of its greater basicity and presence of bromide ion, led to a more complex scheme of reactions. Since sulphide formation was important, either bromosulphonium ion **55** and **56** or ylide formation was of consequence.

Alkoxysulphonium salts **57** prepared by treating a mixture of sulphide, $(-)$-menthol, and pyridine in acetonitrile with t-butyl hypochlorite at $-25°C$ led to optically active sulphonium ylides **58** when treated with sodium dimethyl-malonate[34]. Optically active sulphoxides **59** of known absolute configuration resulted when the alkoxysulphonium salts were hydrolysed in base[39,40]. Since both the hydrolysis and ylide formation proceeded with inversion of configuration at sulphur, this allows absolute configurational assignments to be made to both the alkoxysulphonium salts and the ylides (equations 11 and 12). In addition, it

$$Ar\!-\!\overset{\overset{\displaystyle O\text{-Menthyl}}{|}}{S^+}\!-\!p\text{-Tol} \xrightarrow{\text{NaCH(CO}_2\text{Me)}_2} p\text{-Tol}\!-\!\overset{\overset{\displaystyle C(CO_2Me)_2}{||}}{S}\!-\!Ar \qquad (11)$$

(57)   (58)

$$57 \xrightarrow{\text{NaOH, H}_2\text{O}} p\text{-Tol}\!-\!\overset{\overset{\displaystyle O}{||}}{S}\!-\!Ar \qquad (12)$$

(59)

establishes their approximate optical purity. When Ar was o-anisyl, a 22% optically pure sulphoxide was obtained upon hydrolysis; the optical yield dropped to 14% for Ar = o-tolyl and to zero for Ar = phenyl.

In one instance, transfer of chirality from sulphonium sulphur to carbon was used to assign an absolute configuration to the sulphonium salt (Scheme 5)[21]. $(+)$-1-Adamantylallylethylsulphonium tetrafluoroborate **60**, prepared by reaction

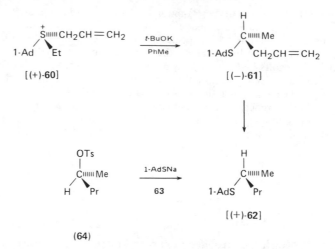

[(+)-60]   [(−)-61]

[(+)-62]

(64)

SCHEME 5

of 1-bromoadamantane with allyl ethyl sulphide and silver tetrafluoroborate and
resolved via its dibenzoyl hydrogen tartrate salt, underwent a [2,3] sigmatropic
rearrangement when treated with potassium *t*-butoxide in toluene for 1 h at −33°C,
yielding (−)-1-adamantyl 1-methyl-3-butenyl sulphide **61** in 48% yield, together
with some 1-adamantyl ethyl sulphide and 1-adamantyl allyl sulphide. The
absolute configuration and enantiomeric purity of **61** were determined as follows.
The sulphide (−)-**61** was reduced to (+)-**62** with diimide. Reaction of sodium
1-adamantanethiolate **63** with (S)-2-pentyl tosylate **64** gave (+)-(R)-**62**. Thus,
(−)-**61** produced from (+)-**60** had the R configuration at sulphur. Consideration of
optical rotations indicated that (−)-**61** contained a minimum of 97% of the R
isomer; there was essentially no loss of optical activity in the transfer of chirality
from sulphur to carbon.

For rearrangement of the ylide to take place, a folded envelope conformation
was assumed to be necessary. Since the absolute configuration of the product **61**
was known, only two ylide conformations appeared likely: **65** and **66**. Sulphur has
the S configuration in **65** and the R configuration in **66**. Of the two, **65** was

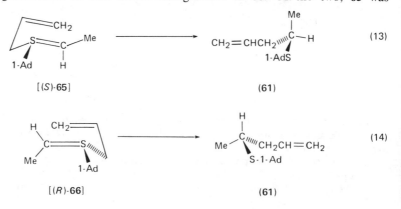

[(S)-65]   (61)   (13)

[(R)-66]   (61)   (14)

believed to minimize non-bonded interactions and was preferred. Therefore, (+)-1-adamantylallylethylsulphonium ion was assigned the S configuration.

Asymmetric induction occurred in the Sommelet rearrangement of optically active benzylsulphonium salts, **67** and **68**, to optically active sulphides, **69** and **70**, but the methods described above have not been applied to this system in an effort to assign absolute configurations (equation 15). Each enantiomeric sulphonium

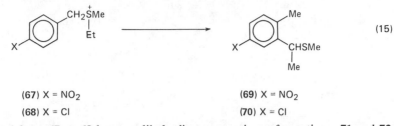

(15)

(**67**) X = NO₂

(**68**) X = Cl

(**69**) X = NO₂

(**70**) X = Cl

ylide derived from **67** or **68** has two likely diastereomeric conformations, **71** and **72**, preceding individual diastereomeric transition states. One transition state leads to

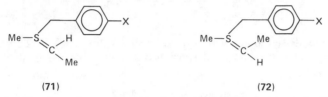

(**71**)   (**72**)

one chiral sulphide, the other to its enantiomer. A not-so-obvious selection of **71** or **72** as the principal sulphide precursor would be required in order to assign an absolute configuration to **67** and **68**[25].

V. CYCLIC SULPHONIUM SALTS

A. Thiiranium and Thiirenium Salts

Thiiranium or episulphonium ions are believed to be intermediates in the addition of sulphenyl halides to alkenes. Anchimeric assistance by the sulphur atom in the solvolyses of β-chlorosulphides also leads to the transient existence of these ions. However, in some cases, moderately stable thiiranium ions such as **73**[41], **74**[41], and **75**[42] have been prepared and isolated. The fact that *cis*-1,2-di-*t*-butylthiirane **76** gave the episulphonium ion **75** upon treatment with methyl fluorosulphonate in methylene chloride at 20°C, whereas the *trans*-sulphide **77** gave no reaction under these conditions, argues for methylation on the diastereotopic sulphide face *trans* to the *t*-butyl groups. *cis*-Methylation of **76** is prevented by the two bulky *t*-butyl groups, which also prevent alkylation of the *trans*-thiirane **77**. However, both **76** and **77** were protonated on sulphur by fluorosulphonic acid at low temperatures. The *cis*-isomer **76** gave diastereomers in the ratio of 4:1 with the isomer derived from *cis* protonation believed to be predominant (**78**, **79**). The proton's small steric requirement allowed protonation of **77**, whereas the methyl group's larger steric bulk prevented methylation to **80**. It can safely be concluded that thiiranium ions are, in general, pyramidal at sulphur.

More recently, episulphonium salts have been synthesized by alkylthiolation of alkenes in methylene chloride at 0°C or in sulphur dioxide at −60°C (equation

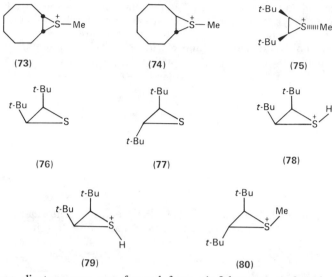

(73)　　　　　(74)　　　　　(75)

(76)　　　　　(77)　　　　　(78)

(79)　　　　　(80)

16). Only one diastereomer was formed from *cis*-2-butene, *cis*-2-cyclooctene, or 2-methyl-2-butene even though existence of a second isomer is possible[43].

$$R_2C=CR_2 + MeSSMe_2^+ SbCl_6^- \xrightarrow{-Me_2S_2} \text{(16)}$$

Thiirenium ions (**81**) are believed to exist since sulphenyl chlorides add *anti* to alkynes and the β-sulphur atom participates in the solvolysis of β-thiovinyl sulphonates[44]. However, a plausible alternative is the open cation **82** in rapid

(81)　　　　　(82)

equilibrium with **81**. Non-empirical SCF-MO calculations indicated that the cyclic ion **81** is located at a potential minimum with an energy equal to or less than that of **82**, also found at a minimum. A rather small barrier separates the two. As is the case for the episulphonium ions, the sulphur atom is pyramidal.

B. Thietanium Salts

The stereochemistry of thietanium ions has not been studied very much, even though the salts are easily prepared by methylation of the corresponding thietanes. For example, *cis*- and *trans*-2,4,-dimethylthietane gave **83** and **84**, respectively[45,46]; 3,3-dimethylthietane gave **85**[47]. The salts are somewhat unstable over a period of time and are susceptible to nucleophilic attack at sulphur and carbon. N.m.r. spectroscopy clearly supports a non-planar structure for thietanium ions. A folded ring is also found in thietanes and their sulphoxides, sulphone, sulphilimine, and sulphoximine derivatives. Methylation of *cis*-2,4-dimethylthietane gave only **83**, one of the two possible diasteromers, whereas 2,2,4-trimethylthietane gave rise to both possible isomers, **86** and **87**, in a 1:1 ratio. These conclusions followed from the

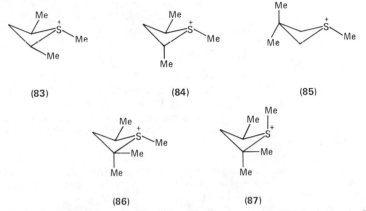

(83)   (84)   (85)

(86)   (87)

n.m.r. spectra. Sulphonium ion **83** gave a clean spectrum with a sharp *S*-methyl signal, but two *S*-methyl absorptions were found for the mixture of **86** and **87**. The *trans* salt **84** apparently exists in one conformation since two *C*-methyl doublets and two α-methine signals appeared in its n.m.r. spectrum[46].

C. Thiolanium Salts

A fair number of five-membered ring sulphonium salts are known and stereochemical studies have been carried out on some of these compounds. 1,2-Dimethylthiolanium perchlorate **88** and 1,3,7-trimethylthiolanium perchlorate **89** have been resolved[22]. The absolute configuration of sulphur in the *S*-methylated penicillinate **32** has been determined[35]. Various studies of pyramidal inversion and base-catalysed hydrogen–deuterium exchange involving the thiolanium ring system have been carried out and are described in the appropriate sections below. Some stereochemistry of sulphonium salts **48** and **49** was covered earlier[30].

Conformational analyses of 1-methylthiolanium ion **90** and the more rigid *trans*-3-methyl-3-thioniabicyclo[4.3.0]nonane ion **91** were achieved by n.m.r. spectroscopy[48,49]. Two limiting conformations exist for the thiolane ring: one is the half-chair or twist-envelope **92** and the other is the envelope form **93**. Cation **91** must of necessity exist with the thiolane ring in the half-chair form, whereas the ring in cation **94** must adopt the envelope conformation. The geminal couping constants for the two pairs of methylene protons alpha to sulphur **91** are 12 and 13 Hz; in **94**, they are both 16.0 Hz. The 13.0 Hz geminal coupling observed for **90** agrees well with the value found for **91** in a half-chair conformation. Further description of **90** and **91** is given in the section on hydrogen–deuterium exchange.

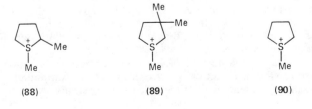

(88)   (89)   (90)

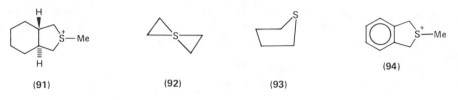

(91)   (92)   (93)   (94)

D. Thianium Salts

Numerous thianium salts have been prepared, usually by alkylation of the parent thians. 1,3,3-Trimethylthianium perchlorate **95** has been resolved and its rate of pyramidal inversion measured[22]. Many thianium ions have been conformationally analysed by proton and carbon-13 n.m.r. spectroscopy, and, in a few instances, also by X-ray crystallography.

1-Methylthianium **96** iodide exists in the chair form in the solid state with the methyl group in the equatorial position[50]. This thiane ring is more folded than the cyclohexane ring, and the S—H(axial) distance is short (only 2.40 Å) compared with the Van der Waals distance of 3.05 Å.

Equatorial disposition of the S-methyl group in **96** is also favoured in solution, but by no means strongly. In fact, a low-temperature [13]C n.m.r. study showed both conformers to be of equal energy. A calculation based on ^{13}C chemical shifts of the S-methyl groups in **96**, **97**, and **98** indicated that 60% of **96** has the S-methyl

(95)   (96)

(97)   (98)

equatorial. An earlier value of about 90% is incorrect[51,52,53]. Thianium perchlorates **97** and **98** were both examined by n.m.r. and X-ray methods[53]. The latter technique established their configurations and also showed that **97** has a more flattened and less folded ring than **98**, perhaps owing to 1-methyl 3,5-hydrogen diaxial interactions. Heating **97** or **98** at 100°C in chloroform established a 3:2 equilibrium ratio favouring **98**. Alkylation of the parent thiane with methyl iodide favoured **98** formation, the ratio of **98** to **97** being 88:12. Carbon-13 n.m.r. (c.m.r.) spectroscopy **97** gave signals for the S-methyl and 3,5-methylene carbons at 16.61 and of 19.04 p.p.m., respectively, downfield from TMS; **98** gave the S-methyl peak at 25.48 p.p.m. and the 3,5-methylene peak at 25.10 p.p.m. The upfield shifts observed for **97** are expected for this more compressed diastereomer based on analogy with methylcyclohexanes.

Barbarella *et al.*[51] investigated several conformationally biased C-alkylated 1-methylthianes, **99–108**, by n.m.r. spectroscopy. These compounds were prepared by treatment of the thianes with trimethyloxonium tetrafluoroborate in methylene

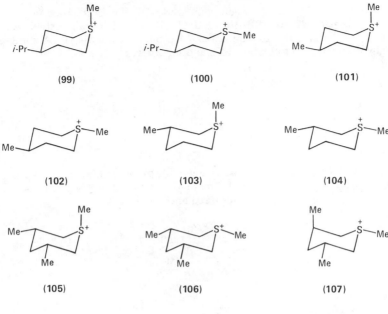

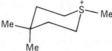

(108)

chloride at 0°C, which led to mixtures containing 80–90% of the equatorial isomers. Only **102** and **106** were obtained pure by fractional crystallization; the others were examined as mixtures. Thermal equilibrium increased the amount of axial isomer in the mixtures to 40–45%. Since the ^{13}C n.m.r. signals for axial S-methyls occur upfield about 8 p.p.m. from their equatorial counterparts[53,54], configurational assignments to **99–107** were possible. Geminal coupling constants for the α-protons were 15 Hz when the S-methyl was axial and 12 Hz when it was equatorial. These values together with the J_{ae} value of 13.8 Hz for **108** were used to calculate the axial–equatorial populations of this ion as 60% axial and 40% equatorial. Calculations based on an S-methyl c.m.r. shift of 22.6 p.p.m. for **108** together with values of 19.0 and 27.5 p.p.m. for the S-methyl shifts of *cis*- and *trans*-4-isopropylthianium ion, respectively, gave the same percentages. Geminal methyl groups at $C_{(4)}$ generally increase the amount of axial isomer in analogous heterocyclic systems.

Barbarella *et al.*[55] also measured the percentage of equatorial isomer at equilibrium (100°C) for the following pairs of diastereomers: **99** and **100**, 57%; **101** and **102**, 59%; **103** and **104**, 57%; and **105** and **106**, 57%.

Although the *cis* isomer **104** exists exclusively in the diequatorial conformation, the *trans* isomer **103** may conceivably exist in two chair forms **103a** and **103e**, separated by an energy barrier on the order of 9 kcal/mol. The isomers will interconvert rapidly at room temperature. However, if a substantial amount of **103e** were present, both $C_{(6)}$ and S-CH$_3$ carbon atoms should undergo anomalous chemical shifts relative to **104** when compared with the shift differences between diastereomeric pairs **99–100**, **101–102**, and **105–106**. All five pairs exhibit Δδ values of about 5.5 p.p.m. for the α-carbons and 7.8 p.p.m. for the *S*-CH₃. Apparently **103** essentially populates conformation **103a** and not **103e**.

(103a)   (103e)

Similar considerations of the diastereomeric 1,2-dimethylthianium cations predict that the *trans* isomer exists with the methyls equatorial **109** while the *cis* isomer populates almost totally conformation **110a** and not **110e**[55].

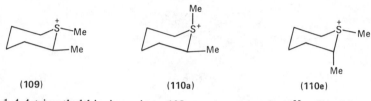

(109)   (110a)   (110e)

The 1,4,4-trimethylthianium ion **108** was re-examined[55]. Consideration of additional ¹³C chemical shifts led to a value for the equatorial population of 35–38%, which is slightly smaller than the previously stated 40%[51].

1,2,2-Trimethylthianium ion also exists mostly with the *S*-CH₃ axial (60%), **111a**, rather than equatorial, **111e**. A twist-boat form such as exists for 1,2,2-trimethyl-cyclohexane did not fit the ¹³C data as well as did conformation **111a**.

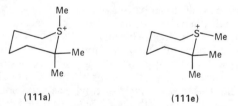

(111a)   (111e)

Eliel and Willer[56,57] obtained the ¹³C.n.m.r. spectra of thirty-four *S*-methylthi-anium salts, the four possible *S*-methyl-1-thiadecalinium salts, six *S*-protonated thianes or thianium salts, six *S*-benzylthianium salts, *cis*- and *trans*-1-ethyl-4-*t*-butyl thianium perchlorate **112**, **113**, and 1-phenylthianium tetrafluoroborate[56,57]. Ions **101–110** plus the 1,3,3-trimethylthianium ion, examined by Barbarella *et al.*, were included among the thirty-four.

All six thianium salts, prepared by adding the parent thiane to fluorosulphonic acid, had their *S*-protons axially oriented. Thianium ion **114** gave a seven-line

(112)   (113)

(114)

spectrum. Six lines were broadened [the line due to $C_{(4)}$ was not], apparently by an incipient exchange process. The equatorial isomer was not detected and free thiane was believed not to be present in the strongly acidic medium. The presence of seven rather than four lines indicated that axial–equatorial interconversions of the methyl groups was slow on the n.m.r. time scale. This means that either protonation occurs exclusively axially, which seems unlikely, or, more likely, that proton exchange leading to equilibration is fast on the laboratory time scale such that only the thermodynamically favoured axial isomer is observed. In contrast, amine salts exchange protons slowly[57].

Eleven pairs of S-methylthianium salts epimeric at sulphur, 1-α- and -β-methyl-*trans*-1-thiadecalinium salts, 1-α- and -β-methyl-*cis*-1-thiadecalinium salts, two pairs of S-benzyl salts, and the *cis–trans* S-ethyl derivatives **112** and **113** were equilibrated at sulphur by heating at 100°C in CD_3CN. The results are given in Table 2 for the process indicated by equation 17. Conformational equilibria were

$$R-\!\!\!\!\!\diagdown\!\!\!\!\!\diagup\!\!\!\!\!S^+ \quad \rightleftharpoons \quad R-\!\!\!\!\!\diagdown\!\!\!\!\!\diagup\!\!\!\!\!\overset{+}{S}\!\!\diagdown R \tag{17}$$

determined by low-temperature ^{13}C n.m.r. spectroscopy at −90°C (−65°C for **128**) for the five salts shown by equations 18–22.

The equilibrium constant for **96** (equation 18) is 1.0 at −90°C, which differs from the value of 1.45 at 100°C obtained using the model compounds **97** and **98** or **105** and **106**. Eliel and Willer speculated that a small entropy difference between

TABLE 2. Configurational equilibria of *C*-alkyl *S*-methylthianium salts. Reprinted with permission from E. L. Eliel and R. L. Willer, *J. Amer. Chem. Soc.*, **99**, 1936 (1977). Copyright by the American Chemical Society

Compound	R (see equation 17)	Anion	K
97, 98	4-*t*-Butyl	ClO_4^-	1.43
101, 102	4-Methyl	PF_6^-	1.31 (1.44)[a]
103, 104	3-Methyl	PF_6^-	1.11 (1.33)[a]
105, 106	*cis*-3,5-Dimethyl	ClO_4^-	1.54 (1.27)[a]
109, 110	2-Methyl	PF_6^-	1.96
115, 116	*trans*-2,5-Dimethyl	PF_6^-	2.25
117, 118	*cis*-2,4-Dimethyl	PF_6^-	2.22
119, 120	*cis*-2,6-Dimethyl	PF_6^-	3.85
121, 122	*trans*-1-Thiadecalin	PF_6^-	2.94
123, 124	2,4,4-Trimethyl	PF_6^-	1.91
125	3,5,5-Trimethyl	PF_6^-	>30
126, 127	2,2,4-Trimethyl	PF_6^-	0.88
128, 129	*cis*-1-Thiadecalin	PF_6^-	0.11

[a]From ref. 55; in D_2O.

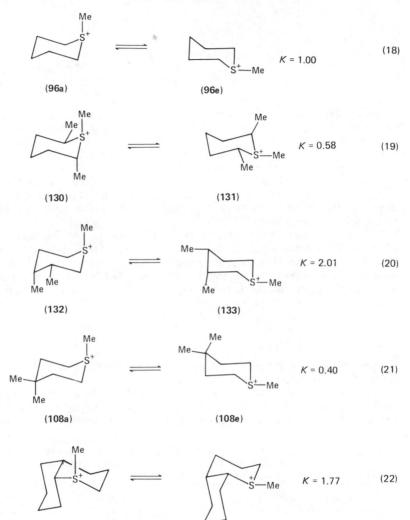

$K = 1.00$ (18)

(96a)   (96e)

$K = 0.58$ (19)

(130)   (131)

$K = 2.01$ (20)

(132)   (133)

$K = 0.40$ (21)

(108a)   (108e)

$K = 1.77$ (22)

(128a)   (128e)

conformers **96a** and **96e** or a solvent effect may be responsible for this 0.3 kcal/mol difference over the 190°C temperature range, rather than effects of the *t*-butyl group in **97** and **98** or the methyl groups in **105** and **106**.

Cook *et al.*[58] methylated 4-hydroxy-4-phenylthiane to form the two diastereomers of **9**. Although their kinetic product ratio was later shown to be in error due to a faulty n.m.r. integration, they did develop a technique for equilibrating the isomers. This consisted of heating the two in methyl iodide–acetonitrile at 100°C.

Methylation of various thianes generally yields much more of the equatorial *S*-methyl isomer than predicted by the relative stability of the axial–equatorial products. That is, the kinetic stereoselectivity exceeds the thermodynamic stereoselectivity. Halfpenny *et al.*[59] tested whether or not methylation of a twist form of the thiane ring contributed substantially to product formation. A small concentration of a highly reactive twist form could invalidate the common assumption that the product ratio reflects the relative amounts of axial to equatorial methylation via chair forms. To that end, they synthesized *cis*- and *trans*-3,5-di-*t*-butylthiane. Although the *cis* isomer exists in the chair form **134**, the *trans* isomer does not; it exists predominantly in a twist form **135**.

(134)

(135)

These two isomers as well as 4-phenylthiane and 4-*t*-butylthiane were methylated using methyl iodide in CD_3CN at 15°C (kinetic control) and at 100°C (thermodynamic control). Rates relative to methyl sulphide were established by competitive methylation. The kinetic rate ratios for methylation by equatorial to axial attack, respectively, were: **134**, 11; 4-phenylthiane, 6.5; and 4-*t*-butylthiane, 3.7. The equilibrium ratios (equatorial to axial) were: **134**, 2.8; 4-phenylthiane, 1.25; and 4-*t*-butylthiane, 1.2. The relative rates for methylation by equatorial and axial attack, respectively, were: methyl sulphide, 1, 1; thiane, 1.6, 0.4; **134**, 2.0, 0.2; **135**, 1.6, 1.6; and 4-phenylthiane, 0.41, 0.06. The sulphide faces of methyl sulphide and **135** are equivalent, so the overall rate of each was divided by two to give the rates indicated. Since **135** was only slightly more reactive than **134**, the authors concluded that contributions from twist forms, in general, to the methylation of thianes are negligible. They also concluded that steric hindrance between the 3,5-methylene groups and the methyl group of methyl iodide in the S_N2 transition state explains the low rate for axial methylation. This hindrance lessens in the product. Thus, greater kinetic than thermodynamic stereoselectivity is accounted for.

A comparison was made with the methylation of 1-methyl-4-phenylpiperidine for which the reverse order of stereoselectivity is found. The authors concluded that the direction of approach to the sulphur atom by the electrophilic methyl group is approximately perpendicular to the $C_{(2)}$—S—$C_{(6)}$ plane (along the axis of the p lone pair of electrons, thus creating the substantial steric interaction with the 3,5-axial hydrogens. This strain is released as the *S*-methyl group adopts the axial position. In contrast, the nitrogen sp³ electron pair lies along the axial axis. Interaction with the 3,5-axial hydrogen atoms is increased, not diminished, as axial product is formed. This explains the reversal of stereoselectivity.

E. 1,3-Dithianium Salts

Conformational analyses of a few 1,3-dithianium salts were carried out by proton n.m.r. spectroscopy[60]. The simplest of these was 1-ethyl-1,3-dithianium tetrafluoroborate, which may exist with the *S*-ethyl group equatorial **136e** or axial **136a**. At room temperature the diastereotopic $C_{(2)}$ methylene protons gave an AB spectrum ($J_{AB} = 13.6$ Hz). At −70°C, a broad singlet replaced these AB peaks. As the temperature was decreased to −102°C, six bands replaced the singlet. These peaks arose from the partial overlap at two AB quartets, one from **136e** and one from **136a**. The quartet with the largest coupling constant ($J_{AB} = 15$ Hz) was assigned to **136a** and that with the smallest ($J_{AB} = 11$–12 Hz) to **136e**. This assignment was based on the smaller couplings observed for the methylene protons in thiane-derived sulphoxides and sulphonium salts with axial electron pairs on sulphur compared to couplings found in their equatorial counterparts. Integration of the signals at $-102°C$ gave an equilibrium constant of 1.5 favouring **136e** and a free energy difference of 0.14 kcal/mol.

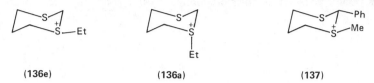

(136e)   (136a)   (137)

1-Methyl-2-phenyl-1,3-dithianium ion (**137**) was also studied by low-temperature n.m.r. spectroscopy. Only one peak for the $C_{(2)}$ proton was observed. The authors concluded that only one conformation existed to any measurable extent; probably this was the diequatorial *trans* isomer.

Some bis-sulphonium salts derived from 1,3-dithianes were also studied. 1,3-Diethyl-1,3-dithianium salts should exist as *cis–trans* isomers. Ring reversal of the *cis* isomer (**138**) leads to interconversion of diaxial with diequatorial conformers, but for the *trans* isomer (**139**) only equivalent conformations are

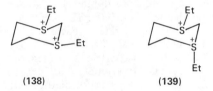

(138)   (139)

interchanged. The n.m.r. spectrum of the 1,3-diethyl-1,3-dithianium $C_{(2)}$ methylene protons consists of an AB quartet due to the *cis* isomer ($J = 11$ Hz) and a singlet due to the *trans* isomer. The small AB coupling constant is in agreement with an expected diequatorial conformation for **138**. The ratio of *cis* to *trans* isomers was 2:1.

1,3-Dimethyl-2-aryl-1,3-dithianium salts were among the bis-sulphonium salts investigated by proton n.m.r. spectroscopy. Only a singlet was observed for the methine proton, suggesting predominance of one isomer in one conformation, perhaps the all-equatorial molecule **140**. When the aryl group was replaced with methyl, however, signals from isomers were detected. 1,2,3-Trimethyl-1,3-dithianium tetrafluoroborates and picrates gave two quartets for the methine protons, two doublets for the $C_{(2)}$ methyl protons, and two singlets for the S-methyl groups. A further description of the conformers giving rise to these signals was not given. When both $C_{(2)}$ protons were replaced with methyl groups to give 1,2,2,3-tetramethyl-

(140)   (141)

1,3-dithianium picrate (**141**), only singlets were observed for the S-methyl and C-methyl protons.

F. Thioxanthenium Salts

Thioxanthenium salts have been studied by several groups[61–64]. The central ring in these salts is boat-shaped, as can be seen from the X-ray structure of a 10-phenylphenoxathiinium salt, a related compound[65].

Six thioxanthenium perchlorates were prepared in the initial study of these compounds[61]. Salt **142a** gave two signals for the methyl protons;

		R^1	R^2	R^3	R
	142a	Me	Me	Ph	H
	142b	H	H	2,5-Xylyl	H
	142c	H	H	Mesityl	H
	142d	H	H	Me	H
	142e	H	H	Me	Cl
	142f	H	H	Et	H
	142g	H	Ph	p-Anisyl	H

(142)

These signals collapsed at 200°C to one singlet, indicating rapid methyl group exchange resulting from rapid pyramidal inversion at sulphur superimposed on an even faster ring reversal.

More recently, a series of 10-substituted-9-phenylthioxanthenium perchlorates were prepared[63]. The 10-p-anisyl salt **142g** crystallized in two forms, m.p. 154 and 210°C, which could be separated mechanically under a microscope. N.m.r. spectroscopy revealed that these two forms were isomers. In trifluoroacetic acid, the p-methoxy group signal appeared at δ5.93 p.p.m. for the 210°C m.p. compound and δ3.87 p.p.m. for the 154°C compound. Although the 9-methine protons for both isomers appeared at δ5.80 p.p.m. in trifluoroacetic acid, separate signals at δ5.82 and δ5.75 p.p.m. were obtained in deuterochloroform. Upon being heated in trifluoro-acetic acid for 10 min at 72°C, the higher melting isomer gave an equilibrium mixture with the lower melting substance in the ratio of 2:1. The same ratio was obtained upon heating the lower melting isomer. Cis–trans isomers are possible for **142g**, with two conformations for each isomer. Each pair of conformers is intercon-vertible by ring reversal, but the cis–trans isomers only by pyramidal inversion at sulphur. The authors suggest that **142g** is the trans isomer and that the two crystalline forms are the two trans conformers. Since the barrier to pyramidal inversion is greater than 20 kcal/mol, whereas the barrier to ring reversal is probably much less, their assignments should be regarded as tentative. Several other 10-aryl and alkyl salts also exist in isomeric forms.

VI. STEREOMUTATION OF SULPHONIUM SALTS

Sulphonium salts can undergo configurational change at sulphur by (1) pyramidal inversion, (2) reversible dissociation into a carbonium ion and neutral sulphide molecule via an S_N1 mechanism, and (3) S_N2 attack at the α-carbon followed by reformation of the salt (Scheme 6). Other processes such as 1,2-elimination followed by recombination of the alkene and sulphide are possible, in principle, but have not been detected.

A. Trialkylsulphonium Salts

Balfe et al.[15] proposed the third of the above processes to account for the racemization of ethylmethylphenacylsulphonium salts (**4**). Equilibration of the

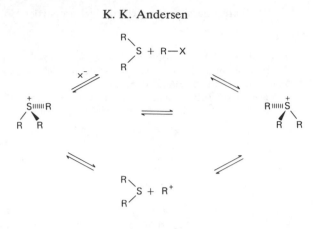

SCHEME 6

cis,trans isomers of **9** was achieved by heating the iodide salts in acetonitrile; the S$_N$2 mechanism was doubtless operative in this case also[58]. In general, sulphonium salts are subject to racemization or epimerization if the counter ion is a good nucleophile such as a halide ion. If this process is to be avoided and the configurational integrity of a salt maintained, the sulphonium cation should be in combination with a non-nucleophilic anion such as tetrafluoroborate, tetraphenylborate, picrate, 2,4,6-trinitrobenzenesulphonate, or hexafluoroantimonate.

Darwish and Tourigny[19] first established that optically active trialkylsulphonium salts may racemize by pyramidal inversion. *t*-Butylethylmethylsulphonium perchlorate **143** racemized 10–15 times faster than it solvolysed in ethanol, acetic acid, water, or acetone. Perchlorate ion is non-nucleophilic. Nucleophilic displacement by the first three solvents mentioned would not be reversible. Thus, only dissociation–recombination of a *t*-butyl cation–ethyl methyl sulphide ion–neutral pair or pyramidal inversion of sulphur would account for the experimental observations. Substituent effects were used to distinguish between these alternatives. Carbocation formation should be facilitated by the added electron-donating methyl group in **144** and lessened by the electron-attracting methoxy group in **145**. Rates of racemization and solvolysis should be increased for **144** and decreased for **145** relative to **143** if carbocation formation is involved in both processes. The relative rates for ethanolysis of **143**, **144**, and **145** at 50°C were

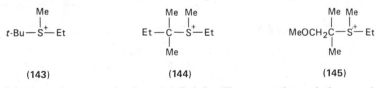

| (143) | (144) | (145) |

1:0.06:6.3, but for racemization 1:1.7:3.8. These results ruled out the S$_N$1 ion–neutral molecule process as the major pathway for racemization and supported instead a process influenced mainly by steric effects rather than electronic effects; that is, they supported the pyramidal inversion mechanism.

Scartazzini and Mislow[20] substantiated this interpretation with their finding that 1-adamantylethylmethylsulphonium perchlorate **146**, a compound which did not solvolyse under the reaction conditions, racemized in acetic acid at 50°C with a first-order rate constant of 8.59×10^{-4} s^{-1}, which was very close to the rate constant

$$\text{1-Adamantyl} - \underset{\underset{\text{Et}}{\overset{|}{\text{S}^+}}}{\overset{\overset{\text{Me}}{|}}{}} - \text{Et}$$

(146)

observed for **144** (4.13×10^{-4} s^{-1}). Racemization of **146**, labelled with deuterium in the ethyl group, in the presence of ethyl methyl sulphide gave no evidence for loss of labelling by exchange of ethyl groups. This finding argued against the possible intervention of an ion–neutral molecule pair. Although adamantyl and *t*-butyl groups differ by a factor of 10^3 in the solvolysis of their halides with the *t*-butyl being faster, their steric bulk is similar. This size factor was believed to be the reason for the similar rates for **143** and **146**, i.e. steric factors dominate the barriers to pyramidal inversion in sulphonium salts.

If one of the alkyl groups attached to sulphur is particularly capable of supporting a positive charge, then racemization via an S$_N$1 process dominated over pyramidal inversion. This was so for *p*-methoxybenzylethylmethylsulphonium perchlorate **147**, which was studied together with three other benzyl salts, **148**, **149**, and **150**, and ethylmethylphenacylsulphonium perchlorate **4**[24]. In methanol at 70°C, the relative rate constants for the loss of optical activity for **148:149:4** were 1:0.99:0.6. Compound **148** underwent racemization 38 times faster than it did methanolysis. It also underwent methanolysis 5 times faster than **149** at 70°C and 33 times faster than **4** at 90°C. These results supported the pyramidal inversion mechanism for racemization and showed again that electron-withdrawing substituents have little effect on the rate.

$$p\text{-MeOC}_6\text{H}_4\text{CH}_2 - \underset{\underset{\text{Et}}{\overset{|}{\text{S}^+}}}{\overset{\overset{\text{Me}}{|}}{}} - \text{Et} \qquad \text{PhCH}_2 - \underset{\underset{\text{Et}}{\overset{|}{\text{S}^+}}}{\overset{\overset{\text{Me}}{|}}{}} - \text{Et} \qquad p\text{-NO}_2\text{C}_6\text{H}_4\text{CH}_2 - \underset{\underset{\text{Et}}{\overset{|}{\text{S}^+}}}{\overset{\overset{\text{Me}}{|}}{}} - \text{Et}$$

(147)   (148)   (149)

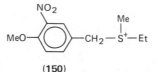

(150)

In contrast, **147** lost optical activity only 1.5 times as fast as it solvolysed at 50°C. The methanolysis was over 1000 times faster for **147** than for **148** and racemization about 15 times faster. Since the electron-withdrawing *p*-nitro group had no effect on the rate of racemization by pyramidal inversion, the *p*-methoxy group must have caused loss of optical activity by another mechanism. Added ethyl methyl sulphide increased the rate of loss of optical activity for **147**, but this increase was not sufficient to account for racemization proceeding via an S$_N$2 nucleophilic displacement reaction. Ethyl methyl sulphide had no effect on the solvolysis rate; this ruled out dissociation to a free benzyl cation followed by recombination with ethyl methyl sulphide.

Salt **150**, which contains both an electron-withdrawing and an electron-donating group, racemized at almost the same rate as **148**. Solvolysis of **150** compared with **147** was decreased because the opposing electronic effects of the *p*-methoxy and *m*-nitro groups cancelled each other. The decrease in racemization rate followed from a change in mechanism as the rate of solvolysis lessened so that pyramidal inversion became dominant. Thus, **150** resembled **148** and **149**. Salt **147**, on the other hand, racemized by formation and recombination of a p-methoxybenzyl cation–ethyl methyl sulphide pair contained in a solvent cage.

In spite of the evidence presented above in support of racemization by pyramidal inversion for sulphonium salt **143**, Brower and Wu[66] concluded that the compound lost optical activity by dissociation, but they suggested that this 'dissociation of t-butyl cation and sulphide may not be quite complete. They determined the activation volumes for the racemization of ethylmethylphenacylsulphonium **4** and t-butylethylmethylsulphonium **143** dibenzoyltartrates in water; salt **4** was also studied in ethanol and methanol. Racemization without bond breakage should proceed with activation volumes near zero, as found for the racemization of 6-nitrodiphenic acid and for phenyl and α-naphthyl p-tolyl sulphoxides. Indeed, the racemization of **4** gave values of 0 ± 2 ml/mol in all three solvents at 60.5°C. However, **143** racemized with a $\Delta V^{\ddagger}$ value of 6.4 ± 1.0 ml/mol at 40°C. Since values of 10 and -10 ml/mol are common for processes involving bond breakage and formation, respectively, the value of 6.4 ml/mol for **143** led to the conclusion that bond scission was involved in racemization. It would be interesting to know $\Delta V^{\ddagger}$ for the racemization of 1-adamantylethylmethylsulphonium ion **146**.

In spite of this puzzling $\Delta V^{\ddagger}$ value for the racemization of **143**, there is good evidence that several S-methyl thiolanium and thianium salts racemize (or epimerize) by pyramidal inversion[22]. Perchlorate salts of optically active 1,2-dimethylthiolanium **88**, 1,3,3-trimethylthiolanium **89**, and 1,3,3-trimethylthianium **95** cations were heated at 100°C in acetic acid, whereupon they lost optical activity by a strictly first-order process with rate constants of 1.25×10^{-4} s^{-1} **88**, 1.04×10^{-4} s^{-1} **89**, and 0.81×10^{-4} s^{-1} **95**. Salts **89** and **95**, recovered after ten half-lives, were completely racemic, but salt **88** with its chiral carbon atom was still optically active although it exhibited laevorotation instead of the initial dextrorotation. This finding was especially significant, for it meant that bond cleavage between the sulphur and tertiary carbon atom, which would have racemized **88** completely, was not occurring. The similar rate constants for **89** and **95** also argue against carbocation formation.

Extrapolation of the rate constants for the racemization of 1-adamantylethyl-methylsulphonium percholate **146** from 34–50 to 100°C indicated that **146** racemized about 2500 times faster than **88**, **89**, or **95**. This may be due to bond-angle strain introduced as the sulphur atom's geometry changes from pyramidal to planar during the configurational inversion process.

Pyramidal inversion in **91** labelled with three deuterium atoms at the α-carbons led to stereomutation, i.e. the lone α-proton exchanged between non-equivalent positions[49]. A rate constant of 2×10^{-5} s^{-1} at 90°C in D$_2$O was obtained, which is in good agreement with the rate constants observed for **88**, **89**, and **95**. Similar thermally induced stereomutations have been used to equilibrate various cyclic sulphonium salts and were mentioned in the section on conformational analysis.

B. Dialkylarylsulphonium Salts

Darwish and Scott[23] measured the rates of racemization for a series of seven dialkylarylsulphonium perchlorates in methanol at 50°C. Salts **151–155** did not differ greatly in their relative rates of racemization: **151**, 1.0; **152**, 1.61; **153**, 1.86; **154**, 4.3; and **155**, 1.19. Electronic effects were rather small. Steric effects, however, were pronounced. Salts in which the larger n-propyl and n-butyl groups replaced the methyl group racemized more rapidly than **151**: **156**, 11.2; **157**, 9.5. An isopropyl substituent had approximately the same effect on racemization as a phenyl group; **160** racemized 1.54 times faster than **151**.

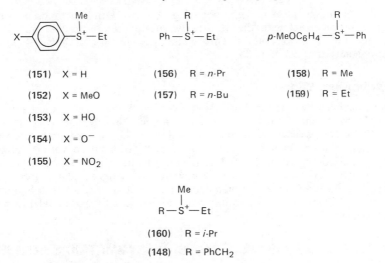

(151) X = H

(152) X = MeO

(153) X = HO

(154) X = O⁻

(155) X = NO₂

(156) R = n-Pr

(157) R = n-Bu

(158) R = Me

(159) R = Et

(160) R = i-Pr

(148) R = PhCH₂

Thermal equilibration of salts **48** and **49** in acetic acid was helpful in assigning configurations; at 110°C the *trans* isomer **48** was favoured over *cis*-**49** by 3:2[30]. No rate measurements were made.

C. Alkyldiarylsulphonium Salts

Two alkyldiarylsulphonium salts have been resolved; they racemized more readily than the dialkylarylsulphonium salts **151**–**155**[23]. Their rates in methanol at 25 and 50°C, respectively, compared with **151** are: **158**, 9.8, 14.2; and **159**, 70.8, 112. The replacement of the methyl group in **158** by an ethyl group **159** increased the rate approximately tenfold, whereas the replacement of a methyl in **151** by a *p*-anisyl group increased the rate about one-hundredfold.

D. Triarysulphonium Salts

No optically active triarylsulphonium salts are known. Attempts to synthesize them by reaction of diarylalkoxysulphonium salts with aryl Grignard[61,67] or organocadmium reagents[29] gave only racemic triarylsulphonium salts (equation 10), even though this type of reaction was successful in preparing optically active diarylalkylsulphonium salts. Compound **142a** was synthesized as a model of triarylsulphonium salts. At 200°C in benzophenone, the two methyl signals coalesced to a single peak; from this and the chemical shift difference at low temperature, an activation energy of 25.4 kcal/mol for pyramidal inversion at 200°C was calculated[61]. Hori *et al.*[63] studied related sulphonium salts **142g** described above. They concluded that a barrier to ring inversion allowed for the existence at room temperature of two conformers of the *trans* isomer. Darwish and Scott[23] estimated that triarylsulphonium salts would racemize in methanol at 25°C with a rate constant of 8 × 10⁻⁴ s⁻¹, which corresponds to an activation energy of about 22 kcal/mol and a half-life of 15 min. When a triarylsulphonium cation is constrained by ring formation as in **142a–g**, the barrier to inversion should increase as was found for **88**, **89**, and **95** compared with **146**[22]. Since barriers to ring inversion are usually low, it may be that the two forms of **142g** are *cis–trans* isomers rather than two conformers. The chemical shifts of the 9-methine protons are identical. This might arise if the 9-phenyl groups are either both pseudo-axial or both pseudo-equatorial with the 10-*p*-anisyl group pseudo-axial in one case and pseudo-equatorial in the other. This would account for the widely separated methoxy signals at δ5.93 and δ3.87.

E. Sulphonium Ylides

Ylides racemize by pyramidal inversion more easily than their parent sulphonium cations. Darwish and Tomlinson[26] first determined this for the ylide **15** derived from ethylmethylphenacylsulphonium perchlorate **4**. The ylide racemized 200 times faster than the cation in methanol at 50°C with an activation enthalpy of 23.3 kcal/mol compared with 29.2 kcal/mol for the salt[68]. Three other closely related ethylmethyl ylides **16–18** had activation enthalpies ranging from 23.6 to 23.9 kcal/mol[27]. Wolfe *et al.*[69] concluded from a study of 1,3-bis-sulphonium salts and their ylides that ylide formation reduced barriers to pyramidal inversion by about 5 kcal/mol.

VII. USE OF SULPHONIUM SALTS IN ASYMMETRIC SYNTHESIS

Even though sulphonium salts are often used in synthesis, primarily as their ylides, they have almost never been employed in their optically active forms for asymmetric synthesis. Trost and coworkers[21,70], and Campbell and Darwish[25] were successful, however, in a few instances, in transferring chirality from sulphur to carbon.

Optically active 1-adamantylallylethylsulphonium tetrafluoroborate **60** underwent a [2,3] sigmatropic rearrangement when treated with potassium *t*-butoxide to give 1-adamantyl 1-methyl-3-butenyl sulphide **61** together with lesser amounts of 1-adamantyl ethyl sulphide and 1-adamantyl allyl sulphide[21]. This reaction was described earlier in the section on absolute configuration.

Sulphonium salt **161** also underwent a [2,3] sigmatropic rearrangement with asymmetric induction, but in this case optically active bases and solvents provided the chiral influence[70]. Lithium (*S*)-(+)-2,2,2-trifluorophenylethoxide in a 1:1 mixture of (*S*,*S*)-(+)-1,4-bis(dimethylamino)-2,3-dimethoxybutane and tetrahydro-furan at −20°C gave sulphide **162** in 48% yield and with 12% enantiomeric

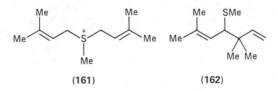

(161)   (162)

purity. If the rearrangement was highly stereospecific as it was for **60**, then this 12% reflects the selectivity of the base for the prochiral methylene groups. Lower optical yields (5%) were obtained in optically active 2,2,2-trifluorophenylethanol–pentane mixtures; *n*-butyllithium–sparteine complex or lithium (−)-menthoxide in tetrahydrofuran gave a racemic product.

Ylide **21** and the related optically active ethylmethyl-*p*-chlorophenacylsulphonium ylide **22** underwent the Sommelet rearrangement, to give achiral sulphides **163** and **164**, respectively, and chiral sulphides **165** and **166**, respectively, the latter two with about 20% enantiomeric purity[25]. Intermediates arising from initial [2,3] sigmatropic rearrangements presumably have the structures shown by **167** and **168**.

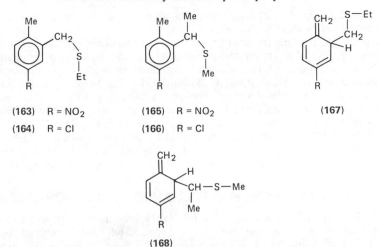

(163) R = NO₂
(164) R = Cl

(165) R = NO₂
(166) R = Cl

(167)

(168)

Structure **168** has two chiral centres and is thus capable of existing as any of four diastereomers, each preceded by a diastereomeric transition state arising from ylides **71** and **72**. Whether small energy differences between these transition states or racemization at sulphur, or both, are responsible for the smaller amount of asymmetric induction compared with the rearrangement of **60** is not clear. It remains to be seen if [2,3] sigmatropic reactions will prove useful in asymmetric induction in various recently studied ring expansions[71].

Processes other than sigmatropic rearrangements have not been successful in asymmetric synthesis. Optically active 1-adamantylethylmethylsulphonium tetra-fluoroborate **146** gave the ylide by abstraction of a methyl proton when treated with *n*-butyllithium. Deuteration of the ylide by deuterofluoroboric acid regenerated sulphonium salt **169** without loss of optical activity. This demonstrated the ylides configurational stability at sulphur, yet treatment of the ylide with benzaldehyde or methyl α-cyanocinnamate gave **170** and **171**, respectively, with very low yields of asymmetric induction[21].

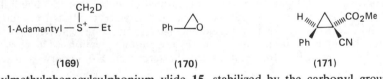

(169)   (170)   (171)

Ethylmethylphenacylsulphonium ylide **15**, stabilized by the carbonyl group, was considerably less reactive than simple ylides such as that derived from **146** and failed to add to cyclohexanone or benzaldehyde, but it did add to chalcone **172** in benzene (17 h at 25°C) to give *meso*-cyclopropane **173** (21% yield) and (±)-cyclopropane **174** (25% yield)[27]. The ease of racemization of **15** doubtless contributed to the formation of racemic product. Similarly, ethylmethyl-*p*-nitro-

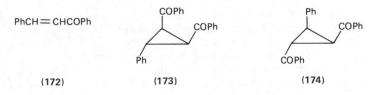

(172)   (173)   (174)

phenacylsulphonium ylide **21** either failed to react with various carbonyl compounds or else produced racemic products[25].

A report of an enantioselective epoxide synthesis by means of trimethylsulphonium ylide and chiral phase-transfer catalysis was shown to be in error[72,73].

Finally, note should be taken of a study on the stereoselectivity in the transfer of the prochiral methyl groups from (*R,S*)-dimethyl-(1-methylpropyl)sulphonium ion **175** to 4-methylbenzenethiolate ion. By specific labelling with ^{14}C, it was shown that the methyls are transferred to the acceptor molecule at different rates, but that this rate difference was mostly a consequence of a surprisingly large ^{12}C/^{14}C isotope effect of 1.16[74]. Consequently, any study in which prochiral methyl groups are transferred selectively and in which ^{14}C is used to follow the process must take into account this large isotope effect in the interpretation of the results.

VIII. BASE-CATALYSED H–D EXCHANGE IN SULPHONIUM SALTS

Protons alpha to sulphur are acidic. Sulphides, sulphoxides, sulphonium salts, and sulphones all have acidic α-hydrogens. Stereoselective H–D exchange has been observed in both sulphoxides and sulphonium salts. Such studies are of interest for several reasons. Firstly, they aid in an understanding of the stereochemistry of exchange processes in general and allow tests of theories which make specific predictions. Secondly, knowledge of factors governing stereoselective ylide formation might prove useful in designing asymmetric syntheses or in understanding possible related enzyme-catalysed reactions.

Hydrogen–deuterium exchange proceeded with stereoselectivity in several thiolanium salts, but without stereoselectivity in homologous 6- and 7-membered rings. 1-Methylthiolanium cation **90** exchanged one pair of its α-methylene protons (those *cis* to the *S*-methyl group) about 30 times faster than the other (*trans*) pair[49,75,76]. 1,3,3-Trimethylthiolanium cation **89** and the thiolanium cation **91** also exchanged their *cis*-α-protons more rapidly than their *trans*-protons. Compounds **89** and **91**, both of which exist in a half-chair conformation, have one *cis*-proton pseudo-axial and one pseudo-equatorial. The pseudo-equatorial proton exchanged the most rapidly of the two. In the *trans* pair, the pseudo-axial proton exchanged more rapidly than the pseudo-equatorial proton. These results do not agree with any theory, such as the gauche effect theory, which proposes that the dihedral angle at $C_{(\alpha)}$—S governs the rate of proton exchange[76]. To be strictly correct, however, the gauche effect theory predicts thermodynamic and not kinetic activity.

1-Methylthianium cation **96**, 1-methylthiopanium cation **176**, *trans*-1,2-dimethylthianium cation **109**, and its *cis* diastereomer **110** underwent exchange

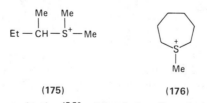

(175)   (176)

without significant stereoselection[47,75]. All of the *S*-methyl protons exchange at about the same rate, and about 100 times faster than the α-protons, for these 5-, 6-, and 7-membered rings. However, the *S*-methyl protons in 1,3,3-thietanium cation **85** are more reactive by a factor of several hundred; the α-protons do not show any special rate enhancement[47]. The reason for the acceleration of proton exchange exocyclic, but not endocyclic, to the ring is unclear.

IX. HETEROSULPHONIUM SALTS

Heterosulphonium salts are treated in Chapter 17, so their stereochemistry will be mentioned only briefly here. Only oxosulphonium salts **177**, alkoxysulphonium salts **178**, and aminosulphonium salts **179** will be discussed.

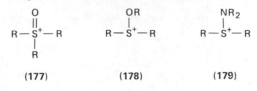

(177)   (178)   (179)

A. Oxosulphonium Salts

(+)-(R)-ethylmethylphenyloxosulphonium salt **180** was prepared by heating a mixture of ethyl phenyl sulphoxide, methyl iodide, and mercury(II) iodide. Treatment of the resulting mercuritriiodide salt with silver (+)-camphor-10-sulphonate gave the sulphonate salt, which was subsequently resolved by fractional recrystallization. Upon passage through an appropriate ion-exchange column, it gave the perchlorate. Sulphonium perchlorate **180** was demethylated by reaction with sodium iodide in acetone to give (+)-(R)-ethyl phenyl sulphoxide **181**. Since the configuration at sulphur remained unchanged during this S_N2 process, the configuration of **180** was established (equation 23)[77,78,79]. Although methylation of dimethyl sulphoxide on sulphur is facile, this is not so for other sulphoxides. Under severe conditions, dialkyl sulphoxides yield trialkylsulphonium salts as the main products[79]. Aryl alkyl sulphoxides react very slowly at sulphur even under forcing conditions.

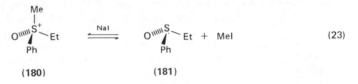

(180)   (181)

Ylides derived from oxosulphonium salts have some use in synthesis, but owing to a lack of a synthetic route to an optically active diarylmethyloxosulphonium salt, no applications to asymmetric synthesis have been reported[80]. Recently, sulphonium salts have been oxidized by treatment with peracids under basic conditions with retention of configuration, thus making optically active oxosulphonium salts accessible[80].

B. Alkoxysulphonium Salts

Treatment of sulphoxides with alkylating agents generally yields alkoxysulphonium salts **178** by alkylation on oxygen rather than oxosulphonium salts **177** by alkylation on sulphur. These compounds are susceptible to reaction with nucleophiles at five positions as shown in Scheme 7[39,40,81].

Johnson and McCants[82] showed that alkoxysulphonium salts react with dilute base (Scheme 7-1, X = O) to regenerate the original sulphoxide in high yield with greater than 90% inversion at sulphur. The small amount of retained product was thought to result from nucleophilic attack on carbon with consequent C—O bond cleavage (Scheme 7-2)[82,83]. Grignard reagents and organocadmium reagents also displace the alkoxy group with inversion at sulphur; these reactions were described

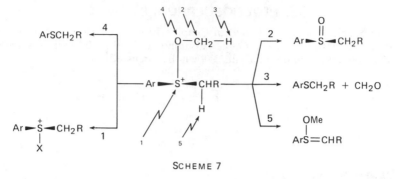

SCHEME 7

previously as methods for obtaining optically active sulphonium salts. Iodide and bromide ions react at carbon (Scheme 7–2) to regenerate the sulphoxide with retained configuration[40]. Chloride ion reacts at both sulphur and carbon whereas fluoride ion prefers sulphur. Mechanisms have been proposed for these reactions[40]. Treatment of alkoxysulphonium salts with amines, triphenylphosphine, sodium thiophenoxide, sodium-p-toluenesulphinate, sodium-p-toluenesulphonamide, and sodium dimethylmalonate have also been studied[34,81].

C. Aminosulphonium Salts

Optically active aminosulphonium salts **179** have been prepared by alkylation of optically active sulphilimines and their rates of racemization by pyramidal inversion have been measured[84]. Methylation of N-p-toluenesulphonyl S-methyl S-phenylsulphilimine lowered the barrier to pyramidal inversion by about 3 kcal/mol. Based on this one example, aminosulphonium salts appear to fall between sulphonium salts and sulphoxides in ease of racemization, with the latter racemizing most slowly.

X. CHIROPTICAL PROPERTIES OF SULPHONIUM SALTS

There is only one study on the chiroptical properties of sulphonium salts[29]. The optical rotatory dispersion and circular dichroism curves for ethylmethyl-p-tolyl- and n-butylmethyl-p-tolylsulphonium salts were measured. Longer wavelength Cotton effects (250–280 nm) correspond to 1_{L_b} secondary u.v. absorption bands; oppositely signed Cotton effects below 250 nm corresponds to the 1_{L_a} primary u.v. bands.

The optical rotatory dispersion curves of (R)- and (S)-benzyl-p-tolylethoxy-sulphonium tetrafluoroborates have been reported[82]. They resemble the curves obtained for the precursor sulphoxides.

XI. REFERENCES

1. K. Freudenberg (Editor), *Stereochemie*, Franz Deuticke, Leipzig and Vienna, 1933.
2. G. Wittig, *Stereochemie*, Akademische Verlagsgesellschaft, Leipzig, 1930.
3. W. H. Mills, *J. Chem. Soc.*, 697 (1941).
4. O. Aschan, *Chem. Ber.*, **32**, 988 (1899).
5. D. Strömholm, *Chem. Ber.*, **33**, 823 (1900).

6. L. Vanzetti, *Gazz. Chim. Ital.*, **30(I)**, 175 (1900).
7. W. J. Pope and S. J. Peachey, *J. Chem. Soc.*, **77**, 1072 (1900).
8. S. Smiles, *J. Chem. Soc.*, **77**, 1174 (1900).
9. W. J. Pope and A. Neville, *J. Chem. Soc.*, **81**, 1552 (1902).
10. J. Meisenheimer, L. Angermann, O. Finn and E. Vieweg, *Chem. Ber.*, **57**, 1745 (1924).
11. J. Smiles, *J. Chem. Soc.*, **87**, 450 (1905).
12. E. Wedekind, *Chem. Ber.*, **58**, 2510 (1925).
13. E. V. Bell and G. M. Bennett, *J. Chem. Soc.*, 86 (1928).
14. G. M. Bennett and W. B. Waddington, *J. Chem. Soc.*, 2832 (1929).
15. M. P. Balfe, J. Kenyon and H. Phillips, *J. Chem. Soc.*, 2554 (1930).
16. F. Mussgnug, *Naturwissenschaften*, **28**, 366 (1940).
17. J. F. Kincaid and F. C. Henriques, Jr., *J. Amer. Chem. Soc.*, **62**, 1474 (1940).
18. R. E. Weston, Jr., *J. Amer. Chem. Soc.*, **76**, 2645 (1954).
19. D. Darwish and G. Tourigny, *J. Amer. Chem. Soc.*, **88**, 4303 (1966).
20. R. Scartazzini and K. Mislow, *Tetrahedron Lett.*, 2719 (1967).
21. B. M. Trost and R. F. Hammen, *J. Amer. Chem. Soc.*, **95**, 962 (1973).
22. A. Garbesi, N. Corsi and A. Fava, *Helv. Chim. Acta*, **53**, 1499 (1970).
23. D. Darwish and C. E. Scott, *Can. J. Chem.*, **51**, 3647 (1973).
24. D. Darwish, S. H. Hui and R. Tomilson, *J. Amer. Chem. Soc.*, **90**, 5631 (1968).
25. S. J. Campbell and D. Darwish, *Can. J. Chem.*, **54**, 193 (1976).
26. D. Darwish and R. Tomilson, *J. Amer. Chem. Soc.*, **90**, 5938 (1968).
27. S. J. Campbell and D. Darwish, *Can. J. Chem.*, **52**, 2953 (1974).
28. K. K. Andersen, *J. Chem. Soc. Chem. Commun.*, 1051 (1971).
29. K. K. Andersen, R. L. Caret and D. L. Ladd, *J. Org. Chem.*, **41**, 3096 (1976).
30. K. K. Andersen, R. L. Caret and I. Karup-Nielsen, *J. Amer. Chem. Soc.*, **96**, 8026 (1974).
31. P. H. Laur, in *Sulfur in Organic and Inorganic Chemistry* (Ed. A. Senning), Vol. 3, Marcel Dekker, New York, 1972, Ch. 24.
32. T. Durst, M. J. LeBelle, R. van den Elzen and K. C. Tin, *Can. J. Chem.*, **52**, 761 (1974).
33. W. H. Pirkle and D. W. House, *Abstracts, 175th National Meeting of the American Chemical Society, Anaheim, California, March 1978*, ORGN 151.
34. M. Moriyama, S. Oae, T. Numata and N. Furukawa, *Chem. Ind. (London)*, 163 (1976).
35. P. M. Denerley and E. J. Thomas, *Tetrahedron Lett.*, 71 (1977).
36. D. K. Herron, *Tetrahedron Lett.*, 2145 (1975).
37. E. Kelstrup, A. Kjaer, S. Abrahamsson and B. Dahlen, *J. Chem. Soc. Chem. Commun.*, 628 (1975).
38. J. W. Cornforth, S. A. Reichard, P. Talalay, H. L. Carrell and J. P. Glusker, *J. Amer. Chem. Soc.*, **99**, 7292 (1977).
39. R. Annunziata, M. Cinquini and S. Colonna, *J. Chem. Soc. Perkin Trans. I*, 1231 (1973).
40. R. Annunziata, M. Cinquini and S. Colonna, *J. Chem. Soc. Perkin Trans. I*, 404 (1975).
41. D. C. Owsley, G. H. Helmkamp and S. N. Spurlock, *J. Amer. Chem. Soc.*, **91**, 3606 (1969).
42. P. Raynolds, S. Zonnebelt, S. Bakker and R. M. Kellog, *J. Amer. Chem. Soc.*, **96**, 3146 (1974).
43. G. Capozzi, O. DeLucchi, V. Lucchini and G. Modena, *Tetrahedron Lett.*, 2603 (1975).
44. I. G. Csizmadia, A. J. Duke, V. Lucchini and G. Modena, *J. Chem. Soc. Perkin Trans. II*, 1808 (1974).
45. B. M. Trost, W. L. Schinski and I. B. Mantz, *J. Amer. Chem. Soc.*, **91**, 4320 (1969).
46. B. M. Trost, W. L. Schinski, F. Chen and I. B. Mantz, *J. Amer. Chem. Soc.*, **93**, 676 (1971).
47. G. Barbarella, A. Garbesi and A. Fava, *Helv. Chim. Acta*, **54**, 2297 (1971).
48. A. Garbesi, G. Barbarella and A. Fava, *J. Chem. Soc. Chem. Commun.*, 155 (1973).
49. G. Barbarella, A. Garbesi, A. Boicelli and A. Fava, *J. Amer. Chem. Soc.*, **95**, 8051 (1973).
50. R. Gerdil, *Helv. Chim. Acta*, **57**, 489 (1974).
51. G. Barbarella, P. Dembeck, A. Garbesi and A. Fava, *Tetrahedron*, **32**, 1045 (1976).
52. J. B. Lambert, C. E. Mixan and D. H. Johnson, *J. Amer. Chem. Soc.*, **95**, 4634 (1973).

53. E. L. Eliel, R. L. Willer, A. T. McPhail and K. D. Onan, *J. Amer. Chem. Soc.*, **96**, 3021 (1974).
54. G. Barbarella, P. Dembech, A. Garbesi and A. Fava, *Org. Mag. Reson.*, **8**, 108 (1976).
55. G. Barbarella, P. Dembech, A. Garbesi and A. Fava, *Org. Mag. Reson.*, **8**, 469 (1976).
56. E. L. Eliel and R. L. Willer, *J. Amer. Chem. Soc.*, **99**, 1936 (1977).
57. R. L. Willer and E. L. Eliel, *Org. Mag. Reson.*, **9**, 285 (1977).
58. M. J. Cook, H. Dorn and A. R. Katritzky, *J. Chem. Soc. B*, 1468 (1968).
59. P. J. Halfpenny, P. J. Johnson, M. J. T. Robinson and M. G. Ward, *Tetrahedron*, **32**, 1873 (1976).
60. I. Stahl and J. Gosselck, *Tetrahedron*, **30**, 3519 (1974).
61. K. K. Andersen, M. Cinquini and N. E. Papanikolaou, *J. Org. Chem.*, **35**, 706 (1970).
62. R. M. Acheson and J. K. Stubbs, *J. Chem. Soc. Perkin Trans. I*, 899 (1972).
63. M. Hori, T. Kataoka and H. Shimizu, *Chem. Lett.*, 1117 (1974).
64. M. Hori, T. Kataoka, H. Shimizu and S. Ohno, *Tetrahedron Lett.*, 255 (1978).
65. A. I. Gusev and Yu. T. Struchov, *Zh. Strukt. Khim.*, **12**, 1120 (1971).
66. K. R. Brower and T. Wu, *J. Amer. Chem. Soc.*, **92**, 5303 (1970).
67. K. K. Andersen and N. E. Papanikolaou, *Tetrahedron Lett.*, 5445 (1966).
68. D. Darwish, *Mech. React. Sulfur Compd.*, **3**, 33 (1968).
69. S. Wolfe, P. Chamberlain and T. F. Garrard, *Can. J. Chem.*, **54**, 2847 (1976).
70. B. M. Trost and W. G. Biddlecom, *J. Org. Chem.*, **38**, 3438 (1973).
71. E. Vedejs, M. J. Arco and J. M. Renga, *Tetrahedron Lett.*, 523 (1978).
72. T. Hiyama, T. Mishima, H. Sawada and H. Nozaki, *J. Amer. Chem. Soc.*, **97**, 1626 (1975).
73. T. Hiyama, T. Mishima, H. Sawada and H. Nozaki, *J. Amer. Chem. Soc.*, **98**, 641 (1976).
74. G. Grue-Sørensen, A. Kjaer, R. Norrestam and E. Wieczorkowska, *Acta Chem. Scand.*, **B31**, 859 (1977).
75. O. Hofer and E. L. Eliel, *J. Amer. Chem. Soc.*, **95**, 8045 (1973).
76. G. Barbarella, A. Garbesi and A. Fava, *J. Amer. Chem. Soc.*, **97**, 5883 (1975); G. Barbarella, P. Dembech, A. Garbesi, F. Bernardi, A. Bottini and A. Fava, *J. Amer. Chem. Soc.*, **100**, 200 (1978).
77. M. Kobayashi, K. Kamiyama, H. Minato, Y. Oishi, Y. Takada and Y. Hattori, *J. Chem. Soc. Chem. Commun.*, 1577 (1971).
78. K. Kamiyama, H. Minato and M. Kobayshi, *Bull. Chem. Soc. Jap.*, **46**, 3895 (1973).
79. M. Kobayashi, K. Kamiyama, H. Minato, Y. Oishi, Y. Takada and Y. Hattori, *Bull. Chem. Soc. Jap.*, **45**, 3703 (1972).
80. K. Ryoke, H. Minato and M. Kobayashi, *Bull. Soc. Chem. Jap.*, **49**, 1455 (1976); M. Kobayashi, K. Okuma and H. Takeuchi, *Abstracts, Eighth International Symposium on Organic Sulphur Chemistry, Portoroz, Yugoslavia, June 1978*, 132.
81. K. Tsumori, H. Minato and M. Kobayashi, *Bull. Chem. Soc. Jap.*, **46**, 3503 (1973).
82. C. R. Johnson and D. McCants, Jr., *J. Amer. Chem. Soc.*, **87**, 5404 (1965).
83. M. Buza, K. K. Andersen and M. D. Pazdon, *J. Org. Chem.*, **42**, 3827 (1978).
84. D. Darwish and S. K. Datta, *Tetrahedron*, **30**, 1155 (1974).

The Chemistry of the Sulphonium Group
Edited by C. J. M. Stirling and S. Patai
© 1981 John Wiley & Sons Ltd.

CHAPTER **11**

Synthesis of sulphonium salts

P. A. LOWE

The Ramage Laboratories, Department of Chemistry and Applied Chemistry, University of Salford, Salford M5 4WT, England

I.	INTRODUCTION	268
II.	FORMATION FROM SULPHIDES	268
	A. Alkylation of Saturated Sulphides	268
	B. Alkylation of Unsaturated Sulphides	272
	C. Alkylation of Aryl Sulphides	273
	D. Alkylation of Cyclic Sulphides	274
	E. Intramolecular Alkylation of Sulphur	279
	F. Arylation of Sulphides	282
III.	FORMATION FROM DI- AND TRISULPHIDES	283
IV.	FORMATION FROM SULPHENYL HALIDES	285
V.	FORMATION FROM SULPHOXIDES	285
	A. From Protonated Sulphoxides	285
	B. From Sulphoxides and Organometallics	286
	C. Alkylation of Sulphoxides and Conversion of the Products into Sulphonium Salts	287
	1. Alkylation of sulphoxides	287
	2. Conversion of alkoxysulphonium salts into alkyl- and arylsulphonium salts	288
VI.	FORMATION OF SULPHUR YLIDES AND THEIR CONVERSION INTO SULPHONIUM SALTS	289
	A. Formation of Sulphonium Ylides	289
	B. Conversion of Sulphonium Ylides into Sulphonium Salts	290
VII.	MODIFICATION OF SULPHONIUM SALTS	292
VIII.	FORMATION FROM HETEROSULPHONIUM SALTS	294
	A. Formation from Chlorosulphonium Salts	294
	B. Formation from Azasulphonium Salts and Related Compounds	295
IX.	MISCELLANEOUS METHODS OF FORMATION	296
X.	METHYLENESULPHONIUM SALTS	297
XI.	FORMATION OF ISOTHIOPHENIUM SALTS	298

XII. FORMATION OF THIOPYRYLIUM SALTS 299
    A. Formation from 1,5-Diketones 299
    B. Formation from Thiopyrans 299
    C. Formation from Pyrylium Salts 301
    D. Formation from 1,3-Dienes and Thiophosgene 301
    E. Formation by Ring Expansion of Thiophens 301
    F. Formation from 2-Chlorovinylmethineammonium Salts . . . 302
    G. Formation from 2-Ketosulphides 302
    H. Formation from Diaryl Sulphides 302
XIII. FORMATION OF DITHIOLIUM SALTS 302
    A. Formation of 1,2-Dithiolium Salts 302
      1. From open-chain compounds 302
      2. From 1,2-dithioles 303
      3. By ring contraction 303
    B. Formation of 1,3-Dithiolium Salts 304
      1. From open-chain compounds 304
      2. From 1,3-dithioles 305
XIV. REFERENCES 305

I. INTRODUCTION

This chapter surveys the methods of formation of sulphonium salts having a tricoordinated sulphur atom attached to three carbon atoms, where the substituents may be alkyl, cycloalkyl, alkenyl, alkynyl, aryl or heteroaryl groups, or where the sulphur atom is part of a monocyclic or bicyclic system. There will also be reference to salts whose structures contain a carbon–sulphur double bond in one or more canonical form, for example methylenesulphonium and isothiouronium salts, and to systems where the sulphur atom is present in an aromatic heterocyclic ring, for example thiophenium, thiopyrylium, and dithiolium salts.

There is increasing interest in the chemistry of heterosulphonium salts in which one or more of the ligands is a heteroatom (halogen, sulphur, oxygen, or nitrogen). Wherever it has so far been possible to convert these salts into the types of sulphonium salts described in the opening sentence, the syntheses of these heterosulphonium salts will be described. Other aspects of the chemistry of these latter salts will be covered in Chapter 15.

Descriptions of some of the earlier preparative methods were given by Goerdeler[1], and many references are to be found in the six-volume series by Reid[2]. The book by Trost and Melvin[3] contains a chapter on the synthesis of sulphonium salts, other references to synthesis are mentioned throughout the text and in the numerous tables, and details of some typical methods of preparation are given. A major portion of the review on sulphur-containing cations by Marino[4] covers the chemistry of sulphonium salts and there is a chapter on sulphonium salts by Stirling in the book edited by Oae[5]. The specialist reviews of the chemistry of sulphimides[6], sulphoximides[7-10], and sulphodiimides[10,11] include reference to the syntheses of related sulphonium compounds.

II. FORMATION FROM SULPHIDES

A. Alkylation of Saturated Sulphides

Dialkyl sulphides have sufficient nucleophilicity for reaction with primary alkyl halides to proceed under mild conditions, frequently simply by mixing the reagents

$$R_2^1S + R^2X \longrightarrow R_2^1R^2S^+ X^- \tag{1}$$

at room temperature (equation 1). Catalysts are often unnecessary although the reaction may be accelerated by the use of polar solvents such as methanol, acetone, acetonitrile, or nitromethane, as the transition state and product are more polar than the starting materials. Alkyl iodides are the most reactive of the simple alkyl halides, the reaction rate decreasing with chain length or chain branching. With activated alkyl groups, such as allyl[12], benzyl[12,13], α-halocarbonyl[14,15], and arylthiomethyl[16], chlorides and bromides are frequently used (equations 2 and 3).

$$Me_2S + CH_2{=}CHCH_2Cl \xrightarrow{H_2O} Me_2\overset{+}{S}CH_2CH{=}CH_2\ Cl^- \tag{2}$$

$$Me_2S + BrCH_2CO_2Et \xrightarrow{Me_2CO} Me_2\overset{+}{S}CH_2CO_2Et\ Br^- \tag{3}$$

Alkylation of a sulphide is reversible: the sulphonium salt can be dealkylated by the halide ion, leading to the following series of reactions (equation 4):

$$\left. \begin{array}{l} R_2^1S + R^2X \rightleftharpoons R_2^1R^2\overset{+}{S}\ X^- \rightleftharpoons R^1R^2S + R^1X \\[2mm] R_2^1S + R^1X \rightleftharpoons R_3^1\overset{+}{S}\ X^- \\[2mm] R^1R^2S + R^2X \rightleftharpoons R^1R_2^2\overset{+}{S}\ X^- \rightleftharpoons R_2^2S + R^1X\ \text{etc.} \end{array} \right\} \tag{4}$$

There is a tendency for the formation of the salt containing alkyl groups having the lowest molecular weight[17]. Thus, reaction of a methanolic solution of dimethyl sulphide with allyl iodide yielded trimethylsulphonium iodide[18], and the reaction between dimethyl sulphide and cyanogen halides resulted in the formation of trimethylsulphonium halide and methylthioacetonitrile (equation 5)[1]. The desired

$$2Me_2S + XCN \longrightarrow Me_3\overset{+}{S}\ X^- + MeSCN \tag{5}$$

product in the above reaction, $Me_2\overset{+}{S}CN\ X^-$, was obtained[19], but by alkylation of the methylthioacetonitrile using a methylating agent possessing a non-nucleophilic anion, in this case trimethyloxonium tetrafluoroborate. This reagent and similar salts[20] have often been used to alkylate all types of sulphides and will be referred to repeatedly throughout this chapter. Other alkylating agents include dialkyl sulphates[21], alkyl sulphonates[22], alkyl fluorosulphonates[23], alkyl trifluoromethane sulphonates (alkyl triflates)[24], dialkoxycarbonium tetrafluoroborates[25], and dialkyl halonium salts[26].

The addition of certain metal salts, e.g. iron(III), mercury(II), and zinc halides, accelerates the reaction[27], although this is more often required with unsaturated or aryl sulphides.

Secondary and tertiary alkyl halides are generally rather unreactive towards dialkyl sulphides; an activating group in the secondary alkyl halide and addition of silver tetrafluoroborate gives the products in quantitative yields (equation 6)[28].

$$R^1CHBrCOR^2 + Me_2S \xrightarrow[Me_2CO]{AgBF_4} Me_2\overset{+}{S}CHR^1COR^2\overset{-}{B}F_4 \tag{6}$$

$R^1, R^2 = Me$
$R^1 = Et, R^2 = H$
$R^1 = n\text{-}Bu, R^2 = H$

The reaction between cycloalkyl halides and dialkyl sulphides in the presence of silver salts generally produces the cycloalkyl sulphides in high yield, which can subsequently be alkylated to give the desired sulphonium salt (equation 7)[29]. In a similar manner, sulphonium salts derived from butyrolactones have been obtained (equation 8)[30].

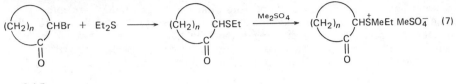

$$n = 3,4,5$$

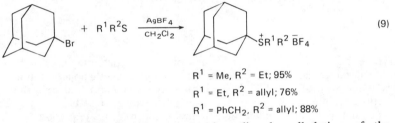

Alkylation of dimethyl sulphide with t-butyl iodide in nitromethane solution gives a low yield of t-butyl dimethylsulphonium iodide[31]; adamantyl bromide, however, reacts readily in the presence of silver tetrafluoroborate (equation 9)[32].

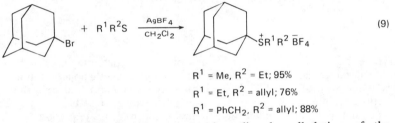

$R^1 = Me, R^2 = Et; 95\%$

$R^1 = Et, R^2 = allyl; 76\%$

$R^1 = PhCH_2, R^2 = allyl; 88\%$

Adamantylsulphonium salts had been prepared earlier by alkylation of the adamantyl alkyl sulphide with methyl iodide in the presence of silver tetrafluoroborate[33].

Bis-sulphonium salts, for example $MeEt\overset{+}{S}CH_2CH_2\overset{+}{S}EtMe$ $2I^-$, have been prepared by dimethylation of the bis-1,2-(ethylthio)ethane[34] and while attempts to alkylate bis-1,2-(ethylthio)ethane and bis-1,3-(ethylthio)propane using phenacyl bromide gave the disulphonium salts, similar reactions with formaldehyde diethylacetal and 1,3-thiolan and 1,4-dithian compounds gave only the monosulphonium salts[35]. However, methylene bis(dialkylsulphonium) tetrafluoroborates have been obtained in yields of 68–99% by alkylating the corresponding sulphides with trialkyloxonium tetrafluoroborates (equation 10)[36-38].

$$R^1SCHR^2SR^1 + R^3\overset{+}{O} \bar{B}F_4 \longrightarrow R^1R^3\overset{+}{S}CHR^2\overset{+}{S}R^1R^3 \quad 2\bar{B}F_4 \qquad (10)$$

Treatment of secondary and tertiary alcohols with sulphuric acid[39] or with acetic–sulphuric acid mixtures[40] in the presence of dialkyl sulphides leads to the formation of sulphonium salts, presumably via the intermediate carbocations. Primary alcohols will react with dialkyl sulphides in the presence of 70% perchloric acid to give the trialkylsulphonium salts, provided that water is removed during the reaction by azeotropic distillation (equation 11)[41].

There was apparently no detectable rearrangement during the reaction.

Dialkyl sulphides, for example dibenzyl sulphide, may be converted directly into the trialkylsulphonium salts by reaction with sulphuric acid[42] or with aluminium halides (equation 12)[43-45].

Dimethyl sulphide (and PhSMe but not Ph₂S) will react with diazoketones in the

$$R^1_2S \;+\; R^2OH \xrightarrow[-H_2O]{70\%\ HClO_4} R^1_2R^2\overset{+}{S}\ ClO^-_4 \tag{11}$$

$R^1 = Et,\ R^2 = Et$ $R^1 = n\text{-}Bu,\ R^2 = benzyl$

$R^1 = n\text{-}Pr,\ R^2 = n\text{-}Pr$ $R^1 = benzyl,\ R^2 = benzyl$

$R^1 = n\text{-}Bu,\ R^2 = n\text{-}Bu$

$$(PhCH_2)_2S \xrightarrow[(H_2SO_4)]{H^+} PhCH_2\overset{+}{\underset{H}{S}}CH_2Ph\ HSO^-_4$$

$$\tag{12}$$

$$(PhCH_2)_3\overset{+}{S}\ HSO^-_4 \xleftarrow{(PhCH)_2S} Ph\overset{+}{C}H_2 \;+\; PhCH_2SH$$

presence of 71% perchloric acid to give the sulphonium salts in excellent yields (equation 13)[46].

$$ArCOCHN_2 \xrightarrow{HClO_4} ArCOCH_2\overset{+}{N_2}\ ClO^-_4 \xrightarrow{R^1R^2S} ArCOCH_2\overset{+}{S}R^1R^2\ ClO^-_4 \tag{13}$$

$R^1 = R^2 = Me$

$R^1 = Me,\ R^2 = Ph$

Dialkyl sulphides will add to an electrophilic alkene (equation 14)[47] or to an activated cyclopropane (equation 15)[48] in the presence of a proton donor. Similar reactions with azirine derivatives are known (equations 16[49] and 17[50]). In the later example the salt undergoes subsequent reaction.

The analogous reactions between oxirans and dialkyl sulphides are unknown, although the desired hydroxyethylsulphonium salts have been obtained both from alkylation of hydroxyethyl sulphides by means of alkyl halides[51,–53] and of trialkyloxonium salts[54], and from the reaction between a ring-opened oxiran and a dialkyl sulphide (equation 18)[55]. 2-Hydroxyethylsulphonium salts have been

$$R^1_2S \;+\; R^2CH\!=\!CHCO_2H \xrightarrow{H^+} R^1_2\overset{+}{S}CHR^2CH_2CO_2H \tag{14}$$

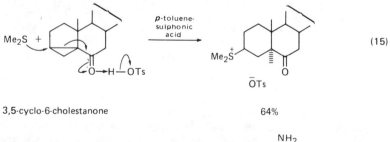

$$\tag{15}$$

3,5-cyclo-6-cholestanone   64%

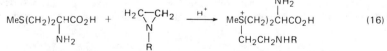

$$\tag{16}$$

$R = CO_2Et,\ CONH_2$

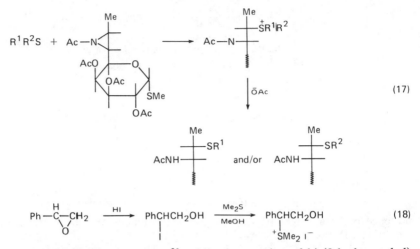

$$(17)$$

$$(18)$$

obtained by hydrolysis of mustard gas[56], while the reaction of bis(2-hydroxyethyl) sulphide with ethylenechlorohydrin yielded the tris(2-hydroxyethyl)sulphonium chloride[57]. Early attempts to prepare chloroalkylsulphonium salts by methylation of the halogenoalkyl sulphides with methyl iodide led only to the formation of trimethylsulphonium salts[58]. Compounds of this type have been obtained, however, by reaction with trimethyloxonium tetrafluoroborate and the salts appear to be stable (equation 19)[38].

$$ClCH_2SMe + Me_3\overset{+}{O}\ \bar{B}F_4 \longrightarrow ClCH_2\overset{+}{S}Me_2\ \bar{B}F_4 \qquad (19)$$

B. Alkylation of Unsaturated Sulphides

Sulphonium salts containing unsaturated ligands, for example the allyl group referred to earlier[12,18] have been obtained by alkylation using allyl halides, and it is possible to introduce propargyl groups in a similar manner (equation 20)[59,60]. This

$$R_2^1S + R^2C\equiv CCH_2Br \longrightarrow R_2^1\overset{+}{S}CH_2C\equiv CR^2\ Br^- \qquad (20)$$

$$R^1 = Alk, R^2 = H, Alk, Ar$$

type of unsaturated sulphonium salt has also been obtained by alkylation of allylic[61] and propargylic[62] sulphides using dimethyl sulphate, and salts containing both of these functional groups have been obtained in this way (equation 21)[60]. Recent

$$PhC\equiv CCH_2SCH_2CH=CH_2 + Me_2SO_4 \longrightarrow PhC\equiv CCH_2\overset{+}{S}CH_2CH=CH_2\ HSO_4^- \qquad (21)$$
$$\underset{Me}{|}$$

interest in the mechanism of terpene biosynthesis has led to the preparation of diallylmethylsulphonium salts by alkylation of diallyl sulphides with trialkyloxonium tetrafluoroborates[63–65]. This sometimes gives rearranged products (equation 22)[64]. Sulphonium salts having the unsaturated group (vinylic or acetylenic) directly

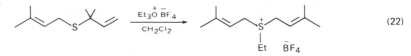

$$(22)$$

attached to the sulphur atom are not as readily obtained owing to the lower nucleophilicity of alkenyl and alkynyl sulphides. An early attempt to prepare dimethylvinylsulphonium iodide by reaction of methyl vinyl sulphide with methyl iodide gave trimethylsulphonium iodide only[66], although subsequent reactions carried out on a solution of the reactants led to the suggestion that the unsaturated salt had indeed been formed initially (see also equation 4). Gosselck *et al.*[67] succeeded in preparing a number of dimethylvinylsulphonium salts in 60–80% yields by reaction of the sulphides with dimethyl sulphate at 90–100°C (equation 23) and also be reaction of dimethylthioacetals under similar conditions (equation 24). Cycloalkenyl sulphides have been converted into the corresponding sulphonium salts using dimethyl sulphate at 40–50°C in yields dependent on the ring size[29].

$$MeSCR^1{=}CHR^2 \ + \ Me_2SO_4 \xrightarrow[\text{1–2 h}]{\text{90–100°C}} Me_2\overset{+}{S}CR^1{=}CR^2 \ Me SO_4^- \qquad (23)$$

R^1 = H, R^2 = Ph R^1 = PhCO, R^2 = p-ClC$_6$H$_4$

R^1 = Ph, R^2 = PhCO R^1 = PhCO, R^2 = p-NO$_2$C$_6$H$_4$

$$R^1 C(SMe)_2 CH_2 R^2 \ + \ Me_2SO_4 \ \longrightarrow \ Me_2\overset{+}{S}CR^1{=}CHR^2 \ MeSO_4^- \qquad (24)$$

R^1 = Me, Ph, p-NO$_2$C$_6$H$_4$, β naphthyl $\qquad$ 60–80%

R^2 = 1-pyridyl, p-NO$_2$C$_6$H$_4$, 2,4-diNO$_2$C$_6$H$_4$, Ph, CO$_2$Et

Alkylation of methyl styryl sulphide with methyl iodide was successful when silver perchlorate was added[68], and trialkyloxonium salts have also been used with considerable success.

Substituted vinylsulphonium salts are readily prepared from the corresponding sulphides and trialkyloxonium salts. For example, bis-1,2-(dialkylthio)ethylene (both *Z* and *E*) react at room temperature with these reagents to give the monosulphonium salts (equation 25)[69]. When R^1 = Et and R^2 = Me, the

$$ \text{(25)} $$

R^1 = Et, PhCH$_2$; R^2 = Me, Et $\qquad$ 80–90%

disulphonium salts were obtained. Irradiation of the *Z*-monosulphonium salt gave the *E*-isomer, which was subsequently converted into the disulphonium salt as before. Further conversions of these and similar unsaturated sulphonium salts are described in Section VII (p. 292). The only example of formation of an alkynylsulphonium salt was by reaction of methyl phenylethynyl sulphide with triethyloxonium tetrafluoroborate (equation 26)[67].

$$PhC{\equiv}CSMe \ + \ Et_3\overset{+}{O} \ \overline{B}F_4 \ \longrightarrow \ PhC{\equiv}\overset{+}{C}SMeEt \ X^- \qquad (26)$$

(isolated as the picrate)

C. Alkylation of Aryl Sulphides

The nucleophilicity of the sulphide group is appreciably lowered when one or both of the R^1 groups in equation 1 is phenyl. Reaction of aryl sulphides with alkyl halides will take place only in the presence of silver[70] or mercury(II) salts[71]. Diphenyl sulphide reacts with a series of alkyl halides in the presence of silver tetrafluoroborate (or perchlorate) to give the sulphonium salts in yields of 41–86% (equation 27)[70]. Diphenyl sulphide, methyl phenyl sulphide and

$$Ph_2S + R^1X \xrightarrow{\text{AgBF}_4} Ph_2\overset{+}{S}R^1 \ X^- \tag{27}$$

(isolated as the tetrafluoroborates or perchlorates)

$$R^1 = \text{Me, Et, } n\text{-Bu, } n\text{-C}_6\text{H}_{13}, n\text{-CH}_2\text{Ph}$$

$$X = \text{Br, I}$$

methyl p-chlorophenyl sulphide react with excess of methyl iodide in the presence of mercury(II) iodide to give the corresponding sulphonium salts in yields of 62–71.5%[71]. Allyldiphenylsulphonium tetrafluoroborate was obtained in 86% yield by the rapid addition of allyl bromide to a slurry of silver tetrafluoroborate and diphenyl sulphide in acetone[72].

Reaction between an alkyl halide and diphenyl sulphide appears to involve an S_N2 displacement by sulphide on a silver-complexed halide[73], as silver halide is not precipitated until the addition of the diphenyl sulphide (equation 28). However,

$$Ph_2S: \longrightarrow \underset{\overset{|}{X\cdots A\overset{+}{g}\,\bar{B}F_4}}{\overset{\overset{R^1}{|}}{CH_2}} \longrightarrow Ph_2\overset{+}{S}CH_2R^1 \ \bar{B}F_4 + AgX \tag{28}$$

repetition of the procedure of Franzen et al.[70] by Rapoport[74] using n-propyl and n-butyl halides gave a mixture of primary and secondary alkyldiphenylsulphonium salts, suggesting the intermediacy of carbocations. n-Alkyl trifluoromethane-sulphonates (triflates) gave unrearranged n-alkyldiphenylsulphonium triflates, albeit in poor yield. Triflate esters of unsaturated alcohols, which are exceptionally reactive, will alkylate diphenyl sulphide in 1–2 min at room temperature; those derived from α-hydroxycarbonyl compounds are less reactive but are still far more effective than the α-halocarbonyl analogues[75].

Methyl and ethyldiphenylsulphonium salts may also be prepared by reaction between diphenyl sulphide and the corresponding trialkyloxonium tetrafluoroborate[76].

D. Alkylation of Cyclic Sulphides

The simplest cyclic sulphonium compounds are the thiiranium (episulphonium) salts. Thiiranium ions have been suggested as intermediates in reactions involving neighbouring group participation, electrophilic addition of sulphenyl halides to alkenes, and in alkylation of thiiranes. There have been a number of reviews on these[77–79] topics and further coverage is given in Chapter 3.

A stable thiiranium salt was first isolated in 70% yield from the alkylation of cyclooctene thioepoxide (cis-1,2-epithiocyclooctane) with trimethyloxonium 2,4,6-trinitrobenzene sulphonate (equation 29)[80]. Subsequently the S-t-butyl salt was obtained in 57% yield in a similar manner[81], but attempts to prepare the analogous cyclohexane compounds were unsuccessful. A rather unstable mono-cyclic S-methylthiiranium salt has been prepared, however, by treating cis-2,3-di-t-butylthiirane with methyl fluorosulphonate (reaction 30)[82]. S-Methylthietanonium salts have been prepared by the treatment of thietanes with

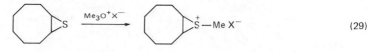

$$X = 2,4,6\text{-trinitrobenzenesulphonate} \tag{29}$$

$$(30)$$

trimethyloxonium tetrafluoroborate at $-30°C$. The salts decomposed on warming and were characterized spectroscopically (equation 31)[83]. Of all reported

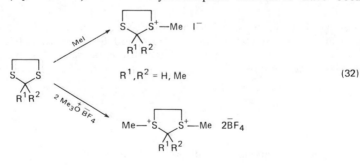

$$(31)$$

alkylations of cyclic sulphides, those concerned with tetrahydrothiophen (thiolan) occur most frequently. Thus, reaction with alkyl halides[84,85], dimethyl sulphate[84], allyl halides[84,86,87], propargyl halides[88,89], and α-halocarbonyl compounds[84] gave good to excellent yields of the salts. Alkylthiomethyl halides required activation by means of silver tetrafluoroborate in order to react[37]; trialkyloxonium salts, alkyl fluorosulphonates, alkyl triflates[90,91], and protonated alkenes[92] have also been used to alkylate tetrahydrothiophen.

1,3-Dithiolans have been alkylated with a variety of reagents to give both mono- and disulphonium salts. Thus, treatment with methyl iodide[93] or phenacyl bromide[35] gave the monosulphonium salt only, whereas reaction with trialkyloxonium tetrafluoroborates gave the bis-sulphonium salts in good to excellent yields (equation 32)[93–95]. Δ³-Dihydrothiophen derivatives have been

$$(32)$$

$$R^1 = R^2 = H; R^1 = H, R^2 = Me$$

treated with trimethyloxonium tetrafluoroborate to give the sulphonium salts in yields up to 90% (reaction 33)[96]. 1,3-Dihydrobenzo(c)thiophen has been reacted with methyl iodide to give the corresponding sulphonium salt, but the reaction was described as being rather sluggish[97].

Thiophen has a much lower nucleophilicity than its hydrogenated derivatives and accordingly proved more difficult to alkylate. The first examples of 1-methylthiophenium salts were obtained by reaction of thiophen with methyl

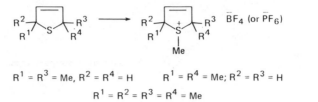

$$R^1 = R^3 = Me, R^2 = R^4 = H \qquad R^1 = R^4 = Me; R^2 = R^3 = H$$
$$R^1 = R^2 = R^3 = R^4 = Me$$

iodide and silver perchlorate and with trimethyloxonium tetrafluoroborate, but yields were not reported[98]. Using methyl iodide and silver tetrafluoroborate in dichloroethane, the S-methyl salts of 3-methyl-, 2,5-dimethyl-, and 2-phenethylthiophen were obtained in low yield (3–12%)[99]. Subsequently, S-methyl- and S-ethylthiophen salts have been obtained in up to 60% yield by this method, while alkylation using methyl and ethyl fluorosulphonates gave >95% yield of the salts derived from 2,3,4,5-tetramethylthiophen, 15% yield from 2,5-dimethylthiophen, but no alkylated product from thiophen[100].

Benzo(b)thiophen and its derivatives are S-alkylated much more readily than the corresponding thiophens. Using methyl, ethyl and even isopropyl iodide in the presence of AgBF$_4$, yields of up to 70% were reported, and yields of 40% using Me$_3$O BF$_4$[99]. Introduction of other electron-donating groups or an additional ring in dibenzothiophen (1) further increased the yields and stabilities of the thiophenium salts, and similar tricyclic systems. Compounds 2,3, and 4, for

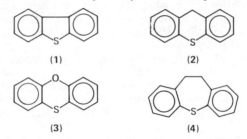

(1)   (2)

(3)   (4)

example, have been S-alkylated in good to excellent yields[101,102]. Monocyclic thian derivatives on alkylation with methyl iodide[103], ethyl iodide[104], and methyl bromoacetate[105] gave mixtures of stereoisomeric salts. An example of the first alkylation is as follows:

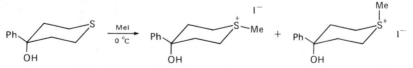

An attempt to alkylate 3,4-dihydro-2H-thiopyran (5) with methyl iodide gave mainly trimethylsulphonium iodide, but the 4-methyl derivative of the isomeric 5,6-dihydro-2H-thiopyran (6) did give a low yield of the sulphonium salt at 24°C; at higher temperatures ring cleavage occurred[106].

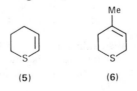

(5)   (6)

Whilst there have been no successful reports of the alkylation of thiopyran itself, 3,5-diphenyl-2H-thiopyran reacted with methyl iodide – silver tetrafluoroborate giving a 90% yield of the thiinium salt[107]:

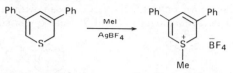

Silver-catalysed alkylations of the corresponding benzo analogues of thiopyran in 56–90% yields have also been reported (equations 34 and 35)[107].

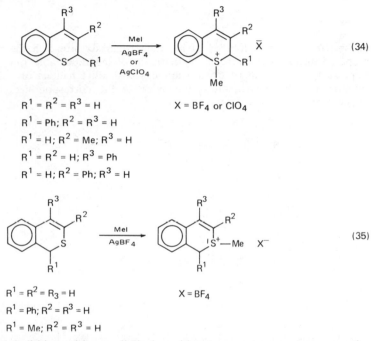

$$R^1 = R^2 = R^3 = H$$
$$R^1 = Ph; R^2 = R^3 = H$$
$$R^1 = H; R^2 = Me; R^3 = H$$
$$R^1 = R^2 = H; R^3 = Ph$$
$$R^1 = H; R^2 = Ph; R^3 = H$$

$$X = BF_4 \text{ or } ClO_4$$

(34)

$$R^1 = R^2 = R_3 = H$$
$$R^1 = Ph; R^2 = R^3 = H$$
$$R^1 = Me; R^2 = R^3 = H$$

$$X = BF_4$$

(35)

Reaction of 1,3-dithian with an allylic bromide at room temperature gave the monosulphonium salt in high yield (equation 36)[108], whereas reaction of some 2-aryl derivatives with trialkyloxonium tetrafluoroborate led to the formation of both mono- and bis-sulphonium salts[109].

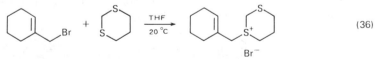

(36)

Monosulphonium salts of 1,3,5-trithiane were obtained on reaction with either aluminium trichloride followed by methyl iodide or with dimethyl sulphate–potassium iodide[43], whilst bis-sulphonium salts were obtained on using trialkyloxonium salts[94].

1,4-Dithian has been both mono- and bisalkylated using methyl iodide and other alkyl halides[110]. In contrast, 1,4-dithiin and its dihydro derivative have to date been converted into monosulphonium salts only. 2,5-Diphenyl-1,4-dithiin reacted with methyl or ethyl iodide in the presence of silver or mercury salts to give the monosulphonium iodides in high yield[111], and reaction of the parent compound with triethyloxonium tetrafluoroborate gave a 56% yield of the mono salt[112]. 2,3-Dihydro-1,4-dithiin gave only a 20% yield of salt with the latter reagent, but higher yields of their benzo analogues (7 and 8) were obtained[112]. Sulphonium salts

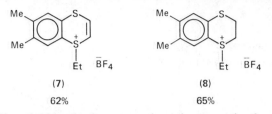

(7)   (8)

62%   65%

derived from cyclic sulphides having more than six atoms in the ring have been prepared during the course of syntheses involving repeatable ring expansions[87,90]. For example, reaction of cyclic sulphides with allyl halides or triflates gave the sulphonium salts, which, after conversion to the corresponding ylides, can then undergo ring expansion (equation 37).

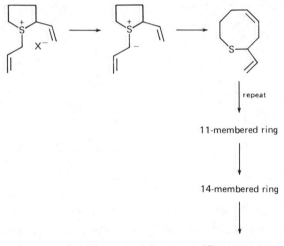

11-membered ring   (37)

14-membered ring

17-membered ring

Formation of sulphonium salts derived from thia-, dithia-, and trithiacyclophanes, e.g. 9 and 10, by means of reactions involving dimethoxycarbonium tetrafluoroborate have been reported by Boekelheide and coworkers[113].

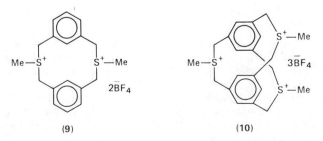

(9)   (10)

Thiacyclophanes have also been converted into their sulphonium salts by means of methyl fluorosulphonate[114].

Until recently, the sulphur atom of penams and cephams had been resistant towards electrophilic attack. By utilizing trimethyloxonium tetrafluoroborate[115,116] or methyl fluorosulphonate[116] both of these important groups of sulphur-containing natural products have now been S-methylated. The reactions in each case were stereoselective; with the penam derivatives only one stereoisomer was produced in nearly quantitative yield (equation 38)[115], whilst with the cepham the product (30% yield) was obtained probably by ring opening of the initially formed sulphonium salt (equation 39)[116].

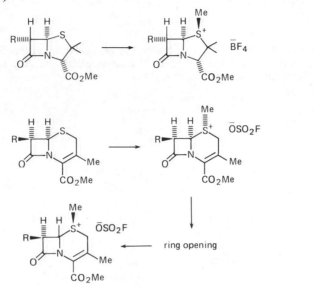

(38)

(39)

E. Intramolecular Alkylation of Sulphur

Cyclic sulphonium salts may be formed when the sulphide group and the alkylating group are present within the same molecule. There have been extensive investigations of many reactions whose mechanisms may involve cyclic sulphonium salts as intermediates and these will be discussed in detail in Chapter 13. In certain cases cyclic sulphonium salts have been isolated from intramolecular reactions such as the solvolysis of halogenylalkyl sulphides $RS(CH_2)_nX$. 2-Halogenoethyl alkyl sulphides cyclise less readily than the 4-halogenobutyl analogues, but a phenyl group on sulphur or on the halogen-bearing carbon atom selectively accelerates three-ring formation[117]. Examples of the formation of three-, four-, five- and six-membered rings are known but, as would be expected, medium-sized rings are not formed in this way. Thiiranium ions were postulated as intermediates in stereospecific[118] isomerization of 2-chloroethyl sulphides, but they were not isolated (equation 40).

A number of relatively stable (several hours at $-30°C$) thiiranium tetrafluoroborates or hexafluoroantimonates have been obtained either by reaction of alkenes with sulphenyl halides (p. 285) in the presence of $AgBF_4$ or $AgSbF_6$, or by formation of the halogenoethyl sulphide followed by treatment with the silver

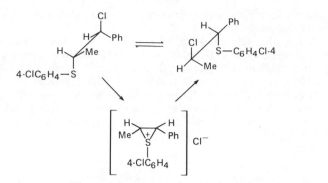

(40)

salts (equations 41 and 42)[119]. The reaction between 1-methylthio-2-chloroisobutyric or isovaleric acid amides with silver *p*-toluenesulphonate led to the formation of stable thiiranium salts (equation 43)[120].

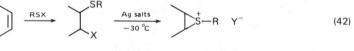

(41)

$$ \text{CICR}^1\text{R}^2\text{CR}^3\text{(SMe)CONHR}^4 \xrightarrow[\text{MeCN}]{\text{AgOTs}} $$

(42)

$$ Y = BF_4, SbF_6 $$

(43)

$$ R^1 = R^2 = H; R^3 = Me, R^4 = H $$
$$ R^1 = R^2 = Me, R^3 = H, R^4 = Ph $$

Stable thietenium salts have been isolated by treatment of 2-phenylmercapto-2-alkyl (or aryl)ethyl ketones with phosphorus oxychloride–perchloric acid (equation 44)[121]. With $R^1 = R^2 = Ph$, the intermediate crystalline perchlorate was

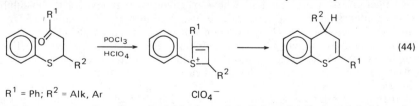

(44)

$$ R^1 = Ph; R^2 = Alk, Ar \qquad ClO_4^- $$

converted by means of sodium hydride into 2,4-diphenyl-1-*S*-phenyl-1-thiacyclo-butadiene, which was characterized spectroscopically and which was re-converted to the thietenium salt on protonation.

Intramolecular alkylation of sulphides by diazoketones (equation 13)[46] gives benzo(*b*)thiophen derivatives (equation 45). When R = Ph, the salt was isolated in almost quantitative yield, but with R = alkyl, the salts decomposed to give the cyclic sulphide.

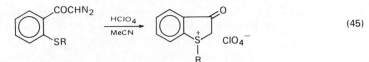

$$(45)$$

Formation of stable bridged cyclic sulphonium salts has been achieved by intramolecular alkylation (equation 46)[122].

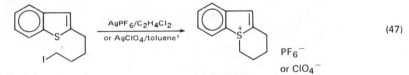

isolated as the picrate

$$(46)$$

(83%)

Intramolecular cyclization of 2-(4-iodobutyl)benzo(b)thiophen on treatment with silver complex salts led to stable crystalline thiophenium salts (equation 47)[123].

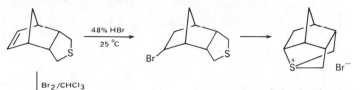

$$(47)$$

Treatment of 4-hydroxymethylthian with concentrated hydrochloric acid at 50°C gave the hygroscopic bicyclic sulphonium chloride, which was subsequently converted into the more stable picrate and chloroplatinate (equation 48)[124]. A similar reaction with the 2-hydroxyethylthian yielded the thiabicyclooctane 1-derivative (equation 49)[125].

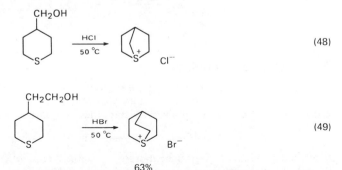

$$(48)$$

$$(49)$$

63%

Prolonged heating of bis(2-hydroxyethyl) sulphide (thiodiglycol) with concentrated hydrochloric acid–zinc chloride at 100°C gave the 1,4-dithian bis-sulphonium salt, isolated as the tetrachlorozincate (equation 50)[126]. Other examples of the preparation of bicyclic sulphonium salts are shown below.

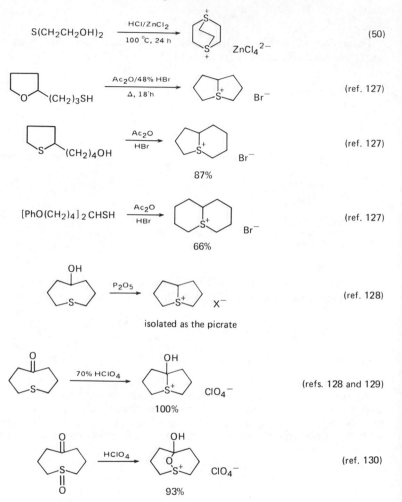

F. Arylation of Sulphides

This is a little used method for formation of sulphonium salts. Triphenylsulphonium tetrafluoroborate was obtained on heating diphenyliodonium tetrafluoroborate with diphenyl sulphide (equation 51)[131]. Reaction of

$$Ph_2\overset{+}{I}\ \overset{-}{BF_4} + Ph_2S \longrightarrow Ph_3\overset{+}{S}\ \overset{-}{BF_4} + PhI \tag{51}$$

benzenediazonium tetrafluoroborate with dimethyl sulphide gave dimethylphenylsulphonium tetrafluoroborate (9%) and trimethylsulphonium tetrafluoroborate (38%). Reaction of the diazonium salt with methyl phenyl sulphide gave a low yield of the dimethylphenylsulphonium tetrafluoroborate[132].

Addition of sulphides (dialkyl, cycloalkyl, and aralkyl, but *not* diaryl) to protonated quinones or quinone imines gives arylsulphonium salts in high yields (equation 52)[133]

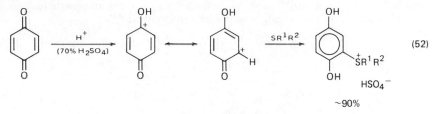

$$\sim 90\%$$

III. FORMATION FROM DI- AND TRISULPHIDES

Alkylation of disulphides using alkyl halides has been shown to yield sulphonium salts (equation 53)[134]. The reaction is slow at room temperature, but when

$$MeSSMe + MeI \longrightarrow Me_3\overset{+}{S}\ I^- \qquad (53)$$

catalysed by metal salts, e.g. mercury(II) iodide, can result in almost quantitative yields (equation 54)[135].

$$PhCH_2SSCH_2Ph + PhCH_2I \xrightarrow[\text{acetone, 25 °C}]{HgI_2} (PhCH_2)_3\overset{+}{S}\ HgI_3^- \qquad (54)$$

$$95\%$$

The first step in the alkylation of a disulphide involves the formation of a disulphide sulphonium (thiosulphonium) salt (equation 55)[136]. The stability of the

$$R^1SSR^1 + R^2I \longrightarrow R^1R^2\overset{+}{S}SR^1\ I^- \qquad (55)$$

$$R^1, R^2 = alkyl$$

alkylated disulphides depends on the choice of the anion; with trialkyloxonium 2,4,6-trinitrobenzene sulphonate[137] or tetrafluoroborate[138], the corresponding thiosulphonium salts have been isolated as stable crystalline compounds. When the counterion is iodide, as in equation 55, further reaction involves attack on the thiosulphonium cation (equation 56)[137].

$$R^1R^2\overset{+}{S}SR^1 + I^- \longrightarrow R^1R^2S + R^1SI$$

$$2R^1SI \longrightarrow R^1SSR^1 + I_2 \qquad\qquad (56)$$

$$R^1R^2S + R^2I \longrightarrow R^1R_2^2\overset{+}{S}\ I^-$$

Alternatively, the sulphide R^1R^2S could be alkylated be means of the thiosulphonium salt, as these salts have been shown to behave as alkylating agents[139], and thiosulphonium salts containing the *t*-butyl group will *t*-butylate dimethyl sulphide (equation 57)[140]. However, thiosulphonium salts can also be used to introduce the thiomethyl group (equation 58)[140].

$$t\text{-Bu}_2\overset{+}{S}SMe\ X^- + Me_2S \longrightarrow t\text{-BuSSMe} + t\text{-Bu}\overset{+}{S}Me_2\ X^- \qquad (57)$$

$$R^1R^2S + Me\overset{+}{S}Me_2\ X^- \longrightarrow R^1R^2\overset{+}{S}SMe\ X^- + Me_2S \qquad (58)$$

Other methods of preparation of thiosulphonium salts include the reaction of a sulphide with an alkylsulphenyl halide[139], on a preparative scale in the presence of $SbCl_5$[141] (equation 59), or with an arylsulphenyl halide in the presence of silver perchlorate (equation 60)[142] and by reaction between chlorodimethylsulphonium hexachloroantimonate and dialkyl sulphides (equation 61)[138].

$$R^1SCl + R^2_2S \xrightarrow[SbCl_5]{CH_2Cl_2} R^1S\overset{+}{S}R^2_2 \ SbCl_6^- \tag{59}$$

$$75-80\%$$

$$ArSCl + R_2S \xrightarrow[AgClO_4]{0\ °C,\ MeCN} Ar\overset{+}{S}SR_2 \ ClO_4^- \tag{60}$$

$$Me_2\overset{+}{S}Cl \ SbCl_6^- + RSSR \longrightarrow Me_2\overset{+}{S}SR \ SbCl_6^- + RSCl \tag{61}$$

It has been shown[81,143] that alkylthiosulphonium salts will add stereospecifically to alkenes and alkynes to give alkylsulphonium salts (equations 62 and 63) and in certain examples the intermediate thiiranium and thiirenium salts have been isolated[141].

$$\overset{E}{MeCH}=CHMe + Me_2\overset{+}{S}SMe \ \overset{-}{X} \xrightarrow{trans\ addition} \begin{array}{c} MeCH-CHMe \ X^- \\ | \qquad | \\ MeS \quad {}^+SMe_2 \end{array} \tag{62}$$

$$erythro$$

$$CH\equiv CH + Me_2\overset{+}{S}SMe \ X^- \longrightarrow Me_2\overset{+}{S}CH\overset{E}{=}CHSMe \tag{63}$$

The reaction with alkenes of a cyclic alkylthiosulphonium salt, obtained by alkylation of the disulphide with a trimethyloxonium salt, gives cyclic sulphonium salts (equations 64 and 65)[144].

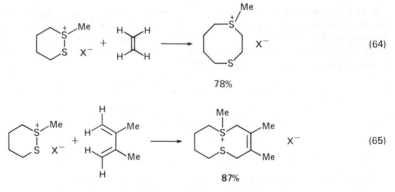

$$78\%$$

$$87\%$$

$$X^- = 2,4,6\text{-trinitrobenzenesulphonate}$$

Thiosulphonium salts derived from trisulphides have been obtained by alkylation using trialkyloxonium salts[145], by reaction between alkylsulphenyl halides and disulphides[141], and by treatment of disulphides with antimony pentachloride[146]. Methyl bis(methylthio)sulphonium hexachloroantimonate will react with alkenes to form stable S-methylthiiranium salts[147] and with alkynes to give S-methylthiirenium salts (equations 66 and 67)[148]. Other aspects of the chemistry of thiosulphonium salts are discussed in Section IV and in Chapter 15.

$$H_2C{=}CH_2 + (MeS)_2\overset{+}{S}Me\ SbCl_6^- \xrightarrow[0\,°C]{CH_2Cl_2} \underset{H}{\overset{H}{\diagdown}}\hspace{-0.5em}\triangle\hspace{-0.9em}\underset{H}{\overset{H}{\diagup}}\ \overset{+}{S}Me\ SbCl_6^- + MeSSMe \qquad (66)$$

85–90%

$$HC{\equiv}CH + (MeS)_2\overset{+}{S}Me\ SbCl_6^- \xrightarrow[0\,°C]{CH_2CH_2} \underset{H}{\overset{H}{\diagup\diagdown}}\hspace{-0.5em}\triangle\ \overset{+}{S}Me\ SbCl_6^- + MeSSMe \qquad (67)$$

100%

IV. FORMATION FROM SULPHENYL HALIDES

Reaction of sulphenyl halides and related compounds with alkenes or alkynes has received considerable attention; the intermediates in these reactions are generally considered to be thiranium (episulphonium) and thiirenium ions (equation 68)[78]

$$\underset{/}{\overset{\diagdown}{}}C{=}C\overset{\diagup}{\underset{\diagdown}{}} + RSCl \longrightarrow \left[\triangle\hspace{-0.9em}\overset{+}{S}{-}R \right] Cl^- \longrightarrow RS{-}\overset{|}{\underset{|}{C}}{-}\overset{|}{\underset{|}{C}}{-}Cl \qquad (68)$$

(see also Chapter 13). The first report of a stable thiiranium salt prepared in this way involved the reaction between cyclooctene and methylsulphenyl 2,4,6-trinitrobenzenesulphonate (equation 69)[81]. Other stable salts have

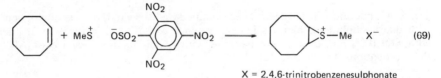

$$X = 2,4,6\text{-trinitrobenzenesulphonate}$$

subsequently been isolated using alkylsulphenyl halides in the presence of silver tetrafluoroborate or antimony pentafluoride[119].

Addition of the methylsulphenyl chloride–antimony pentachloride complex to but-2-yne at $-130\,°C$ (raised to $-80\,°C$) yielded an unstable (decomposition above $-30\,°C$) thiirenium salt. With di(t-butyl)ethyne in liquid sulphur dioxide at $-60\,°C$, however, it was possible to isolate the stable thiirenium salt as the tetrafluoroborate in 60% yield (equation 70)[149]

$$t\text{-BuC}{\equiv}\text{CBu-}t + MeCl + SbCl_5 \xrightarrow[-60\,°C]{liq.\ SO_2} \underset{t\text{-Bu}}{\overset{t\text{-Bu}}{\diagdown\diagup}}\hspace{-0.5em}\triangle\ \overset{+}{S}Me\ SbCl_6^- \qquad (70)$$

V. FORMATION FROM SULPHOXIDES

A. From Protonated Sulphoxides

The earliest report of the formation of sulphonium salts from sulphoxides was of the reaction between diaryl sulphoxides and phenol ethers in the presence of strong acids[150]. The reaction probably proceeds by protonation of the sulphoxide followed by attack of this electrophilic species on an activated benzene nucleus (equation 71). Subsequently, diaryl sulphoxides were reacted with phenols, phenol ethers,

$$Ar_2SO \xrightarrow{H_2SO_4} Ar_2\overset{+}{S}OH \; HSO_4^- \xrightarrow{PhOR} Ar_2\overset{+}{S}-\!\!\!\!\bigcirc\!\!\!\!-OR \quad HSO_4^- \qquad (71)$$

thiophenols, and thiophenol ethers (sulphides) under acidic conditions (H_2SO_4 or P_2O_5[151]. Hydroquinone reacts with alkyl aryl sulphoxides in the presence of $HClO_4$ or with dialkyl sulphoxides in a perchloric acid–phosphorus oxychloride mixture to give yields of better than 90%[152]. Aromatic hydrocarbons reacted with diaryl sulphoxides at 40°C using Lewis acids such as $AlCl_3$ as catalysts[153], although a later report suggested that better yields could be obtained at higher temperatures (equation 72)[154]. Under similar conditions, dibenzothiophen S-oxide reacted with 70% $HClO_4$–$POCl_3$ to give 5-(2-dibenzothienyl)dibenzothiophenium perchlorate (equation 73)[102].

$$Ar_2SO \; + \; Ar'H \xrightarrow[\text{xylene reflux}]{AlCl_3 \; or \; AlBr_3} Ar_2\overset{+}{S} \; + \; Ar'\overset{-}{X} \qquad (72)$$

$$31-61\%$$

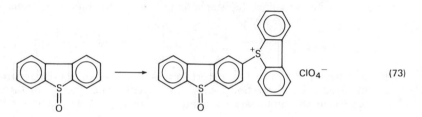

$$(73)$$

Triarylsulphonium salts have also been obtained (in 36% yield) on reaction of benzene with diaryl sulphide dichlorides in the presence of $AlCl_3$[155], in reaction of phenols with $SOCl_2$[156], and by reacting benzene with $SOCl_2$ in the presence of $AlCl_3$ or P_2O_5 (equations 74, 75, and 76)[153].

$$Ar_2SCl_2 \; + \; C_6H_6 \xrightarrow{AlCl_3} Ar_2\overset{+}{S}Ph \; Cl^- \qquad (74)$$

$$3PhOH \; + \; 2SOCl_2 \longrightarrow (HOC_6H_4)_3\overset{+}{S} \; Cl^- \qquad (75)$$

$$3ArH \; + \; SOCl_2 \xrightarrow{AlCl_3} Ar_3\overset{+}{S} \; Cl^- \qquad (76)$$

B. From Sulphoxides and Organometallics

Diaryl sulphoxides react with arylmagnesium bromides to give addition products which with hydrobromic acid give triarylsulphonium bromides in yields of 12–49% (equation 77)[157]:

$$Ar_2SO \; + \; ArMgBr \longrightarrow Ar_2S(OMgBr)Ar \xrightarrow{2HBr} Ar_3\overset{+}{S} \; Br^- \; + \; MgBr_2 \qquad (77)$$

Similar products were obtained by replacing the Grignard reagent with phenyllithium. Triarylsulphonium tetrafluoroborates were obtained either by treatment of the addition complex with HBF_4 or by reaction of the sulphonium bromides with $AgBF_4$[158].

A recent application of this method gave S-aryl derivatives of dibenzothiophen and of 9,9-dimethylthioxanthene (equations 78 and 79)[102].

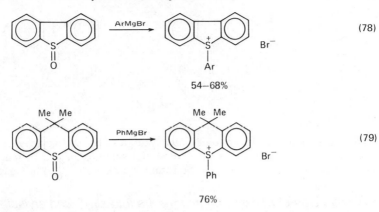

(78)

54–68%

(79)

76%

C. Alkylation of Sulphoxides and Conversion of the Products into Sulphonium Salts

1. Alkylation of sulphoxides

Reaction of dimethyl sulphoxide with methyl iodide gives trimethylsulphoxonium iodide (equation 80)[159,160]. With other alkyl halides, the product is the *O*-alkylated

$$Me_2SO + MeI \longrightarrow Me_3\overset{+}{S}O\ I^- \qquad (ref. 159) \qquad (80)$$

73%

salt, or reaction does not occur. The rate of isomerization of the *O*-methylated to the *S*-methylated salt was shown to be dependent on the nucleophilicity of the counter ion ($I^- > ONO_2^- > OTs^-$)[160]. However, addition of HgI_2 can lead to the formation of sulphoxonium complex halides in low yield, for example on prolonged heating (100 h) of a mixture of *n*-butyl ethyl sulphoxide and methyl iodide (equation 81)[161]. Reaction of methyl phenyl sulphoxide with methyl

$$n\text{-BuSOEt} \xrightarrow[\text{HgI}_2]{\text{MeI}} \left[\underset{\overset{|}{OMe}}{n\text{-BuSEt HgI}_3^-} \right] \longrightarrow n\text{-Bu}\overset{+}{S}O(Me)Et\ HgI_3^- + n\text{-Bu}\overset{+}{S}(Me)Et\ HgI_3^- + I_2 \tag{81}$$

iodide–mercury(II) iodide gave the dimethylphenylsulphoxonium salt in 49% yield[162].

O-Alkylated sulphoxides (alkoxysulphonium salts) may also be obtained by alkylation of sulphoxides with trialkyloxonium salts[163,164], alkoxycarbonium salts[25b], methyl fluorosulphonate[165] and methyl chlorosulphinate[166].

Other methods of formation of alkoxysulphonium salts include reactions of:
(a) sulphides with *t*-butyl or isopropyl hypochlorite (equation 82)[167,168]
(b) chlorosulphonium salts and alcohols[138];
(c) *S*-succinimido salts and alcohols[169];
(d) a sulphide–1-chlorobenzotriazole adduct and alcohols, followed by treatment with $AgBF_4$ (equation 83)[170].

$$R^1R^2S + R^3OCl \xrightarrow{-78\ ^\circ C} R^1R^2R^3\overset{+}{S}Cl \xrightarrow{R^3O^-} R^1R^2\overset{+}{S}OR^3\ Cl^- \tag{82}$$

(isolated as $\bar{S}bCl_6$ or $\bar{B}F_4$ salt)

$$R^1R^2\overset{+}{S}X + R^3OH \xrightarrow{\text{AgBF}_4} R^1R^2\overset{+}{S}OR^3 \ \bar{B}F_4 \qquad (83)$$

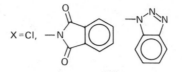

X=Cl,

Further details on the formation of alkoxysulphonium and acyloxysulphonium salts as intermediates in oxidations using dimethyl sulphoxide, and in reactions such as the Pummerer rearrangement, can be found in reviews[4,171] and in Chapter 15.

2. Conversion of alkoxysulphonium salts into alkyl- and arylsulphonium salts

Alkoxysulphonium salts are susceptible to nucleophilic attack by hydroxide[172,164] or alkoxide[173] ions, which causes inversion of configuration at sulphur. A more synthetically useful displacement reaction involves aryl(or alkyl)magnesium halides or alkylcadmium compounds (equation 84)[174]. This reaction leads to inversion of

$$Ar^1_2\overset{+}{S}OEt \ \bar{B}F_4 \xrightarrow{Ar^2MgX} Ar^1_2Ar^2\overset{+}{S} X^- \qquad (84)$$

configuration when optically active alkoxydiaryl- and alkoxyalkylarylsulphonium salts are used, but not with alkoxydialkylsulphonium salts[175]. One advantage of this method of formation over the direct method from a sulphoxide (see Section B) is that lower temperatures are used in the former reaction. The reaction with alkylcadmium compounds is completely stereospecific (equations 85 and 86)[176].

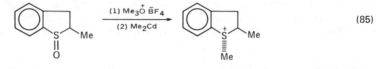

(85)

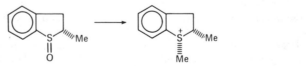

(86)

This method of formation of sulphonium salts has been used to prepare S-aryldibenzothiophens (equation 87)[177,102].

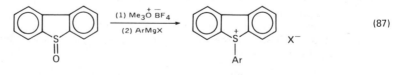

(87)

25–80%

VI. FORMATION OF SULPHUR YLIDES AND THEIR CONVERSION INTO SULPHONIUM SALTS

A. Formation of Sulphonium Ylides

The following methods can be used:

(i) Treatment of an alkylsulphonium salt with a strong base[3].

$$R^1R^2\overset{+}{S}CH_2R^3X^- \xrightarrow[-HX]{base} R^1R^2\overset{+}{S}-\bar{C}HR^3$$

R^1, R^2, R^3 = alkyl or aryl

(ii) Reaction of a sulphoxide with an 'active' methylene compound, generally in the presence of a dehydrating agent, to give a stabilized ylide (equation 88)[178–184].

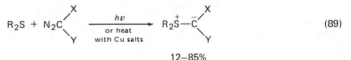

$$(88)$$

R = Me, Ph; X = COMe, CN; Y = COMe, CO_2Et, CN

Dehydrating agents used include Ac_2O, P_2O_5, $SOCl_2$, dicyclohexylcarbodiimide (DCC) and phenyl isocyanate.

(iii) Reaction of a sulphide (or sulphoxide) with a diazo compound bearing strongly electron-accepting conjugative groups, either photolytically[185] or thermally in the presence of copper catalysts (equations 89 and 90)[185,186].

$$R_2S + N_2C\overset{X}{\underset{Y}{\diagdown}} \xrightarrow[\text{with Cu salts}]{hv \text{ or heat}} R_2\overset{+}{S}-\bar{C}\overset{X}{\underset{Y}{\diagdown}}$$

$$(89)$$

12–85%

R = alkyl, aryl; X = Y = COMe, CO_2Me, CO_2Et; X = $COCMe_3$, Y = H

$$Me_2SO + N_2C\overset{X}{\underset{Y}{\diagdown}} \longrightarrow Me_2\overset{+}{S}O-\bar{C}\overset{X}{\underset{Y}{\diagdown}}$$

$$(90)$$

e.g. $Me_2SO + N_2CHCOCMe_3 \xrightarrow[40\ °C]{CuSO_4} Me_2\overset{+}{S}O\bar{C}HCOCMe_3$

49.5%

(iv) Reaction of dialkyl sulphides with iodonium ylides prepared from aryliodosoacetates, and active methylene compounds in the presence of copper(II) ions (equation 91)[187].

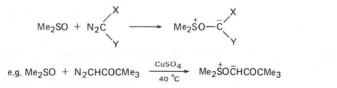

$$(91)$$

60–70%

R = Ph, 4-$NO_2C_6H_4$

X = NO_2, PhCO, CO_2Me

(v) Treatment of selenonium ylides with dialkyl sulphides in the presence of CS_2[188]. The reaction does not occur in the absence of the CS_2, suggesting the formation of an intermediate sulphonium salt (equation 92).

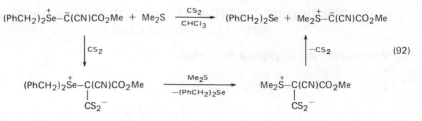

$$\text{(PhCH}_2)_2\overset{+}{\text{Se}}-\overset{-}{\text{C}}\text{(CN)CO}_2\text{Me} + \text{Me}_2\text{S} \xrightarrow[\text{CHCl}_3]{\text{CS}_2} \text{(PhCH}_2)_2\text{Se} + \text{Me}_2\overset{+}{\text{S}}-\overset{-}{\text{C}}\text{(CN)CO}_2\text{Me}$$

$$\text{(PhCH}_2)_2\overset{+}{\text{Se}}-\overset{\text{C(CN)CO}_2\text{Me}}{\underset{\text{CS}_2^-}{|}} \xrightarrow[-\text{(PhCH}_2)_2\text{Se}]{\text{Me}_2\text{S}} \text{Me}_2\overset{+}{\text{S}}-\overset{\text{C(CN)CO}_2\text{Me}}{\underset{\text{CS}_2^-}{|}}$$

(92)

B. Conversion of Sulphonium Ylides into Sulphonium Salts

Reaction of a sulphonium ylide with strong acid is sufficient to convert it into the sulphonium salt. Thus reaction of $\text{Me}_2\overset{+}{\text{S}}\overset{-}{\text{C}}(\text{NO}_2)_2$ with sulphuric acid gives the corresponding sulphonium sulphate, $\text{Me}_2\overset{+}{\text{S}}\text{CH}(\text{NO}_2)_2\ \text{HSO}_4^{-}$[189]. It is often not necessary to isolate the ylide. Reaction of a dialkyl sulphide (but not a diaryl sulphide) with an activated diazo compound in the presence of 71% perchloric acid gives the sulphonium perchlorates directly in 86–100% yield (equation 93)[46]. The

$$\text{R}_2\text{S} + \text{PhCOCHN}_2 \xrightarrow[\text{MeCN}]{\text{HClO}_4} \text{PhCOCH}_2\overset{+}{\text{S}}\text{R}_2\ \text{ClO}_4^{-}$$

(93)

reaction of dimethyl sulphide with benzyne (generated from *o*-bromochlorobenzene and butyllithium) followed by the addition of 20% perchloric acid gave a 77% yield of the sulphonium perchlorate (equation 94)[190]. That the ylide had been formed

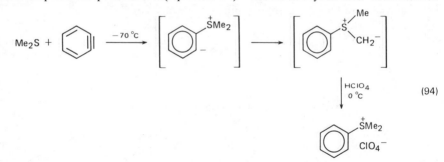

(94)

was shown by quenching the reaction with deutero-acids, the methyl-*d* salt being obtained rather than the phenyl-*d* salt.

Alkylation of a sulphonium ylide can serve as a useful method of preparation of a sulphonium salt that is difficult to obtain directly. Thus, isopropyldiphenylsulphonium salts are conveniently prepared by methylation of diphenyl sulphonium ethylide (equation 95)[76]. The iodide is rather unstable but it

$$\text{Ph}_2\overset{+}{\text{S}}-\overset{-}{\text{C}}\text{HMe} + \text{MeI} \longrightarrow \text{Ph}_2\overset{+}{\text{S}}\text{CHMe}_2\ \text{I}^{-}$$

(95)

can readily be converted into more stable complex salts. Other alkyl iodides have also been used in this reaction. This reaction is not always successful in giving the desired product, however. Early attempts to benzylate dimethylsulphonium methylide resulted only in the isolation of styrene, whilst a subsequent attempt at alkylation of diphenylsulphonium allylide led only to starting material being recovered[72]. Dimethylsulphonium dicyanomethylide reacts with benzyl chloride in the presence of base to give some benzylidene malononitrile, presumably via an intermediate sulphonium salt (equation 96)[191]. Alkylation of an ylide provides a very

$$Me_2\overset{+}{S}-\overset{-}{C}CN_2 \ + \ PhCH_2Cl \ \longrightarrow \ [Me_2\overset{+}{S}-C(CN)_2CH_2Ph\ Cl^-]$$

$$\downarrow \text{base} \qquad\qquad (96)$$

$$PhCH{=}C(CN)_2 \ + \ Me_2S$$

efficient method for the preparation of cyclopropyldiphenylsulphonium salts (and hence cyclopropylides). Whilst these may be obtained by reaction of triphenylsulphonium salts with cyclopropyllithium at $-70°C$, the yields were rather low, and the salts were not isolated[192]. It was shown that diphenyl sulphide will react with 1,3-dihaloalkanes in the presence of silver tetrafluoroborate to give 3-halogenoalkylsulphonium salts. With diiodo- or iodohaloalkanes, the yields were >80%. These salts were treated with base (NaH or t-BuOK) to give the cyclopropyldiphenylsulphonium salts in good to excellent yields (equation 97). It

$$Ph_2S \ + \ I(CH_2)_3X \ \xrightarrow{AgBF_4} \ Ph_2\overset{+}{S}(CH_2)_3X \ \overset{-}{B}F_4 \ \xrightarrow[\text{or } t\text{-BuOK/DMSO}]{NaH/THF} \ Ph_2\overset{+}{S}-\overset{CH_2}{\underset{CH_2}{\overset{|}{CH}}} \ \overset{-}{B}F_4 \qquad (97)$$

X = I,Br,Cl   81%   40–83%

was suggested that the substitution reaction probably proceeds by displacement by the diphenyl sulphide on a silver-complexed halide (cf. p. 000).

An alternative method of formation of cyclopropylsulphonium salts involves the reaction between an acylated sulphonium ylide and vinyldimethylsulphonium bromide, the following mechanism being suggested (equation 98)[193].

$$RCO\overset{-}{C}H\overset{+}{S}Me \ + \ CH_2{=}CH\overset{+}{S}Me_2 \ \ Br^- \ \longrightarrow \ \left[\begin{matrix} RCOCH\overset{+}{S}Me_2 \ Br^- \\ | \\ CH_2\overset{-}{C}H\overset{+}{S}Me_2 \end{matrix} \right]$$

$$\downarrow \qquad\qquad (98)$$

$$\underset{H_2C-CH_2}{\overset{RCO\overset{+}{C}SMe_2 \ Br^-}{\diagup\diagdown}} \ \xleftarrow{-Me_2S} \ \left[\begin{matrix} RCO\overset{-+}{C}SMe_2 \ Br^- \\ | \\ CH_2CH_2\overset{+}{S}Me_2 \end{matrix} \right]$$

Decomposition of sulphur ylides to give alkenes probably takes place via alkylation of the ylide, for example dimethylsulphonium methylide reacts with trimethylsulphonium halides to form ethylene (equation 99)[194], and a similar decomposition occurs with sulphoxonium ylides[195].

$$Me_2\overset{+}{S}Me \ X^- \ + \ Me_2\overset{+-}{S}CH_2 \ \longrightarrow \ Me_2\overset{+}{S}CH_2Me \ \overset{-}{X} \ + \ Me_2S \ \xrightarrow{base} \ Me_2S \ + \ CH_2{=}CH_2 \qquad (99)$$

Reaction between acylated sulphonium ylides and a 1,4-dipole precursor gave a 5-pyrimidylsulphonium salt (equation 100)[196].

Alkylation of a benzoylated sulphonium ylide with trimethyloxonium tetrafluoroborate gave a mixture of the isomeric methoxyvinylsulphonium salts (equation 101)[197].

In contrast, alkylation of a thiocarbonyl ylide gave the (Z)-methylthiovinylsulphonium salt only (equation 102)[198], and acylated dialkylaminosulphonium ylides gave mixtures of the vinylsulphonium salts that were incompletely separated (equation 103)[199].

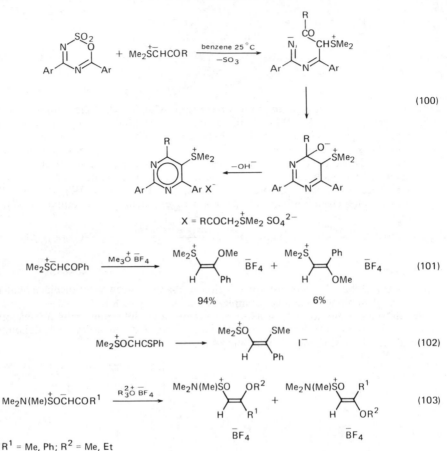

(100)

(101)

94%   6%

(102)

R^1 = Me, Ph; R^2 = Me, Et

(103)

Sulphonium ylides also react with other electrophiles, for example the nitronium ion[189], an arenesulphenyl chloride[189], or thiocyanate[200], giving the corresponding salts (equations 104, 105 and 106). Other synthetic applications of sulphonium ylides are described in Chapter 16.

(104)

(105)

(106)

VII. MODIFICATION OF SULPHONIUM SALTS

Certain sulphonium salts have been prepared by elimination, substitution, or rearrangement reactions on other sulphonium salts, and in some cases these may be at present the only ways in which these compounds can be obtained.

An early example was the preparation of tris(vinylsulphonium) picrylsulphonate, by prolonged reaction of tris(2-chloroethyl)sulphonium chloride with an aqueous sodium hydroxide–sodium carbonate solution at 25°C, followed by treatment with sodium picrylsulphonate. The intermediate bis(2-chloroethyl)vinylsulphonium salt was also isolated (equation 107)[201].

$$(HOCH_2CH_2)_3\overset{+}{S}\ Cl^- \xrightarrow{\ SOCl_2\ } (ClCH_2CH_2)_3\overset{+}{S}\ Cl^-$$

$$\downarrow$$ (107)

$$(CH_2{=}CH)_3\overset{+}{S}\ Cl^- \longleftarrow (ClCH_2CH_2)_2\overset{+}{S}CH{=}CH_2\ Cl^-$$

Conversion of a *cis*-vinylsulphonium salt into the *trans*-isomer has been referred to earlier (p. 273)[69]. Both *cis*- and *trans*-ethylene disulphonium salts react with nucleophiles, such as tertiary amines[202] and sodium methoxide[203], giving monosulphonium salts as the products (equations 108 and 109).

$$R^1R^2\overset{+}{S}CH{=}CH\overset{+}{S}R^1R^2\ 2\bar{B}F_4 + NR_3^3 \longrightarrow R^1R^2\overset{+}{S}CH{=}CH\overset{+}{N}R_3^3\ 2\bar{B}F_4 \qquad (108)$$

$$cis\ \text{and}\ trans$$

$$R^1R^2\overset{+}{S}CH{=}CH\overset{+}{S}R^1R^2\ 2\bar{B}F_4 + NaOMe \longrightarrow R^1R^2\overset{+}{S}CH{=}CHOMe\ \bar{B}F_4 \qquad (109)$$

$$E\ \text{and}\ Z \qquad\qquad\qquad E\ \text{and}\ Z$$
$$\text{stereospecific}$$

$$R^1 = Me;\ R^2 = Et$$

1,4-Bis(dialkylsulphonio)-2-butenes, prepared from Z-1,4-dihalides and dialkyl sulphides, react with an equimolar amount of sodium methoxide to give the Z-1-dialkylsulphonio-1,3-butadiene salts (equation 110)[204]. Subsequent reaction of

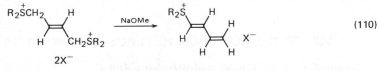

$$R = Me,\ Et$$

these salts with a catalytic amount of an alkoxide gave the isomeric sulphonium salts together with some dialkoxybutylsulphonium salts (equation 111)[205].

$$R_2^1\overset{+}{S}CH{=}CH{-}CH{=}CH_2\ X^- \xrightarrow[R^2OH]{R^2O^-\ (\text{catalytic amount})} R_2^1\overset{+}{S}CH_2CH{=}CHCH_2OR^2\ X^-$$

$$+$$

$$R_2^1\overset{+}{S}CH_2CH(OR^2)CH{=}CH_2\ X^- \qquad (111)$$

$$+$$

$$R^1 = Me,\ R^2 = Me,\ Et \qquad\qquad R_2^1\overset{+}{S}CH_2CH(OR^2)CH_2CH_2OR^2\ X^-$$

Propargylic sulphonium salts isomerize spontaneously in methanolic solution to the corresponding allenic sulphonium salts[59], the reaction being accelerated by triethylamine. Whilst the allenic salts were not isolated in a pure state, their presence was clearly indicated both spectroscopically and by their behaviour towards nucleophiles (equation 112). Allenic sulphonium salts have also been postulated as intermediates in fragmentation[60] and rearrangement reactions[206,207].

$$\overset{+}{Me_2S}CH_2C\equiv CH\ X^- \xrightarrow[Et_3N]{MeOH} \overset{+}{Me_2S}CH=C=CH_2\ X^- \xrightarrow{Nu} \overset{+}{Me_2S}CH_2\overset{|}{\underset{Nu}{C}}=CH_2\ X^- \qquad (112)$$

Nu = OMe, SPh, OPh, NHR^2, NR_2^2

In a manner similar to the preparation of butadienylsulphonium salts (equation 110), butatrienylsulphonium salts have been prepared by rearrangement of 1,4-bis(dialkylsulphonio)-2-butynes with sodium alkoxides at $-40°C$ (equation 113)[208]. These salts were not isolated but their presence was indicated

$$R_2^1SCH_2C\equiv \overset{+}{C}CH_2\overset{+}{S}R_2^1\ 2X^- \xrightarrow[-40\,°C]{NaOR} R_2^1\overset{+}{S}CH=C=C=CH_2\ X^- \qquad (113)$$

R^1 = Me, X = Cl; R^1 = Et, X = Br

spectroscopically and by reaction with methanol and thiophenol. Additionally, reaction with cyclopentadiene yielded a mixture of the stereoisomeric allenic sulphonium salts (equation 114)[209]. The formation of these products is in contrast

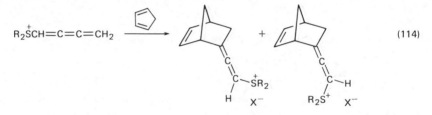

~50:50 (90% overall yield)

to an earlier report of a cycloaddition involving a butatrienyl system which occurred at the central double bond[210].

VIII. FORMATION FROM HETEROSULPHONIUM SALTS

A. Formation from Chlorosulphonium Salts

Chlorodimethylsulphonium salts may be prepared by the following methods[138]:

(1) By reaction between the dimethyl sulphide–antimony pentachloride complex and chlorine (equation 115).

$$Me_2S.SbCl_5 + Cl_2 \longrightarrow \overset{+}{Me_2S}Cl\ \overset{-}{S}bCl_6 \qquad (115)$$

92%

(2) From dimethyl sulphide, sulphuryl chloride and tetrafluoroboric acid (equation 116).

$$Me_2S + SO_2Cl_2 + HBF_4 \longrightarrow \overset{+}{Me_2S}Cl\ \overset{-}{B}F_4 \qquad (116)$$

96%

(3) From dimethyl sulphoxide, antimony pentachloride and thionyl chloride (equation 117).

$$Me_2SO + SbCl_5 + SOCl_2 \longrightarrow Me_2\overset{+}{S}Cl\ \overset{-}{S}bCl_6 \qquad (117)$$

$$91\%$$

These salts have been reacted with carbon nucleophiles to give the corresponding sulphonium salts (equation 118)[138,211]. Other aspects of the chemistry of halosulphonium salts are described in Chapter 15.

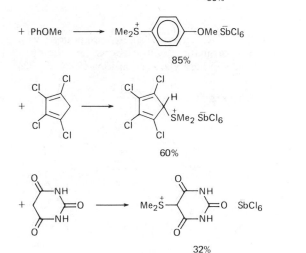

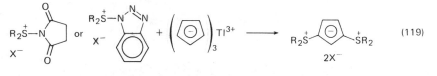

B. Formation from Azasulphonium Salts and Related Compounds

Azasulphonium salts, obtained by the reaction between sulphides and N-chlorosuccinimide[212] or 1-chlorobenzotriazole[170], react with cyclopentadienyl thallium at −20 to −70°C to give bissulphoniocyclopentadienides (equation 119)[213]. Succinimidosulphonium salts have also been reacted with indole and some

R = Me, piperidino, morpholino   20—55% isolated as ClO_4^- salt

of its derivatives to give indolyl-3-diálkylsulphonium salts in quantitative yield (equation 120)[214]. Similar indole derivatives had been obtained previously by alkylation of the 3-thio- and 3-alkylthio- compounds (equation 121)[215].

N-Tosylsulphilimines (N-tosylsulphimides) are obtained by reaction of sulphides

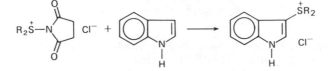

$$R = Me, Et, n\text{-propyl}, n\text{-butyl}$$

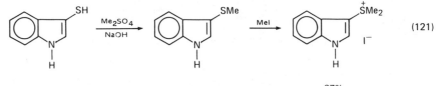

97%

with chloramine-T[6,216]. Reactions of these compounds with either arylmagnesium bromides or with aromatic hydrocarbons in the presence of AlCl$_3$ give triarylsulphonium salts in high yield (equations 122 and 123)[217].

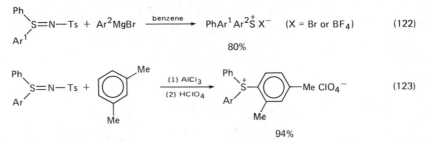

Further aspects of the chemistry of azasulphonium salts are described in Chapter 15.

IX. MISCELLANEOUS METHODS OF FORMATION

It has been noted (p. 283) that alkylation of disulphides using trialkyloxonium salts gives the monosulphonium salts only. A novel method for the preparation of bicyclic bis-sulphonium salts by means of an oxidative coupling reaction requires nitrosyl tetrafluoroborate or nitrosyl hexafluorophosphate as the oxidant (equations 123 and 124)[218]. A cycloaddition reaction between butadiene, 2-methylbutadiene,

$$\text{(123)}$$

$$\text{(124)}$$

or 2,3-dimethylbutadiene and 1,3-dithienium tetrafluoroborate results in the formation of bicyclic sulphonium salts in high yields (equation 125)[219]. Electrochemical methods of formation of sulphonium salts are described in Chapter 14.

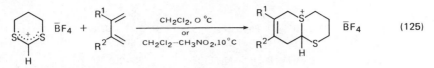

$$R^1 = R^2 = H, 85\%; R^1 = H_1, R_2 = Me, 95\%; R^1 = R^2 = Me, 96\%$$

X. METHYLENESULPHONIUM SALTS

The methylenesulphonium ion structure is considered to be a major resonance contributor to the structure of the α-thiocarbonium ion (11). This ion has been

$$R-\overset{+}{S}=CH_2 \longleftrightarrow R-S-\overset{+}{C}H_2$$

(11)

postulated as an intermediate in the Pummerer rearrangement of sulphoxides to α-substituted sulphides[220,221] (equation 126).

$$R^1SOCHR^2R^3 + HX \longrightarrow \underset{\overset{|}{OH} \ \ X^-}{R^1\overset{+}{S}CHR^2R^3} \overset{-HX}{\longrightarrow} \underset{\overset{|}{OH}}{R^1\overset{+-}{S}CR^2R^3} \longleftrightarrow \underset{\overset{|}{OH}}{R^1S=CR^2R^3}$$

$$\downarrow H^+ (-H_2O) \qquad (126)$$

$$R^1SCXR^2R^3 \overset{X^-}{\longleftarrow} R^1\overset{+}{S}CR^2R^3 \longleftrightarrow R^1\overset{+}{S}=CR^2R^3$$

Methylmethylenesulphonium hexachloroantimonate has been obtained in almost quantitative yield from the reaction between chloromethyl methyl sulphide and antimony pentachloride (equation 127)[138]. References to the reactions of this salt

$$ClCH_2SMe + SbCl_5 \longrightarrow CH_2=\overset{+}{S}Me \ \ \bar{S}bCl_6 \qquad (127)$$

$$97\%$$

and to the possible intermediacy of this and similar ions in reactions are to be found in the review by Marino[4].

Isothiouronium salts may be considered to be derivatives of methylenesulphonium salts, several canonical forms contributing to the hybrid structure 12[222]. They are usually prepared by reaction of thioureas with alkyl, cycloalkyl, aryl, or heterocyclic halides, or with diazonium salts[222–225].

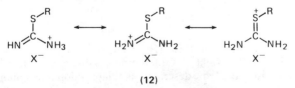

(12)

Other derivatives of methylenesulphonium salts that are becoming of increasing interest include the bis(alkylthio)methyl salts, obtained by reaction of trialkyl o-trithiocarboxylates with one equivalent of trityl tetrafluoroborate (equation 128)[226]. Cyclic analogues (13, 14) of these salts are known. Compound 14 was referred to above, whilst unsaturated derivatives of 13, the 1,3-dithiolium salts, are discussed in Section XIII.

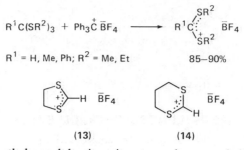

$$R^1C(SR^2)_3 + Ph_3\overset{+}{C}\ \bar{B}F_4 \longrightarrow R^1\overset{SR^2}{\underset{SR^2}{C}}\ \bar{B}F_4 \qquad (128)$$

$R^1 = H, Me, Ph; R^2 = Me, Et \qquad\qquad 85\text{–}90\%$

(13)   (14)

Unsaturated methylenesulphonium ions may be regarded as major canonical contributors to the structures of certain aromatic heterocycles, such as isothiophenium salts and thiopyrylium salts.

XI. FORMATION OF ISOTHIOPHENIUM SALTS

Photolysis of 1,2,3,4,5-pentamethylthiophenium fluorosulphonate at room temperature led to a shift of the 1-methyl group to an adjacent carbon atom, giving the 2,2,3,4,5-pentamethylisothiophenium salt (equation 129)[227]. The product,

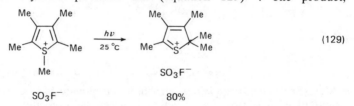

$$\qquad\qquad\qquad\qquad\qquad\qquad\qquad\qquad (129)$$

$SO_3F^- \qquad\qquad\qquad\qquad 80\%$

whose structure was determined by ^1H and ^{13}C n.m.r. spectroscopy, was stable to further photochemical transformation. No other identifiable products were obtained.

The naphtho[1,8-*b,c*]thiophen derivative **15** ($R^1 = Ph$) reacts with trityl perchlorate to give the naphthol[1,8-*b,c*]thienylium perchlorate **16** ($R^1 = Ph$, $R^2 = H$) (equation 130)[228]. Reaction of **15** ($R^1 = Ph$, Br) with triethyloxonium

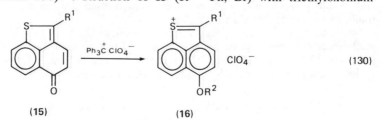

$$\qquad\qquad\qquad\qquad\qquad\qquad\qquad\qquad (130)$$

(15)   (16)

tetrafluoroborate gave the salts **16** ($R^1 = Ph$, $R^2 = Et$) in 71% yield and **16** ($R^1 = Br$, $R^2 = Et$) in 80% yield[229]. Other similar salts, e.g. **17**, have also been prepared. Compounds **16** and **17** are described as thiapseudophenalenium salts[230].

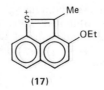

(17)

2-Thiaazulenylium salts, e.g. **19**, have been obtained by treatment of the hydroxy compound **18** with $HClO_4$–CF_3CO_2H or with HCl–$SbCl_5$ (equation 131)[231].

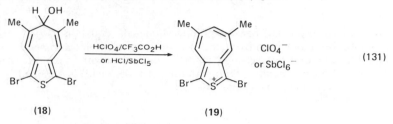

(131)

    (18)   (19)

XII. FORMATION OF THIOPYRYLIUM SALTS

The chemistry of thiopyrylium salts has been reviewed[232,233].

A. Formation from 1,5-Diketones

This, a general method for both monocyclic and ring-fused thiopyrylium salts, is particularly useful for the preparation of aryl-substituted salts. A disproportionation reaction is involved, with the fully reduced thiopyrans being obtained as by-products (equation 132)[234]. By using P_4S_{10} and acid, the yields of

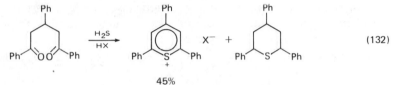

(132)

45%

aryl-substituted salts could be increased to *ca.* 80%[235]. Tetrahydro-1-benzo-thiopyrylium (e.g. **20**) and octahydrothioxanthylium (e.g. **21**) salts may be obtained under similar conditions.

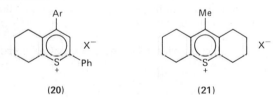

    (20)   (21)

B. Formation from Thiopyrans

Chlorine and iodine, but not bromine, will oxidize 4*H*-thiopyrans to thiopyrylium salts (equation 133)[236]. The thiopyrylium chloride may also be obtained from the

$$\text{thiopyran} \xrightarrow[\text{or } I_2/\text{acetone}/H_2O, \ 25\,^{\circ}C]{Cl_2/CHCl_3, \ -40\,^{\circ}C} \text{thiopyrylium}$$

Cl^- (100%)

I^- (50%)

(133)

same starting material on treatment with PCl_5 (>90% yield)[237]. 5-Hydroxy-5,6-dihydrothiopyran reacts with triphenylmethyl perchlorate by dehydration and hydride ion abstraction to give thiopyrylium perchlorate (equation 134), and treatment of 4-hydroxybenzo[*b*]thiopyran *S*-oxide with perchloric acid

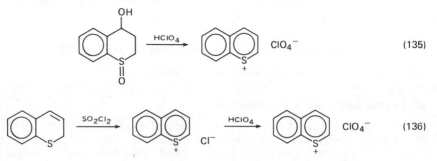

gives the corresponding benzo[b]thiopyrylium salt (equation 135)[238]. The latter salt has also been obtained from the reaction between thiochromene and sulphuryl chloride, followed by exchange of the counter ion with perchloric acid (equation 136)[239]. The corresponding thiopyrylium salt (60% yield) is obtained by reaction of the isothiochromene in a similar manner[239].

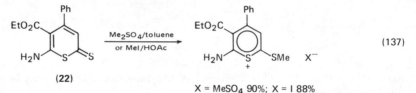

Methylation of 6-aminothiopyran-2-thiones (e.g. **22**) gives the 2-methylthiopyrylium salts in excellent yields (equation 137)[240]. Thiopyrones and

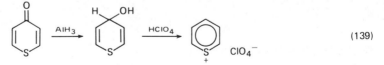

the analogous thiopyranthiones can be converted into thiopyrylium salts, (a) by reduction with aluminium hydrides to the corresponding thiopyranols or thiopyran-thiols, which are then reacted with perchloric or hydriodic acids (equation 139)[237], (b) by reaction with a Grignard reagent followed by dehydration

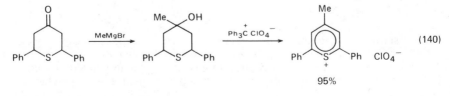

with perchloric acid or triphenylmethyl perchlorate (equation 140)[241], or (c) by reduction with zinc amalgam and hydrochloric acid followed by treatment with triphenylmethyl perchlorate (equation 141)[241]

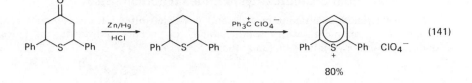

(141)

80%

C. Formation from Pyrylium Salts

Pyrylium perchlorates may be converted into thiopyrylium perchlorates by the successive action of sodium sulphide and perchloric acid (equation 142)[242]

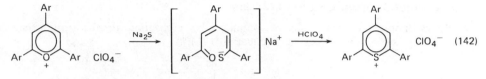

D. Formation from 1,3-Dienes and Thiophosgene

The cycloaddition reaction between 1,4-diphenylbutadiene and thiophosgene gives 2-chloro-3,6-diphenylthiopyrylium chloride in excellent yield (equation 143)[243]

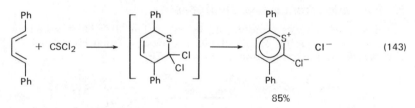

(143)

85%

E. Formation by Ring Expansion of Thiophens

Reaction of thiophen with ethyl diazoacetate followed by acidification gives the thiopyrylium salt (equation 144)[244].

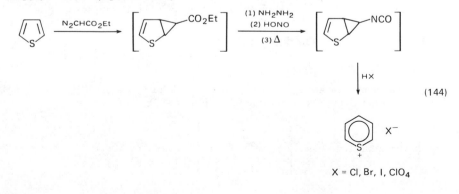

(144)

X = Cl, Br, I, ClO$_4$

F. Formation from 2-Chlorovinylmethineammonium Salts

2-Chlorovinylmethineammonium salts and NN-disubstituted thioacetamides react together on treatment with triethylamine in acetic anhydride to give 2-dialkylaminothiopyrylium salts (equation 145)[245].

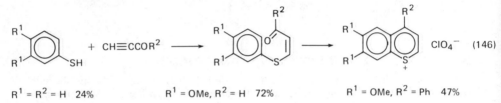

$$R^1 = Ar; NR_2^2 = morpholino, piperidino; R^3 = H, Ph \quad X = ClO_4 \quad 27-69\%$$

G. Formation from 2-Ketosulphides

Benzothiopyrylium salts can be obtained by cyclization of the appropriate 2-ketosulphide with perchloric acid (equation 146)[246].

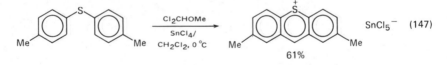

$R^1 = R^2 = H \quad 24\%$ $R^1 = OMe, R^2 = H \quad 72\%$ $R^1 = OMe, R^2 = Ph \quad 47\%$

H. Formation from Diaryl Sulphides

Thioxanthylium salts are obtained by the action of dichloromethyl methyl ether and tin(IV) chloride on diaryl sulphides so designed that electrophilic substitution occurs *ortho* to the bridge atom (equation 147)[247].

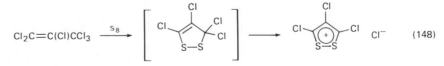

61%

XIII. FORMATION OF DITHIOLIUM SALTS

The chemistry of 1,2-[248,249] and 1,3-dithiolium salts[248,250] has been reviewed.

A. Formation of 1,2-Dithiolium Salts

1. From open-chain compounds

(a) Hexachloropropene reacts with sulphur at 180°C to give 3,4,5-trichloro-1,2-dithiolium chloride (equation 148)[251].

$$Cl_2C = C(Cl)CCl_3 \xrightarrow{S_8} \left[\text{(structure)} \right] \longrightarrow \text{(structure)} \quad Cl^- \quad (148)$$

(b) 1,3-Diketones react with H_2S_2–HCl[252], with P_4S_{10}[253], or with diacetyl disulphide[254] to give good yields of the substituted 1,2-dithiolium salts (equation 149). A 1,2,3-triketone reacts with an H_2S–HCl mixture to give the

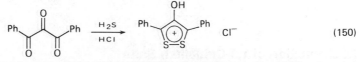

R^1, R^2 = alkyl, aryl

4-hydroxydithiolium salt (equation 150)[255]. Good yields of the salts may often be obtained by the addition of an oxidant such as iodine[256,257], bromine[257], or iron(III) ion[258] to an ethanolic solution of the diketone saturated with H_2S–HCl gases.

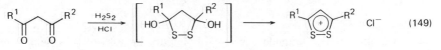

(c) The reaction between malondithiodiamides and iodine, or iron(III) salts and hydrogen peroxide, leads to the formation of 1,2-dithiolium salts (equation 151)[259].

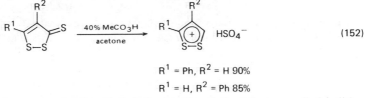

However, as physical chemical studies indicate that the positive charge probably resides on a nitrogen atom, these compounds are more correctly formulated as imminium salts.

2. From 1,2-dithioles

Although the hydrogen abstraction of a 1,2-dithiole to give the 1,2-dithiolium salt appears to be an obvious method of preparation, the method is limited by the instability of the starting materials. A more successful method starts from the 1,2-dithiole-3-ones or the analogous 3-thiones. Treatment of 1,2-dithiole-3-thiones with 40% peracetic acid in acetone[260] or with 30% hydrogen peroxide in glacial acetic acid[261] gives good yields of the 1,2-dithiolium salts (equation 152).

$$ \text{(152)} $$

R^1 = Ph, R^2 = H 90%

R^1 = H, R^2 = Ph 85%

1,2-Dithiole-3-thiones can also be alkylated with dimethyl sulphate or alkyl halides to the corresponding 3-alkylthio-1,2-dithiolium salts[262]. These reagents will not alkylate the 1,2-dithiole-3-ones but alkylation has been accomplished using trialkyloxonium tetrafluoroborates[262].

3. By ring contraction

1,3-Dithiacyclohexenes are converted into 1,2-dithiolium salts by oxidizing agents such as bromine or sulphuryl chloride (equation 153)[264]. Using bromine in benzene,

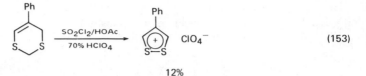

$$(153)$$

12%

the 1,2-dithiolium bromide was obtained in 46–50% yield whilst 2-(4-methoxyphenyl)-5,6-benzodithien was converted into benzo-1,2-dithiolium bromide in 55–60% yield (equation 154)[264].

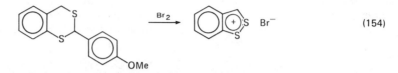

$$(154)$$

B. Formation of 1,3-Dithiolium Salts

1. From open-chain compounds

(a) 1,3-Dithiolium salts were first obtained from the acid-catalysed condensation of benzene-1,2-dithiols with aldehydes, followed by oxidation (equation 155)[265].

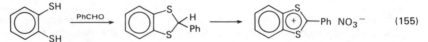

$$(155)$$

Other oxidants include bromine and H_2O_2–quinone. Using carboxylic acids in the presence of phosphorus oxychloride, benzo-1,3-dithiolium salts are obtained in 65–90% yields[265,266]. 2-Amino-1,3-dithiolium salts are obtained in excellent yields from benzene-1,2-dithiols and cyanogen chloride in acidified ethanol (equation 156)[267].

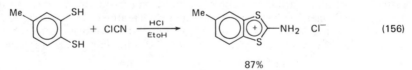

$$(156)$$

87%

(b) Compounds having the general formula **23** are cyclized on treatment with acids (equation 157)[261]. It is also possible to obtain these salts simply by reaction of

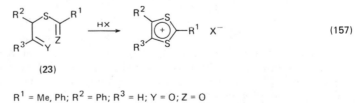

$$(157)$$

(23)

R^1 = Me, Ph; R^2 = Ph; R^3 = H; Y = O; Z = O

R^1 = Me, Ph; R^2 = H, Ph; R^3 = Me, Ph; Y = O; Z = S

an α-substituted ketone with excess of a thioacid in the presence of a strong acid (equation 158)[268]. In a similar manner, α-carboxybenzyl dithiobenzoates are cyclized to 1,3-dithiolium-4-olates with acetic anhydride–triethylamine (equation

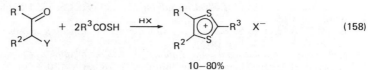

$$(158)$$

10–80%

R^1 = Me, Ph; R^2 = Me, Et, Ph, Ar; R^3 = H, Me, Ph; Y = halogen; X = I, ClO_4

159)[269]. Addition of bromine to *S*-vinyl-*NN*-dialkyldithiocarbamates and subsequent pyrolysis gives 1,3-dithiolium salts in excellent yields (equation 160)[270].

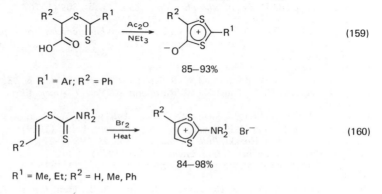

$$(159)$$

85–93%

R^1 = Ar; R^2 = Ph

$$(160)$$

84–98%

R^1 = Me, Et; R^2 = H, Me, Ph

2. From 1,3-dithioles

(a) Hydride-ion abstraction from 1,3-benzodithioles on treatment with trityl tetrafluoroborate gives the 1,3-dithiolium salts in good yield (equation 161)[271]. The

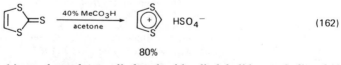

$$(161)$$

75%

same product (86–94% yield) is obtained on treatment of 2-alkoxy-, 2-alkylthio-, or 2-arylthio-1,3-benzothioles with hydrofluoroboric acid in acetic anhydride[271].

(b) Treatment of 1,3-dithiole-2-thiones with 40% peracetic acid in acetone[272], or with 30% hydrogen peroxide in glacial acetic acid[261], gives the 1,3-dithiolium salts in excellent yields (equation 162).

$$(162)$$

80%

(c) 1,3-Dithiole-2-thiones have been alkylated with alkyl halides and dimethyl sulphate[266] to give the 2-alkylthio-1,3-dithiolium salts in high yields, but the corresponding 2-alkoxy compounds have not so far been reported.

XIV. REFERENCES

1. J. Goerdeler, in *Methoden der Organischen Chemie (Houben-Weyl)*, Vol. IX, Georg Thieme Verlag, Stuttgart, 1955, pp. 175–187.

2. E. E. Reid, *Chemistry of Bivalent Sulphur*, Vol. I–VI, Chemical Publishing Co., New York, 1958–65.
3. B. M. Trost and L. S. Melvin, Jr., *Sulphur Ylides, Emerging Synthetic Intermediates*, Academic Press, London, 1975.
4. J. P. Marino, *Sulphur-containing Cations*, in *Topics in Sulphur Chemistry*, Vol 1 (Ed. A. Senning), Georg Thieme Verlag, Stuttgart, 1976, pp. 1–110.
5. C. J. M. Stirling, in *Organic Chemistry of Sulphur* (Ed. S. Oae), Plenum Press, New York, 1977, pp. 473–525.
6. T. L. Gilchrist and C. J. Moody, *Chem. Rev.*, **77**, 409 (1977).
7. C. R. Johnson, *Accounts Chem. Res.*, **6**, 341 (1973).
8. P. D. Kennewell and J. B. Taylor, *Chem. Soc. Rev.*, **4**, 189 (1975).
9. S. L. Huang and D. Swern, *Phosph. Sulph. Relat. Elem.*, 309 (1976).
10. E. Fluck and W. Haubold, *Chem. Ztg.*, **101**, 523 (1977).
11. M. Haacke, *The Chemistry of S,S-Diorgano-Sulphodiimides* in *Topics in Sulphur Chemistry*, Vol. 3 (Ed. A. Senning), Georg Thieme Verlag, Stuttgart, 1976, pp. 185–215.
12. M. J. Hatch, *J. Org. Chem.*, **34**, 2133 (1969).
13. C. R. Hauser, S. W. Kantor and W. R. Brasen, *J. Amer. Chem. Soc.*, **75**, 2660 (1953).
14. A. J. Speziale, C. C. Tung, K. W. Ratts and A. Yao, *J. Amer. Chem. Soc.*, **87**, 3460 (1965).
15. G. B. Rayne, *J. Org. Chem.*, **32**, 3351 (1967); D. A. Rutolo, P. G. Truskier, J. Casanova, Jr., and G. B. Payne, *Org. Prep. Proced.*, **1**, 111 (1969).
16. Y. Hayashi and H. Nozaki, *Bull. Chem. Soc. Jap.*, **45**, 198 (1972).
17. F. E. Ray and I. Levine, *J. Org. Chem.*, **2**, 267 (1937).
18. G. B. Butler and B. M. Benjamin, *J. Amer. Chem. Soc.*, **74**, 1846 (1952).
19. Ref. 3, p. 6.
20. H. Meerwein, S. Hinz, P. Hofmann, E. Kroning and E. Pfeil, *J. prakt. Chem.*, **147**, 257 (1937); H. Meerwein, *Org. Syn.*, Coll. Vol. V, 1080, 1096 (1973); G. K. Helmkamp and D. J. Pettitt, *Org. Syn.*, Coll. Vol. V, 1099 (1973). For a general coverage of these reagents, see H. Perst, *Oxonium Ions in Organic Chemistry*, Verlag Chemie, Academic Press, Weinheim, 1971.
21. F. E. Ray and J. L. Farmer, *J. Org. Chem.*, **8**, 391 (1943).
22. C. G. Swain and E. R. Thornton, *J. Amer. Chem. Soc.*, **83**, 4033 (1961).
23. M. G. Ahmed, R. W. Alder, G. H. James, M. L. Sinnott and M. C. Whiting, *J. Chem. Soc. Chem. Commun.*, 1533 (1968).
24. A. J. Boulton, A. C. G. Gray and A. R. Katritzky, *J. Chem. Soc. B*, 911 (1967); E. Vedejs, D. A. Engler and M. J. Mullins, *J. Org. Chem.*, **42**, 3109 (1977).
25. (a) H. Meerwein, K. Bodenbenner, P. Borner, F. Kunert and K. Wunderlich, *Justus Liebigs Ann. Chem.*, **632**, 38 (1960); (b) S. Kabuss, *Angew. Chem. Int. Ed. Engl.*, **5**, 675 (1966); (c) R. F. Borch, *J. Org. Chem.*, **34**, 627 (1969).
26. G. A. Olah and J. R. DeMember, *J. Amer. Chem. Soc.*, **92**, 2562 (1970).
27. S. Smiles, *J. Chem. Soc.*, **77**, 160 (1900).
28. T. Mukaiyama, K. Hagio, H. Takei and K. Saigo, *Bull. Chem. Soc. Jap.*, **44**, 161 (1971).
29. T. Mukaiyama, T. Adachi and T. Kumamoto, *Bull. Chem. Soc. Jap.*, **44**, 3155 (1971).
30. B. M. Trost and H. C. Arndt, *J. Org. Chem.*, **38**, 3140 (1973).
31. W. H. Saunders, Jr. and S. Asperger, *J. Amer. Chem. Soc.*, **79**, 1612 (1957).
32. B. M. Trost and R. F. Hammen, *J. Amer. Chem. Soc.*, **95**, 962 (1973).
33. R. Scartazzini and K. Mislow, *Tetrahedron Lett.*, 2719 (1967).
34. Ref. 2, Vol. II, p. 70.
35. S. Kato, H. Ishihara, M. Mizuta and Y. Hirabayashi, *Bull. Chem. Soc. Jap.*, **44**, 2469 (1971).
36. C. P. Lillya and P. Miller, *J. Amer. Chem. Soc.*, **88**, 1559 (1966).
37. S. Wolfe, P. Chamberlain and T. F. Garrard, *Can. J. Chem.*, **54**, 2847 (1976).
38. H. Boehme and W. Krack, *Justus Liebigs Ann. Chem.*, 1291 (1977).
39. A. Cascaval and I. Zugravescu, *Bul. Inst. Politeh. Iasi*, **23**, 75 (1977); *Chem. Abstr.*, **88**, 152, 168 (1978); see also Ref. 1, p. 178.
40. A. Cascaval and I. Zugravescu, *Bul. Inst. Politeh. Iasi*, **22**, 75 (1976); *Chem. Abstr.*, **86**, 170,790 (1977).

41. T. W. Milligan and B. C. Minor, *J. Org. Chem.,* **28**, 235 (1963).
42. O. Haas and G. Dougherty, *J. Amer. Chem. Soc.,* **65**, 1238 (1943).
43. S. W. Lee and G. Dougherty, *J. Org. Chem.,* **4**, 48 (1939).
44. D. P. Harnish and D. S. Tarbell, *J. Amer. Chem. Soc.,* **70**, 4123 (1948).
45. E. A. Krulikovskaya, G. L. Ryzhiva and A. G. Dranaeva, *Izv. Vyssh. Uchebn. Zaved. Khim. Khim. Tekhnol.,* **19**, 1699 (1976); *Chem. Abstr.,* **86**, 106,088 (1977).
46. W. T. Flowers, G. Holt and M. A. Hope, *J. Chem. Soc., Perkin Trans. I,* 1116 (1974).
47. A. Schoberl and G. Lange, *Angew. Chem.,* **64**, 224 (1952).
48. N. F. Blau and C. G. Stuckwisch, *J. Org. Chem.,* **27**, 370 (1962).
49. P. A. Capps and A. R. Jones, *J. Chem. Soc. Chem. Commun.,* 320 (1974).
50. B. Bannister, *J. Chem. Soc., Perkin Trans. I,* 274 (1978).
51. Ref. 2, Vol. II, p. 69.
52. T. Durst, R. Viau, R. Van Den Elzen and C. H. Nguyen, *J. Chem. Soc. D,* 1334 (1971).
53. C. R. Johnson, C. W. Schroeck and J. R. Shanklin, *J. Amer. Chem. Soc.,* **95**, 7424 (1973).
54. J. M. Townsend and K. B. Sharpless, *Tetrahedron Lett.,* 3313 (1972).
55. M. Yoshimine and M. J. Hatch, *J. Amer. Chem. Soc.,* **89**, 5831 (1967).
56. W. H. Stein, S. Moore and M. Bergmann, *J. Org. Chem.,* **11**, 664 (1946); see also Ref. 2, Vol. II, pp. 69 and 261.
57. J. S. H. Davies and A. E. Oxford, *J. Chem. Soc.,* 224 (1931).
58. H. Boehme, H. Fischer and R. Frank, *Justus Liebigs Ann. Chem.,* **563**, 54 (1949).
59. G. D. Appleyard and C. J. M. Stirling, *J. Chem. Soc. C,* 1904 (1969).
60. A. Terada and Y. Kishida, *Chem. Pharm. Bull., Tokyo,* **17**, 966 (1969).
61. G. Toennies and J. J. Kolb, *J. Amer. Chem. Soc.,* **67**, 1141 (1945); W. Kirmse and M. Kapps, *Chem. Ber.,* **101**, 1004 (1968).
62. I. Iwai and J. Ide, *Chem. Pharm. Bull., Tokyo,* **12**, 1094 (1964).
63. G. M. Blackburn, W. D. Ollis, J. D. Plackett, C. Smith and I. O. Sutherland, *J. Chem. Soc. Chem. Commun.,* 186 (1968).
64. J. E. Baldwin, R. E. Hackler and D. P. Kelly, *J. Chem. Soc. Chem. Commun.,* 537 (1968); J. E. Baldwin, R. E. Hackler and D. P. Kelly, *J. Amer. Chem.Soc.,* **90**, 4758 (1968).
65. B. M. Trost and W. G. Biddlecom, *J. Org. Chem.,* **38**, 3438 (1973).
66. W. von E. Doering and K. C. Schreiber, *J. Amer. Chem. Soc.,* **77**, 514 (1955).
67. J. Gosselck, L. Béress, H. Schenk and G. Schmidt, *Angew. Chem. Int. Ed. Engl.,* **4**, 1080 (1965).
68. K. Takaki and T. Agawa, *J. Org. Chem.,* **42**, 3303 (1977).
69. H. Braun and A. Amann, *Angew. Chem. Int. Ed. Engl.,* **14**, 755 (1975).
70. V. Franzen, H. J. Schmidt and C. Mertz, *Chem. Ber.,* **94**, 2942 (1961).
71. T. Hashimoto, K. Ohkubo, H. Kitano and K. Fukui, *Nippon Kagaku Zasshi,* **87**, 456 (1966); *Chem. Abstr.,* **65**, 15,259 (1966).
72. J. van der Veen, *Rec. Trav. Chim. Pays Bas,* **84**, 540 (1965).
73. R. W. LaRochelle, B. M. Trost and L. Krepski, *J. Org. Chem.,* **36**, 1126 (1971).
74. Ref. 3, p. 9.
75. C. S. F. Tang and H. Rapoport, *J. Org. Chem.,* **38**, 2806 (1973).
76. E. J. Corey, M. Jautelat and W. Oppolzer, *Tetrahedron Lett.,* 2325 (1967).
77. K. D. Gundermann, *Angew. Chem. Int. Ed. Engl.,* **2**, 674 (1963).
78. W. H. Mueller, *Angew. Chem. Int. Ed. Engl.,* **8**, 482 (1969).
79. G. H. Schmid, *Thiiranium and Thiirenium Salts,* in *Topics in Sulphur Chemistry,* Vol. 3 (Ed. A. Senning), Georg Thieme Verlag, Stuttgart, 1977, p. 101.
80. D. J. Pettitt and G. K. Helmkamp, *J. Org. Chem.,* **28**, 2932 (1963).
81. D. J. Pettitt and G. K. Helmkamp, *J. Org. Chem.,* **29**, 2702 (1964).
82. P. Raynolds, S. Zonnebelt, S. Bakker and R. M. Kellogg, *J. Amer. Chem. Soc.,* **96**, 3146 (1974).
83. B. M. Trost, W. L. Schinski, F. Chen and I. B. Mantz, *J. Amer. Chem. Soc.,* **93**, 676 (1971).
84. M. Umehara, K. Kanai, H. Kitano and F. Fukui, *Nippon Kagaku Zasshi,* **83**, 1060 (1962); *Chem. Abstr.,* **59**, 11,398 (1963).
85. M. Hetschko and J. Gosselck, *Chem. Ber.,* **106**, 996 (1973).

86. Y. Yano, T. Matayoshi, K. Misu and W. Tagaki, *Phosph. Sulph. Relat. Elem.*, **1**, 25 (1976).
87. R. Schmid and H. Schmid, *Helv. Chim. Acta*, **60**, 1361 (1977).
88. J. W. Batty, P. D. Howes and C. J. M. Stirling, *J. Chem. Soc., Perkin Trans. I*, 59 (1973).
89. M. Yashimoto, N. Ishida and Y. Kishida, *Chem. Pharm. Bull., Tokyo*, **19**, 863 (1971).
90. E. Vedejs and J. P. Hagen, *J. Amer. Chem. Soc.*, **97**, 6878 (1975).
91. E. Vedejs, M. J. Mullins, J. M. Renga and S. P. Singer, *Tetrahedron Lett.*, 519 (1978).
92. H. Bosshard, *Helv. Chim. Acta*, **55**, 37 (1972).
93. H. Boehme and W. Krack, *Justus Liebigs Ann. Chem.*, **758**, 143 (1972).
94. I. Stahl, M. Hetschko and J. Gosselk, *Tetrahedron Lett.*, 4077 (1971).
95. T. Oishi, K. Kamemoto and Y. Ban, *Tetrahedron Lett.*, 1085 (1972).
96. B. M. Trost and S. D. Ziman, *J. Amer. Chem. Soc.*, **93**, 3825 (1971).
97. J. Bornstein, J. E. Shields and J. H. Supple, *J. Org. Chem.*, **32**, 1499 (1967); J. von Braun, *Chem. Ber.*, **58**, 2165 (1925).
98. G. C. Brumlik, A. I. Kosak and R. Pitcher, *J. Amer. Chem. Soc.*, **86**, 5360 (1964).
99. R. M. Acheson and D. R. Harrison, *J. Chem. Soc. D*, 724 (1969); R. M. Acheson and D. R. Harrison, *J. Chem. Soc. C*, 1764 (1970).
100. R. F. Heldeweg and H. Hogeveen, *Tetrahedron Lett.*, 75 (1974).
101. R. M. Acheson and J. K. Stubbs, *J. Chem. Soc., Perkin Trans. I*, 899 (1972).
102. M. Hori, T. Kataoka, H. Shimizu and M. Miyagaki, *Yagugaku Zasshi*, **93**, 476 (1973); *Chem. Abstr.*, **79**, 31,798 (1973).
103. M. J. Cook, H. Dorn and A. R. Katritzky, *J. Chem. Soc. B*, 1467 (1968).
104. P. J. Halfpenny, P. J. Johnson, M. J. T. Robinson and M. G. Ward, *Tetrahedron*, **32**, 1873 (1976).
105. E. L. Eliel, R. L. Willer, A. T. McPhail and K. D. Onan, *J. Amer. Chem. Soc.*, **96**, 3021 (1974); E. L. Eliel and R. L. Willer, *J. Amer. Chem. Soc.*, **99**, 1936 (1977).
106. R. F. Naylor, *J. Chem. Soc.*, 2749 (1949).
107. A. G. Hortmann, R. L. Harris and J. A. Miles, *J. Amer. Chem. Soc.*, **96**, 6119 (1974).
108. E. Hunt and B. Lythgoe, *J. Chem. Soc. Chem. Commun.*, 757 (1972).
109. I. Stahl and J. Gosselck, *Tetrahedron*, **29**, 2323 (1973).
110. Ref. 2, Vol. III, p. 68.
111. T. E. Young and R. A. Lazarus, *J. Org. Chem.*, **33**, 3770 (1968).
112. W. Schroth, M. Hassfeld and A. Zschunke, *Z. Chem.*, **10**, 296 (1970).
113. R. H. Mitchell and V. Boekelheide, *Tetrahedron Lett.*, 1197 (1970); V. Boekelheide and R. A. Hollins, *J. Amer. Chem. Soc.*, **92**, 3512 (1970). See also Ref. 3 for further examples.
114. R. Danieli, A. Ricci and J. H. Ridd, *J. Chem. Soc., Perkin Trans II*, 290 (1976).
115. P. M. Denerley and E. J. Thomas, *Tetrahedron Lett.*, 71 (1977).
116. D. K. Herron, *Tetrahedron Lett.*, 2145 (1975).
117. R. Bird and C. J. M. Stirling, *J. Chem. Soc., Perkin Trans. II*, 1221 (1973).
118. G. H. Schmid and V. M. Csizmadia, *Can. J. Chem.*, **50**, 2465 (1972).
119. W. A. Smit, M. Z. Krimer and E. A. Vorob'eva, *Tetrahedron Lett.*, 2451 (1975); E. A. Vorob'eva, M. Z. Krimer and W. A. Smit, *Izv. Akad. Nauk S.S.S.R. Ser. Khim.*, 125 (1975); *Chem. Abstr.*, **85**, 177,148 (1976).
120. O. V. Kil'disheva, M. G. Linkova, L. P. Rasteikiene, V. A. Zabelaite, N. K. Pociute and I. L. Knunyants, *Proc. Acad. Sci. U.S.S.R., Chem. Sect.*, **203**, 1072 (1972); *Chem. Abstr.*, **77**, 88,178 (1972).
121. R. S. Devdhar, V. N. Gogtè and B. D. Tilak, *Tetrahedron Lett.*, 3911 (1974).
122. P. Wilder, Jr., and L. A. Feliu-Otero, *J. Org. Chem.*, **30**, 2560 (1965); P. Wilder, Jr., and L. A. Feliu-Otero, *J. Org. Chem.*, **31**, 4264 (1966).
123. J. A. Cotruvo and I. Degani, *J. Chem. Soc. Chem. Commun.*, 436 (1971).
124. V. Prelog and E. Cerkovnikov, *Justus Liebigs Ann. Chem.*, **537**, 214 (1939).
125. V. Prelog and D. Kohlbach, *Chem. Ber.*, **72**, 672 (1939).
126. E. Deutsch, *J. Org. Chem.*, **37**, 3481 (1972).
127. R. H. Eastman and G. Kritchevsky, *J. Org. Chem.*, **24**, 1428 (1959).
128. C. G. Overberger, P. Barkan, A. Lusi and H. Ringsdorf, *J. Amer. Chem. Soc.*, **84**, 2814 (1962).

129. N. J. Leonard, T. W. Milligan and T. L. Brown, *J. Amer. Chem. Soc.*, **82**, 4075 (1960).
130. N. J. Leonard and C. R. Johnson, *J. Amer. Chem. Soc.*, **84**, 3701 (1962).
131. L. G. Makarova and A. N. Nesmeyanov, *Izv. Akad. Nauk S.S.S.R.*, 617 (1945); *Chem. Abstr.*, **40**, 4686 (1946).
132. M. Kobayashi, H. Minato, J. Fukui and N. Kamigata, *Bull. Chem. Soc., Jap.*, **48**, 729 (1975).
133. H. Bosshard, *Helv. Chim. Acta*, **55**, 32 (1972).
134. Ref. 2, Vol. II, p. 69, and references cited therein.
135. O. Haas and G. Dougherty, *J. Amer. Chem. Soc.*, **62**, 1004 (1940).
136. T. P. Hilditch and S. Smiles, *J. Chem. Soc.*, **91**, 1394 (1907).
137. G. K. Helmkamp, H. N. Cassey, B. A. Olsen and D. J. Pettitt, *J. Org. Chem.*, **30**, 933 (1965).
138. H. Meerwein, K-F. Zenner and R. Gipp, *Justus Liebigs Ann. Chem.*, **688**, 67 (1965).
139. J. K. Kim and M. C. Caserio, *J. Amer. Chem. Soc.*, **96**, 1930 (1974).
140. H. Minato, T. Miura, F. Takagi and M. Kobayashi, *Chem. Lett.*, 211 (1975).
141. G. Capozzi, O. DeLucchi, V. Lucchini and G. Modena, *Synthesis*, 677 (1976).
142. H. Minato, T. Miura and M. Kobayashi, *Chem. Lett.*, 1055 (1975).
143. G. K. Helmkamp, B. A. Olsen and J. R. Koskinen, *J. Org. Chem.*, **30**, 1623 (1965).
144. N. E. Hester, G. K. Helmkamp and G. L. Alford, *Int. J. Sulphur Chem.*, **1**, 65 (1971).
145. P. Dubs and R. Stuessi, Paper presented at the VIIth International Symposium on Organo Sulphur Chemistry, Hamburg, July 12–16th, 1976.
146. R. Weiss and C. Schlierf, *Synthesis*, 323 (1976); R. Weiss, C. Schlierf and K. Schloter, *J. Amer. Chem. Soc.*, **98**, 4668 (1976).
147. G. Capozzi, O. Delucchi, V. Lucchini and G. Modena, *Tetrahedron Lett.*, 2603 (1975).
148. G. Capozzi, O. Delucchi, V. Lucchini and G. Modena, *J. Chem. Soc. Chem. Commun.*, 248 (1975).
149. G. Capozzi, V. Lucchini, G. Modena and P. Scrimin, *Tetrahedron Lett.*, 911 (1977).
150. S. Smiles and R. LeRossignol, *J. Chem. Soc.*, **89**, 696 (1906).
151. F. Kehrmann, S. Lievermann and P. Frumkine, *Chem. Ber.*, **51**, 474 (1918); A. Luttringhaus and K. Hauschild, *Chem. Ber.*, **72**, 890 (1939); F. Krollpfeiffer and W. Hahn, *Chem. Ber.*, **86**, 1049 (1953).
152. S. Ukai and K. Hirose, *Yakugaku Zasshi*, **86**, 187, (1966); *Chem. Abstr.*, **64**, 19,466 (1966).
153. C. Courtot and T. Y. Tung, *C.R. Acad. Sci., Ser. C*, **197**, 1227 (1933).
154. G. H. Wiegand and W. E. McEwen, *J. Org. Chem.*, **33**, 2671 (1968).
155. G. Dougherty and P. D. Hammond, *J. Amer. Chem. Soc.*, **61**, 80 (1939).
156. D. Libermann, *C.R. Acad. Sci., Ser. C*, **197**, 921 (1933).
157. B. S. Wildi, S. W. Taylor and H. A. Potratz, *J. Amer. Chem. Soc.*, **73**, 1965 (1951).
158. K. Fukui, K. Kanai and H. Kitano, *Nippon Kagaku Zasshi*, **82**, 178 (1961); *Chem. Abstr.*, **57**, 9631 (1961).
159. R. Kuhn and H. Trischmann, *Justus Liebigs Ann. Chem.*, **611**, 117 (1958).
160. S. G. Smith and S. Winstein, *Tetrahedron*, **3**, 317 (1958).
161. M. Kobayashi, K. Kamiyama, H. Minato, Y. Oishi, Y. Takada and Y. Hattori, *Bull. Chem. Soc. Jap.*, **45**, 3703 (1972).
162. K. Royke, H. Minato and M. Kobayshi, *Bull. Chem. Soc. Jap.*, **49**, 1455 (1976).
163. H. Meerwein, E. Battenberg, H. Gold, E. Pfeil and G. Willfang, *J. prakt. Chem.*, **154**, 83 (1939).
164. C. R. Johnson and D. McCants, *J. Amer. Chem. Soc.*, **87**, 5404 (1965).
165. H. D. Durst, J. W. Zubrick and G. R. Kieczykowski, *Tetrahedron Lett.*, 1777 (1974).
166. Y. Hara and M. Matsuda, *J. Chem. Soc. Chem. Commun.*, 919 (1974).
167. C. R. Johnson and M. P. Jones, *J. Org. Chem.*, **32**, 2014 (1967).
168. K. Torssell, *Tetrahedron Lett.*, 4445 (1966); K. Torssell, *Acta Chem. Scand.*, **21**, 1 (1967).
169. E. J. Corey and C. U. Kim, *J. Amer. Chem. Soc.*, **94**, 7586 (1972).
170. C. R. Johnson, C. C. Bacon and W. D. Kingsbury, *Tetrahedron Lett.*, 501 (1972).
171. D. Martin and H. D. Hauthal, *Dimethyl Sulphoxide*, Akademie-Verlag, Berlin, 1971 (English translation by E. S. Halberstadt, Van Nostrand Reinhold, New York, 1975).
172. C. R. Johnson, *J. Amer. Chem. Soc.*, **85**, 1020 (1963).

310   P. A. Lowe

173. C. R. Johnson and W. G. Phillips, *J. Org. Chem.*, **32**, 1926 (1967).
174. K. K. Anderson and N. E. Papanikolaou, *Tetrahedron Lett.*, 5445 (1966).
175. K. K. Andersen, *J. Chem. Soc. Chem. Commun.*, 1051 (1971); K. K. Andersen, R. L. Caret and I. Karup-Nielsen, *J. Amer. Chem. Soc.*, **96**, 8026 (1974); K. K. Andersen, R. L. Caret and D. L. Ladd, *J. Org. Chem.*, **41**, 3096 (1976).
176. K. K. Andersen, R. L. Caret and I. Karup-Nielsen, in *Organic Sulphur Chemistry* (Ed. C. J. M. Stirling), Butterworths, London, 1975, p. 333.
177. R. W. LaRochelle and B. M. Trost, *J. Amer. Chem. Soc.*, **93**, 6077 (1971).
178. A. Hochrainer and F. Wessely, *Monatsh. Chem.*, **97**, 1 (1966); A. Hochrainer, *Monatsh. Chem.*, **97**, 823 (1966).
179. H. Nozaki, Z. Morita and K. Kondo, *Tetrahedron Lett.*, 2913 (1966); H. Nozaki, D. Tunemoto, Z. Morita, K. Nakamura, K. Watanabe, M. Takaku and K. Kondon, *Tetrahedron*, **23**, 4279 (1967).
180. W. J. Middleton, E. L. Buhle, J. G. McNally, Jr., and M. Zanger, *J. Org. Chem.*, **30**, 2384 (1965).
181. R. Gompper and H. Euchner, *Chem. Ber.*, **99**, 527 (1966).
182. R. Oda and Y. Hayashi, *Nippon Kagaku Zasshi*, **87**, 1110 (1966); *Chem. Abstr.*, **66**, 85,416 (1967).
183. A. F. Cook and J. G. Moffatt, *J. Amer. Chem. Soc.*, **90**, 740 (1968).
184. D. Martin and H.-J. Niclas, *Chem. Ber.*, **102**, 31 (1969).
185. W. Ando, T. Yagihara, S. Tozune, S. Nakaido and T. Migata, *Tetrahedron Lett.*, 1979 (1969); W. Ando, T. Yagihara, S. Tozune and T. Migata, *J. Amer. Chem. Soc.*, **91**, 2786 (1969).
186. J. Quintana, M. Torres and F. Serratosa, *Tetrahedron*, **29**, 2065 (1973).
187. V. V. Semenov, S. A. Shevelev and A. A. Fainzil'berg, *Izv. Akad. Nauk S.S.S.R. Ser. Khim.*, 2640 (1976); *Chem. Abstr.*, **86**, 139,517 (1977).
188. S. Tamagaki, K. Tamura and S. Kozuka, *Chem. Lett.*, 725 (1977).
189. S. A. Shevelev, V. V. Semenov and A. A. Fainzil'berg, *Izv. Akad. Nauk S.S.S.R. Ser. Khim.*, 2621 (1975); *Chem. Abstr.*, **84**, 58,540 (1976).
190. V. Franzen, H.-I. Joschek and C. Mertz, *Justus Liebigs Ann. Chem.*, **654**, 82 (1962).
191. K. Friedrich and J. Rieser, *Justus Liebigs Ann. Chem.*, 648 (1976).
192. B. M. Trost and M. J. Bogdanowicz, *J. Amer. Chem. Soc.*, **95**, 5298 (1973).
193. G. Schmidt and J. Gosselck, *Tetrahedron Lett.*, 3445 (1969).
194. V. Franzen and H. E. Driesen, *Chem. Ber.*, **96**, 1881 (1963).
195. E. J. Corey and M. Chaykovsky, *J. Amer. Chem. Soc.*, **87**, 1353 (1965).
196. S. Ueda, Y. Hayashi and R. Oda, *Tetrahedron Lett.*, 4967 (1969).
197. S. H. Smallcombe, R. J. Holland, R. H. Fish and M. C. Caserio, *Tetrahedron Lett.*, 5987 (1968).
198. H. Yoshida, T. Yao, T. Ogata and S. Inokawa, *Bull. Chem. Soc. Jap.*, **49**, 3128 (1976).
199. C. R. Johnson and P. E. Rogers, *J. Org. Chem.*, **38**, 1798 (1973).
200. H. Matsuyama, H. Minato and M. Kobayashi, *Bull. Chem. Soc. Jap.*, **51**, 575 (1978).
201. M. A. Stahmann, J. S. Fruton and M. Bergmann, *J. Org. Chem.*, **11**, 704 (1946).
202. H. Braun, A. Amann and M. Richter, *Angew. Chem. Int. Ed. Engl.*, **16**, 471 (1977).
203. H. Braun and A. Amann, *Angew. Chem. Int. Ed. Engl.*, **14**, 756 (1975).
204. H. Braun, N. Mayer and G. Kresze, *Justus Liebigs Ann. Chem.*, **762**, 111 (1972).
205. H. Braun, N. Mayer, G. Strobl and G. Kresze, *Justus Liebigs Ann. Chem.*, 1317 (1973).
206. L. Veniard and G. Pourcelet, *C.R. Acad. Sci., Ser. C*, **273**, 1190 (1971).
207. A. Terada and Y. Kishida, *Chem. Pharm. Bull., Tokyo*, **18**, 490 (1970).
208. H. Braun, G. Strobl and H. Gotzler, *Angew. Chem. Int. Ed. Engl.*, **13**, 469 (1974).
209. H. Braun and G. Strobl, *Angew. Chem. Int. Ed. Engl.*, **13**, 470 (1974).
210. W. Reid and R. Neidhardt, *Justus Liebigs Ann. Chem.*, **739**, 155 (1970).
211. R. Neidlein and B. Stackebrandt, *Justus Liebigs Ann. Chem.*, **914** (1977).
212. E. Vilsmaier and W. Sprugel, *Justus Liebigs Ann. Chem.*, **747**, 151 (1971).
213. K. H. Schlingensief and K. Hartke, *Tetrahedron Lett.*, 1269 (1977).
214. K. Tomita, A. Tereda and R. Tachikawa, *Heterocycles*, **4**, 729 (1976).
215. G. Doyle Daves, Jr., W. R. Anderson, Jr., and M. V. Pickering, *J. Chem. Soc. Chem. Commun.*, 301 (1974).

216. F. G. Mann and W. J. Pope, *J. Chem. Soc.*, **121**, 1052 (1922).
217. P. Manya, A. Sekera and P. Rumpf, *Bull. Soc. Chim. Fr.*, 286 (1971).
218. W. K. Musker and P. B. Roush, *J. Amer. Chem. Soc.*, **98**, 6745 (1976).
219. E. J. Corey and S. W. Walinsky, *J. Amer. Chem. Soc.*, **94**, 8932 (1972).
220. R. Pummerer, *Chem. Ber.*, **43**, 1401 (1910).
221. Ref. 4, p. 21.
222. Ref. 4, p. 16.
223. Ref. 2, Vol, V.
224. M. Bogemann, S. Peterson, O.-E. Schultz and H. Söll, in *Methoden der Organischen Chemie (Houben-Weyl)*, Vol. IX, Georg Thieme Verlag, Stuttgart, 1955, p. 900.
225. D. C. Schroeder, *Chem. Rev.*, **55**, 181 (1955).
226. S. W. Walinsky, *PhD Thesis*, Pennsylvania State University, 1971: *Diss. Abstr.*, 6314 (1972).
227. H. Hogeveen, R. M. Kellogg and K. A. Kuindersma, *Tetrahedron Lett.*, 3929 (1973).
228. D. G. Hawthorne and Q. N. Porter, *Aust. J. Chem.*, **19**, 1909 (1966).
229. R. Neidlein and H. Seel, *Angew. Chem. Int. Ed. Engl.*, **15**, 775 (1976).
230. R. Neidlein, K. F. Cepera and M. H. Salzl, *Chem. Ztg.*, **101**, 558 (1977).
231. M. Hori, T. Kataoka, H. Shimizu and S. Yoshimura, *Yakugaku Zasshi*, **94**, 1429 (1974); *Chem. Abstr.*, **82**, 170,549 (1975).
232. Ref. 4, p. 86.
233. V. G. Kharchenko, S. N. Chalaya and T. M. Konovalova, *Khim. Geterotsikl. Soedin.*, 147 (1975); English Translation, 125 (1976).
234. V. G. Kharchenko, N. M. Kupranets, V. I. Kleimenova, N. V. Polikarpov and A. R. Yakoreva, *J. Org. Chem. U.S.S.R.*, **4**, 1984 (1968).
235. V. G. Kharchenko and V. I. Kleimenova, *J. Org. Chem. U.S.S.R.*, **7**, 618 (1971).
236. E. Molenaar and J. Strating, *Tetrahedron Lett.*, 2941 (1965).
237. I. Degani, R. Fochi and C. Vincenzi, *Tetrahedron Lett.*, 1167 (1963).
238. A. Luttringhaus and N. Engelhard, *Angew. Chem.*, **73**, 218 (1961).
239. A. Luttringhaus and N. Engelhard, *Chem. Ber.*, **93**, 1525 (1960).
240. K. Gewald, M. Buchwalder and M. Peukert, *J. prakt. Chem.*, **315**, 679 (1973).
241. R. Wizinger and H. J. Angliker, *Helv. Chim. Acta*, **49**, 2046 (1966).
242. R. Wizinger and P. Ulrich, *Helv. Chim. Acta*, **39**, 207 (1956).
243. G. Laban and R. Mayer, *Z. Chem.*, **7**, 227 (1967).
244. R. Pettit, *Tetrahedron Lett.*, No, 23, 11 (1960).
245. H. Hartmann, *J. prakt. Chem.*, **313**, 1113 (1971).
246. N. Engelhard and A. Kolb, *Justus Liebigs Ann. Chem.*, **673**, 136 (1964).
247. J. Ashby, M. Ayad and O. Meth-Cohn, *J. Chem. Soc. Perkin Trans I*, 1104 (1973).
248. H. Prinzbach and E. Futterer, *Adv. Heterocycl. Chem.*, **7**, 39 (1966).
249. N. Lozach and J. Vialle, in *The Chemistry of Organic Sulphur Compounds*, Vol. II (Ed. N. Kharasch and C. Y. Meyers), Pergamon Press, Oxford, 1966, p. 257.
250. E. Campaigne and R. D. Hamilton, *Q. Rep. Sulphur Chem.*, **5**, 275 (1970).
251. F. Boberg, *Angew. Chem.*, **72**, 629 (1960); *Justus Liebigs Ann. Chem.*, **679**, 109 (1964).
252. D. Leaver and W. A. H. Robertson, *Proc. Chem. Soc.*, 252 (1960); M. Schmidt and H. Schulz, *Chem. Ber.*, **101**, 277 (1968).
253. H. Behringer and A. Grimm, *Justus Liebigs Ann. Chem.*, **682**, 188 (1965).
254. H. Hartmann, K. Fabian, B. Bartho and J. Faust, *J. prakt. Chem.*, **312**, 1197 (1970).
255. K. Inouye, S. Sato and M. Ohta, *Bull. Chem. Soc. Jap.*, **43**, 1911 (1970).
256. J.-P. Gúemas and H. Quiniou, *C.R. Acad. Sci., Ser. C*, **268**, 1805 (1969).
257. A. R. Hendrickson and R. L. Martin, *J. Org. Chem.*, **38**, 2548 (1973).
258. G. A. Heath, R. L. Martin and I. M. Stewart, *J. Chem. Soc. Chem. Commun.*, 54 (1969).
259. U. Schmidt, *Chem. Ber.*, **92**, 1171 (1959); K. A. Jensen, H. R. Baccaro and O. Buchardt, *Acta Chem. Scand.*, **17**, 163 (1963).
260. E. Klingsberg, *J. Amer. Chem. Soc.*, **83**, 2934 (1961).
261. D. Leaver, W. A. H. Robertson and D. M. McKinnon, *J. Chem. Soc.*, 5104 (1962).
262. B. Bottcher and A. Lüttringhaus, *Justus Liebigs Ann. Chem.*, **557**, 89 (1947); B. Bottcher and F. Bauer, *Chem. Ztg.*, **77**, 135 (1953).

263. J. Faust and J. Fabian, Z. Naturforsch., **24B**, 577 (1969).
264. A Lüttringhaus, M. Mohr and N. Engelhard, Justus Liebigs Ann. Chem., **661**, 84 (1963).
265. W. R. H. Hurtley and S. Smiles, J. Chem. Soc., 1821 (1926).
266. L. Soder and R. Wizinger, Helv. Chim. Acta, **42**, 1733 (1959).
267. R. W. Addor, J. Org. Chem., **29**, 738 (1964).
268. K. Fabian and H. Hartmann, J. prakt. Chem., **313**, 722 (1971).
269. H. Gotthardt and B. Christl, Tetrahedron Lett., 4743 (1968).
270. K. Hiratani, H. Shiono and M. Okawara, Chem. Lett., 867 (1973).
271. J. Nakayama, K. Fujiwara and M. Hoshino, Chem. Lett., 1099 (1975).
272. E. Klingsberg, J. Amer. Chem. Soc., **84**, 3410 (1962); **86**, 5290 (1964).

The Chemistry of the Sulphonium Group
Edited by C. J. M. Stirling and S. Patai
© 1981 John Wiley & Sons Ltd

CHAPTER **12**

Reactivity of sulphonium salts

A. C. KNIPE

*School of Physical Sciences, The New University of Ulster, Coleraine, Co.
Londonderry, N. Ireland BT42 1SA*

I.	INTRODUCTION	314
II.	SUBSTITUTION REACTIONS OF SULPHONIUM SALTS . . .	314
	A. Unimolecular Nucleophilic Displacement of Sulphonium Groups . .	314
	1. Racemization	317
	B. Bimolecular Nucleophilic Displacement of Sulphonium Groups . .	319
	1. Stereochemistry and orientation	319
	2. Solvent dependence	321
	3. Effects of ion pairing	322
	4. Methyl transfer	323
	5. Reversion	324
	6. Miscellaneous nucleophilic displacement reactions . . .	324
	C. Substitution of Aromatic Sulphonium Compounds	324
	1. Nucleophilic aromatic substitution	324
	2. Electrophilic aromatic substitution	325
	D. Nucleophilic Reactions at the Sulphonium Sulphur Atom . .	325
	E. Sulphonium Ions as Intermediates in Substitutions at Sulphur . .	332
III.	ELIMINATION REACTIONS OF SULPHONIUM SALTS . .	334
	A. Unimolecular Elimination of Sulphonium Groups . . .	334
	B. Bimolecular Elimination of Sulphonium Groups . . .	336
	1. Substituent kinetic effects on E2 reactions . . .	336
	a. The 2-phenylethyl system	336
	b. The 1-phenylethyl system	339
	c. Other systems	340
	2. Stereochemistry and orientation of E2 reactions . . .	341
	a. Hofmann *versus* Saytzeff orientation . . .	342
	C. E1cb Reactions of Sulphonium Salts	344
	1. Activation by the sulphonium group	344
	2. Elimination of the sulphonium group	345
	D. The α′,β-Elimination Mechanism	346
	E. α-Elimination Reactions	347
IV.	POLAR ADDITION TO UNSATURATED SULPHONIUM SALTS . .	349

V.   SULPHONIUM YLIDES   355
     A. Formation of Sulphonium Ylides   355
        1. From sulphonium salts   355
           a. Modification of sulphonium ylides by acylation and alkylation   .   .   356
        2. Ylides formed by reaction of carbenes with sulphides   .   .   357
     B. Stability and Structure of Sulphonium Ylides   358
     C. Reactions of Sulphonium Ylides   359
        1. Epoxidation by alkylidene transfer   359
           a. Stereochemistry and mechanism   359
           b. Scope of the epoxidation reaction   363
           c. Epoxidation *versus* cyclopropanation of α,β-unsaturated carbonyl
              compounds   363
        2. Cyclopropanation by alkylidene transfer   366
           a. Stereochemistry and mechanism   366
           b. Scope of the cyclopropanation reaction   368
        3. Rearrangement of sulphur ylides   369
           a. [2,3]-Sigmatropic rearrangement   369
           b. Rearrangement by 1,2-migration   372
        4. Miscellaneous reactions of sulphonium ylides   373
VI. REFERENCES   374

I. INTRODUCTION

Sulphonium salts are known to undergo many novel and synthetically useful reactions but there have been relatively few attempts to review the chemical reactivity of this group. Limited coverage has been given to the biochemistry of sulphonium salts[1], to their addition[2] and elimination[2] reactions, and to the mechanisms of their reactions in general[3,4]; reviews of nucleophilic substitution at tricoordinate sulphur[5] and of the extensive chemistry of the wide range of sulphur containing cations[6] have naturally covered many aspects of the reactivity of sulphonium salts.

In contrast, there has been considerable interest in sulphonium ylides [which are often obtained directly from sulphonium salts by deprotonation at $C_{(\alpha)}$], and the chemistry of these species will be discussed later in this chapter.

In keeping with Marino's classification[6] of sulphur-containing cations, this review will deal only with the reactivity of tricoordinated species but will not include the chemistry of heterosulphonium salts (see Chapter 15) or of cyclic sulphonium salts (see Chapter 13), including thiophenium and thiopyrylium ions.

II. SUBSTITUTION REACTIONS OF SULPHONIUM SALTS

A. Unimolecular Nucleophilic Displacement of Sulphonium Groups

The dependence of rates of solvolysis of *t*-butyl sulphonium salts on solvent[7–12], co-anion[8,10,11,13,14], and pressure[15] have been studied in detail and it is now generally accepted that, in media of high dielectric constant, the mechanism of solvolysis of *t*-butyldimethylsulphonium salts is of the S_N1 type[8,14,16–18]; the considerable sulphur isotope effect[18] is in keeping with a large distortion of the C—S bond in the S_N1 transition state.

Although *t*-butyl chloride is hydrolysed by water *ca.* 40% faster than by D_2O, the rate of hydrolysis of *t*-butyldimethylsulphonium ion is unaffected by this solvent change; this can be attributed to the greater demand of chloride ion than of dimethyl sulphide for electrophilic solvation and is therefore consistent with a heterolytic mechanism for each type of t-butyl substrate[7]. Correspondingly, the co-anion is not involved in the rate-determining step and t-butyldimethylsulphonium perchlorate decomposes unimolecularly at almost the same rate[8] as the analogous sulphonium chloride upon solvolysis in water containing 0–90% of acetone. However, Hyne and coworkers have shown that for solvolysis of t-butyldimethylsulphonium halides in ethanol–water mixtures the dependence of the rate on anion character and on salt concentration increases as the dielectric constant of the medium is lowered[10,11]; this effect, which becomes apparent in the order $I^- > Br^- > Cl^-$ (when the molar fraction of ethanol exceeds 0.55, Table 1), has been attributed to increased participation of ion pairs in a rate-determining process for which k_{ip} must exceed k_+ (equation 1).

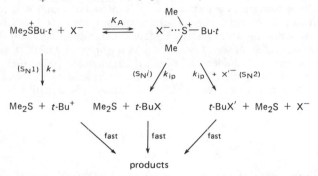

(1)

At high anion concentrations the reaction rate does not reach a limiting value but becomes linearly dependent upon the anion concentration[10]; although this has been attributed to onset of a competing mechanism, whereby the anion attacks the ion pair (k_{ip}'), an alternative explanation in terms of a normal salt effect has not been ruled out. A wide range of anion types (X^-, Cl^-, NO_3^-, ClO_4^-, and AcO^-) has been studied[13] in order to determine the dependence on anion nucleophilicity; the values k_{ip}' and k_{ip} have been evaluated from the expression $k_{obs} = k_{ip} \pm k_{ip}' [X^-]$, which is obtained from the general expression

$$k_{obs} = (k_i + k_{ip}K_a [X^-] + k_{ip}'K_A [X^-]^2)/(1 + K_A [X^-])$$

where $K_A[X^-] \gg 1$ and k_+. While k_{ip}' does generally depend on the nucleophilicity of X^-, the acetate anion is anomalous and more nucleophilic than expected; this is also reflected in an unusually high k_{ip} value for acetate and has been attributed to

TABLE 1. Rates of solvolysis[11] of t-butyldimethyl-sulphonium salts (0.001 M) in ethanol–water mixtures at 78.4°C

Molar fraction EtOH	$10^4 k$ (s^{-1})		
	I^-	Br^-	Cl^-
0.000	3.55	3.52	3.54
0.552	5.81	5.77	5.76
0.855	7.73	6.68	6.08

enhancement of the number of favourable configurations for attack by this 'ambident' nucleophile. In contrast, almost identical values of k_{ip} are obtained for NO_3^- and ClO_4^-, which are presumably too weak to assist sulphonium ion decomposition by nucleophilic attack within the ion pair.

Darwish and Tourigny[14] have also explored the effect of the counter ion (X^-) on the product distribution and rate of solvolysis of t-butyl(and t-amyl)ethylmethylsulphonium salts in ethanol, acetic acid, and $AcOH–Ac_2O$ mixtures[14]; a higher yield of alkene is obtained when $X = Br$ than when $X = ClO_4$, and the results have been interpreted by an ion-pair mechanism.

Attempts have been made to correlate rates of solvolysis of t-butyldimethylsulphonium halides in binary mixtures of water with alcohols[8,12], N,N-dimethylformamide, N-methylformamide, dioxan, and ethylene glycol[12]; no straightforward solvent parameter can qualitatively equate the increase in rate which generally accompanies a decrease in solvent polarity and it is apparent that specific microscopic solvent–solute interactions are important in the activation process[12]. Thus, while the Grunwald–Winstein relationship, $\log(k/k^0) = my$, can be used to correlate rates of reaction in ethanol ($m = -0.062$), it proves to be unsatisfactory for solvent amides, for which a larger nucleophilic role is evident; it has been suggested that quaternized ammonium intermediates may well be involved (equation 2). Kosower Z values, which are themselves sensitive to microscopic

$$R_2NC{\overset{O}{\underset{H}{\diagdown}}} + t\text{-Bu}\overset{+}{S}Me_2 \xrightarrow[-Me_2S]{} R_2\overset{+}{N}{\overset{\overset{O}{\|}{C}\diagup H}{\underset{t\text{-Bu}}{\diagdown}}} \longrightarrow \text{elimination} + \text{substitution} \quad (2)$$

products

solvation effects, generally provide a better means of correlating rates of unimolecular solvolysis of sulphonium salts although anomalous behaviour has been discerned for bifunctional solvents such as dioxan.

Results of an investigation of the pressure dependence of rates of solvolysis of t-butylsulphonium iodide[15] in ethanol–water mixtures (60°C) suggest that the behaviour of the partial molar volumes of the reactant and the transition state is remarkably similar; this is in contrast with the corresponding situation for solvolysis of t-butyl chloride. For the sulphonium salt the overall volume of activation ($\Delta V\ddagger$) passes through a maximum at $ca.$ 0.2–0.3 molar fraction of ethanol, whereas a minimum in $\Delta V\ddagger$ is observed for t-butyl chloride; this reversal of behaviour parallels that previously noted[19] for $\Delta H\ddagger$ which exhibits corresponding minimum and maximum values, respectively. The reversal is also evident in the sign of $\Delta V\ddagger$, which is negative for solvolysis of t-butyl chloride but positive for solvolysis of the corresponding sulphonium salt; in the latter case increased electrostriction on activation is apparently insufficient to outweigh the volume increase due to bond stretching and it has been argued that charge dispersal between S and C atoms, rather than charge transfer from S to C, predominates.

Rates and activation parameters for solvolysis of α-phenylethyl- and benzyldimethylsulphonium bromide in ethanol–water mixtures have been compared with those for solvolysis of t-butyldimethylsulphonium iodide and the corresponding alkyl chlorides[20]; for reactions in 0.204 molar fraction ethanol the relative rates of solvolysis within the halide and sulphonium salt series are equally affected by structural changes, and stabilization of the incipient carbocation is apparently the dominant influence in each case. The activation parameters for reactions of the sulphonium salts are remarkably insensitive to structural change, and throughout the range of solvents used $E_a = 32.4–33.0$ kcal mol^{-1} and $\Delta S\ddagger = 15–18$ cal °C^{-1}.

The sulphur kinetic isotope effects for hydrolysis of substituted-benzyl dimethylsulphonium tosylates ($k^{32}/k^{34} \approx 1.01$ for a range of substituents from p-MeO to m-Br) give no indication of mechanistic change throughout the series, even though the p-methoxy derivative reacts very much faster than expected; the results suggest that the reaction proceeds via a S_N2-like transition state in which the degree of nucleophilic participation by solvent depends on the nucleophilic capability of the ring substituent[21].

A unimolecular decomposition of a triple ion consisting of two sulphonium cations and one halide ion has been proposed to account for the kinetics of decomposition of benzyldimethylsulphonium halides in chloroform (to give the benzyl halide and dimethyl sulphide as the sole products) for which the reaction rate has been correlated with the anion nucleophilicity[22].

The rates of solvolysis of a series of sulphonium salts **1**, which may be

Rates of solvolysis of **1** in 0.75 molar fraction
EtOH–H$_2$O at 60°C

X	n	$10^4 k$ (s^{-1})
CO$_2$H	2	13.5
CO$_2$H	3	5.4
CO$_2$H	4	3.2
CO$_2$Et	3	6.5
CO$_2$CH$_2$CF$_3$	3	9.6
CH$_3$	3	2.7
CH$_3$	2	2.4
H	1	0.54

considered as models for the biologically important sulphonium salt S-adenosylmethionine, have been investigated in an attempt to detect a possible anchimeric role (equation 3) for the carboxyl group of methionine in transmethylation reactions[23]. Although it has been argued that t-butyl transfer is assisted by interaction between the carboxyl group and the incipient carbocation, this need not necessarily involve covalent bond formation.

1. Racemization

Optically active tertiary sulphonium salts have been known for many years[24-26] and there has been considerable interest in the mechanism of their racemization[27-30]. t-Butylethylmethylsulphonium perchlorate racemizes $ca.$ 15-fold faster than it solvolyses in a variety of solvents[28], and it has been argued that racemization in ethanol occurs by pyramidal inversion about the central sulphur atom (equation 3). An alternative mechanism. whereby heterolytic carbon–sulphur

$$R^2\text{\tiny{IIIIII}}\underset{R^3}{\overset{R^1}{S^+}} \rightleftharpoons ^+S\text{\tiny{IIIIII}}\underset{R^3}{\overset{R^1}{R^2}}$$

(3)

bond cleavage ($cf.$ equation 5) yields a t-butyl cation–ethyl methyl sulphide ion–molecule pair (which may either return to racemic sulphonium salt or react with the solvent), is supported by measurements of activation volume[31] ($\Delta V = +6.4$ ml mol^{-1}) in water at 40°C but has been ruled out for reaction in ethanol on the basis of substituent effects; thus, although the relative rates of ethanolysis (1:0.06:6.3) of **2:3:4** are consistent with a mechanism of heterolytic cleavage (under the influence of electron-withdrawing **3** or electron-donating **4**

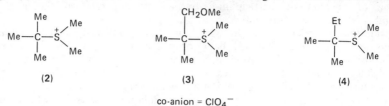

(2)   (3)   (4)

co-anion = ClO$_4^-$

substituents), the relative rates of racemization (1:1.7:3.8) can best be attributed to increase in non-bonded interactions in the ground state, relative to the transition state, of a concerted pyramidal inversion. Likewise, racemization of optically active 1-adamantyl ethylmethylsulphonium perchlorate in acetic acid is believed to occur by pyramidal inversion since no evidence for C—S bond heterolysis has been found[29]; this salt fails to solvolyse under the racemization conditions and perchlorate ion is unlikely to promote inversion by a displacement mechanism, such as that proposed to account for racemization of *l*-phenacylethylmethylsulphonium halides[26] (equation 4).

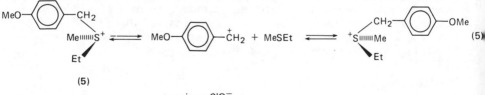

$$\qquad\qquad\qquad\qquad\qquad\qquad\qquad\qquad\qquad\qquad\qquad\qquad (4)$$

Although inversion mechanisms are also believed[27] to account of racemization of benzylethylmethyl-, *p*-nitrobenzylethylmethyl-, and phenacylethylmethylsulphonium perchlorates ($\Delta V^{\ddagger} \approx 0$ for reaction[31] in H$_2$O, MeOH, and EtOH at 60°C), there is evidence to suggest that the *p*-methoxybenzylethylmethyl system **5** racemizes principally by C—S bond heterolysis (equation 5).

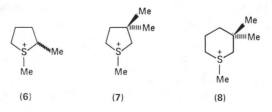

(5)

(5)

co-anion = ClO$_4^-$

High ring strain during pyramidal inversion is believed[32] to account for the slow rates of racemization of the cyclic sulphonium salts **6–8**. Enantiomers of triarylsulphonium ions can only be isolated if their unusually rapid pyramidal inversion[33] is precluded by ring strain effects as in **9** or **10**.

(6)   (7)   (8)

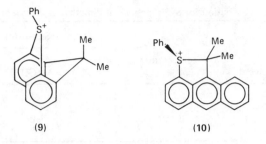

(9)   (10)

B. Bimolecular Nucleophilic Displacement of Sulphonium Groups

1. Stereochemistry and orientation

Although by 1953 it was well established that concerted displacement of halide ion reaction of ionic nucleophiles with alkyl halides generally occurred with net inversion of configuration, there was no reason to believe that this would necessarily be the case for S_N2 displacement reactions of different charge type. It was subsequently concluded[34] that the formation of 1-phenylethyl bromide by nucleophilic displacement of dimethyl sulphide from dimethyl 1-phenylethylsulphonium bromide must occur with inversion at the benzylic carbon atom, as evidenced by the stereochemical results depicted in equation 6. Almost

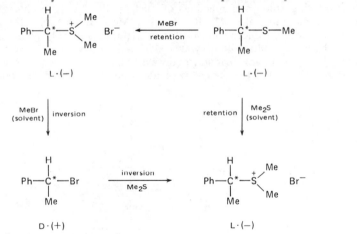

(6)

quantitative stereochemical inversion has likewise been shown to accompany the S_N2 reaction between azide ion and dimethyl 1-phenylethylsulphonium chloride in 80% ethanol[35a]:

$$PhCH(Me)\overset{+}{S}Me_2Cl^- + NaN_3 \longrightarrow PhCH(Me)N_3 + Me_2S + NaCl$$

and it has been established that this substrate also undergoes Walden inversion upon reaction (in acetonitrile) with the neutral nucleophile thiourea[35a]:

$$S{:}C(NH_2)_2 + MeCH(Ph)\overset{+}{S}Me_2 \longrightarrow MeCH(Ph)S\overset{+}{C}(NH_2)_2 + Me_2S$$

Although the preceding examples of nucleophilic attack on dimethyl 1-phenylethylsulphonium ion proceed with predominant attack at the secondary

TABLE 2. Product distribution for S_N2 reactions of dimethyl-1-phenylethylsulphonium ion

| | | | | Substitution (%) | | |
Nucleophile	Solvent	Temperature (°C)	10^4k ($M^{-1} s^{-1}$)	At MeCHPH	At Me	Reference
$(H_2N)_2C{=}S$	MeCN	59.8	10.2	>99	<1	35a
N_3^-	80% EtOH	55.2	12	>99	<1	35c
EtO^-	EtOH	34.9	7.6^a	84	16	35b
MeS^-	EtOH	25.0	Large	42	58	35b

^{a}For substitution at MeCHPh.

carbon atom, this is not always the case; thus, reactions with EtO^- and MeS^- in ethanol occur with 16% and 58% displacement of methyl 1-phenylethylsulphide, respectively[35b]. The latter estimate was based on stereochemical analysis of the displaced sulphide which may be obtained from the optically active substrate either with 100% inversion (by attack of MeS^- at the benzylic carbon) or 100% retention (by attack of MeS^- at the methyl carbon). The collective results (Table 2) can be rationalized if it is assumed that attack on the 1-phenylethyl moiety is most likely to be preferred in those cases where C—leaving group bond breaking is well advanced in the transition state.

An interesting study[36] of reactions of nucleophiles (Y^-), in chloroform, with a series of sulphonium salts (**12**; R = Et, Bu, Me_2CH, etc.; X = Br, Cl, PhS, H) which bear electron-withdrawing substituents [these salts were generated *in situ* by reaction between sulphonium ylides **11** and appropriate electrophiles] has revealed three competing reaction pathways (equation 7); when R is (+)-2-octyl, X = Br, and $Y^- = Br^-$, (−)-2-octyl bromide is obtained with 59% optical purity. In

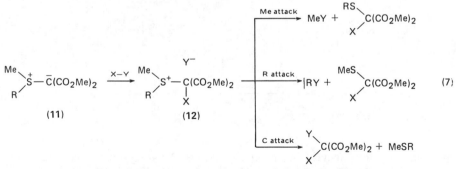

order to account for the relative yields of methylated and alkylated products (which reflect relative rates of substitution which are quite different from those usually observed for S_N2 reactions), in particular the marked preference for benzylation rather than methylation, it has tentatively been suggested that the electrophilic carbon may develop considerable positive charge during the reaction. This is consistent with an earlier report that the neutral and alkaline hydrolysis of (substituted-benzyl)methylphenylsulphonium tosylates occurs with almost exclusive cleavage of the benzyl—sulphur bond and decreases in rate with substituent electron withdrawal[37].

Sulphonium salts which contain at least one benzyl group react with sulphides by benzyl transfer; rates of reaction of sulphides with tribenzylsulphonium perchlorate are dependent on steric effects, although the basicity of the sulphide is also important; for reactions of one sulphide with a range of sulphonium salts, relief of strain in the sulphonium salt appears to be the most important rate-determining factor[38].

The distributions of products of thermal decomposition of sulphonium salts[39–44] have been investigated; although sulphonium hydroxides may decompose by competing pathways to give each of the alternative products of elimination and substitution of the sulphonium group, it has been established that the mode of reaction of an individual alkyl group is, to a first approximation, an intrinsic property of the group and insensitive to the other groups present[40] (see also Tables 12–17).

The kinetics of thermal decomposition of solid benzyl dimethyl sulphonium salts are determined by the surface area exposed as well as by the nature of the counter ion[41,42].

2. Solvent dependence

In 1935 it was shown[45] that the rates of reaction of a series of trialkylsulphonium hydroxides ($R_3\overset{+}{S}$ $\overline{O}H$; R = Me, Et, i-Pr, and t-Bu), in aqueous ethanol, increase as the proportion of ethanol decreases; thus, for R = Me at 100°C, k_2 = 0.133, 5.44, 64 and 2670 M^{-1} h^{-1} in water, 60%, 80%, and 100% ethanol, respectively. $Me_3\overset{+}{S}$ $\overline{O}H$ behaves similarly, although predominant E2 reaction (86–100%) occurs, and where R = i-Pr the reaction mechanism changes from S_N1 (k^1 = 0.00788 s^{-1}) in water to E2 (k_2 = 6.76 M^{-1} s^{-1}) in 60% ethanol. A detailed study of the trimethylsulphonium bromide, chloride, phenoxide, and carbonate has established that for reaction in ethanol there is apparently a point of mechanistic change as the basicity of the anion increases from phenoxide to carbonate; thus, the bromide and chloride undergo unimolecular decomposition ($k^{100°C}$ = 0.279 and 0.263 h^{-1}, respectively) at a rate which is close to that for reaction of the carbonate for which there are signs of concentration dependence ($k^{100°c}$ – 0.242 and 0.301 h^{-1}, at 0.0362 and 0.171 M, respectively); in contrast, the kinetics of reaction of the sulphonium phenoxide are of first order in both anion and cation.

In a recent and more detailed study, the kinetics of decomposition of a wide range of salts $R^1R^2R^3\overset{+}{S}$ $\overline{O}H$, in aqueous methanol and ethanol, have been correlated with dielectric permeability according to an equation for ion–dipole interactions[46]. The change in chemical potential which corresponds to transfer of the transition state ($Me_2\overset{+}{S}$ ... CH_3 ... Br^-) from DMF to ethanol has been estimated[47] as 2.12 kcal mol^{-1} at 25°C; however, for this anionic displacement reaction it is still not clear to what extent the large decrease in rate constant, which accompanies the change from dipolar aprotic to protic solvents, is caused by transition state solvation ($k^{25°C}$ = 1.72 × 10^{-8} and 4.90 × 10^{-10} M^{-1} s^{-1} for reaction in ethanol and DMF, respectively).

For reaction of sodium azide with (substituted-benzyl)dimethyl sulphonium p-toluenesulphonates at 60°C it has been found[48] that where the ring substituent is p-Me, H, and m-Cl, respectively, a change from water to 80% dioxan causes the second-order rate constant to increase by 115, 202, and 241-fold. The solvent dependence of S_N2 reactions of Ingold's 'charge type 4' (neutral reagent, positive substrate) was probed for the first time in 1960 when it was found that reaction of triethylamine with $Me_3\overset{+}{S}$ NO_3^- follows second-order kinetics and goes to completion in water, methanol, ethanol, and nitromethane[49]; the rate of reaction is only moderately accelerated by decrease of solvent polarity (Table 3) and the

TABLE 3. Comparison of the rates of reaction of triethylamine
with $Me_3\overset{+}{S}NO_3^-$ in different solvents at 44.6°C

Rate	Solvent			
	H_2O	MeOH	EtOH	$MeNO_2$
$10^6 k_2$	6.53	41.3	66.7	775
Relative rate[a]	1	6	10	119

[a]Relative to water.

activation parameters reveal a concomitant decrease in entropy, which is outweighted by a favourable decrease in the activation enthalpy.

Sulphur isotope effects for reactions of Me_3S^+ with Br^- (1.36%), PhS^- (1.20%), EtO^- (0.96%), and PhO^- (0.96%) in ethanol indicate that increasing the basicity of the nucleophile leads to a more reactant-like transition state[50]; the isotope effect for reaction in $EtO^-/EtOH$ decreases almost linearly with mole-% DMSO and reaches a value of 0.35% at 65% DMSO. These results are in keeping with the Swain–Thornton rule and with Thornton's contention that basicity is the most appropriate measure of the electron-releasing ability of a nucleophile.

3. Effects of ion pairing

Although reactions of trimethyl- and tribenzylsulphonium salts follow first-order kinetics, in acetone–water (9:1) at 50–100°C it has been found that this is not attributable to rate-determining carbocation formation or nucleophilic displacement by solvent; the reactions involve bimolecular anionic nucleophilic displacement for which the first-order behaviour is a consequence of an exactly compensating salt effect[51]. An alternative explanation whereby the anionic displacement is considered to occur within an ion pair was ruled out for reaction of tribenzylsulphonium chloride in 90% acetone, since conductivity measurements indicated that ion pair formation is unimportant in this medium; direct displacement by solvent molecules is also ineffective since it has been established that the corresponding perchlorate fails to react at a measurable rate, even at 100°C in 90% acetone, but that good second-order kinetics are observed upon addition of lithium chloride.

Hughes et al.[52] subsequently reported that not all salts $Me_3S\ X^-$ decompose at the same rate in ethanol; thus, at 100°C the iodide (0.01 M) is ca. nine times more reactive than the bromide, which is ca. 4.4 times more reactive than the chloride. This is in contrast with the previously envisaged unimolecular decomposition of the sulphonium ion, which may nonetheless explain the very much slower and nearly identical rates of reaction of the perchlorate and fluoroborate salts, at the same concentration. A unimolecular decomposition of $Me_3\overset{+}{S}\ X^-$, where X = perchlorate, picrate, or arenesulphonate, has also been reported for reactions in aqueous solution[37]; ethanolysis at 100°C ($k = 17.8 \times 10^{-8}\ s^{-1}$), occurs ca. 20-fold faster than hydrolysis ($k = 0.81 \times 10^{-8}\ s^{-1}$), and this is in marked contrast with the 20,000-fold increase in the rate of decomposition of $Me_3\overset{+}{S}\ OH$ which occurs on changing from water to ethanol[45].

Reactions of Me_3S^+ with X^- (where X = EtO, CN, I, SCN, N_3, Cl, or F) in ethanol and methanol at 100°C have been found to proceed by the S_N2 mechanism, and hydrolysis and methanolysis of Me_3S^+ is also believed to occur by a bimolecular displacement for which methanol is a better nucleophile than water[53].

The marked influence of ion association, in ethanol, on the anionic nucleophilic displacement reaction has been equated (where X = Br) by the relationship $k_2 = k_2^0 \alpha^2$ (where α is the fraction of dissociated ions at ionic strength μ), but it has not been possible, by inclusion of formal activity coefficients, to distinguish kinetically between mechanisms which involve separated ions and those which involve ion pairs.

Sneen *et al.*[54] have studied the competitive substitutions (by solvent water and added nucleophile) of benzyldimethylsulphonium ion (I⁻, 140°C) and its *p*-methoxy derivative (N₃⁻, 60°C); the reactions exhibit borderline behaviour between S_N1 and S_N2 and it has been argued that the competing nucleophilic displacements occur by attack on a common ion–dipole assemblage **13** rather than by a combination of S_N1 and S_N2 components (equation 8).

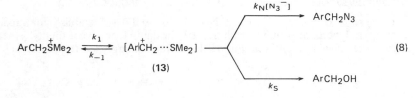

$$\text{(8)}$$

4. Methyl transfer

There has been particular interest in the kinetics and mechanism of methyl transfer from sulphonium compounds to various nucleophiles, since this is a common process in biological systems[55–58]. For reactions of a series of substituted-phenyldimethylsulphonium perchlorates it has been found that anionic nucleophiles in water generally react less rapidly than amine nucleophiles but that the Hammett rho values obtained for reaction with hydroxide ion (1.60 in water at 78.8°C), pyrrolidine or *n*-butylamine (1.74 in acetonitrile at 25°C) are similar[55]. For reaction of *p*-nitrophenyldimethylsulphonium ion with various amines in acetonitrile, a Brönsted coefficient of $\beta = 0.36$ has been determined and, by appropriate comparison, it has been concluded that transmethylation from simple sulphonium compounds is similar to that from methyl iodide.

By investigation of reactions of a series of methylsulphonium salts bearing appropriately positioned nucleophilic moieties it was subsequently established[56] that there is a strict requirement for a linear transition state in intramolecular transmethylation reactions; such restriction may therefore account for the steric controls which appear to be operative in enzyme-catalysed transmethylation processes (e.g. where *S*-adenosyl-L-methionine serves as the methyl donor).

The transfer of diastereotopic methyl groups from a monochiral dimethyl-sulphonium ion **14a** to a nucleophile (*p*-thiocresolate ion) has been studied by C-14 labelling. The large rate differences observed for transfer of ¹²C and ¹⁴C methyl groups from tosylates **14b** and **14c** are, however, a consequence of an unusually large isotope effect ($k^{12}/k^{14} \approx 1.16$), there being only slight diastereotopic preference[58,59].

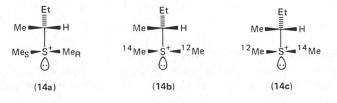

5. Reversion

The term 'reversion' has been introduced to indicate the reverse of the reaction by which sulphonium salts are formed from sulphides by alkylating agents:

$$R^1SR^2 + R^3X \rightleftharpoons R^1R^2R^3\overset{+}{S} X^-$$

The reversion process, which occasionally leads to unexpected products upon attempted alkylation of sulphides[60], has been investigated kinetically[61–63], and is believed to account for racemization of phenacylsulphonium salts[26,30]. The reversion process displays the usual regioselectivity expected of a nucleophilic displacement[64] and is correspondingly rare for arylsulphonium halides[65].

6. Miscellaneous nucleophilic displacement reactions

Deuterium labelling techniques have been used[66] to establish that reaction of trimethylsulphonium bromide with $LiAlH_4$ in diethylene glycol diethyl ether occurs by a simple displacement and not via the corresponding ylide (equation 9).

$$H_3\overline{Al}-H \qquad D_3C-\overset{+}{\underset{\downarrow}{S}}(CD_3)_2 \longrightarrow H_3Al + HCD_3 + S(CD_3)_2 \qquad (9)$$

An interesting example of intramolecular nucleophilic displacement of a sulphonium group by the π electrons of a non-activated double bond has been reported[67]; thus, *endo*-bicyclo[3.2.1]octan-2-ol **16** is obtained upon hydrolysis of **15** (equation 13).

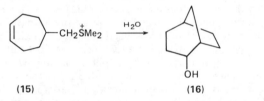

$$(10)$$

(15)   (16)

Kice and Favstritsky[68] have established that the facile displacement of Me_2S from dimethylmethylthiosulphonium ion by dimethyl sulphide:

$$Me_2S + Me\overset{+}{S}SMe_3 \longrightarrow Me_3S\overset{+}{S}Me + Me_2S$$

occurs much more rapidly than the following exchange of thiomethoxide ion:

$$MeS^- + MeSSMe \longrightarrow MeSSMe + MeS^-$$

C. Substitution of Aromatic Sulphonium Compounds

It would be inappropriate, in this chapter, to review the directive effects of the sulphonium group (see Chapter 9), but the following examples have been included in order to illustrate the stability of such groups to the usual conditions of aromatic substitution.

1. Nucleophilic aromatic substitution

The sulphonium group, in the 4-position, is a better activating substituent than a 2-nitro group for amine-induced displacement of chloride from chlorobenzenes[69].

Upon reaction of acetone with *m*-nitrophenylsulphonium fluoroborate in ethanolic potassium hydroxide an anionic σ complex **17** is formed (equation 11), which

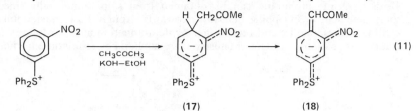

(11)

(17)   (18)

subsequently gives **18** on standing[70]. Nonetheless, the sulphonium group is a very effective leaving group in nucleophilic aromatic substitution reactions for which the order of ease of displacement $Me_2S^+ > Me_3N^+ > F > NO_2 > Cl > NR_2$ has been determined[71] by comparison of rates of reaction of the corresponding *p*-nitrophenyl derivatives with MeONa–MeOH; this represents a general order cation > dipole > neutral group (although fluorine is in an anomalous position), and it has been suggested that electrostatic interactions may be important in stabilizing the rate-determining transition state, for addition of the anion; further, it has been suggested that reaction of the sulphonium substrate, relative to the ammonium analogue, may be facilitated by the ability of sulphur to expand its octet and therby stabilize the Meisenheimer intermediate.

2. Electrophilic aromatic substitution

The electrostatic effects of the sulphonium group on electrophilic aromatic halogenations[72,73] and nitrations[73–77] have been investigated. For silver-catalysed chlorination and bromination of dimethylphenylsulphonium and -selenonium methylsulphate the *m-/p*-ratios (15–27) are similar to those for nitration reactions, whereas the *o-/p*-ratio for bromination and chlorination (1.2–6.1) is greater than that for nitration[76] (0.4–0.6); this has tentatively been ascribed[72] to halogenation via an undefined intermediate which favours electrophilic substitution *ortho* to the cationic centre.

The overall rate of nitration[75] of dimethylphenylsulphonium methylsulphate by nitric acid (10^{-2} M) in concentrated sulphuric acid, relative to that of benzene, is 3.98×10^{-9} and the partial rate factors $f_o = 2.6 \times 10^{-10}$, $f_m = 1.12 \times 10^{-8}$ and $f_p = 9.44 \times 10^{-10}$ have been determined; relative to phenyl trimethylammonium ion, the f_m and f_p values are lower by *ca.* 8-fold and 25-fold, respectively, and this may be a consequence of greater transmission of the electrostatic effect of the positive charge facilitated by π(dp) overlap of the 3d orbitals of sulphur.

For the series $Ph(CH_2)_n\overset{+}{Z}Me_2$ (where $n = 0–2$ and $Z = S$ or Se), the rate of nitration increases and the percentage of *meta*-substitution decreases as the positive pole is removed further from the ring, and also when Z is changed from S to Se[74]. Dimethyl(2-thienyl)sulphonium salts undergo nitration and bromination primarily at the 4- and 5-positions[73].

D. Nucleophilic Reactions at the Sulphonium Sulphur Atom

Many interesting and synthetically important reactions of sulphonium compounds involve their conversion into sulphurane intermediates, which may then undergo a wide variety of subsequent reactions. Sulphuranes are a class of compound in which sulphur has expanded its formal valence shell from eight to ten electrons by either a double (π-sulphurane) or single (σ-sulphurane) bond formation.

Both types of sulphurane can be obtained from sulphonium salts since these display ambident electrophilic behaviour towards strongly basic nucleophiles (e.g. alkyllithium compounds) and react either by deprotonation at C_α or by S—Nu bond formation (equation 12). The chemistry of π-sulphuranes (sulphonium ylides) will

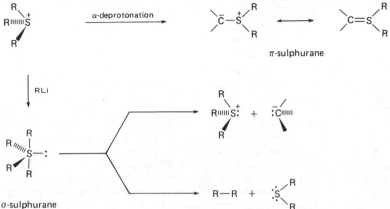

π-sulphurane

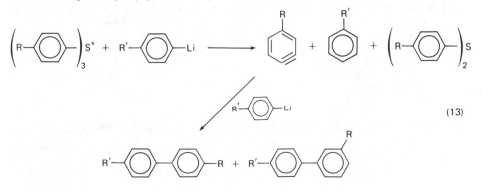

be considered in a later section of this chapter (and in Chapter 16); reactions which are believed to involve intermediate σ-sulphuranes will be reviewed in this section.

σ-Sulphuranes may return to the octet state either by heterolytic cleavage, or by a fragmentation whereby a dialkyl sulphide is released upon coupling of two substituents (equation 12). Such reactions are believed, for example, to account for the formation[78] of styrene upon reaction of triphenylsulphonium ion with vinyllithium[79,80], and for the formation of allylbenzene, biphenyl and n-butyl phenyl sulphide upon reaction of allyldimethylsulphonium ion with n-butyllithium. Coupled products also form in those cases where ylide generation is impossible; thus, treatment of triphenylsulphonium salt with phenyllithium[81] gives rise to biphenyl and diphenyl sulphide. There has been considerable interest in the reactions of triarylsulphonium salts with aryl lithiums[79–86]. Although it is generally accepted[81,82,85] that the formation of sulphides and biaryl compounds proceeds predominantly by formation of a thermally unstable tetraaryl sulphurane, it has been suggested by Franzen et al.[82] that a benzyne intermediate may be involved in a minor pathway (equation 13). This has been tested for reaction of tri-*p*-tolylsulphonium ion with phenyllithium since the intermediate aryne (3,4-dehydrotoluene) should give a mixture of *pp'*- and *mp'*-bitolyl, with the latter predominating[82]; the reaction has been conducted under a variety of conditions[79,80,82–86] but only in some cases[83,84] has *mp'*-bitolyl been detected. Correspondingly, C-14 labelling experiments[85] have confirmed that reaction of labelled phenyllithium with triphenylsulphonium tetrafluoroborate proceeds predominantly via tetraphenylsulphurane; thus, half of the original label appears in each of the products (i.e. as singly labelled PhPh and unlabelled PhSPh, or *vice versa*). An alternative S$_N$Ar reaction would be expected to form predominantly singly labelled biphenyl and unlabelled diphenyl sulphide since label exchange between Ph*Li and Ph$_3$S$^+$ is not significant under the experimental conditions.

The alternative addition–elimination route (S$_N$Ar reaction) has also been discussed for reactions of *S*-phenyldibenzothiophenium fluoroborate **19** with phenyl lithium and vinyllithium; thus, while the quantitative formation of 2-phenylthio-*o*-terphenyl in the former case could be attributed to preferential formation of an adduct **20** (in which the negative charge can be delocalized over two rings), it is unreasonable then to argue that formation of **21** (in which only one ring participates in delocalization) predominates in the latter case and accounts for the formation of styrene and dibenzothiophen as the major products (equation 14). However, the intermediacy of a σ-sulphurane **22** can account for all of the results.

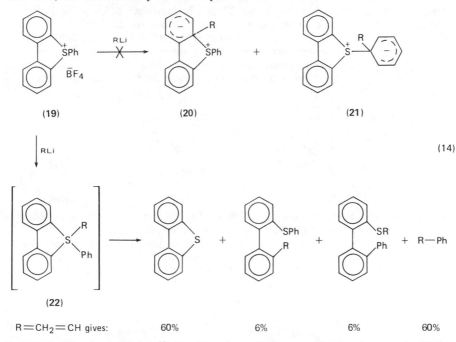

R=CH$_2$=CH gives: 60%   6%   6%   60%

It has also been established[80] that insertion of electron-withdrawing groups X (on the phenyl ring of **19**) or Y (on phenyllithium) facilitates the formation of dibenzothiophen and the product (XC$_6$H$_5$C$_6$H$_5$Y) of aryl coupling; aryl–biphenyl coupling occurs exclusively when neither aryl group bears electron-withdrawing substituents.

Although the scope and synthetic utility of this novel coupling procedure have been investigated[79], the mechanism for collapse of the intermediate sulphurane remains controversial. Homolytic cleavage followed by radical coupling has, however, been ruled out on the following grounds:

(i) the results indicate that aryl groups are preferentially involved in coupling and that alkyl–alkyl coupling does not occur, even though alkyl radicals are known to be more stable than aryl radicals;

(ii) the geometric integrity of the vinyl group is retained when propenyllithium is used; this would require a coupling rate of $>10^{12}$ s^{-1} since vinyl radicals generally have an inversion frequency of *ca.* 10^{10} s^{-1}.

The mode of decomposition of sulphuranes may be complicated by stereochemical considerations and Trost and coworkers have, by analogy to phosphorus chemistry[83], made the following structural assumptions[80,82] in their efforts to rationalize the products of addition of organolithiums to five membered ring sulphonium salts:

(1) pentacoordinated sulphur exists as a trigonal bipyramid;
(2) the electron pair prefers a basal orientation;
(3) the five membered ring prefers the apical–basal orientation; and
(4) groups enter and leave from an apical position.

Thus, the observation that terphenyl **23** is obtained by addition of Ar'Li to *S*-Ar-dibenzothiophenium fluoroborate, whereas terphenyl **25** is the product of addition of ArLi to the *S*-Ar'-dibenzothiophenium salt, has been attributed to preferential (kinetic control) formation of the intermediate sulphuranes **24a** and **24b**, respectively; it is assumed that coupling between the biphenyl group and apical aryl groups is precluded by transition state strain energy. The stereochemical

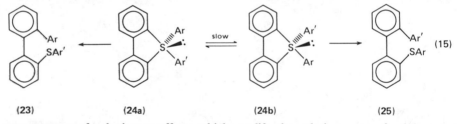

| (23) | (24a) | (24b) | (25) |

(15)

consequences of substituent effects which modify the relative rates of sulphurane isomerization and coupling reactions have been investigated[80,82].

Stereospecific fragmentation of sulphurane intermediates has also been invoked[87] to account for the formation of *cis*- and *trans*-1,2-dimethylcyclopropane **28a,b** from *cis*- and *trans*-1,2,4-trimethylthietanonium fluoroborates **26a,b** respectively, upon reaction with *n*-butyllithium at $-78°C$.

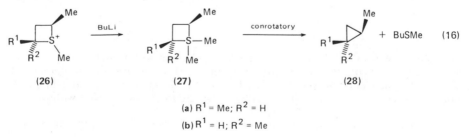

| (26) | (27) | (28) |

(16)

(a) R^1 = Me; R^2 = H
(b) R^1 = H; R^2 = Me

Upon reaction of 2,5-dihydrothiophenium salts with *n*-butyllithium[88], products of both elimination and fragmentation are obtained, in a ratio which is apparently a function of steric hindrance to approach of the nucleophile towards sulphur; thus, fragmentation increases (at the expense of elimination) in the order **29c** < **29a** < **29b**. Formation of **30** from **29c** may arise by an E2 process whereas formation of **32** from both **29a** and **29b** probably occurs via the ylide **31**. The competing fragmentation is remarkably stereospecific and, since **29a** gives *cis,trans*-hexa-2,4-diene **34** almost exclusively, this would require that the proposed sulphurane intermediate **33** decompose by disrotatory elimination of butyl methyl sulphide.

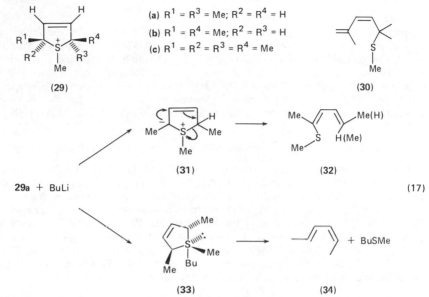

(a) $R^1 = R^3 = Me$; $R^2 = R^4 = H$

(b) $R^1 = R^4 = Me$; $R^2 = R^3 = H$

(c) $R^1 = R^2 = R^3 = R^4 = Me$

(29)

(30)

29a + BuLi

(31)   (32)   (17)

(33)   (34)

The stoichiometry of thiophen formation upon reaction of sulphur ylides with cyclopropenium salts has also been rationalized[86] by way of an intermediate sulphurane. The formation of mixtures of aromatic hydrocarbons, biaryls, methyl aryl ethers, diaryl sulphides, and high-molecular-weight products upon reaction of triarylsulphonium salts with alcoholic alkoxide bases[88-90] is apparently a consequence of free-radical reactions which are initiated by homolysis of an intermediate sulphurane.

The products of pyrolysis to triarylsulphonium halides (PhArAr'SX) are also believed to arise by intramolecular decomposition of an intermediate quadrivalent sulphur compound[91]; the product proportions suggest that decomposition is governed by relief of steric strain: thus, of the three possible pairs of products (PhArS + Ar'X, etc.) from 2,5-dimethylphenylphenyl-*p*-tolylsulphonium halide, 2,5-dimethylhalobenzene and phenyl *p*-tolyl sulphide predominate. Further, the relative amounts of 2,5-dimethylhalobenzenes fall in the order I > Br > Cl and in each case the amount of halobenzene exceeds that of the *p*-halotoluene; these observations are inexplicable in terms of a bimolecular S_NAr reaction and a free-radical mechanism is unlikely since no biaryls were detected. Tetracovalent sulphur intermediates have also been implicated in reactions of sulphonium ions with dialkyl sulphides[92,93].

Walden inversion at sulphonium sulphur has also been reported[94]. Thus, the stereoisomeric alkoxysulphonium salts **36a** and **36b** react with dimethylcadmium to give **35a** and **35b**, respectively, with complete inversion of configuration at sulphur; inversion also occurs upon reaction of **36a** with MeMgBr although reaction of **36b** is complicated by its partial isomerization to **36a** under these conditions. Reactions

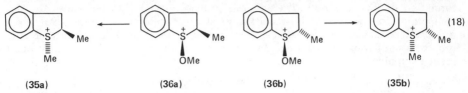

(35a)   (36a)   (36b)   (35b)

of alkoxides with alkoxysulphonium salts are often preceded by exchange of the alkoxy moieties; thus, whereas treatment of $[Me_2SOMe]^+\bar{B}F_4$ (labelled with ^{14}C in the OMe group) with NaH–DMSO gives, by elimination, labelled formaldehyde, only 1% of the expected label is found in the formaldehyde obtained[95] upon reaction of the salt with NaOMe–MeOH. This can be attributed to consecutive alkoxide exchange and elimination, as envisaged to account for the isolation of 2-methylpropene and dimethyl sulphoxide upon reaction of $[Me_2SOMe]^+\bar{B}F_4$ with t-BuONa–t-BuOH:

$$Me_2\overset{+}{S}OMe + t\text{-BuO}^- \longrightarrow Me_2SOBu\text{-}t \longrightarrow Me_2S{=}O + Me_2C{=}CH_2$$

Deuterium-labelling studies subsequently established[96] that the eliminative formation of carbonyl products proceeds by intramolecular fragmentation of the intermediate sulphur ylide (α',β-elimination mechanism); thus, when methylphenyltrideuteriomethoxysulphonium fluoroborate **37a** is treated with NaH–THF, phenyl monodeuteriomethyl sulphide **38a** is formed in high yield. When unlabelled **37** is treated with NaOMe–MeOD there is negligible incorporation of deuterium in the methyl phenyl sulphide obtained, whereas reaction of **37b** with NaH–THF leads to the dideuteriomethyl sulphide **38b**. Apparently the ylide acts as an internal base more rapidly than it re-forms **37** by re-protonation.

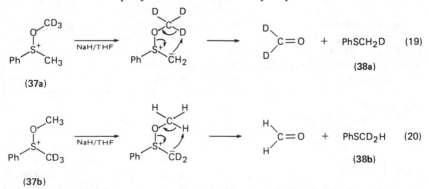

(37a)

(37b)

However, the ultimate course of the alkoxide reaction can be altered[95] by increasing the acidity of the hydrogen on the carbon atoms adjacent to the positive sulphur. For example the O-methyl salt **39** (R = Ph) of dibenzyl sulphoxide forms α-methoxy dibenzyl sulphide when treated with methoxide, and the 2,4-DNP derivative of benzaldehyde is obtained if the reaction mixture is treated with 2,4-DNP–acid; this has been interpreted according to equation 21 since (i) an

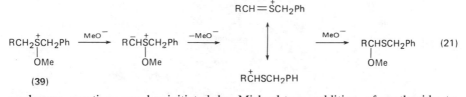

$$RCH_2\overset{+}{S}CH_2Ph \xrightarrow{MeO^-} \overline{RC}H\overset{+}{S}CH_2Ph \xrightarrow{-MeO^-} \qquad \xrightarrow{MeO^-} RCHSCH_2Ph \quad (21)$$

(39) with OMe, OMe, and OMe substituents; center shows $RCH=\overset{+}{S}CH_2Ph$ and $R\overset{+}{C}HSCH_2PH$

analogous reaction can be initiated by Michael-type addition of methoxide to methoxyphenylvinylsulphonium fluoroborate, (ii) in cases where more than one pathway is available, substituents (R) which are able to stabilize a carbocation favour the production of such 'Pummerer products', and (iii) in a double-label experiment, crossover products indicated that the alkoxy group and the S-containing fragment become solvent separated. The migration of a group from a dialkylsulphonium sulphur atom always proceeds to the least substituted α-carbon (which, in most cases, is the site of the least stable carbonium ion) and it has been argued that this reflects the difference of acidities of the α-protons since the rate of ylide formation may kinetically control the product distribution.

Deuterium-labelling techniques have also been employed to determine the mechanism of borohydride reduction of alkoxysulphonium fluoroborates whereby the salts, in alcohols or THF, are efficiently converted into the parent sulphides[97]. All possible ylide mechanisms have been discounted since the methyl phenyl sulphide obtained upon reaction of 37a with NaBH$_4$–THF incorporates only trace amounts of deuterium, and unlabelled 38 is obtained from unlabelled 37 with NaBD$_4$–MeOH. It has been concluded that the reaction proceeds by hydride displacement of methoxide to give the conjugate acid of the product sulphide:

$$\overline{B}H_4 + Ph\overset{+}{S}OMe(Me) \longrightarrow Ph\overset{+}{S}H(Me) + MeO^- \longrightarrow PhSMe + MeOH$$

It is possible that the results could also be accommodated by fragmentation of an intermediate sulphurane, PhSOMe(Me)(H), although additional products might well be expected.

The hydrolysis of optically active alkoxysulphonium salts to the corresponding sulphoxides occurs with inversion of configuration[98]; further, the dimethyl sulphoxide obtained[99] upon hydrolysis of dimethylmethoxysulphonium perchlorate in ^{18}O-enriched water is rich in ^{18}O. These results are consistent with a mechanism (equation 22) whereby the sulphoxide is formed by elimination of alcohol from an

$$Me_{\text{\tiny IIIII}}\overset{\cdot\cdot}{S}{}^+\!\!-OMe + H^{18}O^- \rightleftharpoons Me_{\text{\tiny IIIII}}\underset{\underset{18}{OH}}{\overset{OMe}{\underset{|}{\overset{|}{S}}}}\!-: \rightleftharpoons Me-\overset{\overset{18}{O}}{\underset{||}{S}}-Me + MeOH \quad (22)$$

intermediate σ-sulphurane. Likewise, Tang and Mislow[100] observed that the base-catalysed hydrolysis of both cis- and trans-1-ethoxy-3-methylthietanium ions 40 and 41 proceeds with complete inversion; they concluded that pseudo-rotation (e.g. about the lone pair of electrons as a pivot), which would be expected to occur much less readily than that observed in phosphoranes, is not a complicating factor (i.e. if the thietanium ring adopts the anticipated a,e arrangement and hydroxide attack is from an axial direction).

Nucleophilic displacements at sulphonium centres are a common feature of oxidations promoted by certain heterosulphonium salts. For example, succinimidodimethyl sulphonium cation has been used to convert catechols into the corresponding quinones via eliminative oxidation of an intermediate

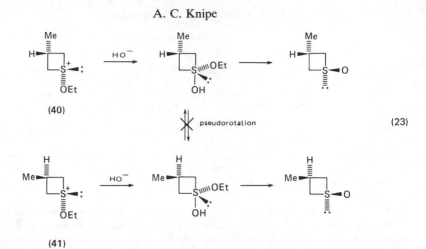

(40)

pseudorotation   (23)

(41)

aryloxydimethylsulphonium cation[101]. Likewise, alcohols can be oxidized to carbonyl compounds by base-induced decomposition of the corresponding alkoxydimethylsulphonium salt, formed by reaction with trifluoroacetoxy-dimethylsulphonium trifluoroacetate[102]. Sulphonium salts have also been used as catalysts for oxidation of tetralin[103] and cumene[104], polymerization of isocyanates[105], dehydration of aldoximes[106], and promotion of carbamate reactions[107].

E. Sulphonium Ions as Intermediates in Substitutions at Sulphur

Sulphonium ion intermediates have often been proposed to account for the stereochemistry of nucleophilic substitution at tricoordinate sulphur[5]; many such examples involve heterosulphonium ions and will therefore be commented upon only briefly since this interesting class of onium compounds is given detailed coverage in Chapter 15.

Many stereochemical transformations of sulphoxides and related compounds involve conversion to sulphonium intermediates by O-alkylation reactions[108]; subsequent hydrolysis of alkoxysulphonium salts generally proceeds with inversion[98,100,109], as illustrated by the following Walden inversion cycles (equations 24 and 25), which probably involve σ-sulphurane intermediates. Furthermore, hydrolysis of **42** in $^{18}OH_2$ (equation 26) forms the ketosulphoxide **43** in which the

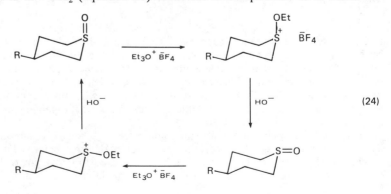

(24)

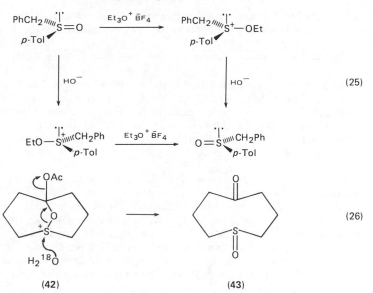

(25)

(26)

(42)   (43)

sulphonyl oxygen has undergone complete exchange with the solvent, whereas the keto-oxygen remains unenriched[99]; this has been attributed to rear-side attack of solvent at sulphonium sulphur.

Racemization of many simple sulphoxides can be achieved at room temperature in concentrated hydrochloric acid[110]; this probably involves pre-equilibrium protonation followed by formation of a halosulphonium ion intermediate:

$$R_2S{=}O + H^+ \rightleftharpoons R_2S{=}\overset{+}{O}H$$

$$R_2S{=}\overset{+}{O}H + H^+ + X^- \rightleftharpoons X{-}\overset{+}{S}R_2 + H_2O$$

Several workers[110–113] have implicated this common equilibrium sequence in order to account for a range of concurrent reduction, racemization and oxygen-exchange reactions of sulphoxides. Thus, a second halide ion may reduce the halosulphonium ion to the sulphide:

$$X{-}\overset{+}{S}R_2 + X^- \rightleftharpoons R_2S + X_2$$

by a process which goes to completion when iodide is used as the reducing agent; with chloride and bromide ions the equilibrium lies further to the left and the sulphoxide may therefore undergo predominant racemization and oxygen exchange, rather than reduction. Racemization may be a consequence either of the reversal of reaction above or of the alternative halide exchange processes[108] (equation 27).

$$Hal{-}\overset{\underset{\displaystyle R'}{|}}{\underset{}{\overset{\displaystyle R}{|}}}S{-}Hal \rightleftharpoons Hal{-}\overset{+}{S}\overset{\displaystyle R}{\underset{\displaystyle R'}{}} + Hal^- \rightleftharpoons \overset{\displaystyle R}{\underset{\displaystyle R'}{}}S^+{-}Hal + Hal^- \quad (27)$$

The rates of both reduction and racemization of sulphoxides have generally been found[112–114] to be first order in both halide ion and substrate, but second order in hydrogen ion concentration; thus, reduction, for which the stoichiometry is

$$RR^1SO + 2H^+ + 2I^- \longrightarrow RR^1S + H_2O + I_2$$

apparently involves the second halide ion subsequent to rate-determining formation of halosulphonium ion by the sequence in equation 28.

$$>S=O \overset{H^+}{\rightleftharpoons} >S\overset{+}{=}\overset{}{O}H \overset{X^-}{\rightleftharpoons} X-\overset{|}{\underset{|}{S}}-OH \overset{H^+}{\rightleftharpoons} X-\overset{|}{\underset{|}{S}}-\overset{+}{O}H_2 \overset{-H_2O}{\rightleftharpoons} X-\overset{+}{S}< \quad (28)$$

While it has been established that for phenyl p-tolylsulphoxide in hydrochloric acid the anion-dependent[115] rates of ^{18}O exchange and racemization are identical[110,116], this is not the case for reactions of asymmetric sulphoxides in all mineral acid; thus, in sulphuric acid the rate of oxygen exchange is generally independent of the anion concentration, and the mechanism may vary from A-1 $(k_{rac} = k_{exch})$[117–119] to A-2 $(k_{exch}/k_{rac} \approx 0.5)$[119,120] as the acid concentration is reduced.

Halosulphonium ion intermediates have also been implicated in anchimerically assisted reduction and racemization reactions of sulphoxide[121–126], oxidation of sulphones by t-butyl hypochlorite, chlorinolyses of sulphides[127,128] or arenesulphenyl chlorides[129], and reduction of N-aryl sulphonylsulphimides with iodide ion in aqueous perchloric acid[130,131]. Oxysulphonium salts are almost certainly intermediates in the racemization and oxygen-exchange reactions of sulphoxides, catalysed by carboxylic acids[132–134] or by acetic anhydride[135–140], and in the Pummerer reaction[95,141–148] (equation 29) (whereby sulphoxides which bear α-hydrogen atoms

$$R'-\overset{O}{\underset{\|}{S}}-CH_2R \overset{Ac_2O}{\rightleftharpoons} R'-\overset{OAc}{\underset{|}{S^+}}-CH_2R \overset{-H^+}{\underset{slow}{\longrightarrow}} R'-\overset{OAc}{\underset{|}{S^+}}-\bar{C}HR$$

$$\downarrow -AcO^- \qquad (29)$$

$$R'-\underset{\underset{OAc}{|}}{S}-CHR \overset{AcO^-}{\longleftarrow} R'-\overset{+}{S}=CHR$$

may be converted into α-substituted sulphides upon reaction with acids, anhydrides, and acyl halides).

Oxysulphonium ions also feature in both A-1 and A-2 mechanisms of acid-catalysed ring opening of episulphoxides[149–152], the stereomutation[153] and reduction[154] reactions of cyclic sulphoxides, and acid hydrolyses of cyclic and open-chain sulphites[155–158] (e.g. equation 30), sulphinic acid derivatives[159] and aryl sulphinylsulphones[160–162]

$$\left[\begin{array}{c}O\\O\end{array}\right]S=O \overset{H^+}{\rightleftharpoons} \left[\begin{array}{c}O\\O\end{array}\right]S\overset{+}{=}\overset{}{O}H \overset{H_2O}{\underset{slow}{\longrightarrow}} \left[\begin{array}{c}O\\O\end{array}\right]\overset{OH}{\underset{\overset{+}{O}H_2}{S}} \overset{}{\underset{fast}{\longrightarrow}} \left[\begin{array}{c}OH\\OH\end{array}\right] \quad (30)$$

III. ELIMINATION REACTIONS OF SULPHONIUM SALTS

A. Unimolecular Elimination of Sulphonium Groups

Ingold and coworkers, in their pioneering work on the fundamental mechanisms of elimination and substitution reactions of alkyl halides, also reported extensive investigations of unimolecular (and bimolecular) reactions of analogous sulphonium compounds[16,17,163–168]; these results have been summarized admirably[169,170a]. More recently there has been considerable interest in the effects of counter ions on the

TABLE 4. Molar percentage ($\pm 1.0\%$) of alkene (2-methyl-propene) from solvolysis[171] of t-BuX in different solvents at 75°C

X	H_2O	EtOH	AcOH
Cl	7.6	44.2	73
Br	6.6	36.0	69.5
I	6.0	32.3	—
$\overset{+}{S}Me_2ClO_4^-$	6.5	17.8	11.7
$\overset{+}{O}H_2$	4.7	—	—

distribution of products obtained during unimolecular solvolyses of t-alkyl derivatives[14,171,172], and in the activation volumes[173,174] for such reactions.

A striking dependence of elimination to substitution ratio on both leaving group and solvent has been reported for t-butyl and t-amyl compounds[171]. It is apparent from the results in Table 4 that, in water, the percentage of alkene formed upon solvolysis of t-BuX is small and almost equal for the leaving groups X = Cl, Br, I, $\overset{+}{S}Me_2$ and $\overset{+}{O}H_2$. A change from water to less dissociating solvents, ethanol and acetic acid, produces a progressive increase in both the proportion of elimination (which decreases in the order Cl > Br > I > $\overset{+}{S}Me_2$) and its dependence upon the nature of the leaving group. These variations have been ascribed to participation of the counter ion at the ion-pair stage of the E1 reaction; further, by taking the distribution of products from t-BuSMe$_2$ ClO$_4^-$ to be a guide to that from the dissociated t-butyl cation, it has been estimated that in ethanol and acetic acid, respectively, at least 73% and 95% of the elimination from t-BuCl involves the counter chloride ion.

The effect of the leaving group on the proportions of products obtained upon ethanolysis of 2-pentyl derivatives (PrCHYMe; Y = Br, $\overset{+}{S}Me_2Br^-$, $\overset{+}{S}Me_2I^-$, $\overset{+}{S}Me_2ClO_4^-$, $\overset{+}{N}Me_3Br^-$, and $\overset{+}{N}Me_3I^-$) has been investigated more recently[172]. A change of leaving group from bromine to any of the onium halides leads to a substantial decrease in the proportion of 2-pentyl ethyl ether, relative to isomeric pentenes, even though the composition of the pentene mixture remains unaltered; a further change to the sulphonium perchlorate produces no further change in the ether to alkene ratio but causes a substantial change in the composition of the alkene mixture. Thus, the ether to alkene ratio is dependent upon the charge type of the leaving group while the proportions of the isomeric pentenes apparently depend on the counter ion of the intermediate carbocation. This observation cannot be adequately explained by invoking a series of intermediate ion pairs unless it is further argued that the ion pair is born into the solvation environment left behind by its precursor. The product proportions obtained upon ethanolysis of 2-methyl-2-butyl derivatives are also dependent upon the leaving group.

Volumes of activation have been measured for E1 and E2 reactions of neutral and ionic substrates[173]. For unimolecular solvolysis of t-amyl chloride in ethanol–water (4:1) at 34.2°C, the value $\Delta V^{\ddagger} = -18$ cm^3 mol^{-1} can be attributed to the largely ionic transition state; this is in marked contrast with the large positive value, $\Delta V^{\ddagger} = 14$ cm^3 mol^{-1} obtained for solvolysis of the corresponding sulphonium iodide at 53.8°C.

For the bimolecular elimination reaction between 2-bromobutane and methoxide ion at 47.8°C, the activation volume ($\Delta V^{\ddagger} = -10$ cm^3 mol^{-1}) can best be interpreted if it is assumed that bond rupture is not far advanced[173]; again, the value is in contrast with that for the analogous bimolecular elimination reaction of the ionic

TABLE 5. Activation volumes for unimolecular and bimolecular reactions of halide and onium substrates[173]

Reaction	Solvent	Temperature (°C)	$\Delta V^{\ddagger}$ (cm^3 mol^{-1})
t-AmCl solvolysis	EtOH (80%)	34.2	-18
	MeOH (80%)	29.8	-16
t-Am$\overset{+}{S}$Me$_2$ I$^-$ solvolysis	EtOH (80%)	53.8	$+14$
2-Bromobutane/NaOEt	EtOH	47.8	-10
t-Am$\overset{+}{N}$Me$_3$ I$^-$/KOH	EtOH	84.8	$+15$
Et$_3$$\overset{+}{S}Br^-$/NaOH	H$_2$O	109.0	$+11$
Et$_3$$\overset{+}{S}Br^-$/NaOMe	MeOH	150.9	$+15$
Et$_4$N/NaOMe	MeOH	104.8	$+20$

substrate triethylsulphonium bromide ($\Delta V^{\ddagger} = 15$ cm^3 mol^{-1} at 50.9°C). For each of the substrates investigated (Table 5) the activation volumes are approximately equal for the competing substitution and elimination reactions; further, the failure to discern any difference in volumes of activation for bimolecular elimination reactions of t- and sec-amyl substrates tends to support Ingold's view[170b] that Hofmann orientation is governed by inductive, rather than steric, effects.

B. Bimolecular Elimination of Sulphonium Groups

1. Substituent kinetic effects on E2 reactions

a. *The 2-phenylethyl system*. There has been considerable interest in the base-induced E2 reactions of 2-phenylethylsulphonium salts. Measurements of the effects of isotopic substitution of α-hydrogen[175], β-hydrogen[176,177], sulphur[18,178–180], β-carbon[181,182], and the solvent[183,184] have been combined with studies of the kinetic influence of aryl substituents[177,185,186] in order to gain insight into the structure of the transition state for bimolecular elimination reactions of sulphonium salts.

For reaction of 2-arylethyldimethylsulphonium bromides with sodium hydroxide in water containing 0–85 mol-% DMSO, the following observations have been reported[177]. As the solvent is gradually enriched in DMSO the Hammett rho value undergoes a small, sharp increase and then remains almost constant ($>20\%$ DMSO) while the β-deuterium isotope effect k_H/k_D passes through a maximum; the rate increases by *ca.* 10^7-fold on going from water to 85% DMSO and there is a large reduction in both the enthalpy and entropy of activation. Although the variations of isotope effect and rho value are not marked (Table 6), it has been suggested that a more 'reactant-like' transition state is induced upon addition of DMSO. This view is supported by the corresponding trend in sulphur isotope effects[179] since it has been found that the rate acceleration is accompanied by a decrease in the sulphur isotope effect from 0.7% in water to 0.1% in DMSO. Approach to a more reactant-like transition is the trend to be expected upon increase in the basicity of the hydroxide ion and decrease in the stability of the ionic reactants with increase in concentration of the dipolar aprotic solvent.

A preliminary report[18] of a small (*ca.* 0.15%) ^{34}S isotope effect for E2 reaction of 2-phenylethyldimethylsulphonium bromide in aqueous alkali appeared to support a transition state with unexpectedly little C—S bond rupture, but this figure was later revised[178] to 0.63% (i.e. *ca.* 30–40% of the experimental and theoretical

TABLE 6. β-Deuterium isotope effects and
Hammett rho values for E2 reaction of dimethyl-
2-phenylethylsulphonium bromide in aqueous
DMSO at 50°C

DMSO (mol-%)	k_H/k_D	ρ
0.0	5.03	2.11
19.4	5.29	2.536
26.3	5.55	—
38.0	5.73	2.556
50.2	5.61	2.605
60.3	5.63	2.593
71.8	4.79	—

maximum value[18]), which parallels the behaviour of the nitrogen isotope effect for
the analogous trimethylammonium compounds.

The dependence of the β-deuterium isotope effect on the nature of the aryl
substituent has been investigated[176] for hydroxide ion-promoted 1,2-elimination of
2-phenylethyl bromides and 2-phenylethylsulphonium salts in 50.2 mol-%
DMSO–H_2O (Table 7); the intention was to test whether k_H/k_D also passes
through a maximum when the substituent is varied at this critical solvent
composition (cf. the solvent dependence[177]). All of the isotope effects are higher
than expected from theoretical considerations[187] and this may be a consequence of
proton tunnelling for which there is some evidence, particularly in the case of the
p-methoxy compounds. For both leaving groups a systematic increase in the
magnitude of the isotope effect is observed as the electron-withdrawing power of
the substituent (Y) increases, up to Y = p-acetyl; in the case of the phenylethyl
bromide there is a subsequent decrease in k_H/k_D when Y = p-NO_2 but this
attainment of a maximum could not be confirmed for the sulphonium series since it
was not possible to measure the isotope effect for the corresponding p-nitro
derivative.

Since the E2 reaction of dimethyl(2-phenylethyl)sulphonium ion with hydroxide
ion in water has been shown to be near the E1cB end of the E2 spectrum of
transition states[177], the increase in k_H^H/k_D^D as the substituent is changed from
Y = p-MeO to Y = p-Ac in 50.2 mol-% DMSO–H_2O must indicate a decreasing
extent of proton transfer; this is consistent with the prediction[188] that
electron-withdrawing substituents at $C_{(\beta)}$ will result in transition states which are
more reactant-like.

TABLE 7. β-Deuterium isotope effects
(k_H/k_D) for reaction[176] of YC_6H_4-β-CH_2-
α-CH_2X with NaOH in 50.2 mol-%
DMSO–H_2O at 20°C

Y	X = $\overset{+}{S}Me_2$	X = Br
H	7.4 ± 0.1	8.5 ± 0.2
p-Cl	8.1 ± 0.2	8.8 ± 0.2
p-Br	8.4 ± 0.1	9.2 ± 0.1
p-Ac	9.6 ± 0.5	9.7 ± 0.3
p-NO_2	—	8.7 ± 0.3

Although 2-phenylethyl(1,1-d_2-dimethyl)sulphonium bromide exchanges $ca.$ 75% of the deuterium label before it reacts in aqueous alkali at 80°C, it has been estimated that the secondary deuterium kinetic isotope effect for the β-elimination process is negligible[175]. It has correspondingly been argued that for reaction of 2-phenylethyl bromide, which displays an appreciable secondary isotope effect ($k_H/k_D = 1.17$) upon elimination in EtONa–EtOH, transition-state formation requires a larger electron transfer from carbon to halogen than is the case in C—S heterolysis.

A study of β-^{13}C isotope effects has been made in an attempt to determine the contribution which the tunnel effect makes to β-deuterium isotope effects for reaction of 2-phenylethyldimethylsulphonium and 2-phenylethyltrimethylammonium ions with sodium hydroxide in DMSO–water mixtures[181,182]. The ^{12}C/^{13}C and $(H/D)_{(\beta)}$ isotope effects depend on solvent composition in distinctly different ways, the former passing through a maximum and the latter passing through a shallow minimum (ammonium salt) or rising curvilinearly (sulphonium salt) with increase in the DMSO concentration; tunnelling cannot account for this dichotomy, although it generally makes an important contribution to the magnitude of both hydrogen and carbon isotope effects. Since the tunnelling contribution is relatively constant the variations in isotope effect are definitely a function of the structure of the reactant or of the base–solvent system; thus, the maximum k_H/k_D exhibited by each of the onium substrates as the proportion of DMSO is varied is still believed to be a consequence of a three-centre proton transfer process in which the proton is approximately half-transferred in the transition state.

Isotope effects (k_{DO^-}/k_{HO^-}, at 80.45°C) for E2 reactions of DO$^-$ in D$_2$O, $versus$ HO$^-$ in H$_2$O, have been determined for 2-phenylethyltrimethylammonium salts (1.79 ± 0.04) and related compounds[183,184]; these large effects indicate that proton transfer to the base is extensive in the transition state. It has been estimated that, even after allowance for a substantial solvent isotope effect (say 10%), the experimental isotope effect for the sulphonium ion is indicative of a transition state in which the proton is almost half-transferred between substrate and base; this is reasonably consistent with conclusions based on the corresponding primary deuterium isotope effects $(k_H/k_D)_{(\beta)}$ for reaction[189] in water at 80°C (4.14), 50°C (5.05) or 30°C (5.93), and also with the large Hammett rho values of 2.21 and 2.03 which have been determined[186] for reaction of this substrate at 30 and 60.55°C, respectively.

Hammett rho values have also been determined[185] for E2 reactions of 2-phenylethyldimethylsulphonium bromides (ρ = 2.64) and 2-phenylethyl bromides (ρ = 2.15) in EtONa–EtOH. The greater sensitivity of the sulphonium series to substituent effects suggests that more incipient negative charge develops on C$_{(\beta)}$ during reaction of these onium substrates. This also supports the Hughes-Ingold explanation of the empirical rules of orientation in elimination reactions which requires that the onium substrates (which obey the Hofmann rule) are more sensitive to electrostatic effects than are alkyl halides (which follow the Saytzeff rule). The activation enthalpies for reaction of these onium salts are 3–5 kcal higher than for the corresponding bromides, but a compensating entropic advantage ensures that the former substrates are the more reactive. As expected for reactions between oppositely charged ions, reaction rates for the sulphonium salts with hydroxide ion in water are $ca.$ 10^3-fold less than are those in ethanol[186], although the enthalpies of activation are nearly the same in the two solvents (Table 8). Correlations between the substituent induced chemical shifts for α- and β-methylene protons and rate constants for E2 reactions of aryl-substituted 2-phenylethyl bromides and 2-phenylethyldimethylsulphonium bromides have been discerned[190].

TABLE 8. Effect[185,186] of solvent on E2 reactions of p-XC$_6$H$_4$CH$_2$CH$_2$$\overset{+}{S}$(CH$_3$)$_2$ Br$^-$ at 30.05°C in EtO$^-$–EtOH and HO$^-$–H$_2$O

X	Solvent	k_2 (10^5 M^{-1} s^{-1})	$\Delta H^{\ddagger}$ (kcal mol^{-1})	$\Delta S^{\ddagger}$ (cal °C^{-1} mol^{-1})
H	EtOH	500	23.3	7.7
H	H$_2$O	0.347	24.0	−4.2
Me	EtOH	232	25.0	12.2
Me	H$_2$O	0.145	24.6	−4.2
McO	EtOH	111	25.4	11.6
MeO	H$_2$O	0.099	24.2	−6.2
Cl	EtOH	2440	22.2	7.2
Cl	H$_2$O	1.19	23.1	−4.9
MeCO	EtOH	34,000	12.3	−19.7
MeCO	H$_2$O	146	12.3	−30.9

b. *The 1-phenylethyl system*. In contrast with the near quantitative E2 reactions of 2-phenylethyldimethylsulphonium bromide in NaOEt–EtOH, the 1-phenylethyl isomer **44** (X = H) affords four products **45–48** by five competing pathways (equation 31)[191]. Primary deuterium isotope effects of 5.9 and 3.5 have been

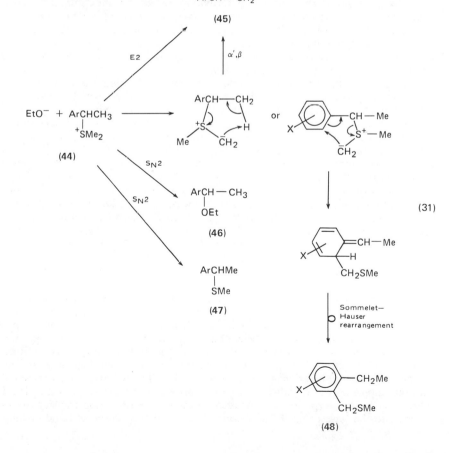

TABLE 9. Hammett correlations and isotope effects for reaction of 1-arylethyldimethylsulphonium bromides **44** with NaOEt–EtOH at 35°C (see equation 31)[191]

Reaction product	k_H/k_D	ρ	r
45	5.6^a	0.95 ± 0.19	0.94
46	0.95	-0.63 ± 0.41	0.67
47	1.26	0.95 ± 0.11	0.98
48	0.93	4.84 ± 0.38	0.99

^{a}Weighted average of k_H/k_D for E2 (5.9) and α',β (3.5) reactions.

determined for the E2 and α', β-components, respectively, by comparing the deuterium content of the methyl sulphide from β-d_3 and α-d_7 substrates. By analysis of the product distribution (where X = p-Me, p-F, p-Br, and p-Cl), and subsequent determination of the partial rates, it has also been possible to determine the Hammett rho values for the competing reactions (Table 9). Only for formation of **45**, which involves $C_{(\beta)}$—H bond breaking, is k_H/k_D large, facilitation of this process by the inductive effect of substituents X is believed to account for the rho value observed. The large rho value for formation of **48** by the Sommelet Hauser rearrangement is consistent with the ylide mechanism depicted in equation 31.

   c. *Other systems.* Although most attention has been given to E2 reactions of 2-phenylethylsulphonium salts, there have been complementary studies of the kinetic effects of isotopic substitution of trialkylsulphonium ions: Deuterium and sulphur isotope effects have been *theoretically* calculated for E2 reactions of hydroxide ion with ethyldimethylsulphonium ion[187]; the deuterium isotope effect goes through a maximum when the proton is half-transferred and is not strongly affected by the extent of weakening of the C—S bond. Similarly, the sulphur isotope effect increases with increasing extent of the C—S bond weakening and does not depend much on the extent of proton transfer. It is apparent, however, that the sulphur isotope effect may remain small until the C—S bond is extensively weakened and that k_H/k_D may be strongly influenced by the extent of coupling of the proton transfer with other atomic motions; the calculations make it clear that it is foolish to draw even semiquantitative conclusions from single measurements of isotope effects, although systematic changes with reactant structure or basicity of the medium may be more informative. Rather inconclusive deductions have been made from the observation[192] that for E2 reactions of t-butyldimethylsulphonium ion with ethoxide in 97% ethanol (at 24°C) the $^{32}S/^{34}S$ isotope effect is 0.72%, while for the corresponding S_N1–E1 reaction a value of 1.03% is obtained (at 40°C); the latter figure is considerably less than that for solvolysis in water (1.8%), and this has not yet been explained.

   The controversial suggestion has been made that bimolecular elimination reactions promoted by weak bases which are, however, good carbon nucleophiles may sometimes proceed by an E2C mechanism[193b] in which leaving group departure is assisted by a nucleophile–$C_{(\alpha)}$ interaction; this has prompted investigation of the relative reactivity of benzenethiolate and alkoxide ions in effecting E2 reactions[194–196]. Rates of reaction of benzenethiolates with t-butyl chloride and t-butyldimethylsulphonium iodide in ethanol have been correlated[194] with the appropriate pK_a values to give the Brönsted parameters in Table 10. The value β has been interpreted as a measure of the degree of proton transfer from the substrate to the base in the E2 transition state, or of the degree of bond making in

TABLE 10. Brönsted parameters for reactions of substituted benzene thiolates in ethanol[194]

Substrate	Temperature (°C)	Reaction type	β	Log G
t-BuCl	45.0	E2	0.17	−2.63
t-BuŠMe$_2$	25.0	E2	0.46	1.41
t-BuŠMe$_2$	25.0	S$_N$2	0.27	0.52

the S$_N$2 transition state. It is clear that increasing the electronegativity and tightness of binding of the leaving group results in a shift of E2 transition state character from the E1 type (chloride) to a central type (sulphonium salt).

In keeping with other E2/S$_N$2 reactions, the S$_N$2 reaction of the sulphonium salt (to give PhSMe and t-BuSMe) is less sensitive to the basicity of the nucleophile than is the competing E2 reaction. Benzenethiolate is 8.3 times more reactive than ethoxide in dehydrochlorinating t-BuCl but this trend is reversed for elimination in corresponding sulphonium ion, for which the ratio is 0.31:1; this has again been attributed to a shift of E2 transition state character from E1 type, to a central or E1cb-type for which the sensitivity to base strength apparently outweighs other factors which may determine nucleophilic reactivity. Similar trends have been reported[195] for E2 reactions of benzyldimethylcarbinyl derivatives (PhCH$_2$CMe$_2$X) for which the EtS$^-$/MeO$^-$ reactivity ratio (α) decreases as the leaving group ability decreases from X = Cl (6.5) to SMe$_2$ (0.8) and then SO$_2$Me (0.05).

2. Stereochemistry and orientation of E2 reactions

Investigations[197,198] of the stereochemistry of E2 reactions of secondary substrates have revealed an interesting dichotomy between the behaviour of onium and neutral substrates as the base–solvent system is varied. Thus, while the ratio of cis to trans olefin obtained in base-catalysed E2 reactions of 5-nonyl bromide increases on going from MeO$^-$–MeOH to EtO$^-$–EtOH to t-BuO$^-$–t-BuOH, the opposite trend is found for 5-nonyltrimethylammonium and 5-nonyltrimethylsulphonium ions (Table 11). It has been suggested that this is a consequence of reaction of onium substrates by a E1cb-like mechanism since this should be favoured in EtO$^-$−EtOH (relative to t-BuO$^-$−t-BuOH) and for ammonium ions more than for sulphonium ions (in any one solvent).

Hofmann elimination reactions of cis and trans isomers of the activated 2-phenylcyclohexyl onium system, in EtONa–EtOH at 25°C, occur preferentially by

TABLE 11. Stereochemistry of alkenes obtained[198] by base-catalysed elimination reactions: PrCH$_2$CHXCH$_2$Pr → PrCH=CHCH$_2$Pr

X	Base–solvent	cis-Non-4-ene in the cis/trans mixture (%)
Me$_3$N̈	t-BuO$^-$–t-BuOH	2.6
	EtO$^-$–EtOH	74
	MeO$^-$–MeOH	81
Me$_2$Š	t-BuO$^-$–t-BuOH	9
	EtO$^-$–EtOH	64

trans elimination[199]. The relative rate of *trans* elimination from the *cis* isomer to *cis* elimination from the *trans* isomer is 133 for 2-phenylcyclohexyltrimethylammonium ion and 383 for reaction of the dimethylsulphonium ion; these ratios are small in comparison with those found for alkyl halides, and a comparison of the rates of reaction with those for corresponding 2-phenylethyl onium salts suggests that this may be because the *trans* elimination is abnormally slow in the case of the ammonium compound whereas *cis* elimination (apparently by a multi-stage, E1cb or α', β mechanism) of the sulphonium salt is abnormally fast.

The contrasting behaviour of onium and halide substrates with regard to preference for *syn* or *anti* elimination or for formation of *cis*- or *trans*-alkenes prompted extensive investigation of the kinetics and stereochemistry of reaction a wide range of cyclic and acyclic onium substrates. The emphasis has been on ammonium salts, for which detailed comment would be inappropriate in this text. The reader is referred to authoritative texts[193c,200] on elimination reactions, in which relevant studies (particularly those of Bartsch, Zavada and coworkers during the period 1965–77) have been reviewed.

Illustrative of developments during this period is the realization that the stereochemistry of E2 reactions of cycloalkylonium ions in RO⁻–ROH (whereby *cis*-alkenes and *trans*-alkenes are formed by *anti* and *syn* elimination mechanisms, respectively) is much less dependent upon ion-pair effects than for reactions of analogous cycloalkyl halides[200a]. The increase in proportion of *trans*-alkene formed from cycloalkylammonium and cycloalkylsulphonium salts, as the solvent–base system is varied in the series MeOK–MeOH, EtOK–EtOH, *i*-PrOK–*i*-PrOH, *t*-BuOK–*t*-BuOH, apparently reflects an increase in proportion of *syn* elimination via an E1cb-like transition state which is favoured by increased proton affinity of the base and decreased solvating power of the solvent[200b,201].

a. *Hofmann* versus *Saytzeff orientation*. Ingold and coworkers studied the kinetics of reaction of $RR'CHCH_2SMe_2$ and $RR'CHCH_2NMe_3$ (where R = H, Me, Et, *i*-Pr, *t*-Bu, and R' = H, Me) with EtO⁻–EtOH and *t*-BuO⁻–*t*-BuOH in an attempt to decide whether the Hofmann orientation[193d], which is typical of E2 reaction of onium substrates, is due to steric or to electronic effects[202]. In contrast with claims of Brown *et al.*, it has been argued that inductive and electromeric effects cause the familiar kinetic and orientational pattern, and that steric hindrance only becomes a complicating factor above determinable thresholds of molecular complexity. The following general features are observed for the reactions depicted in equation 32.

$$RR'CHCH_2\overset{+}{S}Me_2 + AlkO^- \longrightarrow \begin{cases} RR'CHCH_2SMe + MeOAlk \\ RR'CHCH_2OAlk + SMe_2 \\ RR'C{=}CH_2 + HOAlk + SMe_2 \end{cases} \quad \begin{matrix} \\ S_N2 \\ E2 \end{matrix} \quad (32)$$

k^{S_N2} is approximately independent of alkyl structure of the substrate (less true for reactions with *t*-BuO⁻ than with EtO⁻), whereas k^{E2} is markedly dependent on the alkyl structure and follows the Hofmann pattern of variation; k^{E2} spans the ranges 10^2 and 10^{4-5} for reaction of the sulphonium and ammonium salts, respectively. The spread of elimination rates is greater for reactions in *t*-BuO⁻–*t*-BuOH and less pronounced for sulphonium than for ammonium salts which generally give higher yields of elimination product; the latter observation has been ascribed to greater β-H activity of the ammonium salts, (relative to the sulphonium salt) as a consequence of the greater electronegativity of the ammonium (relative to the sulphonium) group. An attempt has been made to interpret these results on a theoretical basis, and also to predict the proportions of alkenes to be expected from pyrolyses of sulphonium (and ammonium) hydroxides (see Tables 12–17).

TABLE 12. Second-order rate consts [k_2 (M^{-1} s^{-1})] and alkene proportions (%) for reactions[202] of RR'CHCH$_2$ṠMe$_2$ with EtO$^-$–EtOH at 64.08°C

Alkyl-ṠMe$_2$	R	R'	$10^5 k_2$	Alkene (%)	$10^5 k_2$(S$_N$2)	$10^5 k_2$(E2)
Me	—	—	480	—	480	—
Et	H	H	383	20.5	304	79
n-Pr	Me	H	381	7.6	352	29
n-Bu	Et	H	406	5.2	385	21
Isopentyl	i-Pr	H	431	3.8	415	16
3,3-Dimethylbutyl	t-Bu	H	361	0.12	361	0.43
Isobutyl	Me	Me	503	2.0	493	10

TABLE 13. Second-order rate consts [k_2 (M^{-1} s^{-1})] and alkene proportions (%) for reactions[202] of RR'CHCH$_2$ṄMe$_3$ with t-BuO$^-$–t-BuOH at 72.85°C

Alkyl-ṄMe$_3$	R	R'	$10^5 k_2$	Alkene (%)	$10^5 k_2$(S$_N$2)	$10^5 k_2$(E2)
Et	H	H	30,200[a]	99.8	—	30,100
n-Pr	Me	H	513	72.8	140	373
n-Bu	Et	H	274	31.9	186	87.7
Isopentyl	i-Pr	H	211	10.7	188	22.5
3,3-Dimethylbutyl	t-Bu	H	181	0.42	180	0.72

[a]Extrapolated from results at lower temperature.

TABLE 14. Second-order rate consts[202] [k_2 (M^{-1} s^{-1})] and alkene proportions (%) for reactions[202] of RR'CHCH$_2$ṠMe$_2$ with t-BuO$^-$–t-BuOH at 34.86°C

Alkyl-ṠMe$_2$	R	R'	$10^5 k_2$	Alkene (%)	$10^5 k_2$(S$_N$2)	$10^5 k_2$(E2)
Et	H	H	835	32.5	564	271
n-Pr	Me	H	402	8.2	369	33
n-Bu	Et	H	350	5.5	331	19
Isopentyl	i-Pr	H	343	2.3	335	7.9
3,3-Dimethylbutyl	t-Bu	H	270	1.2	267	3.2

TABLE 15. Second-order rate consts [k_2 (M^{-1} s^{-1})] and alkene proportions (%) for reactions[202] of RR'CHCH$_2$ṄMe$_2$ with EtO$^-$–EtOH at 104.22°C

Alkyl-ṄMe$_3$	R	R'	$10^5 k_2$	Alkene (%)	$10^5 k_2$(S$_N$2)	$10^5 k_2$(E2)
Me	—	—	20.3	—	20.3	—
Et	H	H	101.8	70.1	30.5	71.3
n-Pr	Me	H	27.6	18.7	22.4	5.16
n-Bu	Et	H	26.6	10.6	23.8	2.82
n-Decyl	C$_8$H$_{17}$	H	17.3	15.0	14.7	2.60
Isopentyl	i-Pr	H	26.4	4.0	25.3	1.06
3,3-Dimethylbutyl	t-Bu	H	24.0	0.35	24.0	0.084
Isobutyl	Me	Me	27.1	6.2	25.4	1.68

TABLE 16. Proportions (%) of alkene from pyrolyses[202] of $RR'R''\overset{+}{S}$ OH^-

R	R'	R''	Found	Calculated
Et	Me	Me	27	27
n-Pr	Me	Me	8	10
n-Bu	Me	Me	4	8
n-Hexyl	Me	Me	19	~7
n-Decyl	Me	Me	21.9	~7
Isobutyl	Me	Me	2	2
i-Pr	Me	Me	63	81
t-Bu	Me	Me	100	100
Et	Et	Me	55	52
Et	Et	Et	86	80
n-Pr	n-Pr	Me	18	27
n-Pr	n-Pr	n-Pr	36	57
i-Pr	i-Pr	i-Pr	100	100

TABLE 17. Proportions (%) of lower homologues and proportions (%) of alk-1-ene isomers from pyrolyses[202] of $RR'R''S$ OH^-

R	R'	R''	Found	Calculated
Et	Et	n-Pr	85	85
Et	n-Pr	Me	80	73
Et	n-Pr	n-Pr	63	60
Et	i-Pr	Me	25	7
Et	i-Bu	Me	93	90
Et	i-Bu	i-Bu	94	80
n-Pr	n-Pr	n-Bu	74	73
n-Pr	n-Bu	n-Bu	40	40
i-Pr	i-Bu	Me	94	99
s-Bu	Me	Me	73	73
1-Methylbutyl	Me	Me	87	80
t-Pentyl	Me	Me	86	85

C. E1cb Reactions of Sulphonium Salts

1. Activation by the sulphonium group

There have been relatively few reports of elimination reactions which are activated by a sulphonium group in the β-position[203-209]. These reactions, in general, proceed by the E1cb mechanism, e.g. as illustrated by base-induced elimination of phenoxide ion from aryl ethyl ethers which bear activating sulphonium, acetyl, and sulphonyl groups in the β-position[206-208]. Thus, rates of elimination from β-phenoxyethyldimethylsulphonium iodide and methyl β-phenoxyethyl sulphoxide in aqueous hydroxide are ca. 1.5-fold less than for reaction of the corresponding bis-β-deuterio-substrates in D_2O[207]; this corresponds (upon correction for solvent and secondary isotope effects) to a primary isotope effect of about unity and is inconsistent with a concerted E2 mechanism. Further, for reaction in EtONa–EtOH the effect of nuclear substitution[207] of the leaving group of 2-aryloxyethyl sulphones (ρ = 1.5), sulphoxides (ρ = 1.2), and sulphonium salts is insensitive to the nature of the activating group and is consistent with a pre-equilibrium E1cb process in which there is a moderate degree of extension of the bond to the leaving group. Nuclear substitution of the activating group gives Hammett plots for which the rho values are large and positive (1.7 and 2.1 for aryl sulphoxides and aryl sulphones, respectively); these values are also consistent with those expected for control of acid dissociation of the $C_{(\beta)}$—H bond combined with a small composite effect on k_2 (equation 33). The results have been interpreted in terms of a pre-equilibrium mechanism, where k_{-1} [BH] $\gg k_2$, and $k_{obs} = k_1 k_2 / k_{-1}$ [BH].

$$XCH_2CH_2OAr + B^- \underset{k_{-1}}{\overset{k_1}{\rightleftharpoons}} X\bar{C}HCH_2OAr + BH$$

$$\downarrow k_2 \qquad\qquad (33)$$

$$XCH_2CH_2OY \xleftarrow[YOH]{base} XCH=CH_2 + ArO^-$$

The alkaline fission of several compounds of the general structure $YCH_2CH_2\overset{+}{S}Me_2$ I^- has been found to occur by one or more of the reactions[204] in equations 34–36.

$$YH + HC\equiv CH + SMe_2 + HI \quad \text{Double Elimination} \qquad (34)$$

$$YCH_2CH_2\overset{+}{S}Me_2\ I^- \longrightarrow YCH=CH_2 + SMe_2 + HI \quad \text{Single Elimination} \qquad (35)$$

$$YCH_2CH_2SMe + MeI \quad \text{Dissociation} \qquad (36)$$

Single elimination (Y = MeS or PhCONH) and dissociation (Y = PhCH$_2$NH) reactions are well known, but a claim[204] that double elimination (which predominates when Y = PhCOO or PhO) occurs by a concerted mechanism, rather than by the more probable[207] stepwise mechanism, has been revised[205] (at least for the 2-aroyloxy derivatives). Thus, it has been shown[205] that alkaline fission of 2-aroyloxydimethylsulphonium iodides proceeds in the following stages:

(i) $ArCO_2(CH_2)_2\overset{+}{S}Me_2\ I^- + NaOH \rightleftharpoons ArCO_2Na + CH_2=CH\overset{+}{S}Me_2\ I^- + H_2O$

(ii) $CH_2=CH\overset{+}{S}Me_2\ I^- + NaOH \longrightarrow HC\equiv CH + SMe_2 + NaI + H_2O$

The first step (at 25°C) occurs much more rapidly than the second and is reversible since the sulphonium substituent activates the vinyl group towards anionic addition. The first step, which is subject to a negative primary salt effect, is activated by electron-attracting aryl substituents, there being no significant *ortho*-effect.

A sulphonium ylide has also been implicated in the mechanism of hydrolysis of *o*-nitrophenyl-α-dimethylsulphonium acetate; an intermediate ketene, apparently formed by an E1cb reaction, reacts with either the lyate species (to produce the carboxylic acid) or with added nucleophiles (to form acyl derivatives).

2. Elimination of the sulphonium group

Stirling and coworkers have exploited[210] the reversible E1cb mechanism of reaction of β-substituted sulphones (PhSO$_2$CH$_2$CH$_2$Z) in EtONa–EtOH in order to determine the leaving-group abilities of a wide range of groups Z; this is feasible since values of k_1/k_{-1} (cf. equation 33) can be estimated and changes in k_{obs} give relative values of k_2. The results show the following striking features: (a) the range

TABLE 18. Relative rates of $(Elcb)_R$ reaction of $PhSO_2CH_2CH_2Z$ with EtONa–EtOH at 25°C, which are a good indication of the leaving group ability of Z

	Z							
	$\overset{+}{S}MePh$	$\overset{+}{N}Me_2Ph$	$\overset{+}{P}Ph_3$	OPh	OMe	SPh	SO_2Ph	SOPh
$Log(k^Z/k^{OPh})$	+6.25	+5.17	+4.40	0	−3.91	−1.22	+0.30	−0.15

of leaving group ability is very large (*ca.* 10^6), (b) positively charged leaving groups (including sulphonium) depart very rapidly, irrespective of the type of atom connection, and (c) for neutral leaving groups, reactivity shows no correlation with the pK_a of the acid ZH, the C—Z bond strength, or the nucleophilicity of Z^- towards MeI, but is broadly dependent on the group of the periodic table to which the connecting atom belongs. The results in Table 18 illustrate the trends observed.

D. The α′,β-Elimination Mechanism

Sulphonium or ammonium salts which possess α-hydrogen(s) on one alkyl group and β-hydrogen(s) on another may undergo α′, β-elimination whereby the conjugate base, obtained by removal of an α-proton, promotes intramolecular β-elimination[193e,211]. While the α′, β-mechanism can be observed with very hindered

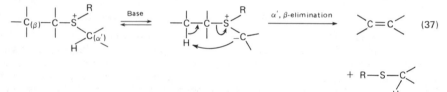

$$\qquad (37)$$

substrates or with organometallic bases, it is never predominant for reactions of simple onium salts in aqueous hydroxide solutions. The unusually high *trans/cis* ratios (3.5 and 5.0, respectively) obtained for the pent-2-enes obtained from 2- and 3-pentyldimethylsulphonium ions with *t*-BuOK–*t*-BuOH have, however, been attributed to operation of the α′, β-mechanism; by investigation of the deuterium content of the dimethyl sulphide obtained upon reaction of 3-pentyl-2,2,4,4-d_4-dimethylsulphonium iodide with BuOK–BuOH, *t*-BuOK–*t*-BuOH and *t*-BuOK–95% DMSO–5% *t*-BuOH it has been found that the percentages of elimination which proceed by the α′, β-route are 1.7, 65.2 and 73, respectively[212].

The only prior examples of α′, β-elimination from sulphonium salts involved tritylsodium as the base under aprotic conditions[213,214]; thus, tritylsodium reacts (a) with cyclooctyldimethylsulphonium iodide to give *cis*-cyclooctene by a *syn*-α′, β-elimination mechanism[213], (b) with $(MeCD_2)_3S^+$ to form ethene by predominant attack on the methylene group[214], and (c) with **49** to form triphenylmethane which is 79% deuterated[213] even though the β-hydrogen is activated by a phenyl group (equation 38). In contrast, the negligible deuterium content of the dimethyl

$$PhCH_2CD_2\overset{+}{S}(CD_3)_2Br^- \xrightarrow{Ph_3\bar{C}\,Na^+} Ph_3CD + PhCH{=}CD_2 + CD_3SCHD_2 \qquad (38)$$

$$\text{(49)} \qquad\qquad\qquad \text{(79\% deuterated)}$$

TABLE 19. The α′,β mechanism in eliminations from β-perdeuterated sulphonium iodides $(R\overset{+}{S}Me_2I)^{216}$

R	Base/solvent	Temperature (°C)	% α′,β-
2-Propyl	n-BuOK/n-BuOH	35	0.9 ± 0.2
2-Propyl	t-BuOK/t-BuOH	35	21.2 ± 2.0
3-Pentyl	n-BuOK/n-BuOH	35	1.7 ± 0.3
3-Pentyl	t-BuOK/t-BuOH	35	65.2 ± 0.6
3-Pentyl	t-BuONa/t-BuOH	35	60.0 ± 2.0
3-Pentyl	t-BuOK/5% t-BuOH–95% DMSO	35	73.0 ± 0.2
Cyclopentyl	NaOH/H₂O	90	0.0 ± 0.1
Cyclopentyl	n-BuOK/n-BuOH	35	2.6 ± 1.0
Cyclopentyl	t-BuOK/t-BuOH	35	82.4 ± 0.4
Cyclohexyl	n-BuOK/n-BuOH	96	0.0 ± 0.1
Cyclohexyl	t-BuOK/t-BuOH	40	4.3 ± 0.2
2-Phenylethyl	NaOH/H₂O	96	0.0
2-Phenylethyl	NaOH/39% H₂O–61% DMSO	30	1.0 ± 0.0
2-Phenylethyl	NaOH/16% H₂O–84% DMSO	30	12.2 ± 0.6
2-Phenylethyl	NaOH/16% H₂O–84% DMSO	60	9.2 ± 0.8

sulphide obtained upon reaction of $PhCD_2CH_2\overset{+}{S}Me_2$ with aqueous hydroxide[215] indicates that this base system is unable to promote the α′, β-mechanism in this case.

The results in Table 19 (which were obtained by tracer studies) suggest that the structure of the sulphonium salt also affects the propensity toward α′, β-elimination[216]. The trend in α′, β-elimination 3-pentyl > 3-propyl is believed to be a consequence of the relative ease of the competing E2 reactions, whereas the trend cyclopentyl > 3-pentyl > cyclohexyl has been ascribed to conformational influences on the rate of ylide fragmentation; it has correspondingly been concluded that the five-membered cyclic transition state is most easily formed when the C—S and $C_{(\beta)}$—H bonds are mostly nearly eclipsed (cyclopentyl), and least easily formed when they are most nearly staggered (cyclohexyl). The intermediate behaviour of the 3-pentyl derivative, which has relatively free rotation about the $C_{(\alpha)}$—$C_{(\beta)}$ bond, is much closer to that of the cyclopentyl than of the cyclohexyl derivative.

It is clear that an increase in base strength of the reaction medium favours the α′, β-more than the E2 mechanism, and this has been attributed to the concomitant increase in concentration of the ylide, rather than to a solvent effect on the rate of its subsequent intramolecular decomposition; the rate of ylide formation is unlikely to be an important factor since it has been established that, even for reaction of 3-pentyldimethylsulphonium iodide under the relatively mild conditions of sodium deuteroxide in D_2O, the α′-hydrogens (methyl) are ca. 98% exchanged within one half-life of the elimination reaction. That ion pairs are involved in t-BuOH is suggested by the significantly greater proportion of α′, β-elimination obtained with t-BuOK than with t-BuONa in the case of 3-pentyldimethylsulphonium iodide.

E. α-Elimination Reactions

Carbenes are commonly generated by thermal or photochemical decomposition of diazoalkanes, and there is mounting evidence for carbene formation by photolytic α-elimination reactions of the isoelectronic sulphur ylides[78,217–220]. Thus, upon irradiation of dimethylsulphonium phenacylide in cyclohexene, a mixture of 7-benzoylnorcarane and acetophenone is obtained in nearly the same ratio as that obtained upon photolysis of diazoacetophenone under the same conditions (equation 39)[217,221]. Likewise, the formation of cyclopropene as the major product

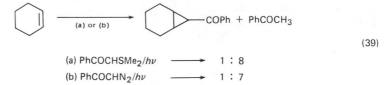

$$(39)$$

(a) PhCOCHSMe$_2$/$h\nu$ $\longrightarrow$ 1 : 8
(b) PhCOCHN$_2$/$h\nu$ $\longrightarrow$ 1 : 7

of photodecomposition of diphenylsulphonium allylide can be attributed to cyclization of the intermediate vinyl carbene (equation 40)[78].

$$Ph_2S\diagdown\diagup \xrightarrow[-78\,°C]{h\nu} Ph_2S + \overset{..}{\diagdown}\diagup \longrightarrow \triangledown \qquad (40)$$

In contrast, there is little evidence[219,222,223] for thermal generation of carbenes from sulphonium ylides, although cyclopropanation has been achieved in low yield by copper-catalysed thermolysis of dimethylsulphonium phenacylide in the presence of cyclohexene[217,220]. The results of a detailed investigation of reactions of p-nitrobenzylsulphonium ions in aqueous alkali have been interpreted by implicating p-nitrophenylcarbene in the production of pp'-dinitrostilbene **51** and its derivatives[224–227]. The formation of cis- and trans-**51** in 99% yield upon reaction of p-nitrobenzyldimethylsulphonium ion in aqueous NaOH is first order in both hydroxide ion and sulphonium ion, and is preceded by rapid exchange[224] of the α-hydrogen atoms of the substrate; these results suggest that the reaction proceeds by reversible formation of the ylide, but the sulphur isotope effect[225] ($k^{32}/k^{34} = 1.0066 \pm 0.0008$) is smaller than that expected for subsequent elimination of dimethyl sulphide to form the intermediate carbene. It has therefore been suggested[225,227] that the reaction *may* proceed via steps 1–4 (equation 41) such that the isotope effect becomes the mean of those for steps 2 and 4.

(1) $\quad ArCH_2\overset{+}{S}Me_2 + HO^- \underset{}{\overset{fast}{\rightleftarrows}} Ar\overset{-}{C}H\overset{+}{S}Me_2 + H_2O$

(2) $\quad Ar\overset{-}{C}H\overset{+}{S}Me_2 \xrightarrow{slow} Ar\overset{..}{C}H + Me_2S$

(3) $\quad Ar\overset{..}{C}H + Ar\overset{-}{C}H\overset{+}{S}Me_2 \rightleftarrows \underset{\underset{Ar}{|}}{Ar\overset{-}{C}H\overset{+}{C}H\overset{+}{S}Me_2}$

$$(50)$$

$$(41)$$

(4) $\quad$ **50** $\xrightarrow{fast}$ ArCH=CHAr

$$(51)$$

$$Ar = p\text{-}NO_2C_6H_4$$

The nature of the substituent on the phenyl ring is critical[224] and it has been found that unsubstituted, m-chloro- and p-methylbenzyldimethylsulphonium ions do not yield stilbenes but give high yields of the corresponding alcohols when treated with base[223]. The corresponding m-nitro compound forms the benzyl alcohol (5%) and mm'-dinitrostilbene oxide[224] when the reaction is performed under air or nitrogen; alternative schemes (equations 42a and 42b) have been tentatively

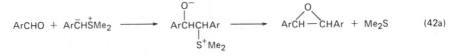

$$\text{ArCHO} + \text{ArC}^-\overset{+}{\text{H}}\text{SMe}_2 \longrightarrow \underset{\underset{\overset{|}{\text{S}^+\text{Me}_2}}{\overset{|}{\text{ArCHCHAr}}}}{} \longrightarrow \text{ArCH}\overset{\text{O}}{-}\text{CHAr} + \text{Me}_2\text{S} \qquad (42a)$$

$$\text{or ArCHO} + \text{Ar}\overset{..}{\text{CH}} \longrightarrow \text{ArCH}\overset{\text{O}}{-}\text{CHAr} \qquad (42b)$$

proposed to account for formation of the latter product, it being assumed that the aldeyhde arises through involvement of the corresponding carbene in an intermolecular oxydation–reduction process.

The nature of the leaving group is also important and, in contrast with the sulphonium analogue, p-nitrobenzyltrimethylammonium ion is unreactive even in refluxing aqueous hydroxide. p-Nitrobenzyltriphenylphosphonium bromide forms triphenylphosphine oxide (88–97%) and p-nitrotoluene (56%) (presumably by initial attack of HO$^-$ on phosphorus), together with pp'-dinitrostilbene in a yield of only 0.5%[227].

The dependence of stilbene formation on the nature of the R group in p-NO$_2$C$_6$H$_4$CH$_2$SMeR X$^-$ has also been investigated[226] for reactions conducted at 60°C in aqueous NaOH under nitrogen, with the results given in Table 20. Since the yield of the stilbene is relatively insensitive to the electronic nature of the aryl leaving group R, it is probably that steric, rather than electronic, effects account for the progressively higher yields obtained when R is changed from Me to i-Pr and then Ar.

IV. POLAR ADDITION TO UNSATURATED SULPHONIUM SALTS

In 1955 Doering and Schreiber reported an investigation of hitherto unknown reactions of nucleophiles with vinyl-sulphonium and -ammonium salts[228]. The former salts had been prepared by a rapid hydroxide-induced dehydrobromination of 2-bromoethyldimethylsulphonium ion; this reaction is in contrast with the much slower S$_N$2 displacement reaction of 2-bromoethyltrimethylammonium ion, to give choline, under the same conditions. The vinyl sulphonium salt **52** reacts very rapidly with a variety of nucleophiles (equation 43) to give 2-substituted-ethyldimethylsulphonium ions, whereas it was impossible to add bases to vinyltrimethylammonium ion. For example, reaction of ethanol with **52** is complete within a few minutes at room temperature in the presence of a trace

TABLE 20. Dependence of stilbene formation on the nature of the R group in p-NO$_2$C$_6$H$_4$CH$_2$SMeR X$^-$

Yield of p-nitrobenzyl alcohol (%)	Yield of pp'-dinitrostilbene (%)	R	X$^-$
19	19	Ph	ClO$_4^-$
10	16	p-ClC$_6$H$_4$	ClO$_4^-$
12	21	p-MeC$_6$H$_4$	ClO$_4^-$
<0.5	48	i-Pr	Br$^-$
0	99	Me	TsO$^-$

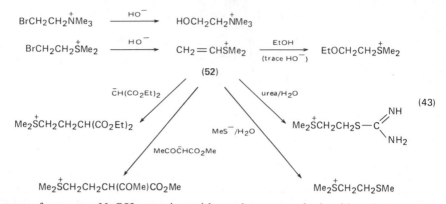

amount of aqueous NaOH; reaction with weak aqueous hydroxide solution gives $Me_2\overset{+}{S}CH_2CH_2OCH_2CH_2\overset{+}{S}Me_2$, presumably via $HOCH_2CH_2\overset{+}{S}Me_2$. The contrasting behaviour of vinyl-sulphonium and -ammonium salts has been attributed to the ability of the sulphonium moiety to stabilize an α-carbanion by d-orbital participation.

The reactivity of vinylsulphonium salts is markedly dependent on the nature of substituents on the vinyl group[229]. Thus, nucleophilic attack on $C_{(\beta)}$ is hindered in the case of $PhCH{=}CH\overset{+}{S}Me_2$ which reacts, in aqueous alkali, to give phenylacetylene (14.7%) and styrene oxide (13.3%); the epoxide is formed by intramolecular reaction of the addition product (i.e. a 2-hydroxyalkylsulphonium salt). The alkyne-forming elimination reaction is more pronounced when the β-hydrogen is activated, as in $PhCO\overset{.}{C}H{=}CPh\overset{+}{S}Me_2$ which forms $PhCOC{\equiv}CPh$ (74%) in aqueous alkali at 0°C; $Me_2\overset{+}{S}CPh{=}CHCO_2Et$ behaves similarly[230]. The additional carbanion stabilizing group on $C_{(\alpha)}$ ensures that $ArCH{=}C(COPh)\overset{+}{S}Me_2$ reacts with ethoxide via $ArCH(OEt)\overset{-}{C}(COPh)\overset{+}{S}Me_2$ which, during work-up, is converted into ArCHO and $Me_2\overset{+}{S}CH_2COPh$.

Initial carbanionic addition of the sodium salt of an active methylene compound $H_2CR^3R^4$ (where $R^4 = CN$ and $R^3 = Ar$ or CN) to α,β-disubstituted vinyl dimethylsulphonium salts $Me_2\overset{+}{S}CR^1{=}CHR^2$ forms the basis of a useful

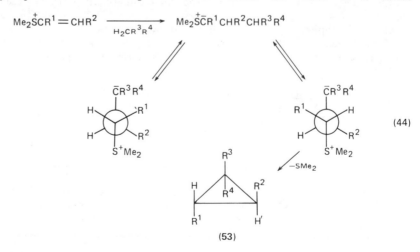

stereospecific synthesis of cyclopropanes according to the sequence in equation 44[231,232]. In each case the cyclopropane **53**, in which H and H′ have a *trans* orientation, is formed stereospecifically. This work has been extended[233] to include reactions of $Ph_3\overset{+}{P}CHR^1$ with $R^2CH{=}CHSR_2$.

Lithium enolates of acyclic ketones[234] also add, as carbon nucleophiles, to vinylsulphonium salts, whereupon the adducts may cyclize to give cyclopropanes (cf. equation 44); upon reaction of dimethylstyrylsulphonium perchlorate with the lithium enolates of cyclopentanone and cyclohexanone, the intermediate ylides react by alternative routes to give spirocyclopropanes and 8*a*-hydroxy-2-thiadecalins, respectively[235].

Stirling and coworkers have investigated nucleophilic addition–displacement reactions of allenic sulphonium salts with carboxylate ions[236], carbonyl-stabilized carbanions[237–239] and a range of monodentate nucleophiles[240,241]; recent interest has focused on reactions in which initial nucleophilic addition is followed by intramolecular substitution[237–238,241], addition–elimination[237,241], or ylide rearrangement[237]. Thus, in neutral or alkaline solution, dimethylprop-2-ynylsulphonium bromide **54** isomerizes readily to its allenic isomer **55**, which undergoes base-catalysed nucleophilic addition of carboxylic acid in Et_3N–DMSO; amines and alcohols react rapidly with the carbonyl group of the resulting adducts (activated vinyl esters) **56** to give amides and esters, respectively. This reaction sequence has been incorporated in a novel synthetic scheme whereby alkynylsulphonium salts are prepared (from sulphide and alkynyl bromide) and used, *in situ*, as acylation catalysts.

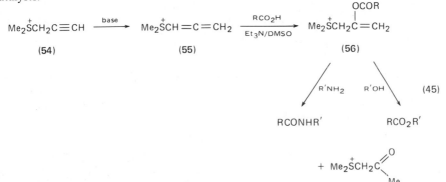

The salt **54** also reacts with alcohols, thiols, and amines (LH) to give $Me_2\overset{+}{S}CH_2C(L){=}CH_2$ adducts[240], presumably via the allene **55**; under the relatively strong basic conditions employed, this kinetic product subsequently isomerizes to the thermodynamically more stable isomer $Me_2\overset{+}{S}CH{=}C(L)Me$, which may then react with L′H by 1,2-addition or by dealkylation to give $Me_2\overset{+}{S}CH_2C(L)(L')Me$ or $Me\overset{+}{S}CH{=}C(L)Me$, respectively. Thus, reaction of **55** with $Ph\overset{-}{S}Na^+$ in ethanol gives $Me_2\overset{+}{S}CH_2C(SPh)_2Me$ and $MeSCH{=}C(SPh)Me$. Addition of cyanide or arenesulphinate ions[241] to **55** gives allenic sulphonium salts which are susceptible to attack of a further anion at the electrophilic double bond; this occurs with displacement of the sulphonium group (presumably by an S_N' pathway) and may be followed by isomerization or by further Michael addition (equation 46).

Reactions of **55** with bidentate sulphur nucleophiles have also been investigated[241]; with disodium ethane-1,2-disulphinate and acetic acid (proton donor) the cyclic bis-sulphone **57** is obtained, apparently by an addition–S_N' sequence (equation 47).

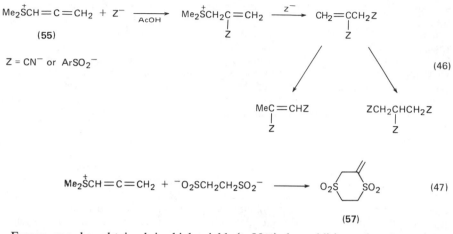

$$Me_2\overset{+}{S}CH=C=CH_2 + Z^- \xrightarrow[AcOH]{} Me_2\overset{+}{S}CH_2\underset{Z}{C}=CH_2 \xrightarrow{Z^-} CH_2=\underset{Z}{C}CH_2Z$$

(55)

Z = CN⁻ or ArSO₂⁻   (46)

$$MeC=\underset{Z}{CHZ} \qquad ZCH_2\underset{Z}{CHCH_2}Z$$

$$Me_2\overset{+}{S}CH=C=CH_2 + {}^-O_2SCH_2CH_2SO_2{}^- \longrightarrow \quad (47)$$

(57)

Furans can be obtained in high yield (>80%) by addition of a β-oxoester, β-oxosulphone, or β-diketone in EtONa–EtOH to an ethanolic solution of **54**; the probable reaction mechanism is depicted in equation 48.

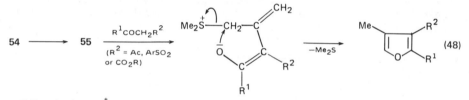

$$54 \longrightarrow 55 \xrightarrow[(R^2 = Ac, ArSO_2 \\ or\ CO_2R)]{R^1COCH_2R^2} \qquad \xrightarrow{-Me_2S} \qquad (48)$$

Likewise, $Me_2\overset{+}{S}CH=C=CHPh\ Br^-$ reacts with sodium acetoacetate to give 4-benzyl-3-carbethoxy-2-methylfuran (63%). In contrast, the adduct obtained upon reaction of $Me_2\overset{+}{S}C(Me)=C=CH_2\ Br^-$ with sodium acetoacetate is believed to cyclize by an intramolecular S_N2' mechanism (equation 49).

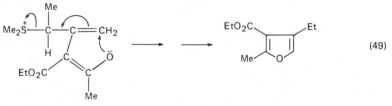

(49)

For reaction of the unsubstituted allene **55** with methylacetoacetate, an elegant isotopic labelling experiment has been conducted in order to determine whether the cyclization step follows an S_N or S_N' mechanism[241]. The p.m.r. spectrum of the furan obtained from [α-¹³C]-labelled **55** showed satellites of the $C_{(5)}$ proton signal whereas no satellite signals of the $C_{(4)}$ methyl group were observed; the c.m.r. spectrum correspondingly showed powerful intensification of the signal assigned to $C_{(5)}$. This is consistent with the S_N mechanism, with which the rival S_N' mechanism is apparently able to compete only when there is steric hindrance (e.g. by a methyl group) at the α-position (see equation 50).

Addition of a cyano-stabilized enolate to **54** gives a cyanofuran, but heterocyclic products are not obtained when the stabilizing group is dimethylsulphonio or

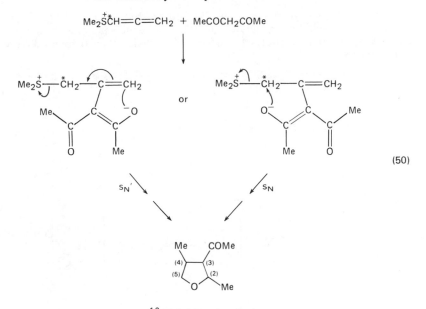

$$(50)$$

$(^{13}C \text{ label found at } C_{(5)})$

nitro[241]; thus, the conjugate base of $PhCOCH_2\overset{+}{S}Me_2$ becomes degraded by ethoxide ion (equation 51) more rapidly than it reacts with **54**.

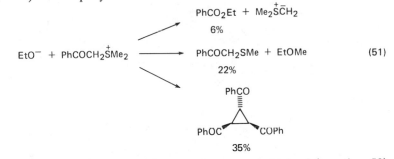

$$(51)$$

It is noteworthy that cyclopropanes, but not furans, are obtained (equation 52) upon reaction of active methylene compounds with dimethylvinylsulphonium bromide[238,239]. This dichotomy may reflect a preference of the allylic centre to

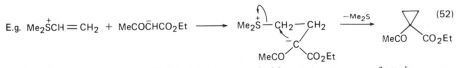

$$(52)$$

react with oxygen nucleophiles but is more probably a consequence of strain energy in the transition state for formation of a cyclopropane ring which bears a methylene substituent.

Allylic adducts **58** of malonic esters with allenic sulphonium salts have been isolated and treated with EtONa–EtOH, whereupon the corresponding ylides undergo [2,3]-sigmatropic rearrangement[237]; rearrangement occurs directly when the malonyl carbon atom bears an alkyl group, but when hydrogen is present a 1,3-prototropic shift first occurs to give the conjugated tautomer which rearranges regiospecifically and entirely through an *S*-methyl rather than an *S*-ethyl group (equation 53).

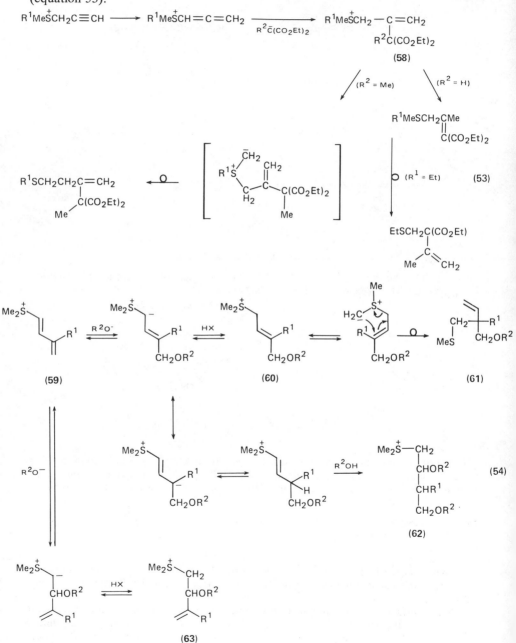

[2,3]-Sigmatropic rearrangement of ylides obtained upon addition of alcoholic alkoxides to 1,3-butadienylsulphonium salts has also been reported[242]; **59** gives **60**, **61**, and **63** with catalytic amounts of alkoxide, whereas at high alkoxide concentrations the rearrangement product **62** predominates.

V. SULPHONIUM YLIDES

Numerous reviews have dealt with the chemistry of ylides in general[243-251] or sulphur ylides in particular[252-263]; developments in the chemistry of the ylides of sulphur selenium, and tellurium continue to be reviewed periodically[264,265]. The following section deals primarily with the chemistry of sulphonium ylides, $R_2\bar{S}\bar{C}R_2'$, where R = alkyl or aryl and R' = H, alkyl, or aryl (unstabilized ylides); however, comparison will frequently be made with the behaviour of ylides where $C_{(\alpha)}$ bears a group R' which has a conjugating heteropolar multiple bond (stabilized ylides). Comparison with oxosulphonium ylides, $R_2\bar{S}(=O)\bar{C}R_2'$, will also be made since there is continuing interest in the contrasting behaviour of analogous sulphonium and oxosulphonium ylides. The chemistry of ylides derived from heterosulphonium salts is covered in Chapter 15, and synthetic applications of sulphonium ylides are considered in detail in Chapter 16.

A. Formation of Sulphonium Ylides

1. From sulphonium salts

The ability of sulphur to stabilize an adjacent carbanion is well documented for sulphides and, whatever may be the precise cause, stabilization is further enhanced in the case of sulphonium salts[228,266-268]. Ammonium salts are typically *ca.* 2.6 pK units less acidic than sulphonium analogues and the greater stability of the sulphonium ylides (π-sulphranes) formed by α-deprotonation of the latter compounds cannot be ascribed to electrostatic effects alone; delocalization of electron density into 'low-lying' d-orbitals[269] or into a combination of s- and p-orbitals have variously been invoked to account for the stability of sulphonium ylides which are correspondingly considered to be best represented by a combination of hybrids **64** and **65**.

$$\begin{array}{ccc} \diagdown \diagup \\ S=C \\ \diagup \diagdown \end{array} \quad \longleftrightarrow \quad \begin{array}{ccc} \diagdown \diagup \\ \overset{+}{S}-\bar{C} \\ \diagup \diagdown \end{array} \qquad (55)$$

$$\text{(64)} \qquad\qquad\qquad \text{(65)}$$

Recognition of the stability of sulphur ylides has prompted considerable interest in the base-promoted reactions of sulphonium salts which bear α-hydrogen atoms, and in the ability of α,β-unsaturated sulphonium salts to act as Michael acceptors (see p. 349). Although sulphur ylides have been known since the isolation of dimethylsulphonium fluorenylide[270] in 1930, the chemistry of such species has been widely exploited only since *ca.* 1960, and their application to organic synthesis (see Chapter 16) has been encouraged mainly by the work of groups led by Trost[252], Corey[271], and Franzen[222,272].

Ylides are commonly generated by deprotonation of a sulphonium salt[273] which is either pre-formed or generated *in situ*. Aqueous sodium hydroxide is usually sufficiently basic to generate the ylide, as evidenced by the complete deuteration of trimethylsulphonium iodide upon reaction with NaOD–D$_2$O for 3 h at 62°C; this mild condition is often preferred to those which induce essentially irreversible deprotonation and the possibility of competing side reactions. Alkoxide bases in the corresponding alcohol[217,221,274–279], NaH in DMSO[271,279–285], or nBuLi[285,286] are, however, commonly employed for irreversible generation of stabilized (e.g. by carbonyl[217,221,274–278], cyano[286], sulphonyl[287], or sulphonium[288,289] groups) sulphonium ylides or oxosulphonium ylides, which have often been isolated as air-sensitive compounds[279,281]. In the case of non-stabilized ylides it is necessary to use stronger bases such as dimsyl sodium[222,271,272,290–292], an organo-lithium[78,222,271,293], or lithium dialkylamide[294–297] in order to achieve irreversible ylide formation. Careful choice of reagent is necessary if competing side reactions, such as E2 elimination or σ-sulphurane formation, are to be avoided; it is preferable to use the highly basic t-BuLi or lithium dialkylamide in preference to the more nucleophilic n-alkyl- or aryllithium, in order to reduce the proportion of products derived from intermediate σ-sulphuranes (equation 56)[78,295].

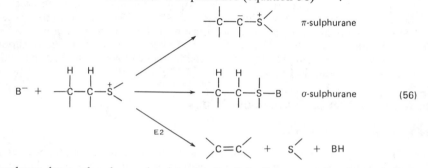

Procedures have also been developed whereby ylides may be formed from sulphonium salts which are generated *in situ*; thus, since triarylsulphonium salts undergo substituent exchange upon reaction with alkyllithium, they can be used to generate diarylalkylides: cyclopropyl[298] and n-butyl[79,80] diphenylsulphonium ylides **66** and **67** have been obtained in yields of *ca.* 20% and 40%, respectively, as

$$Ph_3\overset{+}{S}\ \overset{-}{B}F_4 \ + \ \triangleright\!\!-\!Li \ \xrightarrow{\ THF\ } \ Ph_2\overset{+}{S}\!-\!\overset{-}{\triangleleft} \qquad (57a)$$

$$(66)$$

$$Ph_3\overset{+}{S}\ \overset{-}{B}F_4 \ + \ n\text{-BuLi} \ \longrightarrow \ Ph_2\overset{+}{S}\overset{-}{C}HPr\text{-}n \qquad (57b)$$

$$(67)$$

evidenced by the products obtained upon subsequent quenching with cyclohexanone. However, in view of the relatively poor yields obtained, such a procedure is adopted only when the sulphonium salt cannot be pre-formed by conventional methods. Arylsulphonium ylides have also been formed *in situ* by reaction between sulphides and benzyne[299]; this procedure is applicable to formation of methylides[82] and allylides[300,301] but is unsatisfactory in those cases where the intermediate zwitterion may fragment by intramolecular elimination[82] (equation 58).

a. *Modification of sulphonium ylides by acylation and alkylation.* Since ylides behave as carbon nucleophiles, they can generally be modified by alkylation or acylation reactions with suitable electrophiles[286,302–307]; equation 59 illustrates the range of stabilized ylides which can be formed in this manner[302]. The ambient

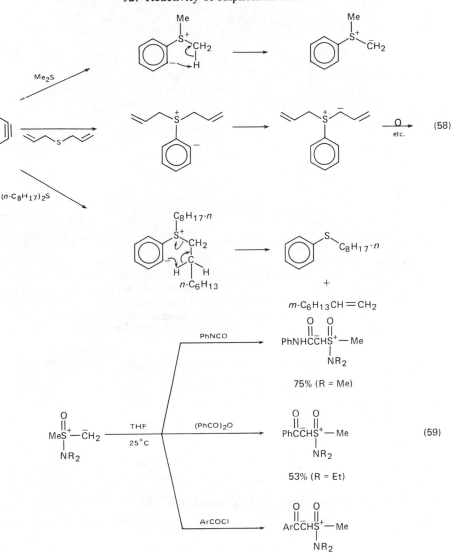

(58)

(59)

75% (R = Me)

53% (R = Et)

65% (Ar = p-ClC$_6$H$_4$; R = Me)

nucleophilicity of ylides which are conjugatively stabilized by α-substituents may, however, complicate such reactions[221,277,307] (equation 60). Non-stabilized ylides present no such problems and, for example, ylide **70** can conveniently be prepared[296] *in situ* from **68** via the rather unstable salt **69**.

2. Ylides formed by reaction of carbenes with sulphides

Since it was recognized that the products of certain reactions[308–315] of sulphides with carbenes were those to be expected from decomposition of ylide

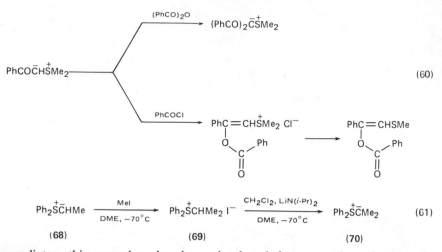

(60)

$$Ph_2\overset{+}{S}\overset{-}{C}HMe \xrightarrow[\text{DME, } -70°C]{\text{MeI}} Ph_2\overset{+}{S}CHMe_2 \ I^- \xrightarrow[\text{DME, } -70°C]{CH_2Cl_2, \ LiN(i\text{-Pr})_2} Ph_2\overset{+}{S}\overset{-}{C}Me_2 \quad (61)$$

(68)   (69)   (70)

intermediates, this procedure has been developed for generation of ylides, in particular those which are susceptible to [2,3]-sigmatropic rearrangement[316–322]. Stable ylides (e.g. **71**) have been isolated in high yield[323] by this technique, which is applicable to dialkyl or diaryl sulphides and is sometimes highly stereoselective[324]. Ylides generated from the reaction of episulphides with carbenes undergo an unusual α-fragmentation to give the alkenes derived from the episulphides[325].

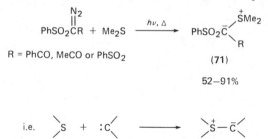

R = PhCO, MeCO or PhSO$_2$

(71)

52–91%

(62)

i.e. $\overset{\diagdown}{\underset{\diagup}{}}S + :C\overset{\diagup}{\underset{\diagdown}{}} \longrightarrow \overset{\diagdown}{\underset{\diagup}{}}\overset{+}{S}-\overset{-}{C}\overset{\diagup}{\underset{\diagdown}{}}$

B. Stability and Structure of Sulphonium Ylides

Whereas it is often possible to isolate, crystallize[320,321,326–328], and even distil[258,274,275,329] sulphur ylides which are stabilized by the conjugating influence of carbonyl, sulphonyl, cyano, or nitro substituents, unstabilized ylides (i.e. those which bear only alkyl, aryl, or vinyl substituents) are usually generated and reacted at *ca.* $-70°$C; approximate half-lives ($t_{1/2}$) have been reported for dimethylsulphonium methylide[271] (min/10°C), diphenylsulphonium ethylide[295] (5 min/20°C), -allylide (30 min/$-15°$C), and -cyclopropylide[330] (2.5 min/25°C). Many investigators have attempted to establish the importance of charge delocalization into the orbitals of sulphur in determining the stability and geometry of sulphonium ylides[268,279,331–342]. It is clear from n.m.r. spectra of **72**[342] and **73**[341], crystal structures of **74**[343] and **75**[344,345], resolution of **76**[346], and generation of the optically active non-stabilized ylide **77** from the corresponding sulphonium salt[347,348] that in each case the sulphur atom is in an asymmetric tetrahedral environment; it is therefore unlikely that there is a high degree of C—S double bond character in such species, although it must still be significant relative to the analogous ammonium ylides (less stable).

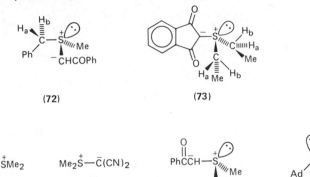

(72)　　　　　　　(73)

(74)　　　　　　(75)　　　　　　(76)　　　　　　(77)

(Ad = 1-adamantyl)

Enantiomers of optically active sulphonium salts are often sufficiently stable to pyramidal inversion[349] (about an asymmetric sulphonium centre) to permit their use in asymmetric synthesis.

C. Reactions of Sulphonium Ylides

Sulphur ylides are isoelectronic with diazo compounds and display similar chemical reactivity. However, the marked zwitterionic character of the ylides makes them particularly reactive towards polar multiple bonds and towards Michael acceptors, with which they react almost exclusively by initial nucleophilic addition. The reaction of sulphur ylides with carbonyl compounds has been studied intensively, and widely used for the synthesis of epoxides; the competing cyclopropanation of α, β-unsaturated carbonyl compounds, via initial 1,4-addition of sulphur ylides, has also received considerable attention. The efficacy of these alkylidene transfer reactions will be discussed, with particular reference to the contrasting behaviour of sulphonium and oxosulphonium ylides where appropriate.

1. Epoxidation by alkylidene transfer

a. *Stereochemistry and mechanism.* The stereochemistry of epoxidation is consistent with a nucleophilic addition–1,3-elimination mechanism; this is illustrated by the stereochemistry of the cyclopropanol 79 obtained[330] upon addition of the cyclopropylide 78 to acetone, followed by subsequent nucleophilic ring opening of the epoxide by n-BuLi (equation 63) (the experiment was actually performed with mixtures of cis- and trans-78). The result suggests that the stereochemistry about the ylidic carbon is retained in the addition step and that inversion occurs at this centre during the subsequent intramolecular nucleophilic displacement of sulphide by the neighbouring alkoxide anion. The stereochemistry of the cyclization step is further supported by the observations[350–352] depicted by equations 64–66, which also establish that the formation of betaines from non-stabilized ylides must be essentially irreversible.

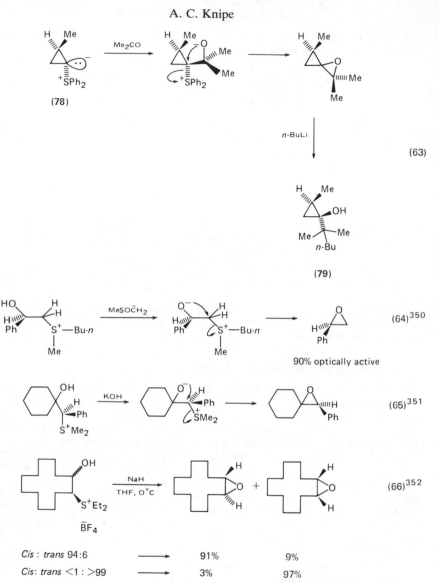

(63)

(78)

(79)

(64)[350]

90% optically active

(65)[351]

(66)[352]

Cis : trans 94 : 6   ⟶   91%   9%

Cis : trans <1 : >99   ⟶   3%   97%

While it has been found that the betaine formed upon reaction of formaldehyde with dimethylsulphonium benzylide undergoes 1,3-elimination faster than it reverts to ylide and aldehyde[353], it is clear that the converse is true for corresponding betaines produced from oxosulphonium ylides and other stabilized ylides[350]. This is dramatically demonstrated by the observation[229] that, following its independent generation, betaine **80** decomposes to give only benzaldehyde and the stabilized ylide **81**; the epoxide **82** is not formed (equation 67).

There is a distinct dichotomy[271,354,355] between sulphonium methylides and oxosulphonium methylides in the stereochemistry of their addition to conformationally stable cyclohexanone ring systems; thus, dimethylsulphonium ylide

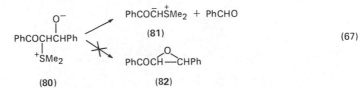

(67)

(80)   (81)   (82)

reacts with 4-*t*-butylcylcohexanone[271] and *trans*-2-decalone[354] predominantly by axial attack, whereas oxosulphonium methylides react by almost exclusive equatorial attack[271,354,356]. It has been suggested that sulphonium methylide reacts by a four-centre reaction for which the symmetry-allowed 2*a* + 2*s* mode requires the transition state configuration **83**; it has correspondingly been argued that steric

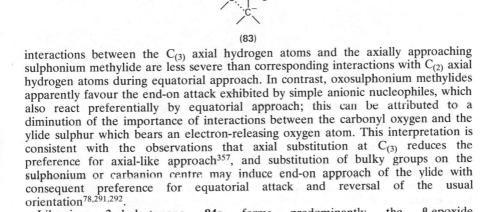

(83)

interactions between the $C_{(3)}$ axial hydrogen atoms and the axially approaching sulphonium methylide are less severe than corresponding interactions with $C_{(2)}$ axial hydrogen atoms during equatorial approach. In contrast, oxosulphonium methylides apparently favour the end-on attack exhibited by simple anionic nucleophiles, which also react preferentially by equatorial approach; this can be attributed to a diminution of the importance of interactions between the carbonyl oxygen and the ylide sulphur which bears an electron-releasing oxygen atom. This interpretation is consistent with the observations that axial substitution at $C_{(3)}$ reduces the preference for axial-like approach[357], and substitution of bulky groups on the sulphonium or carbanion centre may induce end-on approach of the ylide with consequent preference for equatorial attack and reversal of the usual orientation[78,291,292].

Likewise, 3-cholestanone **84a** forms predominantly the β-epoxide (**85a:86a** = 32:1) and the α-epoxide (**85a:86a** = 1:2) upon reaction with dimethyloxosulphonium methylide and dimethylsulphonium ylide, respectively (equation 68)[358]; the preference for axial approach of dimethyl sulphonium

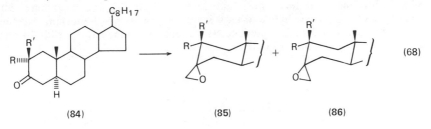

(84)   (85)   (86)

(a) R = R' = H; (b) R = Me, R' = H

methylide is, however, reversed in the case of **84b** (**85b:86b** = 3:1), which is more sterically hindered at its α-face. Norpinone **87** is converted exclusively into the α-epoxide **88** upon reaction with either of the methylides (equation 69).

In contrast, the *endo*-approach of dimethyloxosulphonium methylide to 5-norbornan-2-one **89** is anomalous with respect to that for other carbanions, including dimethylsulphonium methylide, for which a less sterically hindered

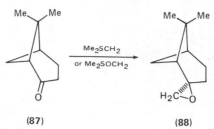

(87)   (88)

(69)

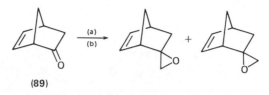

(89)

(70)

(a) = Me$_2$SOCH$_2$ $\longrightarrow$ 2.3 : 1

(b) = Me$_2$SCH$_2$ $\longrightarrow$ 1 : 15.7

exo-approach is preferred (equation 70); this has been attributed[359] to an end-on approach of the oxosulphonium ylide which is anchimerically assisted by interaction of the sulphonium group and the proximate double bond; analogous reactions of 2-norbornen-7-one[359,360], 1-methyl 5-norbornen-2-one[361], and 1-methyl-bicyclo-[4.2.1]non-7-en-9-one[362] have been reported.

Enantioselective synthesis of 2-phenyloxirane from benzaldehyde has been achieved by using Me$_2$S=CH$_2$ in the presence of catalytic amounts of an optically active β-oxido-quaternary ammonium zwitterion, under phase-transfer conditions[363]; the asymmetric induction apparently originates from dipole–dipole interaction of the chiral catalyst with the otherwise achiral ylide.

With the exception of ketones which are very sterically hindered towards carbonyl addition, or which are readily converted into the corresponding enolate anion[271,364,365], deprotonation at the α-carbon atom does not usually compete with methylene transfer; thus, epoxidation 2,2,6,6-d_4-cyclohexanone **90** by dimethylsulphonium methylide occurs with essentially no deuterium exchange[366] (equation 71), and 3-cycloheptenone **91** undergoes methylene transfer to give **92** with no double bond migration (equation 72)[367].

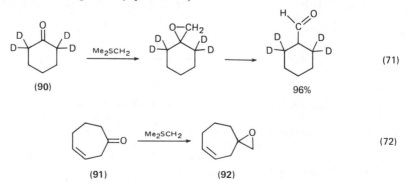

(90)   96%

(71)

(91)   (92)

(72)

b. *Scope of the epoxidation reaction.* Although oxosulphonium methylides and the more reactive dimethylsulphonium methylide react with practically all aldehydes and ketones, yields may often be improved by incorporating substituents which increase the electrophilicity of the carbonyl carbon. Whereas α,β-unsaturation causes oxosulphonium methylides to form cyclopropanes, rather than epoxides, simple double bonds (not conjugated to the carbonyl group) are unreactive[368]. Base-stable groups such as enol ethers[369] and acetals are generally stable under the conditions of methylene transfer and hydroxyl, thio, or amino groups may be present, although they may subsequently become involved in intramolecular decomposition of the epoxides[370-376]. Epoxidation usually occurs much more rapidly than nucleophilic addition of the ylide to ester, amide or nitrile functions which may therefore be present also[377-381].

Stabilized ylides react only with the very electrophilic carbonyl centres of α,β-diketones[281,382], aromatic aldehydes[221,383], or α-haloaldehydes[382] and are particularly unreactive towards hindered ketones[285]. Ylides stabilized by more than one conjugating substituent (e.g. dicyano[384-386], dinitro[387,388] or dicarbomethoxy[389] groups) are even more unreactive and are of limited synthetic utility.

The epoxidation reactions of cyclopropylides[390] have been used to considerable synthetic advantage in cyclopentanone annelation[291,391] γ-butyrolactone annelation[292], secoalkylation[392] and geminal alkylation[393-395]. For example, cyclobutanones can be prepared in one step by acid-catalysed rearrangement (equation 73) of oxaspiropentanes, **(93)**, which are readily formed by the

$$\text{Ph}_2\text{S}^+\text{---}\triangleleft + \begin{matrix}\text{R}^1\\\text{R}^2\end{matrix}\text{C}=\text{O} \longrightarrow \begin{matrix}\text{R}^1\\\text{R}^2\end{matrix}\triangle\!\!\!\triangleright \xrightarrow{} \text{R}^1\text{---}\square\text{R}^2\text{O} \tag{73}$$

(93)

spiroannelation reaction of diphenylsulphonium propylide with aldehydes and ketones[391,396,397]. The cyclobutanones can subsequently be oxidized to γ-butyrolactones (which are appropriately substituted at the α-position) or ring enlarged to cyclopentanones.

c. *Epoxidation* versus *cyclopropanation of α,β-unsaturated carbonyl compounds.* Whereas both sulphonium and oxosulphonium methylides react with aldehydes and ketones to form epoxides, oxosulphonium methylides form cyclopropanes in preference to epoxides, upon reaction with α,β-unsaturated ketones (equations 74–76)[398-400].

The cyclopropanation reaction is in contrast with that of carbenes (which prefer to attack double bonds which bear electron-donating groups) and occurs only upon

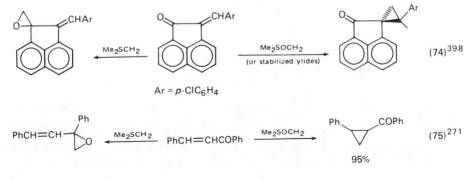

$$\text{Ar} = p\text{-ClC}_6\text{H}_4$$

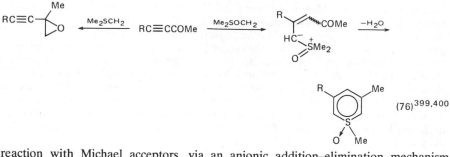

(76)[399,400]

reaction with Michael acceptors, via an anionic addition–elimination mechanism (equation 77). The dichotomy has been attributed to contrasting kinetic and

$$\text{(77)}$$

Y = carbanion stabilizing group

thermodynamic control of carbonyl reactions of the sulphonium and oxosulphonium ylides, respectively. Thus, whereas for the sulphonium methylide 1,3-elimination (epoxidation) of the intermediate betaine **94** occurs much faster than reversion of carbonyl addition ($k_2 > k_{-1}$) (equation 78), the reverse is true for oxosulphonium

$$\text{(78)}$$

(94)

and other stabilized (e.g. see also equation 67) ylides. The stabilized and oxosulphonium ylides therefore have the opportunity to form cyclopropanes by conjugate addition under thermodynamically controlled conditions. It has been established that Michael addition of dimethyloxosulphonium methylide to α,β-ethylenic ketones is slow relative to the subsequent intramolecular cyclization to give the corresponding cyclopropyl ketone[401] (equation 79).

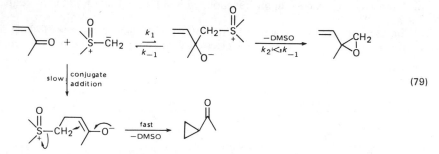

(79)

Thermodynamic control (with consequence cyclopropanation) can, however, be achieved in the case of unstabilized sulphonium ylides which incorporate structural features capable of either hindering the epoxidation step (k_2) or enhancing the reversion of carbonyl addition (k_{-1}); thus, isopropylides[402] and cyclopropylides[290] (for which S$_N i$ reaction at the tertiary carbon of the corresponding betaine is unfavourable) and allylides[78] ('partially stabilized' ylides) can be used to effect preferential cyclopropanation (equation 80). Substitution at $C_{(\beta)}$ may, nonetheless, prevent the Michael addition from competing effectively[402] (equation 81). The

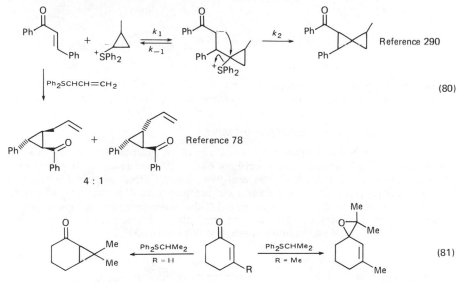

(80)

(81)

α-carbanion stabilizing influence of alkynyl[403], phenyl[404], and phenylthio[307] groups may also facilitate cyclopropanation reactions of otherwise unstabilized sulphonium ylides.

A stereospecific formation of the cyclopropane **96** has also been achieved by intramolecular reaction of the unstabilized ylide derived from **95** (equation 82); the alternative carbonyl addition would proceed by an unfavourable 7-membered transition state[405].

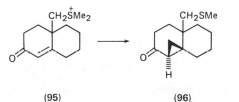

(82)

(95)   (96)

Although cyclopropanation, rather that epoxidation, occurs preferentially upon reaction of oxosulphonium ylides with α,β-unsaturated ketones, both reactions may occur if the reaction is conducted at elevated temperature and in the presence of excess of ylide. Thus, 3,6-dihydro-2*H*-pyrans and pent-2-ene-1,5-diols have been formed via acid hydrolysis of the cyclopropyl epoxides, which can be obtained either by treating α,β-unsaturated ketones with dimethyloxosulphonium methylide or by treating cyclopropyl ketones with diemthylsulphonium methylide[406].

In many cases ylides which are markedly stabilized by α-substituents are too unreactive to form epoxides by addition to unconjugated carbonyl groups and may therefore react only with α,β-unsaturated carbonyl compounds where a conjugate 1,4-addition route is available. The predominant product (e.g. **97** > **98**, equation 83) usually[275,281,407] bears the ylide stabilizing group and the activating group of the

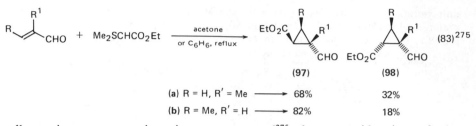

(a) R = H, R' = Me ——→ 68%   32%

(b) R = Me, R' = H ——→ 82%   18%

alkene in a *trans* orientation, as expected[275] from consideration of the conformational stabilities of the intermediate zwitterions (see equations 83 and 86).

2. Cyclopropanation by alkylidene transfer

a. *Stereochemistry and mechanism.* Formation of an intermediate zwitterion can best account for the lack of stereospecificity observed upon reaction of ylides with *cis*- and *trans*-alkenes; thus, both **99a** and **99b** give the same mixture (*ca.* 1:1) of isomeric cyclopropanes **100a** and **100b** upon reaction with dimethyloxosulphonium methylide[408] (equation 84), and only *trans*-1,2-dicarbomethoxycyclopropane **102** is obtained upon reaction of dimethylmaleate or fumarate **101** with phenyldimethylaminooxosulphonium methylide (equation 85)[283].

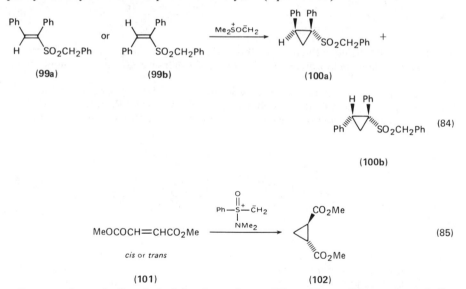

For reaction of dimethylsulphonium phenacylide with arylidenemalononitrile, care has been taken to establish that the cyclopropyl product is the same as that obtained from the proposed carbanion intermediate **103** when it is generated by an independent route (equation 86)[409].

The stereochemistry of the product of such stepwise cyclopropanations (e.g. see equation 80) is largely determined by the influence of non-bonded interactions on the relative stability, and rate of cyclization, of conformers of the intermediate zwitterion[78,217,410]. This is well illustrated by the formation[410] of an optically active cyclopropane upon reaction (equation 87) of optically active adamantlyethyl

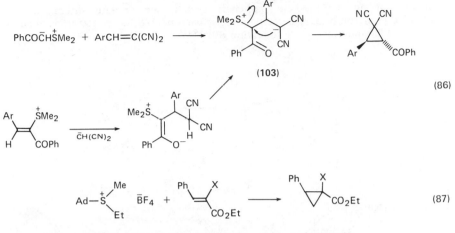

(86)

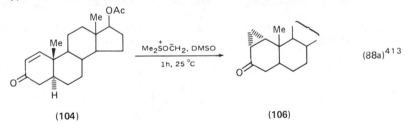

(87)

optically active   X = CN or CO$_2$Et   optically active

sulphonium methyide (in which the sulphonium centre is asymmetric) with dimethyl benzalmalonate; this, and related examples[350,411], naturally require that the sulphonium group is intimately involved in the cyclization step. Optically active cyclopropanes have also been prepared by methylene transfer reactions of the optically active NN-dimethylamino-p-tolyloxosulphonium methylide with aldehydes and ketones[412].

Steric approach control usually determines the stereochemistry of cyclopropanation of cycloalkenones. This is illustrated by the formation[413,414] of **106** and **107** upon reaction of Mc$_2$SOCH$_2$ at the least hindered face (α and β, respectively) of the steroidal enones **104** and **105** (equations 88a and 88b). When

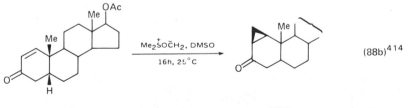

steric factors are relatively insignificant, stereoelectronic control may predominate[415–418]. For example, formation of an axial C—C bond at the site of attack on cyclohexenones can be attributed to a stereoelectronic preference for formation of an intermediate half-chair cyclohexane enolate rather than its alternative boat conformer; correspondingly, carvone **108** forms **109** upon reaction with oxosulphonium methylide (equation 89)[415,416].

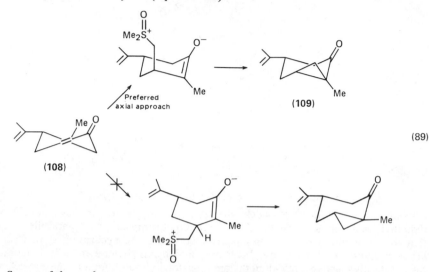

(89)

b. Scope of the cyclopropanation reaction. Excellent cyclopropanation yields have been achieved by reaction of oxosulphonium methylides with α,β-unsaturated ketones[271,419-425], nitriles, isonitriles[426], sulphones[408,427], sulphonamides[428], sulphonates[428], and nitro compounds[429]. Alkyl substitution at both $C_{(\alpha)}$ and $C_{(\beta)}$ of the Michael acceptor may retard cyclopropanation either by sterically hindering the addition, or by destabilizing the intermediate carbanion[430,431]; conversely, it is advantageous to substitute $C_{(\alpha)}$ with electron-withdrawing substituents[432].

Isolated double and triple bonds and also hydroxyl, sulphydryl, amino, imine, nitrile, isonitrile, nitro, carbonyl, sulphonyl, and sulphoxide groups are usually stable to the conditions of cyclopropanation by sulphonium ylides, which have therefore gained popularity in natural product syntheses. Cyclopropanation by

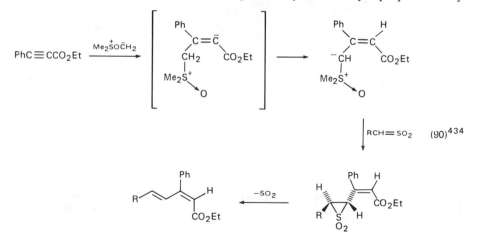

alkylidene, rather than methylene, transfer is generally accomplished[280,433] by using aryldimethylaminoxosulphonium alkylides since simple oxosulphonium alkylides are generally inaccessible and saturated sulphonium alkylides may form either cyclopropanes or epoxides, depending on the structure of both the Michael acceptor and the ylide[78,294,402].

Addition of sulphur ylides to acetylenic Michael acceptors is generally followed by isomerization of the intermediate zwitterion (to form a stabilized ylide, which may be used for subsequent reactions[434–436]) rather than by cyclization to form a highly strained ring[437] (see equations 76 and 90).

Reactions of sulphonium ylides with alkynes have been used to form furans[436,438].

3. Rearrangement of sulphur ylides

Since this topic has recently been reviewed elsewhere[252] and is given further attention in Chapter 16, only an outline of the general reaction types will be presented here.

Rearrangements are a predominant feature of the chemistry of allylsulphonium ylides and may occur either by a symmetry-allowed [2,3]-sigmatropic rearrangement to give homoallytic sulphides, or by a non-concerted 1,2-shift (equations 91a and 91b).

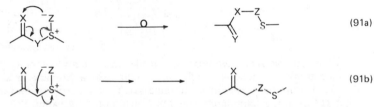

$$(91a)$$

$$(91b)$$

a. *[2,3]-Sigmatropic rearrangement*. [2,3]-Sigmatropic rearrangement apparently occurs by a suprafacial–suprafacial process[439] with complete allylic inversion. Thus, **111** and **113** are obtained upon rearrangement of the ylides derived from **110** and **112**, respectively (equations 92a and 92b).

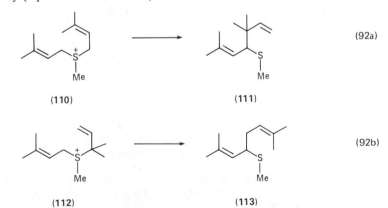

Inversion may also proceed regiospecifically (e.g. when the migration terminus is a cyclohexyl ring carbon[440], as in equation 93) and with a high degree of transfer of chirality from sulphur to carbon[410] (e.g. equation 94).

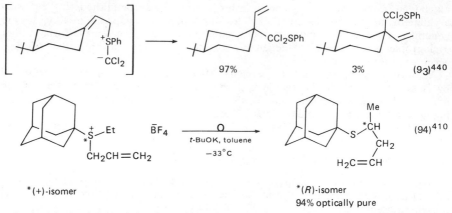

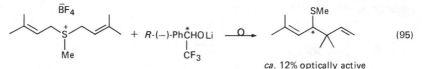

*(+)-isomer

*(R)-isomer
94% optically pure

Optical induction can also be achieved upon rearrangement of an optically active ylide formed simply by the action of a chiral base on a prochiral sulphonium salt[441] (equation 95).

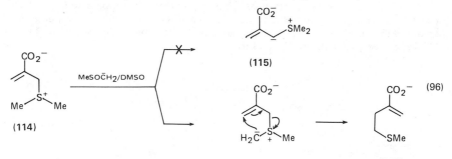

[2,3]-Sigmatropic rearrangements are facile and usually compete effectively with alkylidene transfer reactions of allylides which bear a primary or secondary alkyl group. Thus, only the homoallylic sulphide **116** is obtained[442] when formation of the allylide **115** from the sulphonium salt **114** is attempted (equation 96).

[2,3]-Sigmatropic rearrangements also occur when the unsaturation is due to an alkyne[443,444] or allene[443,445,446] grouping (see equation 97), and need not necessarily involve only carbon skeletons[447] (e.g. β-ketosulphonium salts may rearrange with the oxygen atom at the migration terminus; equation 98). Likewise, the Pummerer reaction (equation 99) may be considered to be a rearrangement of an intermediate acyloxy-ylide[140]; azasulphonium salts also undergo analogous rearrangement[448–450] (equations 100a and 100b).

Repeatable [2,3]-sigmatropic shifts have recently been used as the basis of a noval 'ring-growing' procedure for macrocycle synthesis[451]; this is illustrated by the sequence in equation 101.

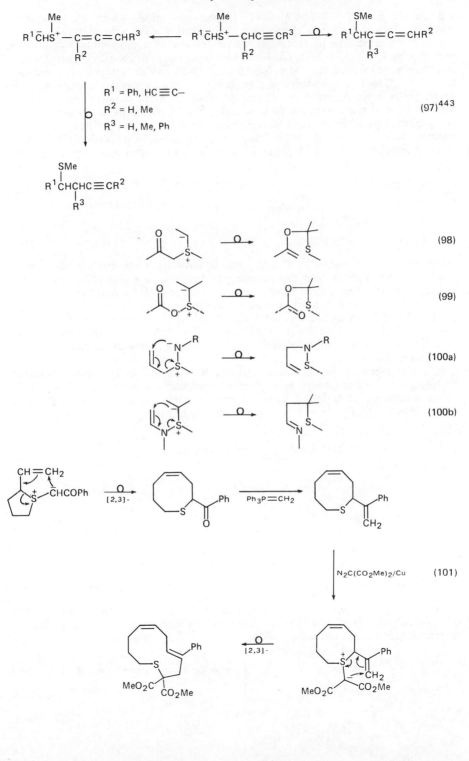

$R^1 = Ph, HC\equiv C-$

$R^2 = H, Me$

$R^3 = H, Me, Ph$

(97)[443]

(98)

(99)

(100a)

(100b)

$[2,3]-$

$Ph_3P=CH_2$

$N_2C(CO_2Me)_2/Cu$ (101)

$[2,3]-$

b. *Rearrangement by 1,2-migration*. Sulphonium ylides which are unable to undergo intramolecular elimination reactions may thermolyse by a 1,2-rearrangement. The first 'Stevens rearrangement' of a sulphonium salt or ylide was reported in 1932 and involved the conversion of phenacylmethylbenzyl-sulphonium bromide, in MeONa–MeOH, into α-thiomethyl-β-phenylpropio-phenone[452]. This result was subsequently verified[435] and analogous rearrangements of actual ylides were reported by Schollkopf *et al.*[453]. More detailed studies of the original Stevens rearrangement have revealed that it occurs in competition with a [2,3]-sigmatropic rearrangement and a Sommelet rearrangement[455–457] and apparently proceeds by a homolytic dissociation and migration of the benzyl radical. Rearrangement of the ylides **117** in THF (equation 102) is characterized by a very low ρ-value

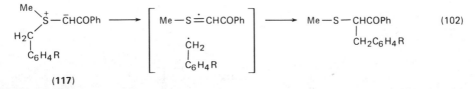

(102)

(**117**)

as the R group is varied[458]; CIDNP effects have been observed in the corresponding n.m.r. spectra, and the coupling by-products are also consistent with a homolytic mechanism. Similar results have been reported[459] for reactions in diphenyl ether at 130°C and results of crossover experiments have been rationalized by a mechanism of rearrangement within a caged radical pair. The observation that the α-monodeuteriobenzyl analogue of **117** (R = H) undergoes rearrangement with 36 ± 15% retention of optical activity requires that 44 ± 20% of the radicals remain in the cage[459].

Rearrangement of the stabilized ylides **118** (equation 103) exhibit migration preference when one group on sulphur is particularly capable of stabilizing the radical intermediate[460]; thus, **118b** and **118c** form **119b** and **119c**, respectively, upon thermolysis at 115°C. The α',β-elimination mechanism is preferred by **118d**.

$$(MeO_2C)_2C{=}S{\Big\langle}{\genfrac{}{}{0pt}{}{R^1}{R^2}} \quad \xrightarrow{\Delta} \quad (MeO_2C)_2C{\Big\langle}{\genfrac{}{}{0pt}{}{S{-}R^1}{R^2}}$$

(**118**)   (**119**)

(103)

(**a**) $R^1 = R^2 = Me$

(**b**) $R^1 = Me$; $R^2 = CH_2Ph$

(**c**) $R^1 = Et$; $R^2 = CH_2COPh$

(**d**) $R^1 = Me$; $R^2 = Et$ $\xrightarrow{\Delta}$ $(MeO_2C)_2CHSMe + CH_2{=}CH_2$

CIDNP effects have also been observed[461] during reaction of benzyne with dibenzyl sulphide (equation 104) and the products have been attributed to Stevens rearrangement of an intermediate ylide.

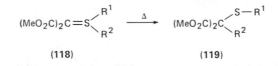

(104)

4. Miscellaneous reactions of sulphonium ylides

Many 'unsaturated' functional groups undergo nucleophilic additions of sulphur ylides. Aziridine synthesis by methylene transfer to imines is less effective than the analogous epoxidation of the corresponding carbonyl compounds[272,378,462], although 3-phenylazabicyclobutane and its 2-methyl and 2,2-dimethyl derivatives have been formed in 40–70% yields by this procedure[463] (equation 105).

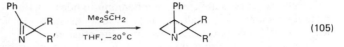

$$(105)$$

Ethylenimines have been obtained[464] in moderate yield upon treatment of ArCH=NN=CHAr with Me_2S=CHPh or by stereospecific ethylidene transfer to benzalaniline[280]. N-Iminoaziridine derivatives have also been prepared by methylene transfer to azines[465–467].

Stabilized ylides react with arylazoethylenes to form 4,5-dihydropyrazolines[468] and with 2-diazoacenaphthenone to form hydrazones[469]; they also react as typical nucleophiles with azides[470–472], from which Δ^2-1,2,3-triazolines can also be obtained[473]. Dimethyloxosulphonium methylide and β-oxosulphonium salts $R_2\overset{+}{S}CH_2COR'$ have been converted into more stable ylides [$Me_2S(O)$=CHCONHAr and R_2S=C(COR')CONHAr, respectively] by reaction with aryl isocynates[474–477], and reactions with ketenes have been used to synthetic advantage[478,479].

Dimethylsulphonium methylide reacts with α-nitroso ketones to form 5-hydroxy-2-isoxazolines[480] (equation 106), and nitrosobenzene reacts with

$$R^1COCR^2\text{=}NOH \xrightarrow[Me_2\overset{+}{S}\overset{-}{C}H_2]{} \qquad (106)$$

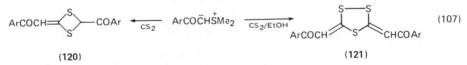

9-dimethylsulphonium fluorenylide to give N-phenylfluorenone ketoxime (a nitrone) rather than an oxazirane[481,482]; *trans*-5-acyl-2-isoxazolines can be obtained by reaction of keto-stabilized sulphonium salts with α-chlorooximes or with the isomeric nitrosochlorides[483].

Sulphonium ylides (R^1CH=SR^2R^3, where R^1 = Ph, p-ClC$_6$H$_4$, CO$_2$Et; R^2 = Me; R^3 = Me, Et) also react with CS$_2$ to form R^1C(=SR^2R^3)CSCH$_2R^1$, which may subsequently rearrange to $R^1C(SR^2)$=C(SR^3)CH$_2R^1$ or $R^1C(SR^3)$=C(SR^2)CH$_2R^1$ on standing; in contrast, 1,3-dithietane **120** and the 1;2,4-trithiole **121** are obtained upon reaction of the ylide (where R^1 = ArCO; R^2 = R^3 = Me) with CS$_2$ alone or in ethanol, respectively[484] (equation 107).

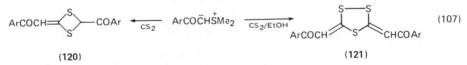

ArCOCH=⟨S⟩—COAr ⟵$_{CS_2}$ ArCO$\overset{+}{C}$H$\overset{-}{S}$Me$_2$ $\xrightarrow{CS_2/EtOH}$ ArCOCH⟨S—S⟩CHCOAr (107)

(120)   (121)

A wide range of intermolecular displacement reactions of ylides may also give rise to novel products, particularly in the case of stabilized ylides which may act as ambident nucleophiles[485,486], or of sulphonium ylides which bear neighbouring leaving groups which are subject to displacement by carbanionic S$_N$i reactions[287]; no attempt will be made to review such examples. Intramolecular Michael additions of sulphonium ylides have also been used to advantage[487].

Reaction of acyl-substituted sulphonium ylides with diphenylthiirene dioxide apparently proceeds by initial Michael addition[488], and 2-pyrones[489] and phenols[490] are formed by similar reactions with diphenylcyclopropenone; reaction of diphenylcyclopropenone with dimethyloxosulphonium methylide has also been studied[491]. Sulphonium and oxosulphonium ylides promote ring expansion of aziridines and azetidines via ring opening and S_Ni reaction of the intermediate azomethine ylide[492,493].

Reactions of stabilized sulphonium ylides with quinazolinium[494] and phenylpyrylium[494,495] salts, and with aromatic cyclopropenium[496] and tropylium[497,498] carbocations, have been investigated.

Alkylidene insertion within the C—B bond has been achieved by reaction of trialkylboranes with functionally substituted ylides[449], and o- and p-nitrotoluenes (o-:p- = 15) have been obtained by S_NAr reaction between nitrobenzene and dimethyloxosulphonium methylide[500].

There has been considerable interest[501–505] in the structure and stability of complexes of sulphur ylides with Pd(II), Pt(II), and Hg(II), and it has recently been shown that dimethylphenylphosphine can be used to liberate dimethyloxosulphonium (but not dimethylsulphonium) ylide from its complex with trimethylgold(III); both complexes are stable towards ketones[506].

Coordination complexes bearing sulphonium ligands were prepared for the first time in 1978, by (i) S-alkylation of a coordinated sulphide[507], and (ii) displacement[508] of H_2O from $[(NH_3)_5RuH_2O]^+(PF_6)_2$ by $Me_3\overset{+}{S}$.

VI. REFERENCES

1. S. K. Shapiro and F. Schlenk, *Adv. Enzymol.*, **22**, 237 (1960).
2. C. J. M. Stirling, *Int. J. Sulfur Chem., Part C*, **6**, 41 (1971).
3. C. J. M. Stirling, *Sulphonium Salts*, in *Organic Chemistry of Sulfur*, (Ed. S. Oae), Plenum Press, New York, pp. 473–525; V. Franzen, *The Chemistry of Sulphonium Compounds*, in *Organic Sulphur Compounds*, Vol. 2 (Ed. N. Kharasch and C. Y. Meyers), Pergamon Press, Oxford.
4. S. Oae, *Kagaku (Kyoto)*, **19**, 795 (1964); *Chem. Abstr.*, **64**, 1911 (1966).
5. J. G. Tillett, *Chem. Rev.*, **76**, 747 (1976).
6. J. P. Marino, *Topics Sulfur Chem.*, **1**, 1 (1976).
7. C. G. Swain, R. Cardinaud and A. D. Ketley, *J. Amer. Chem. Soc.*, **77**, 934 (1955).
8. C. G. Swain and L. E. Kaiser, *J. Amer. Chem. Soc.*, **80**, 4092 (1958).
9. J. B. Hyne and R. Wolfgang, *J. Phys. Chem.*, **64**, 699 (1960).
10. J. B. Hyne and J. W. Abrell, *Can. J. Chem.*, **39**, 1657 (1961).
11. J. B. Hyne, *Can. J. Chem.*, **39**, 1207 (1961).
12. J. B. Hyne and J. H. Jensen, *Can. J. Chem.*, **41**, 1679 (1963).
13. J. B. Hyne and J. H. Jensen, *Can. J. Chem.*, **40**, 1394 (1962).
14. D. Darwish and G. Tourigny, *J. Amer. Chem. Soc.*, **94**, 2191 (1972).
15. C. S. Davis and J. B. Hyne, *Can. J. Chem.*, **50**, 2270 (1972).
16. E. D. Hughes and C. K. Ingold, *J. Chem. Soc.*, 1571 (1933).
17. K. A. Cooper, E. D. Hughes, C. K. Ingold, G. A. Maw and B. J. MacNulty, *J. Chem. Soc.*, 2049 (1948).
18. W. H. Saunders, Jr., and S. Ašperger, *J. Amer. Chem. Soc.*, **79**, 1612 (1957).
19. J. B. Hyne, *J. Amer. Chem. Soc.*, **82**, 5129 (1960).
20. J. B. Hyne and H. S. Golinkin, *Can. J. Chem.*, **41**, 3139 (1963).
21. M. P. Friedberger and E. R. Thornton, *J. Amer. Chem. Soc.*, **98**, 2861 (1976).
22. M. N. Islam and K. T. Leffek, *J. Chem. Soc. Perkin Trans. II*, 958 (1977).
23. J. B. Hyne and J. H. Jensen, *Can. J. Chem.*, **43**, 57 (1965).
24. W. J. Pope and S. J. Peachey, *J. Chem. Soc.*, 1072 (1900).
25. S. Smiles, *J. Chem. Soc.*, 1174 (1900).

26. M. P. Balfe, J. Kenyon and H. Phillips, *J. Chem. Soc.*, 2554 (1930).
27. D. Darwish, S. H. Hui and R. Tomlinson, *J. Amer. Chem. Soc.*, **90**, 5631 (1968).
28. D. Darwish and G. Tourigny, *J. Amer. Chem. Soc.*, **88**, 4303 (1966).
29. R. Scartazzini and K. Mislow, *Tetrahedron Lett.*, 2719 (1967).
30. J. F. Kincaid and F. C. Henriques, *J. Amer. Chem. Soc.*, **62**, 1474 (1940).
31. K. R. Brower and T. L. Wu, *J. Amer. Chem. Soc.*, **92**, 5303 (1970).
32. A. Garbesi, N. Corsi and A. Fava, *Helv. Chim. Acta*, 1499 (1970).
33. K. K. Andersen, M. Cinquini and N. E. Papaniko Laou, *J. Org. Chem.*, **35**, 706 (1970).
34. S. Siegel and A. F. Graefe, *J. Amer. Chem. Soc.*, **75**, 4521 (1953).
35. (a) H. M. R. Hoffmann and E. D. Hughes, *J. Chem. Soc.*, 1259 (1964); (b) H. M. R. Hoffmann, *J. Chem. Soc.*, 823 (1965); (c) S. H. Harvey, P. A. T. Hoye, E. D. Hughes and C. Ingold, *J. Chem. Soc.*, 800 (1960).
36. H. Matsuyama, H. Minato and M. Kobayashi, *Bull. Chem. Soc. Jap.*, **48**, 3287 (1975).
37. C. G. Swain, W. D. Burrows and B. J. Schowen, *J. Org. Chem.*, **33**, 2534 (1968).
38. D. Van Ooteghem, R. Deveux and E. J. Goethals, *Int. J. Sulfur Chem.*, **8**, 31 (1973).
39. C. K. Ingold, J. A. Jessop, K. I. Kuriyan and A. M. M. Mandour, *J. Chem. Soc.*, 533 (1933).
40. C. K. Ingold and I. K. Kuriyan, *J. Chem. Soc.*, 991 (1933).
41. W. D. Burrows, *J. Org. Chem.*, **33**, 3507 (1968).
42. W. D. Burrows and J. H. Cornell, *J. Org. Chem.*, **32**, 3840 (1967).
43. E. Goethals and P. de Radzitsky, *Bull. Soc. Chim. Belg.*, **73**, 546 (1964).
44. E. Goethals and P. de Radzitzky, *Bull. Soc. Chim. Belg.*, **73**, 579 (1964).
45. J. L. Gleave, E. D. Hughes and C. K. Ingold, *J. Chem. Soc.*, 236 (1935).
46. A. A. Sosunova, A. P. Kilimov and V. V. Smirnov, *Zh. Obshch. Khim.*, **45**, 1533 (1975); *Chem. Abstr.*, **83**, 130933 (1975).
47. I. P. Evans and A. J. Parker, *Tetrahedron Lett.*, 163 (1966).
48. C. G. Swain, T. Rees and L. J. Taylor, *J. Org. Chem.*, **28**, 2903 (1963).
49. E. D. Hughes and D. J. Whittingham, *J. Chem. Soc.*, 806 (1960).
50. R. T. Hargreaves, A. M. Katz and W. H. Saunders, Jr., *J. Amer. Chem. Soc.*, **98**, 2614 (1976).
51. C. G. Swain, L. E. Kaiser and T. E. C. Knee, *J. Amer. Chem. Soc.*, **80**, 4089 (1958).
52. E. D. Hughes, C. K. Ingold and Y. Pocker, *Chem. Ind. (London)*, 1282 (1959).
53. Y. Pocker and A. J. Parker, *J. Org. Chem.*, **31**, 1526 (1966).
54. R. A. Sneen, G. R. Felt and W. C. Dickason, *J. Amer. Chem. Soc.*, **95**, 638 (1973).
55. J. K. Coward and W. D. Sweet, *J. Org. Chem.*, **36**, 2337 (1971).
56. R. Lok and J. K. Coward, *Bioorg. Chem.*, **5**, 169 (1976).
57. B. A. Bolto and J. Miller, *J. Org. Chem.*, **20**, 558 (1955).
58. G. Grue-Sørensen, A. Kjaer and E. Wieczorkowska, *J. Chem. Soc. Chem. Commun.*, 355 (1977).
59. G. Grue-Sørensen, A. Kjaer, R. Norrestam and E. Wieczorkowska, *Acta Chem. Scand.*, **B31**, 859 (1977).
60. S. R. Reymenn, *Chem. Ber.*, **7**, 1288 (1874).
61. H. A. Taylor and W. C. M. Lewis, *J. Chem. Soc.*, 665 (1922).
62. R. F. Corran, *Trans. Faraday Soc.*, **23**, 605 (1927).
63. A. C. Knipe and C. J. M. Stirling, *J. Chem. Soc. B*, 1218 (1968).
64. K. von Auers, *Chem. Ber.*, **53**, 2285 (1920).
65. G. H. Wiegand and W. E. McEwen, *J. Org. Chem.*, **33**, 2671 (1968).
66. S. Asperger, D. Stefanovic, D. Hegedic, D. Pavlovic and L. Klasinc, *J. Org. Chem.*, **33**, 2526 (1968).
67. C. Chuit and H. Felkin, *C. R. Acad. Sci., Ser. C*, **264**, 1412 (1967).
68. J. L. Kice and N. A. Favstritsky, *J. Amer. Chem. Soc.*, **91**, 1751 (1969).
69. L. Y. Ignatov, O. A. Ptitsyna and O. A. Reutov, *Dokl. Akad. Nauk SSSR*, **231**, 874 (1976); *Chem. Abstr.*, **86**, 105393 (1977).
70. O. A. Ptitsyna, L. Y. Ignatov and O. A. Reutov, *Izv. Akad. Nauk SSSR, Ser. Khim.*, 2403 (1976); *Chem. Abstr.*, **86**, 89326 (1977).
71. B. A. Bolto and J. Miller, *Aust. J. Chem.*, **9**, 74 (1956).
72. H. M. Gilow, R. B. Camp, Jr., and E. C. Clifton, *J. Org. Chem.*, **33**, 230 (1968).

73. L. I. Belen'kii, N. S. Ksenzhek and L. Y. Gol'dfarb, *Khim. Geterotsikl. Soedin.*, 310 (1972); *Chem. Abstr.*, **77**, 48122 (1972).
74. H. M. Gilow, M. De Shazo and W. C. Van Cleave, *J. Org. Chem.*, **36**, 1745 (1971).
75. N. C. Marziano, E. Maccarone and R. C. Passerini, *Tetrahedron Lett.*, 17 (1972).
76. H. M. Gilow and G. L. Walker, *J. Org. Chem.*, **32**, 2580 (1967).
77. J. W. Baker and W. G. Moffitt, *J. Chem. Soc.*, 1722 (1930).
78. R. W. La Rochelle, B. M. Trost and L. Krepski, *J. Org. Chem.*, **36**, 1126 (1971).
79. B. M. Trost, R. La Rochelle and R. C. Atkins, *J. Amer. Chem. Soc.*, **91**, 2175 (1969).
80. R. W. La Rochelle and B. M. Trost, *J. Amer. Chem. Soc.*, **93**, 6077 (1971).
81. G. Wittig and H. Fritz, *Justus Liebigs Ann. Chem.*, **577**, 39 (1952).
82. V. Franzen, H. I. Joschek and C. Mertz, *Justus Liebigs Ann. Chem.*, **654**, 82 (1962).
83. Y. H. Khim and S. Oae, *Bull. Chem. Soc. Jap.*, **42**, 1968 (1969).
84. B. K. Ackerman, K. K. Andersen, I. K. Nielsen, N. B. Peynircioglu and S. A. Yeager, *J. Org. Chem.*, **39**, 964 (1974).
85. D. Harrington, J. Weston, J. Jacobus and K. Mislow, *J. Chem. Soc. Chem. Commun.*, 1079 (1972).
86. B. M. Trost and H. C. Arndt, *J. Amer. Chem. Soc.*, **95**, 5288 (1973).
87. B. M. Trost, W. L. Schinski and I. B. Mantz, *J. Amer. Chem. Soc.*, **91**, 4320 (1969).
88. J. W. Knapezyk, G. H. Wiegand and W. E. McEwen, *Tetrahedron Lett.*, 2971 (1965).
89. J. W. Knapczyk and W. E. McEwen, *J. Amer. Chem. Soc.*, **91**, 145 (1969).
90. J. W. Knapczyk, C. C. Lai, W. E. McEwen, J. L. Calderon and J. J. Lubinkowski, *J. Amer. Chem. Soc.*, **97**, 1188 (1975).
91. G. H. Wiegand and W. E. McEwen, *Tetrahedron Lett.*, 2639 (1965).
92. H. Minato, T. Miura, F. Takagi and M. Kobayashi, *Chem. Lett.*, 211 (1975).
93. R. Tanikaga, K. Nakayama, K. Tanaka and A. Kaji, *Chem. Lett.*, 395 (1977).
94. K. K. Andersen, R. I. Caret and I. Karup-Nielsen, *J. Amer. Chem. Soc.*, **96**, 8026 (1974).
95. C. R. Johnson and W. G. Phillips, *Tetrahedron Lett.*, 2101 (1965).
96. C. R. Johnson and W. G. Phillips, *J. Org. Chem.*, **32**, 1926 (1967).
97. C. R. Johnson and W. G. Phillips, *J. Org. Chem.*, **32**, 3233 (1967).
98. C. R. Johnson and D. McCants, *J. Amer. Chem. Soc.*, **87**, 5404 (1965).
99. N. J. Leonard and C. R. Johnson, *J. Amer. Chem. Soc.*, **84**, 3701 (1962).
100. R. Tang and K. Mislow, *J. Amer. Chem. Soc.*, **91**, 5644 (1969).
101. J. P. Marino and A. Schwartz, *J. Chem. Soc. Chem. Commun.*, 812 (1974).
102. K. Omura, A. K. Sharma and D. Swern, *J. Org. Chem.*, **41**, 957 (1976).
103. K. Ohkubo, T. Aoji and K. Yoshinaga, *Bull. Chem. Soc. Jap.*, **50**, 1883 (1977).
104. W. J. M. Van Trilborg, *Tetrahedron*, **31**, 2841 (1975).
105. J. E. Kresta, C. S. Shen and K. C. Frisch, *Makromol. Chem.*, **178**, 2495 (1977).
106. T.-L. Ho and C. M. Wong, *Synth. Commun.*, **5**, 423 (1976).
107. J. E. Kresta, C. S. Shen and K. C. Frisch, *Makromol. Chem.*, **178**, 2127 (1977).
108. S. G. Smith and S. Winstein, *Tetrahedron*, **3**, 317 (1958).
109. C. R. Johnson, *J. Amer. Chem. Soc.*, **85**, 1020 (1963).
110. K. Mislow, T. Simmons, J. T. Melillo and A. L. Ternay, *J. Amer. Chem. Soc.*, **86**, 1452 (1964).
111. S. Allenmark, *Acta Chem. Scand.*, **19**, 1 (1965).
112. D. Landini, G. Modena, F. Montanari and G. Scorrano, *J. Amer. Chem. Soc.*, **92**, 7168 (1970).
113. R. A. Strecker and K. K. Andersen, *J. Org. Chem.*, **33**, 2234 (1968).
114. D. Landini, F. Montanari, G. Modena and G. Scorrano, *J. Chem. Soc. Chem. Commun.*, 86 (1968).
115. I. Ookuni and A. Fry, *J. Org. Chem.*, **36**, 4097 (1971).
116. H. Yoshida, T. Numata and S. Oae, *Bull. Chem. Soc. Jap.*, **44**, 2875 (1971).
117. S. Oae, T. Kitao and Y. Kitaoka, *Bull. Chem. Soc. Jap.*, **28**, 543 (1965).
118. S. Oae and N. Kunieda, *Bull. Chem. Soc. Jap.*, **41**, 696 (1968).
119. N. Kunieda and S. Oae, *Bull. Chem. Soc. Jap.*, **46**, 1745 (1973).
120. N. Kunieda and S. Oae, *Bull. Chem. Soc. Jap.*, **42**, 1324 (1969).
121. S. Allenmark, *Ark. Kemi*, **26**, 37 (1967).

122. S. Allenmark and H. Johnsson, *Acta Chem. Scand.*, **21**, 1672 (1967).
123. S. Allenmark and C. E. Hagberg, *Acta Chem. Scand.*, **22**, 1461 (1963).
124. S. Allenmark and C. E. Hagberg, *Acta Chem. Scand.*, **22**, 1694 (1968).
125. S. Allenmark and H. Johnsson, *Acta Chem. Scand.*, **23**, 2902 (1969).
126. S. Allenmark and C. E. Hagberg, *Acta. Chem. Scand.*, **24**, 2225 (1970).
127. H. Kwart, R. W. Body and D. M. Hoffman, *J. Chem. Soc. Chem. Commun.*, 765 (1967).
128. H. Kwart and P. S. Strilko, *J. Chem. Soc. Chem. Commun.*, 767 (1967).
129. E. N. Givens and H. Kwart, *J. Amer. Chem. Soc.*, **90**, 378 (1968).
130. C. Dell'Erba and D. Spinelli, *Ric. Sci.*, **34**, 456 (1964).
131. C. Dell'Erba, G. Guanti, G. Leandri and G. P. Corollo, *Int. J. Sulphur Chem.*, **8**, 261 (1973).
132. S. Oae, M. Yokoyama and M. Kise, *Bull. Chem. Soc. Jap.*, **41**, 1221 (1968).
133. E. Johnsson, *Acta Chem. Scand.*, **21**, 1277 (1967).
134. N. Kunieda and S. Oae, *Bull. Chem. Soc. Jap.*, **41**, 1025 (1968).
135. R. F. Watson and J. F. Eastham, *J. Amer. Chem. Soc.*, **87**, 664 (1965).
136. N. Kunieda, K. Sakai and S. Oae, *Bull. Chem. Soc. Jap.*, **42**, 1090 (1969).
137. M. Kise and S. Oae, *Bull. Chem. Soc. Jap.*, **43**, 1804 (1970).
138. S. Oae and M. Kise, *Bull. Chem. Soc. Jap.*, **43**, 1416 (1970).
139. S. Oae and M. Kise, *Tetrahedron Lett.*, 2261 (1968).
140. S. Oae and M. Kise, *Tetrahedron Lett.*, 1409 (1967).
141. R. Pummerer, *Chem. Ber.*, **42**, 2282 (1909).
142. S. Iriuchijima, K. Maniwa and G. Tsuchihashi, *J. Amer. Chem. Soc.*, **97**, 596 (1975).
143. W. E. Parham and M. D. Bhavsar, *J. Org. Chem.*, **28**, 2686 (1963).
144. S. Iriuchijima, K. Maniwa and G. Tsuchihashi, *J. Amer. Chem. Soc.*, **96**, 4280 (1974).
145. L. Horner and P. Kaiser, *Justus Liebigs Ann. Chem.*, **626**, 19 (1959).
146. S. Oae, T. Kitao, S. Kawamura and Y. Kitaoka, *Tetrahedron*, **19**, 817 (1963).
147. F. G. Bordwell and B. M. Pitt, *J. Amer. Chem. Soc.*, **77**, 572 (1955).
148. J. Kuszmann, P. Sohar and G. Y. Horvath, *Tetrahedron*, **27**, 5035 (1971).
149. G. D. Hartzell and J. N. Paige, *J. Amer. Chem. Soc.*, **88**, 2616 (1966).
150. G. E. Manser, A. D. Mesure and J. G. Tillett, *Tetrahedron Lett*, 3153 (1968).
151. K. Kondo, A. Negishi and G. Tsuchihashi, *Tetrahedron Lett.*, 3173 (1969).
152. K. Kondo, A. Negishi and I. Ojima, *J. Amer. Chem. Soc.*, **94**, 5786 (1972).
153. L. Sagramora, A. Garbesi and A. Fava, *Helv. Chim. Acta*, **55**, 675 (1972).
154. R. Curci, F. Di Furia, A. Levi and G. Scorrano, *J. Chem. Soc. Perkin Trans. II*, 408 (1975).
155. C. A. Bunton, P. B. D. de la Mare and J. G. Tillett, *J. Chem. Soc.*, 4754 (1958).
156. C. A. Bunton, P. B. D. de la Mare and J. G. Tillett, *J. Chem. Soc.*, 1766 (1959).
157. P. A. Bristow, M. Khowaja and J. G. Tillett, *J. Chem. Soc.*, 5779 (1965).
158. C. A. Bunton and G. Schwerin, *J. Org. Chem.*, **31**, 842 (1966).
159. J. L. Kice and K. Ikura, *J. Amer. Chem. Soc.*, **90**, 7378 (1968).
160. J. L. Kice and G. Guaraldi, *J. Amer. Chem. Soc.*, **89**, 4113 (1967).
161. J. L. Kice and G. Guaraldi, *J. Org. Chem.*, **31**, 3568 (1966).
162. J. L. Kice, *Progr. Inorg. Chem.*, **17**, 147 (1972).
163. K. A. Cooper, E. D. Hughes, C. K. Ingold and B. J. MacNulty, *J. Chem. Soc.*, 2038 (1948).
164. K. A. Cooper, M. L. Dhar, E. D. Hughes, C. K. Ingold, B. J. MacNulty and L. I. Woolf, *J. Chem. Soc.*, 2043 (1948).
165. E. D. Hughes, C. K. Ingold and G. A. Maw, *J. Chem. Soc.*, 2072 (1948).
166. E. D. Hughes, C. K. Ingold, G. A. Maw and L. I. Woolf, *J. Chem. Soc.*, 2077 (1948).
167. E. D. Hughes, C. K. Ingold and L. I. Woolf, *J. Chem. Soc.*, 2084 (1948).
168. E. D. Hughes, C. K. Ingold and A. M. M. Mandour, *J. Chem. Soc.*, 2090 (1948).
169. M. L. Dhar, E. D. Hughes, C. K. Ingold, A. M. M. Mandour, G. A. Maw and L. I. Woolf, *J. Chem. Soc.*, 2093 (1948).
170. C. K. Ingold, *Structure and Mechanism in Organic Chemistry*, Cornell University Press, New York, 1953: (a) pp. 420–472; (b) p. 31.
171. M. Cocivera and S. Winstein, *J. Amer. Chem. Soc.*, **85**, 1702 (1963).

172. I. N. Feit and D. G. Wright, *J. Chem. Soc. Chem. Commun*, 776 (1975).
173. K. R. Brower and J. S. Chen, *J. Amer. Chem. Soc.*, **87**, 3396 (1965).
174. K. R. Brower, *J. Amer. Chem. Soc.*, **85**, 1401 (1963).
175. S. Asperger, N. Ilakovac and D. Pavlovic, *J. Amer. Chem. Soc.*, **83**, 5032 (1961).
176. L. F. Blackwell and J. L. Woodhead, *J. Chem. Soc. Perkin Trans. II*, 234 (1975).
177. A. F. Cockerill, *J. Chem. Soc. B*, 964 (1967).
178. W. H. Saunders, Jr., A. F. Cockerill, S. Asperger, L. Klasinc and D. Stefanovic, *J. Amer. Chem. Soc.*, **88**, 848 (1966).
179. A. F. Cockerill and W. H. Saunders, Jr., *J. Amer. Chem. Soc.*, **89**, 4985 (1967).
180. D. M. Hegedic, *Indian J. Chem., Sect. B*, **15**, 283 (1977).
181. J. Banger, A. Jaffe, A.-C. Lin and W. H. Saunders, Jr., *Faraday Symp. Chem. Soc.*, **10**, 113 (1975).
182. J. Banger, A. Jaffe, A.-C. Lin and W. H. Saunders, Jr., *J. Amer. Chem. Soc.*, **97**, 7177 (1975).
183. L. J. Steffa and E. R. Thornton, *J. Amer. Chem. Soc.*, **85**, 2680 (1963).
184. L. J. Steffa and E. R. Thornton, *J. Amer. Chem. Soc.*, **89**, 6149 (1967).
185. W. H. Saunders, Jr., and R. A. Williams, *J. Amer. Chem. Soc.*, **79**, 3712 (1957).
186. W. H. Saunders, Jr., C. B. Gibbons and R. A. Williams, *J. Amer. Chem. Soc.*, **80**, 4099 (1958).
187. A. M. Katz and W. H. Saunders, Jr., *J. Amer. Chem. Soc.*, **91**, 4469 (1969).
188. E. R. Thornton, *J. Amer. Chem. Soc.*, **89**, 2915 (1967).
189. W. H. Saunders, Jr., and D. H. Edison, *J. Amer. Chem. Soc.*, **82**, 138 (1960).
190. L. F. Blackwell, P. D. Buckley and K. W. Jolley, *Aust. J. Chem.*, **27**, 2283 (1974).
191. F. L. Roe and W. H. Saunders, *Tetrahedron*, **33**, 1581 (1977).
192. W. H. Saunders, Jr., and S. E. Zimmerman, *J. Amer. Chem. Soc.*, **86**, 3789 (1964).
193. W. H. Saunders and A. F. Cockerill, *Mechanisms of Elimination Reactions*, Wiley–Interscience, New York, 1973: (a) p. 31; (b) p. 194; (c) pp. 140–151; (d) pp. 136–199; (e) pp. 31–37, 197.
194. D. J. McLennan, *J. Chem. Soc. B*, 709 (1966).
195. J. F. Bunnett and E. Baciocchi, *J. Org. Chem.*, **32**, 11 (1967).
196. J. F. Bunnett and E. Baciocchi, *Proc. Chem. Soc.*, 238 (1963).
197. J. Zavada and J. Sicher, *Collect. Czech. Chem. Commun.*, **30**, 388 (1965).
198. J. Zavada and J. Sicher, *Proc. Chem. Soc.*, 96 (1963).
199. S. J. Cristol and F. R. Stermitz, *J. Amer. Chem. Soc.*, **82**, 4692 (1960).
200. B. Capon, M. J. Perkins, C. W. Rees and A. R. Butler (Editors), *Organic Reaction Mechanisms*, Wiley, New York: (a) 1973, p. 361; (b) 1967, p. 114; (c) 1965, p. 90; 1966, p. 103; 1968, p. 140; 1969, p. 155; 1970, p. 147; 1971, p. 135; 1972, p. 139; 1974, p. 399; 1975, p. 389; 1976, p. 435; 1977, p. 449.
201. J. Zavada and J. Sicher, *Collect. Czech. Chem. Commun.*, **32**, 3701 (1967).
202. D. V. Banthorpe, E. D. Hughes and C. Ingold, *J. Chem. Soc.*, 4054 (1960).
203. E. D. Hughes, *Chem. Ind. (London*, 191 (1951).
204. C. W. Crane and H. N. Rydon, *J. Chem. Soc.*, 766 (1947).
205. P. Mamalis and H. N. Rydon, *J. Chem. Soc.*, 1049 (1953).
206. J. Crosby and C. J. M. Stirling, *J. Amer. Chem. Soc.*, **90**, 6869 (1968).
207. J. Crosby and C. J. M. Stirling, *J. Chem. Soc. B*, 679 (1970).
208. R. P. Redman and C. J. M. Stirling, *J. Chem. Soc. D*, 633 (1970).
209. B. Holmquist and T. C. Bruice, *J. Amer. Chem. Soc.*, **91**, 3003 (1969).
210. D. R. Marshall, P. J. Thomas and C. J. M. Stirling, *J. Chem. Soc. Chem. Commun.*, 940 (1975).
211. M. Anteunis, *Meded. Vlaam. Chem. Ver.*, **25**, 175 (1963); *Chem. Abstr.*, **60**, 9109 (1964).
212. J. K. Borchardt, R. Hargreaves and W. H. Saunders, Jr., *Tetrahedron Lett.*, 2307 (1972).
213. V. Franzen and H. J. Schmidt, *Chem. Ber.*, **94**, 2937 (1961).
214. V. Franzen and C. Mertz, *Chem. Ber.*, **93**, 2819 (1960).
215. W. H. Saunders, Jr., and D. Pavlovic, *Chem. Ind. (London)*, 180 (1962),

216. W. H. Saunders, S. D. Bonadies, M. Braunstein, J. K. Borchardt and R. T. Hargreaves, *Tetrahedron*, **33**, 1577 (1977).
217. B. M. Trost, *J. Amer. Chem. Soc.*, **89**, 138 (1967).
218. T. Kunieda and B. Witkop, *J. Amer. Chem. Soc.*, **93**, 3487 (1971).
219. A. W. Johnson, V. J. Hruby and J. L. Williams, *J. Amer. Chem. Soc.*, **86**, 918 (1964).
220. B. M. Trost, *J. Amer. Chem. Soc.*, **88**, 1587 (1966).
221. A. W. Johnson and R. T. Amel, *J. Org. Chem.*, **34**, 1240 (1969).
222. V. Franzen, H. J. Schmidt and C. Mertz, *Chem. Ber.*, **94**, 2942 (1961).
223. C. G. Swain and E. R. Thornton, *J. Org. Chem.*, **26**, 4808 (1961).
224. I. Rothberg and E. R. Thornton, *J. Amer. Chem. Soc.*, **85**, 1704 (1963).
225. C. G. Swain and E. R. Thornton, *J. Amer. Chem. Soc.*, **83**, 4033 (1961).
226. I. Rothberg and E. R. Thornton, *J. Amer. Chem. Soc.*, **86**, 3302 (1964).
227. I. Rothberg and E. R. Thornton, *J. Amer. Chem.Soc.*, **86**, 3296 (1964).
228. W. v. E. Doering and K. C. Schreiber, *J. Amer. Chem. Soc.*, **77**, 514 (1955).
229. J. Gosselck, G. Schmidt, L. Beress and H. Schenk, *Tetrahedron Lett.*, 331 (1968).
230. J. Gosselck, L. Beress, H. Schenk and G. Schmidt, *Angew. Chem. Int. Ed. Engl.*, **4**, 1080 (1965).
231. J. Gosselck, H. Ahlbrecht, F. Dost, H. Schenk and G. Schmidt, *Tetrahedron Lett.*, 995 (1968).
232. J. Gosselck, L. Beress and H. Schenk, *Angew. Chem. Int. Ed. Engl.*, **5**, 596 (1966).
233. R. Manske and J. Gosselck, *Tetrahedron Lett.*, 2097 (1971).
234. K. Takaki and T. Agawa, *J. Org. Chem.*, **42**, 3303 (1977).
235. K. Takaki, H. Takahashi, Y. Ohshiro and T. Agawa, *J. Chem. Soc. Chem. Commun.*, 675 (1977).
236. G. D. Appleyard and C. J. M. Stirling, *J. Chem. Soc. C*, 1904 (1969).
237. G. Griffiths, P. D. Howes and C. J. M. Stirling, *J. Chem. Soc. Perkin Trans. I*, 912 (1977).
238. J. W. Batty, P. D. Howes and C. J. M. Stirling, *J. Chem. Soc. Perkin Trans. I*, 65 (1973).
239. J. W. Batty, P. D. Howes and C. J. M. Stirling, *J. Chem. Soc. D*, 534 (1971).
240. J. W. Batty, P. D. Howes and C. J. M. Stirling, *J. Chem. Soc. Perkin Trans. I*, 59 (1973).
241. B. S. Ellis, G. Griffiths, P. D. Howes, C. J. M. Stirling and B. R. Fishwick, *J. Chem. Soc. Perkin Trans. I*, 286 (1977).
242. H. Braun, N. Mayer, G. Strobl and G. Kresze, *Justus Liebigs Ann. Chem.*, 1317 (1973).
243. I. Fleming, *Chem. Ind. (London)*, 449 (1975).
244. G. Wittig, *Accounts Chem. Res.*, **7**, 6 (1974).
245. G. Wittig, *J. Organomet. Chem.*, **100**, 279 (1975).
246. P. A. Lowe, *Chem. Ind. (London)*, 1070 (1970).
247. R. F. Hudson, *Chim. Ind. (Milan)*, **54**, 335 (1972).
248. R. F. Hudson, *Chem. Brit.*, **7**, 287 (1971).
249. H. G. Heal, *J. Chem. Educ.*, **35**, 192 (1958).
250. A. W. Johnson, *Ylide Chemistry*, Academic Press, New York, 1966.
251. D. Lloyd, *Chem. Scr.*, **8A**, 14 (1975).
252. B. M. Trost and L. S. Melvin, *Sulfur Ylides – Emerging Synthetic Intermediates*, Academic Press, New York, 1975.
253. R. Oda, *Kagaku (Kyoto)*, **21**, 1044 (1966); *Chem. Abstr.*, **70**, 10704 (1969).
254. Y. Kishida, *Yuki Gosei Kagaku Kyokai Shi*, **26**, 1037 (1968); *Chem. Abstr.*, **70**, 76915 (1969).
255. K. Kondo, *Yuki Gosei Kagaku Kyokai Shi*, **31**, 1011 (1973); *Chem. Abstr.*, **81**, 135210 (1974).
256. W. Ando, *Yuki Gosei Kagaku Kyokai Shi*, **29**, 899 (1971); *Chem. Abstr.*, **77**, 74788 (1972).
257. M. Takaku and H. Nozaki, *Kagaku No Ryoiki*, **22**, 697 (1968); *Chem. Abstr.*, **69**, 95538 (1968).
258. H. Koenig, *Fortschr. Chem. Forsch.*, **9**, 487 (1968); *Chem. Abstr.*, **69**, 35056 (1968).

259. J. C. Bloch, *Justus Liebigs Ann. Chem.*, **10**, 419 (1965).
260. A. Hochrainer, *Oesterr. Chem.-Ztg.*, **67**, 297 (1966); *Chem. Abstr.*, **66**, 10664 (1967).
261. C. Agami, *Bull. Soc. Chim. Fr.*, 1021 (1965).
262. E. Block, *J. Chem. Educ.*, **48**, 814 (1971); E. Block, *Reaction of Organosulfur Compounds*, Academic Press, New York, 1978.
263. See also reviews listed in refs. 264 and 265.
264. A. W. Johnson, in *Organic Compounds of Sulphur, Selenium and Tellurium*, (Ed. D. H. Reid), Specialist Periodical Reports, The Chemical Society, London: 1970, Vol. 1, p. 248; 1973, Vol. 2, p. 288; 1975, Vol. 3, p. 322.
265. E. Block and M. Haake, in *Organic Compounds of Sulphur, Selenium and Tellurium*, (Ed. D. R. Hogg), Specialist Periodical Reports, Chemical Society, London, 1977, Vol. 4, p. 78.
266. K. Tatsumi, Y. Yoshioka, K. Yamaguchi and T. Fueno, *Tetrahedron*, **32**, 1705 (1976).
267. W. von E. Doering and A. K. Hoffmann, *J. Amer. Chem. Soc.*, **77**, 521 (1955).
268. N. F. Blaw and C. G. Stuckwisch, *J. Org. Chem.*, **22**, 82 (1957).
269. J. I. Musher, *Adv. Chem. Ser.*, **110**, 44 (1972).
270. C. K. Ingold and J. A. Jessop, *J. Chem. Soc.*, 713 (1930).
271. E. J. Corey and M. Chaykovsky, *J. Amer. Chem. Soc.*, **87**, 1353 (1965).
272. V. Franzen and H. E. Driesen, *Chem. Ber.*, **96**, 1881 (1963).
273. N. Kunieda, Y. Fukiwara, J. Nokami and M. Kinoshita, *Bull. Chem. Soc. Jap.*, **49**, 575 (1976).
274. D. A. Rutolo Jr., P. G. Truskier, J. Casanova, Jr., and G. B. Payne, *Org. Prep. Proced.*, **1**, 111 (1969).
275. G. B. Payne, *J. Org.Chem.*, **32**, 3351 (1967).
276. J. Casanova, Jr., and D. A. Rutolo, Jr., *J. Chem. Soc. Chem. Commun.*, 1224 (1967).
277. H. Nozaki, K. Nakamura and M. Takaku, *Tetrahedron*, **25**, 3675 (1969).
278. J. T. Lumb, *Tetrahedron Lett.*, 579 (1970).
279. K. W. Ratts and A. N. Yao, *J. Org. Chem.*, **3**, 1185 (1966).
280. C. R. Johnson and E. R. Janiga, *J. Amer. Chem. Soc.*, **95**, 7692 (1973).
281. B. M. Trost and H. C. Arndt, *J. Org. Chem.*, **38**, 3140 (1973).
282. H. Nozaki, D. Tunemoto, Z. Morita, K. Nakamura, K. Watanabe, M. Takaku and K. Kondo, *Tetrahedron*, **23**, 4279 (1967).
283. C. R. Johnson, M. Haake and C. W. Schroeck, *J. Amer. Chem. Soc.*, **92**, 6594 (1970).
284. C. R. Johnson and P. E. Rogers, *J. Org. Chem.*, **38**, 1793 (1973).
285. J. Adams, L. Hoffman, Jr., and B. M. Trost, *J. Org. Chem.*, **35**, 1600 (1970).
286. D. Jeckel and J. Gosselck, *Tetrahedron Lett.*, 2101 (1972).
287. P. R. H. Speakman, *J. Chem. Soc. C*, 2180 (1968).
288. C. P. Lillya and P. Miller, *J. Amer. Chem. Soc.*, **88**, 1559 and 1560 (1966).
289. C. P. Lillya and E. P. Miller, *Tetrahedron Lett.*, 1281 (1968).
290. B. M. Trost and M. J. Bogdanowicz, *J. Amer. Chem. Soc.*, **95**, 5307 (1973).
291. B. M. Trost and M. J. Bogdanowicz, *J. Amer. Chem. Soc.*, **95**, 5311 (1973).
292. B. M. Trost and M. J. Bogdanowicz, *J. Amer. Chem. Soc.*, **95**, 5321 (1973).
293. Y. Hayasi, M. Takaku and H. Nozaki, *Tetrahedron Lett.*, 3179 (1969).
294. E. J. Corey, K. Achiwa and J. A. Katzenellenbogen, *J. Amer. Chem. Soc.*, **91**, 4318 (1969).
295. E. J. Corey and W. Oppolzer, *J. Amer. Chem. Soc.*, **86**, 1899, (1964).
296. E. J. Corey, M. Jautelat and W. Oppolzer, *Tetrahedron Lett.*, 2325 (1967).
297. E. J. Corey, K. Lin and M. Jautelat, *J. Amer. Chem. Soc.*, **90**, 2724 (1968).
298. B. M. Trost, R. W. LaRochelle and M. J. Bogdanowicz, *Tetrahedron Lett.*, 3449 (1970).
299. T. Otsubo and V. Boekelheide, *Tetrahedron Lett.*, 3881 (1975).
300. G. M. Blackburn, W. D. Ollis, J. D. Plackett, C. Smith and I. O. Sutherland, *J. Chem. Soc. Chem. Commun.*, 186 (1968).
301. H. Hellmann and D. Eberle, *Justus Liebigs Ann. Chem.*, **662**, 188 (1963).
302. C. R. Johnson and P. E. Rogers, *J. Org. Chem.*, **38**, 1798 (1973).
303. H. Nozaki, D. Tunemoto, S. Matubara and K. Kondo, *Tetrahedron*, **23**, 545 (1967).
304. K. Kondo and D. Tunemoto, *J. Chem. Soc. Chem. Commun.*, 1361 (1970).

305. S. Kato, H. Ishihara, M. Mizuta and Y. Hirabayashi, *Bull. Chem. Soc. Jap.*, **44**, 2469 (1971).
306. I. Stahl and J. Gosselck, *Tetrahedron Lett.*, 989 (1972).
307. Y. Hayasi and H. Nozaki, *Bull. Chem. Soc. Jap.*, **45**, 198 (1972).
308. P. Y. Johnson, E. Koza and R. E. Kohrman, *J. Org. Chem.*, **38**, 2967 (1973).
309. W. Kirmse and M. Kapps, *Chem. Ber.*, **101**, 994 and 1004 (1968).
310. I. Ojima and K. Kondo, *Bull. Chem. Soc. Jap.*, **46**, 1539 (1973).
311. I. Ojima and K. Kondo, *Chem. Lett.*, 119 (1972).
312. K. Kondo and I. Ojima, *J. Chem. Soc. Chem.Commun.*, 62 (1972).
313. J. H. Robson and H. Shechter, *J. Amer. Chem.Soc.*, **89**, 7112 (1967).
314. W. E. Parham and R. Konros, *J. Amer. Chem. Soc.*, **83**, 4034 (1961).
315. W. E. Parham and S. H. Gwen, *J. Org. Chem.*, **31**, 1694 (1966).
316. W. Ando, *J. Org. Chem.*, **42**, 3365 (1977).
317. Huynh-Chanh, V. Ratovelomanana, and S. Julia, *Bull. Soc. Chim. Fr.*, 710 (1977).
318. P. A. Grieco, D. Boxler and K. Hiroi, *J. Org. Chem.*, **38**, 2572 (1973).
319. M. Yoshimoto, S. Ishihara, E. Nakayama and N. Soma, *Tetrahedron Lett.*, 2923 (1972).
320. W. Ando, S. Kondo, K. Nakayama, K. Ichibori, H. Kohoda, H. Yamato, I. Imai, S. Nakaido and T. Migita, *J. Amer. Chem. Soc.*, **94**, 3870 (1972).
321. W. Ando, M. Yamada, E. Matsuzaki and T. Migita, *J. Org. Chem.*, **37**, 3791 (1972).
322. H. Matsuyama, H. Minato and M. Kobayaski, *Bull. Chem. Soc. Jap.*, **46**, 1512 (1973).
323. W. Illger, A. Liedhegener and M. Regitz, *Justus Liebigs Ann. Chem.*, **760**, 1 (1972).
324. D. C. Appleton, D. C. Bull, J. McKenna, J. M. McKenna and A. R. Walley, *J. Chem. Soc. Chem. Commun.*, 140 (1974).
325. Y. Hata, M. Watanabe, S. Inoue and S. Oae, *J. Amer. Chem. Soc.*, **97**, 2553 (1975).
326. Z. Yoshida, S. Yondeda and M. Hazama, *J. Org. Chem.*, **37**, 1364 (1972).
327. J. Diekmann, *J. Org. Chem.*, **30**, 2272 (1965).
328. R. Gompper and H. Euchner, *Chem. Ber.*, **99**, 527 (1966).
329. N. E. Miller, *Inorg. Chem.*, **4**, 1458 (1965).
330. B. M. Trost and M. J. Bogdanowicz, *J. Amer. Chem. Soc.*, **95**, 5298 (1973).
331. K. W. Ratts, *J. Org. Chem.*, **37**, 848 (1972).
332. W. G. Phillips and K. W. Ratts, *J. Org. Chem.*, **35**, 3144 (1970).
333. A. W. Johnson and R. T. Amel, *Can. J. Chem.*, **46**, 461 (1968).
334. S. Fliszar, R. F. Hudson and G. Salvadori, *Helv. Chim. Acta*, **46**, 1580 (1963).
335. F. G. Bordwell and P. S. Bouten, *J. Amer. Chem. Soc.*, **78**, 87 (1956).
336. F. G. Bordwell and P. S. Bouten, *J. Amer. Chem. Soc.*, **78**, 854 (1956).
337. C. Kissel, R. J. Holland and M. C. Caserio, *J. Org. Chem.*, **37**, 2720 (1972).
338. M.-H. Whangbo, S. Wolfe and F. Bernadi, *Can. J. Chem.*, **53**, 3040 (1975).
339. J. I. Musher, *Tetrahedron*, **30**, 1747 (1974).
340. F. Bernardi, I. G. Csizmadia, A. Mangini, H. B. Schlegel, M.-H. Wangbo and S. Wolfe, *J. Amer. Chem. Soc.*, **97**, 2209 (1975).
341. A. F. Cook and J. G. Moffatt, *J. Amer. Chem. Soc.*, **90**, 740 (1968).
342. K. W. Ratts, *Tetrahedron Lett.*, 4707 (1966).
343. A. T. Christensen and E. Thom, *Acta Crystallogr., Sect B*, **27**, 581 (1971).
344. A. T. Christensen and W. G. Witmore, *Acta Crystallogr., Sect. B*, **25**, 73 (1969).
345. J. P. Schaefer and L. L. Reed, *J. Amer. Chem. Soc.*, **94**, 908 (1972).
346. D. Darwish and R. L. Tomilson, *J. Amer. Chem. Soc.*, **90**, 5938 (1968).
347. G. Barbarella, A. Garbesi, A. Boicelli and A. Fava, *J. Amer. Chem. Soc.*, **95**, 8051 (1973).
348. A. Garbesi, G. Barbarella and A. Fava, *J. Chem. Soc. Chem. Commun.*, 155 (1973).
349. S. J. Campbell and D. Darwish, *Can. J. Chem.*, **52**, 2953 (1974).
350. C. R. Johnson, C. W. Schroeck and J. R. Shanklin, *J. Amer. Chem. Soc.*, **95**, 8424 (1973).
351. T. Durst, R. Viaw, R. Van Der Elzen and C. H. Nguyen, *J. Chem. Soc. Chem. Commun.*, 1334 (1971).
352. J. M. Townsend and K. B. Sharpless, *Tetrahedron Lett.*, 3313 (1972).
353. Y. Masao and M. J. Hatch, *J. Amer. Chem. Soc.*, **89**, 5831 (1967).
354. R. G. Carlson and N. S. Behn, *J. Org. Chem.*, **32**, 1363 (1967).

355. C. E. Cook, R. C. Corley and M. E. Wall, *Tetrahedron Lett.*, 891 (1965).
356. C. R. Johnson and G. F. Kateker, *J. Amer. Chem. Soc.*, **92**, 5753 (1970).
357. S. R. Landor, P. W. O'Connor, A. R. Tatchell and I. Blair, *J. Chem. Soc. Perkin Trans. I*, 473 (1973).
358. J. D. Ballantine and P. J. Sykes, *J. Chem. Soc. C*, 731 (1970).
359. R. S. Bly, C. M. Du Bose, Jr., and G. B. Konizer, *J. Org. Chem.*, **33**, 2188 (1968).
360. R. K. Bly and R. S. Bly, *J. Org. Chem.*, **28**, 3165 (1963).
361. H. L. Goering and C.-S. Chang, *J. Org. Chem.*, **40**, 3276 (1975).
362. W. Carruthers and M. I. Qureshi, *J. Chem. Soc. C*, 2238 (1970).
363. T. Hiyama, T. Mishima, H. Sawada and H. Nozaki, *J. Amer. Chem. Soc.*, **97**, 1626 (1975).
364. B. Holt, J. Howard and P. A. Lowe, *Tetrahedron Lett.*, 4937 (1969).
365. W. E. Parham and L. J. Czuba, *J. Amer. Chem. Soc.*, **90**, 4030 (1970).
366. J. B. Galper and B. M. Babior, *Biochim. Biophys. Acta*, **158**, 289 (1968).
367. P. G. Gassman and E. A. Armour, *Tetrahedron Lett.*, 1431 (1971).
368. B. C. Clark, Jr., and D. J. Goldsmith, *Org. Prep. Proced. Int.*, **4**, 113 (1972).
369. R. E. Ireland, D. R. Marshall and J. W. Tilley, *J. Amer., Chem. Soc.*, **92**, 4754 (1970).
370. P. Bravo, G. Gaudiano and M. G. Zubiani, *J. Heterocylc. Chem.*, **7**, 967 (1970).
371. M. Chaykovsky, L. Benjamin, R. I. Fryer and W. Metlesics, *J. Org. Chem.*, **35**, 1178 (1970).
372. M. C. Sacquet, B. Graffe and P. Maitte, *Tetrahedron Lett.*, 4453 (1972).
373. P. Bravo, G. Gaudiano and A. Umani-Ronchi, *Tetrahedron Lett.*, 679 (1969).
374. B. Holt and P. A. Lowe, *Tetrahedron Lett.*, 683 (1966).
375. P. Bravo, G. Gaudiano and C. Ticozzi, *Gazz, Chim. Ital.*, **103**, 95 (1973).
376. P. Bravo, G. Gaudiano and A. Umani-Ronchi, *Gazz. Chim. Ital.*, **100**, 652 (1970).
377. H. Koenig and H. Metzger, *Z. Naturforsch.*, **18B**, 976 (1963).
378. H. König, H. Metzger and K. Seelert, *Chem. Ber.*, **98**, 3724 (1965).
379. K. Mashimo and Y. Sato, *Tetrahedron Lett.*, 905 (1969).
380. K. Mashimo and Y. Sato, *Chem. Pharm. Bull.*, **18**, 353 (1970).
381. H. O. House, S. G. Boots and V. K. Jones, *J. Org. Chem.*, **30**, 2519 (1965).
382. G. B. Payne, *J. Org. Chem.*, **33**, 3517 (1968).
383. K. W. Ratts and A. N. Yao, *J. Org. Chem.*, **31**, 1689 (1966).
384. W. J. Middleton, E. L. Buhle, J. G. McNally, Jr., and M. Zanger, *J. Org. Chem.*, **30**, 2384 (1965).
385. J. Rieser and K. Friedrich, *Justus Liebigs Ann. Chem.*, 648 (1976).
386. K. Wallenfels, K. Friedrich and J. Rieser, *Justus Liebigs Ann. Chem.*, 656 (1976).
387. S. A. Shevelev, V. V. Semenov and A. A. Fainzil'berg, *Izv. Akad. Nauk SSSR, Ser. Khim.*, 139 (1977).
388. S. A. Shevelev, V. V. Semenov and A. A. Fainzil'berg, *Izv. Akad. Nauk SSSR, Ser. Khim.*, 2032 (1976); *Chem. Abstr.*, **86**, 43337 (1977).
389. H. Matsuyama, H. Minato and M. Kobayashi, *Bull. Chem. Soc. Jap.*, **46**, 3158 (1973).
390. See also B. M. Trost, *Accounts Chem. Res.*, **7**, 85 (1974).
391. B. M. Trost, K. Hiroi and N. Holy, *J. Amer. Chem. Soc.*, **97**, 5873 (1975).
392. B. M. Trost and M. J. Bogdanowicz, *J. Amer. Chem. Soc.*, **94**, 4777 (1972).
393. B. M. Trost and M. Preckel and L. M. Leichter, *J. Amer. Chem. Soc.*, **97**, 2224 (1975).
394. B. M. Trost and M. Preckel, *J. Amer. Chem.Soc.*, **95**, 7862 (1973).
395. B. M. Trost and M. J. Bogdanowicz, *J. Amer. Chem. Soc.*, **95**, 2038 (1973).
396. B. M. Trost and M. J. Bogdanowicz, *J. Amer. Chem. Soc.*, **93**, 3773 (1971).
397. M. J. Bogdanowicz, J. Mitchell and B. M. Trost, *Tetrahedron Lett.*, 887 (1972).
398. O. Tsuge and I. Shinkai, *Bull. Chem. Soc. Jap.*, **43**, 3514 (1970).
399. C. Dumont, M. Vidal and P. Arnaud, *C. R. Acad. Sci., Ser. C*, **266**, 1085 (1968).
400. A. G. Hortmann and R. L. Harris, *J. Amer. Chem. Soc.*, **93**, 2471 (1971).
401. F. Rocquet and A. Sevin, *Bull. Soc. Chim. Fr.*, 881 (1974).
402. E. J. Corey and M. Jautelat, *J. Amer. Chem. Soc.*, **89**, 3912 (1967).
403. M. Yoshimoto, N. Ishida, Y. Kishida, *Chem. Pharm. Bull.*, **20**, 2593 (1972).
404. M. Hetschko and J. Gosselck, *Chem. Ber.*, **106**, 996 (1973).
405. R. S. Matthews and T. E. Meteyer, *J. Chem. Soc. D*, 1576 (1971).

406. J. A. Donnelly, S. O'Brien and J. O'Grady, *J. Chem. Soc. Perkin Trans. I*, 1674 (1974).
407. Y. Tamura and T. Miyamoto, *J. Chem. Soc. Perkin Trans. I*, 1125 (1974).
408. R. M. Dodson, P. D. Hammen, E. H. Jancis and G. Klose, *J. Org. Chem.*, **36**, 2698 (1971).
409. G. Schmidt and J. Gosselck, *Tetrahedron Lett.*, 2623 (1969).
410. B. M. Trost and R. F. Hammen, *J. Amer. Chem. Soc.*, **95**, 962 (1973).
411. C. R. Johnson and C. W. Schroeck, *J. Amer. Chem. Soc.*, **95**, 7418 (1973).
412. C. R. Johnson and C. W. Schroeck, *J. Amer. Chem. Soc.*, **90**, 6852 (1968).
413. D. E. Evans, G. S. Lewis, P. J. Palmer and D. J. Weyell, *J. Chem. Soc. C*, 1197 (1968).
414. R. Wiechert, O. Engelfried, V. Kerb, H. Laurent, H. Müller and G. Schulz, *Chem. Ber.*, **99**, 1118 (1966).
415. F. Rocquet, A. Sevin and W. Chodkiewicz, *Tetrahedron Lett.*, 1049 (1971).
416. M. Narayanaswamy, V. M. Sathe and A. S. Rao, *Chem. Ind. (London)*, 921 (1969).
417. J. E. Heller, A. S. Dreiding, B. R. O'Connor, H. E. Simmons, G. L. Buchanan, R. A. Raphael and R. Taylor, *Helv. Chim. Acta*, **56**, 272 (1973).
418. F. Rocquet and A. Sevin, *Bull. Soc. Chim. Fr.*, 888 (1974).
419. C. M. Agami and C. Prevost, *Bull. Soc. Chim. Fr.*, 2299 (1967).
420. C. Agami and J. Aubouet, *Bull. Soc. Chim. Fr.*, 1391 (1967).
421. R. M. Roberts, R. G. Landolt, R. N. Greene and E. W. Heyer, *J. Amer. Chem. Soc.*, **89**, 1404 (1967).
422. Y. Bessière-Chrétien and M. M. El Gaïed, *Bull. Soc. Chim. Fr.*, 2189 (1971).
423. W. G. Dauben, G. W. Shaffer and E. J. Deviny, *J. Amer. Chem. Soc.*, **92**, 6273 (1970).
424. W. G. Dauben and R. E. Wolf, *J. Org. Chem.*, **35**, 374 (1970).
425. C. Agami, *C. R. Acad. Sci., Ser. C*, **264**, 1128 (1967).
426. U. Schöllkopf, R. Harms and D. Hoppe, *Justus Liebigs Ann. Chem.*, 611 (1973).
427. W. E. Truce and V. V. Badiger, *J. Org. Chem.*, **29**, 3277 (1964).
428. W. E. Truce and C. T. Goralski, *J. Org. Chem.*, **33**, 3849 (1968).
429. J. Asunskis and H. Schechter, *J. Org. Chem.*, **33**, 1164 (1968).
430. W. G. Dauben, L. Schutte, G. W. Shaffer and R. B. Gagosian, *J. Amer. Chem. Soc.*, **95**, 468 (1973).
431. S. R. Landor and N. Punja, *J. Chem. Soc. C*, 2495 (1967).
432. Y. Sugimura, N. Soma and Y. Kishida, *Tetrahedron Lett.*, 91 (1971).
433. For a general review, see C. R. Johnson, *Accounts Chem. Res.*, **6**, 341 (1973).
434. J. Ide and Y. Yura, *Tetrahedron Lett.*, 3491 (1968).
435. A. Terada and Y. Kishida, *Chem. Pharm. Bull.*, **18**, 505 (1970).
436. M. Higo and T. Mukaiyama, *Tetrahedron Lett.*, 2565 (1970).
437. C. Kaiser, B. M. Trost, J. Beeson and J. Weinstock, *J. Org. Chem.*, **30**, 3972 (1965).
438. Y. Hayasi, M. Kobayashi and H. Nozaki, *Tetrahedron*, **26**, 4353 (1970).
439. R. B. Woodward and R. Hoffmann, *The Conservation of Orbital Symmetry*, Academic Press, New York, 1969, p. 106.
440. G. Andrews and D. Evans, *Tetrahedron Lett.*, 5121 (1972).
441. B. M. Trost and W. G. Biddlecom, *J. Org. Chem.*, **38**, 3439 (1973).
442. L. S. Melvin, Jr., *PhD Thesis*, University of Wisconsin, (1973); see Ref. 252 and Chapter 7, ref. 19.
443. G. Pourcelot, L. Veniard and P. Cadiot, *Bull. Soc. Chim. Fr.*, 1275 (1975).
444. A. Terada and Y. Kishida, *Chem. Pharm. Bull.*, **18**, 991 (1970).
445. P. A. Grieco, M. Meyers and R. S. Finkelhor, *J. Org. Chem.*, **39**, 119 (1974).
446. D. Michelot, G. Linstrumelle and S. Julia, *J. Chem. Soc. Chem. Commun.*, 10 (1974).
447. J. E. Baldwin, R. E. Hackler and D. P. Kelly, *J. Chem. Soc. Chem. Commun.*, 537 and 538 (1968).
448. A. S. F. Ash and F. Challenger, *J. Chem. Soc.*, 2792 (1952); P. A. Brescoe, F. Challenger and P. S. Duckworth, *J. Chem. Soc.*, 1755 (1956).
449. P. G. Gassman, T. J. van Bergen and G. Gruetzmacher, *J. Amer. Chem. Soc.*, **95**, 6508 (1973).
450. P. G. Gassman and T. J. van Bergen, *J. Amer. Chem. Soc.*, **95**, 2718 (1973).
451. E. Vedejs and J. P. Hagen, *J. Amer. Chem. Soc.*, **97**, 6878 (1975).
452. T. Thomson and T. S. Stevens, *J. Chem. Soc.*, 69 (1932).

453. H. Bohme and W. Krause, *Chem. Ber.,* **82**, 426 (1949).
454. U. Schollkopf, G. Ostermann and G. Schossig, *Tetrahedron Lett.,* 2619 (1969).
455. E. B. Ruiz, *Acta Salamant. Cienc.,* **2**, 64 (1958); *Chem. Abstr.,* **54**, 7623 (1960).
456. Y. Hayashi and R. Oda, *Tetrahedron Lett.,* 5381 (1968).
457. K. W. Ratts and A. N. Yao, *Chem. Ind. (London),* 1963 (1966); *J. Org. Chem.,* **33**, 70; (1968).
458. U. Schollkopf, J. Schossig and G. Ostermann, *Justus Liebigs Ann. Chem.,* **737**, 158 (1970).
459. J. E. Baldwin, W. K. Erickson, R. E. Hackler and R. M. Scott, *J. Chem. Soc. Chem. Commun.,* 576 (1970).
460. W. Ando, T. Yahihara, S. Tozune, I. Imai, J. Suzuki, T. Toyama, S. Nakaido and T. Migita, *J. Org. Chem.,* **37**, 1721 (1972).
461. H. Iwamura, M. Iwamura, T. Nishida, M. Yoshida and J. Nakayama, *Tetrahedron Lett.,* 63 (1971).
462. A. J. Speziale, C. C. Tung, K. W. Ratts and A. Yao, *J. Amer. Chem. Soc.,* **87**, 3460 (1965).
463. A. G. Hortman and D. A. Robertson, *J. Amer. Chem. Soc.,* **94**, 2758 (1972).
464. S. Hillers, A. V. Eremeev, D. A. Tikhomirov and E. Liepins, *Khim. Geterotsikl. Soedin,* 426 (1975); *Chem. Abstr.,* **83**, 28023 (1975).
465. W. S. Johnson, H. A. P. de Jongh, C. E. Coverdale, J. W. Scott and U. Burckhardt, *J. Amer. Chem. Soc.,* **89**, 4523 (1967).
466. R. Huisgen, R. Sustmann and K. Bunge, *Tetrahedron Lett.,* 3603 (1966); see, however, B. C. Elmes, *Tetrahedron Lett.,* 4139 (1971).
467. B. C. Elmes, *Tetrahedron Lett.,* 4139 (1971).
468. C. P. Dalla, *Ann. Chim. (Rome),* **63**, 895 (1973); *Chem. Abstr.,* **83**, 9891 (1975).
469. O. Tsuge and M. Koga, *Org. Prep. Proced. Int.,* **7**, 173 (1975).
470. E. Van Loock, G. L'abbe and G. Smets, *Tetrahedron Lett.,* 1693 (1970).
471. G. L'abbe, E. Van Loock and G. Smets, *J. Org. Chem.,* **36**, 2520 (1971).
472. E. Van Loock, G. L'abbe and G. Smets, *Tetrahedron,* **28**, 3061 (1973).
473. G. Gaudiano, C. Ticozzi, A. Umani-Ronchi and P. Bravo, *Gazz. Chim. Ital.,* **97**, 1411 (1967).
474. H. Metzger and H. Koenig, *Z. Naturforsch.,* **18B**, 987 (1963).
475. K. Kondo, Y. Liu and D. Tunemoto, *J. Chem. Soc. Perkin Trans. I,* 1279 (1974).
476. Y. Nakajima, M. Muroi, J. Oda and Y. Inouye, *Agric. Biol. Chem.,* **37**, 277 (1973).
477. H. Wittmann and F. A. Petio, *Z. Naturforsch.,* **29B**, 765 (1974).
478. T. Mukaiyam, M. Higo and T. Sakashita, *Tetrahedron Lett.,* 3697 (1971).
479. R. A. Ruden, *J. Org. Chem.,* **39**, 3607 (1974).
480. P. Bravo, G. Gaudiano and C. Ticozzi, *Gazz. Chim. Ital.,* **102**, 395 (1972).
481. A. W. Johnson, *Proc. N. Dakota Acad. Sci.,* **16**, 72 (1962); *Chem. Abstr.,* **60**, 4070 (1964).
482. A. W. Johnson, *J. Org. Chem.,* **28**, 252 (1963).
483. P. Bravo, G. Gaudiano, P. P. Ponti and C. Ticozzi, *Tetrahedron,* **28**, 3845 (1972).
484. Y. Hayashi, T. Akazawa, K. Yamamoto and R. Oda, *Tetrahedron Lett.,* 1781 (1971).
485. H. Nozaki, M. Takaku and Y. Hayasi, *Tetrahedron Lett.,* 2303 (1967).
486. H. Nozaki, M. Takaku, Y. Hayasi and K. Kondo, *Tetrahedron,* **24**, 6563 (1968).
487. J. Ide and Y. Kishida, *Chem. Pharm. Bull.,* **16**, 793 (1968).
488. Y. Hayasi, H. Nakamura and H. Nozaki, *Bull. Chem. Soc. Jap.,* **46**, 667 (1973).
489. Y. Hayasi and H. Nozaki, *Tetrahedron,* **27**, 3085 (1971).
490. Y. Tamura, T. Miyamoto, H. Kiyokawa and Y. Kita, *J. Chem. Soc. Perkin Trans. I,* 2053 (1974).
491. L. Salisbury, *J. Org. Chem.,* **40**, 1340 (1975).
492. M. Vaultier, R. Danion-Bougot, D. Danion, J. Hamelin and R. Carrie, *Tetrahedron Lett.,* 1923 (1973).
493. M. Vaultier, R. Danion-Bougot, D. Danion, J. Hamelin and R. Carrie, *J. Org. Chem.,* **40**, 2990 (1975).
494. G. G. DeAngelis and H. J. Hess, *Tetrahedron Lett.,* 1451 (1969).
495. A. R. Katritzky, S. Q. A. Rizvi and J. W. Suwinski, *Heterocycles,* **3**, 379 (1975).

496. B. M. Trost, R. C. Atkins and L. Hoffman, *J. Amer. Chem. Soc.*, **95**, 1285 (1973).
497. Y. Sugimura, K. Iino, I. Kawamoto and Y. Kishida, *Tetrahedron Lett.*, 4985 (1972).
498. Y. Sugimura, K. Iino, I. Kawamoto and Y. Kishida, *Chem. Lett.*, 1985 (1972).
499. J. J. Tufariello, L. T. C. Lee and P. Wojtkowski, *J. Amer. Chem. Soc.*, **89**, 6804 (1967).
500. V. J. Traynelis and J. V. McSweeney, *J. Org. Chem.*, **31**, 243 (1966).
501. P. Bravo, G. Fronza, C. Ticozzi and G. Gaudiano, *J. Organomet. Chem.*, **74**, 143 (1974).
502. H. Koezuka, G. Matsubayashi and T. Tanaka, *Inorg. Chem.*, **14**, 253 (1975).
503. E. T. Weleski, J. L. Silver, M. D. Jansson and J. L. Burmeister, *J. Organomet. Chem.*, **75**, 365 (1975).
504. H. Koezuka, G. Matsubayashi and T. Tanaka, *Inorg. Chem.*, **15**, 417 (1976).
505. M. Seno and S. Tsuchiya, *J. Chem. Soc. Dalton Trans.*, 751 (1977).
506. J. P. Fackler and C. Paparizos, *J. Amer. Chem. Soc.*, **99**, 2363 (1977).
507. R. D. Adams and D. F. Chodosh, *J. Amer. Chem. Soc.*, **100**, 812 (1978).
508. C. A. Stein and H. Taube, *J. Amer. Chem. Soc.*, **100**, 336 (1978).